GIANT MOLECULES

Second Edition

GIANT MOLECULES

Essential Materials for Everyday Living and Problem Solving

SECOND EDITION

Charles E. Carraher, Jr.

A JOHN WILEY & SONS, INC., PUBLICATION

Published by John Wiley & Sons, Inc., Hoboken, New Jersey.
Published simultaneously in Canada.

For general information on our other products and services please contact our Customer Care Department within the U.S. at 877-762-2974, outside the U.S. at 317-572-3993 or fax 317-572-4002.

Wiley also publishes its books in a variety of electronic formats. Some content that appears in print, however, may not be available in electronic format.

Library of Congress Cataloging-in-Publication Data:

Carraher, Charles E.
Giant molecules : essential materials for everyday living and problem solving. – 2nd ed. / Charles E. Carraher, Jr.
p. cm.
Rev. ed. of: Giant molecules / Raymond B. Seymour, Charles E. Carraher. ©1990.
Includes index.
ISBN 0-471-27399-6 (cloth)
1. Polymers. 2. Plastics. I. Seymour, Raymond Benedict, 1912- Giant molecules. II. Title.
QD381.S47 2003
668.9–dc21 2003009073

Printed in the United States of America

10 9 8 7 6 5 4 3 2 1

CONTENTS

PREFACE

Today, a scientific and technological revolution is occurring, and at its center are giant molecules. This revolution is occurring in medicine, communication, building, transportation, and so on. Understanding the principles behind this revolution is within the grasp of each of us, and it is presented in this book.

Giant molecules form the basis for life (human genome, proteins, nucleic acids), what we eat (complex carbohydrates, straches), where we live (wood, concrete), and the society in which we live (tires, plants, paint, clothing, biomaterials, paper, etc.). This text introduces you to the world of giant molecules, the world of plastics, fibers, adhesives, elastomers, paints, and so on, and also provides you with an understanding of why different giant molecules perform in the way they do. Giant molecules lend themselves to a pictorial presentation of the basic principles that govern their properties. This pictorial approach is employed in this text to convey basic principles and to show why different giant molecules behave in a particular manner; we use visual aids such as drawings, pictures, figures, structures, and so on. This text allows us to understand why some giant molecules are suitable for long-term memory present in the human genome while others are strong, allowing their use in bullet-resistant vests, others are flexible and used in automotive dashboards and rubber bands, others are good adhesives used to form space age composites, others are strong and flexible forming the cloths we wear, and so on.

This text is written so that those without any previous science training will be able to understand the world of giant molecules. Thus, the book begins with essential general basics, moving rapidly to material that forms the basics that enables the presentation of general precepts and fundamentals that apply to all materials and especially giant molecules. The initial two steps are accomplished in the first two chapters, and the remainder of the book considers materials concepts, fundamentals, and application. These basics are covered in a broad-brush manner but emphasize the fundamentals that are critical to the success of dealing with and understanding the basics of materials composed of giant molecules.

The book is arranged so that the earlier chapters introduce background information needed for later chapters. Basic concepts are interwoven and dispersed with illustrations that reinforce these basic concepts in practical and applied terms introduced throughout the text. The material is presented in an integrated, clear, and concise manner that combines basics/fundamentals with brief/illustrative applications.

Each chapter has a

- Glossary
- Bibliography
- Questions and answers section

A grouping of appropriate electronic sites is included.

This book is written for two different audiences. The first audience is the technician that wants to know about plastics, paints, textiles, rubbers, adhesives, fabrics and fibers, and composites. The second audience is those students required to include a basic science course in their college/university curriculum. This book can act as the basis of that course and as an alternative to a one-semester course in geology, chemistry, physics, and biology. Furthermore, it may have use in precollege (high school) trade schools and as an alternative advanced elective to fulfill a science requirement in high school.

CHARLES E. CARRAHER, JR.

The Society of Plastics Engineers is dedicated to the promotion of scientific and engineering knowledge of plastics and to the initiation and continuation of educational programs for the plastics industry. Publications, both books and periodicals, are major means of promoting this technical knowledge and of providing educational materials.

This 2nd Edition of Giant Molecules contains enough easily read basic science to permit the nonscientist to understand the structure and use of all polymers. The Society of Plastics Engineers, through its Technical Volumes Committee, has long sponsored books on various aspects of plastics and polymers. The final manuscripts are reviewed by the Committee to ensure accuracy of technical content. Members of this Committee are selected for outstanding technical competence and include prominent engineers, scientists, and educators.

In addition, the Society publishes Plastics Engineering Magazine, Polymer Engineering and Science, Journal of Vinyl and Additive Technology, Polymer Composites, proceedings of its Annual Technical Conference and other selected publications. Additional information can be obtained from the Society of Plastics Engineers, 14 Fairfield Drive, Brookfield, CT, 06804 - *www.4spe.org*.

Executive Director & CEO
Society of Plastics Engineers

MICHAEL R. CAPPELLETTI

1

THE BUILDING BLOCKS OF OUR WORLD

Giant Molecules: Essential Materials for Everyday Living and Problem Solving, Second Edition, by Charles E. Carraher, Jr.
ISBN 0-471-27399-6

1.1 INTRODUCTION

Science in the broadest sense is our search to understand what is about us. The quest is marked by observation, testing, inquiring, gathering data, explaining, questioning, predicting, and so on. Four major sciences have evolved, yet today's areas of inquiry generally require contributions from more than one. Thus subdisciplines such as biochemistry have developed, and geophysical combinations and other areas of study have also developed: chemical engineering, geography/geology, medical biology, patient law, medical technology, medical physics, and so on. In general terms the four major areas of science can be briefly described as follows:

Biology or Biological Sciences: Study of living systems.

Chemistry: Study of the chemical and physical properties and changes of matter.

Geology: Study of the earth.

Physics: Study of the fundamental components and regularities of nature and how they fit together to form our world.

Mathematics is the queen of science dealing with quantities, magnitudes, and forms and their relationship to one another and to our world.

Engineering deals with design and construction of bridges, highways, computers, biomedical devices, industrial robots, roads, and so on. Giant molecules are used in these endeavors. The design and construction of plants that process prepolymer starting materials as well as this effort of engineering the polymers themselves, along with the machinery used in polymer processing, are also part of the assignment.

This chapter presents a brief overview of some of the science that is essential for an appreciation of the science of giant molecules.

We will be concerned with matter—that is, anything that has mass and occupies space. The term mass is used to describe a quantity of matter. However, in most cases, we will refer to weight instead of mass. Weight, unlike mass, varies with the force of gravity. For example, an astronaut in orbit may be weightless but his or her mass is the same as it was on the earth's surface.

1.2 SETTING THE STAGE

Polymers exist as essential materials for sophisticated objects such as computers and the space shuttle and as simple materials such as rubber bands and plastic spoons. They may be solids capable of stopping a bullet, or they may be liquids such as silicon oils offering a wide variety of flow characteristics.

We not only run across polymers in our everyday lives, but also have questions involving them. When mixing an epoxy adhesive (glue) it gets warm. Why? The dentist stuck a "blue light" into my mouth when I was having a cavity filled. What was happening? When I looked at the filaments in my rug I noticed they were star-shaped and hollow. How did they do this? Information in this book will allow you to better understand giant molecules that make up the world in which you live and to have a reasonable answer and explanation to observations such as those made above.

This initial chapter begins to lay the framework to understanding the giant molecule. It introduces you to atoms, elements, compounds, the periodic table, balanced equations, and so on, all essential topics that allows you to appreciate the wonderful world of the giant molecule that is about you.

Please enjoy the trip.

1.3 BASIC LAWS

All science is based on the assumption that the world about us behaves in an orderly, predictable, and consistent manner. The scientist's aim is to discover and report this behavior. It is an adventure we hope you will share with us in this course.

The scientific method involves making observations, looking for patterns in the observations, formulating theories based on the patterns, designing ways to test these theories, and, finally, developing "laws."

Observations may be qualitative (it is cool outside) or quantitative (it is 70°F outside). A *qualitative* observation is general in nature *without* attached *units*. A *quantitative* observation is more specific in *having units* attached. Gathering quantitative observations can be referred to as gathering measurements, collecting data, or performing an experiment. Patterns are often seen only after numerous measurements are made. Such patterns may be expressed by employing a mathematical relationship. Younger children like balloons; but with other children about, they often resort to hiding the balloons—sometimes in the refrigerator. Later they notice that the balloons became smaller in the refrigerator. Thus the volume of the balloon, V, is directly related to temperature, T. This is expressed mathematically as

$$V \propto T$$

Our theory then is that as temperature increases, the volume of the balloon increases. This may also be called a hypothesis. We can test this hypothesis by further varying the temperature of the balloon and noting the effect on volume. We can then construct a model from which other hypotheses can be formed and other measurements performed.

Continuing with the balloon (made out of giant molecules) example, we can construct a model that says that pressure, the force per unit area, which is acting to expand the balloon, is due to gaseous particles—that is, molecules. This model can also be called a theory that resulted from interpretation, or speculation.

Eventually, a theory that has been tested in many ways over a long period is elevated to the status of a "law." We have a number of "laws" that are basic to the sciences. The following are some of these.

1. The world about us behaves in an orderly, predictable, and consistent manner. Thus, copper wire conducts an electric current yesterday, today, and tomorrow; under usual conditions, water will melt near 0°C (32°F) yesterday, today, and tomorrow, and so on. We also hope that the orderly, predictable, and consistent behavior is explainable and knowable.

2. Mass/energy cannot be created or destroyed. This is called the Law of Conservation of Mass/Energy. It was originally described by Antoine Lavoisier around 1789 and referred to only as the conservation of mass. Later, Albert Einstein extended this to show that mass and energy were related by the famous equation

$$E = mv^2$$

where E is energy, m is mass, and v is velocity. Thus, while the total mass/energy is conserved, they are convertible as described by the Einstein equation.

Lavoisier was born in Paris in 1743. His father wanted him to become a lawyer, but Lavoisier was fascinated by science. He wrote the first modern chemistry textbook, *Elementary Treatise on Chemistry*, in 1789. To help support his scientific work, he invested in a private tax-collecting firm and married the daughter of one of the company's executives. His connection to the tax collectors proved fatal, for eventually the French revolutionaries demanded his execution. On May 8, 1794, Lavoisier was executed on the guillotine.

3. A given compound always contains the same proportion of elements by weight and the same number of elements. Thus water molecules always contain one oxygen atom and two hydrogen atoms. Another compound that contains two oxygen atoms and two hydrogen atoms is not water, but rather is a different compound called hydrogen peroxide, often used as a disinfectant in water. This observation is a combination of two laws: first, the Law of Definite Proportions, described by the Frenchman Joseph Proust (1754–1826), and second, the Law of Multiple Proportions, initially described by the Englishman John Dalton (1766–1844). In fact, Dalton was the first to describe what compounds, elements, and chemical reactions were. Briefly, the important aspects are as follows:

(a) Each element is composed of tiny particles called atoms.
(b) The atoms of the same element are identical; atoms of different elements differ from the atoms of the first element.
(c) Chemical compounds are formed when atoms combine with each other.
(d) Each specific chemical compound contains the same kind and number of atoms.
(e) Chemical reactions involve reorganization of the atoms.

John Dalton was a poor, humble man. He was born in 1766 in the village of Eaglesfield in Cumberland, England. His formal education ended at age 11, but he was clearly bright and, with help from influential patrons, began a teaching career at a Quaker school at the age of 12. In 1793 he moved to Manchester, taking up the post as tutor at New College.

He left in 1799 to pursue his scientific studies full time. On October 12, 1803, he read his now famous paper, "Chemical Atomic Theory," to the Literary and Philosophical Society of Manchester. He went on to lecture in other cities in England and Scotland. His reputation rose rapidly as his theories took hold, which laid the foundation for today's understanding of the world around us.

4. Electrons are arranged in ordered, quantized energy levels about the nucleus, which is composed of neutrons and protons. Most of us are familiar with a rainbow. The same colors can be obtained by passing light through a prism, resulting in a continuous array called a spectrum. If elements are placed between the continuous light source and the prism, certain portions of the spectrum are blank and produce a discontinuous spectrum. Different discontinuous spectra were found for different elements.

Eventually, this discovery led to an understanding that the electrons of the same elements resided in the same general energy levels and that they accepted only the specific energy (the reason for the blank spots in the spectrum) that permitted the electrons to jump from one energy level to another. These energy levels are called quantum levels. We live in a quantized universe in which movement, acceptance of energy, and emission of energy are all done in a discontinuous, quantized manner. Fortunately, the size of these allowable quantum levels decreases as the size of the matter in question increases, as is the case in atomic structure. Thus, at the atomic level the world behaves like it is quantizied, but at our everyday level it behaves as if it were continuous.

1.4 MATTER/ENERGY

As far as we know, the universe is composed of matter/energy and space. Space, as presently understood, is contained within three dimensions. Energy may be divided according to form (magnetic, radiant, light), magnitude (ultraviolet, infrared, microwave), source (chemical energy, coal, oil, light, sugar, moving water, wind, nuclear), or activity (kinetic or potential). Briefly, *kinetic energy* is energy in action—the lighting of a light bulb by a battery. *Potential energy* is energy at rest—a charged battery not being discharged. Potential energy can be converted to kinetic energy and, conversely, kinetic into potential. Thus a book on a shelf represents potential energy. If the book is pushed from the bookshelf, the potential energy is converted into kinetic energy.

Matter/energy is conserved as described in the Law of Conservation of Matter/Energy. Matter can be described in terms of its physical state as solid, liquid, or gas. As shown in Figure 1.1, a *solid* has a fixed volume and a fixed shape and does not assume the shape and volume of its container. A *liquid* has a fixed volume but not a fixed shape. It takes the shape of the portion of the container it occupies. A *gas* has neither a fixed volume nor shape. Some materials are solids, liquids, or gases depending on temperature or the time scale we use. Thus, glass acts like a solid at room temperature but begins to flow when heated to about 750°F, then acting

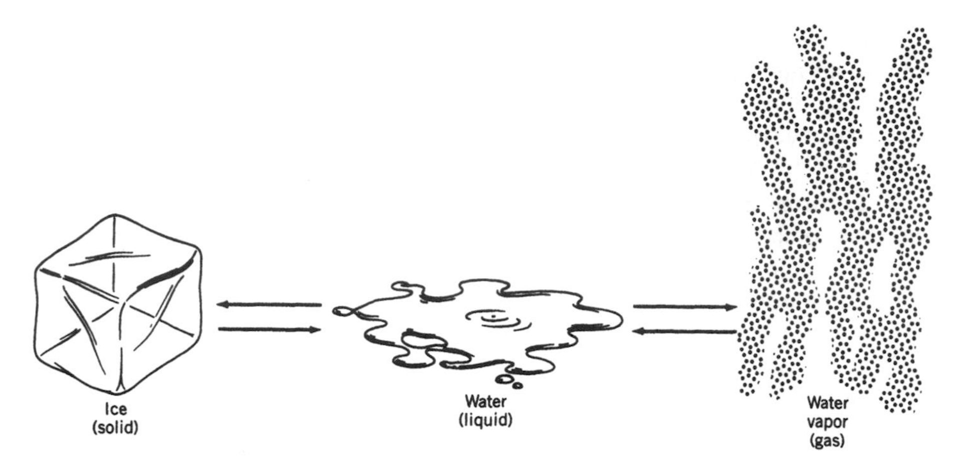

Figure 1.1. Water undergoing changes in state. *From left to right:* Solid to liquid (melting) and liquid to gas (vaporization, boiling). *From right to left:* Gas to liquid (condensation) and liquid to solid (freezing).

like a liquid. Glass acts like a solid when hit by a ball, but acts like a slow-flowing liquid when viewed over a period of a thousand years.

Most non-cross-linked matter undergoes transitions from solid to liquid to gas as temperature is increased or from gas to liquid to solid as temperature is decreased. These transitions are given names such as melting or freezing points. Thus, water below 0°C is solid, it melts (melting point) at 0°C (32°F), and it boils (temperature of evaporation or boiling point) at 100°C (212°F). In turn, water above 212°F is a gas that condenses to a liquid at 212°F and freezes at 32°F.

Boiling, freezing, and melting are all physical changes. A *physical change* does not alter the chemical composition. Water can be broken into its elements of hydrogen and oxygen, however, and such a process is called a *chemical change* since the chemical composition of the matter is changed.

Physical properties are properties that can be measured without changing the chemical composition of the matter. Your height, color of hair, and weight are all physical properties. Other physical properties are density, color, boiling point, and freezing point.

Physical properties can be extensive or intensive. An *extensive property* is one that depends on the amount of matter present. Thus, mass is an extensive property. *Intensive properties* do not depend on the amount of matter present. Density, boiling point, and color are intensive properties.

Chemical properties are properties that matter exhibits when its chemical composition changes. The reaction of an iron nail with oxygen to form rust is a chemical reaction, and the fact that iron reacts with oxygen is a chemical property of iron.

Matter can also be divided into components. Heterogeneous matter includes sidewalk cement, window glass, and most natural materials. Homogeneous matter or solutions include carbonated beverages, sugar in water, and brass (an alloy of zinc and copper). Examples of compounds include water, polyethylene, and table salt (NaCl). Some elements are iron (Fe), carbon (C), aluminum (Al), and copper (Cu).

1.5 SYMBOLS FOR THE ELEMENTS

The ancient Greeks represented their four elements by triangles and barred triangles, that is, fire = △, water = ▽, air = 🜁, and earth = 🜃. Although none of these is an element, the triangle is still used as a symbol for heat or energy in chemical equations. The ancient Babylonians and medieval alchemists represented these elements by using variations of the moon and other celestial bodies.

John Dalton used circles as symbols for elements in the eighteenth century. His symbols for some of the common elements were: oxygen = ○, hydrogen = ⊙, nitrogen = ⦶, carbon = ●, and sulfur = ⊕. This cumbersome system of symbols was displaced early in the nineteenth century by Jöns J. Berzelius, who used the capitalized initial letter of the name of each element. To avoid redundancy, he used a second lowercase letter to distinguish carbon (C) from calcium (Ca), and so on. Some symbols, such as Na for sodium and Fe for iron, were derived from the Latin names, which, in these examples, are natrium and ferrum, respectively.

Notice that the chemical symbol for all of the elements begins with a capital letter. For some elements a second, always small, letter is added. Only a few elements play a dominant role in synthetic and biological giant polymers. These are carbon (C), hydrogen (H), nitrogen (N), oxygen (O), chlorine (Cl), phosphorus (P), and sulfur (S). Additional elements are important in inorganic giant molecules, with silicon (Si) being the most important.

1.6 ELEMENTS

Even in ancient times, many philosophers believed that all matter was composed of a limited number of substances or elements. According to the early Chinese philosophers, there were four elements, namely, earth, solids such as wood, yin, and yang. The ancient Greek philosophers believed that all material forms consisted of various combinations of earth, air, fire, and water. The ancient Babylonians identified seven metallic elements, and many newly discovered substances were also called elements by philosophers during the Middle Ages.

An element is now defined as a substance consisting of identical atoms. There are 110 or more known elements, but we are interested in only a handul of these, namely, hydrogen, carbon, oxygen, nitrogen, and a few others.

Only a few of the over 100 elements are common in nature. These can be remembered using the mnemonic "P. Cohn's CAFE"—that is, phosphorus, carbon, oxygen, hydrogen, nitrogen, sulfur, calcium (Ca), and iron (Fe).

1.7 ATOMS

Some ancient Greek philosophers, such as Aristotle, maintained that matter was continuous, but 2400 years ago Democritus insisted that all matter was discrete—that is, made up of indivisible particles. He named these particles *atomos*, after the Greek word meaning indivisible. Over 23 centuries later, this concept for matter was adopted by John Dalton, who coined the word atom.

According to Dalton's theory, all matter consists of small, indestructible solid particles (atoms) that are in constant motion. These atoms, which are the building units of our universe, are characteristic for each element, such as oxygen (O), hydrogen (H), carbon (C), and nitrogen (N).

The scientists of the early nineteenth century did not recognize the difference between an atom and a molecule, which is a combination of atoms. This enigma was solved by Amedeo Avogadro and his student Stanislao Cannizzaro. These Italian scientists, who coined the term molecule from the Latin name *molecula* or little mass, showed that, under similar conditions of temperature and pressure, equal volumes of all gases contained the same number of molecules. They showed that simple gases, such as oxygen, hydrogen, and nitrogen, existed as diatomic molecules, which could be written as O_2, H_2, and N_2.

The atoms of these gases are unstable and combine spontaneously to produce stable molecules, which are the smallest particles of matter that can exist in a free state. Although the oxygen (O_2), hydrogen (H_2), and nitrogen (N_2) molecules are diatomic, most compounds consist of polyatomic molecules. For example, water (HOH), which is written H_2O, is a triatomic molecule, ammonia (NH_3) is a tetraatomic molecule, and methane (CH_4) is a pentaatomic molecule. Chemical formulas show the relative number and identity of atoms in each specific molecule or compound.

1.8 CLASSICAL ATOMIC STRUCTURE

Each atom consists of a dense, positively charged nucleus that is surrounded by a less dense cloud of negatively charged particles. The magnitude of each of these positively charged nuclear particles, called protons (after the Greek word *protos* or first), is equal to the magnitude of the negatively charged particles, called electrons (after the Greek word for amber). Thus, all neutral atoms contain an equal number of + and − charged particles. The mass of a proton is about 1840 times that of the electron, and the diffuse cloud occupied by the electrons has a diameter that is about 100,000 times that of the nucleus.

The nucleus may also contain dense neutral particles called neutrons (from the Latin word *neuter*, meaning neither), which have a mass similar to that of the positively charged protons. A hydrogen atom consists of one proton and one electron, whereas the oxygen atom consists of eight protons, eight neutrons, and eight electrons. These atoms have mass numbers of 1 and 16, respectively. The mass number is equal to the sum of the number of protons and neutrons in an atom. We will not be concerned with other atomic particles such as neutrinos, mesons, quarks, and gluons, and except for its contribution to mass, we can disregard the neutron.

It is generally accepted that electric current results from the flow of electrons, but the actual existence of these negatively charged atomic particles was not recognized until their presence was observed by J. J. Thomson in 1897. The neutron was discovered by James Chadwick in 1932. The proton, which was discovered by Ernest Rutherford in 1911, is simply the hydrogen atom without an electron. It is the positively charged building unit for the nuclei of all elements.

The presently accepted model for the atom is based on many discoveries made by a host of scientists. Many of these investigators were recipients of Nobel prizes. Obviously, their many contributions cannot be discussed in depth in this book nor learned in an introductory science course. You may find it advantageous to scan much of the description of atomic structure and read it more carefully after you have read some of the subsequent chapters.

In the early part of the twentieth century, Henry Moseley showed that x rays with characteristic wavelengths were produced when metallic elements were bombarded by electrons. He assigned atomic numbers to these elements based on the wavelength of the x rays. The atomic number is equal to the number of protons, which, since the atom has a neutral charge, is also equal to the number of electrons in each atom. The atomic numbers are 1 for hydrogen, 7 for nitrogen, and 8 for oxygen. The mass atomic weights for these atoms are about 1.00, 14.01, and 16.00, respectively. The difference between the atomic weight and atomic number is the average number of neutrons present in the each atom.

Niels Bohr proposed an atomic model in which the electrons traveled in relatively large orbits around the compact nucleus and the energy of these electrons was restricted to specific energy levels called quantum levels. The lowest energy level was near the nucleus, but under certain conditions an electron could pass from one energy level to another; this abrupt change is called a "quantum jump."

Remember, the number of protons is the atomic number and it tells what the element is. Thus, the element with 12 protons is carbon. The element with one proton is hydrogen, and so on. If the atom is neutral, the atomic number, number of protons, is the same as the number of electrons. Electrons are important since it is the outer or valence electrons that form the bond between two atoms and thus connect these two atoms. It is the sharing of electrons that allow the creation of giant molecules.

Figure 1.2 contains an illustration of an atom of carbon containing within the nucleus six positively charged protons (solid circles) and six neutrons. About and outside the nucleus are six negatively charged electrons, with two of the electrons being inner electrons and four of the electrons being further out. It is these outer four electrons that are involved in bonding as carbon forms different compounds. The electrons travel about the nucleus at a speed of about one-third the speed of light. Because they are near the speed of light, electrons behave as both solids and waves.

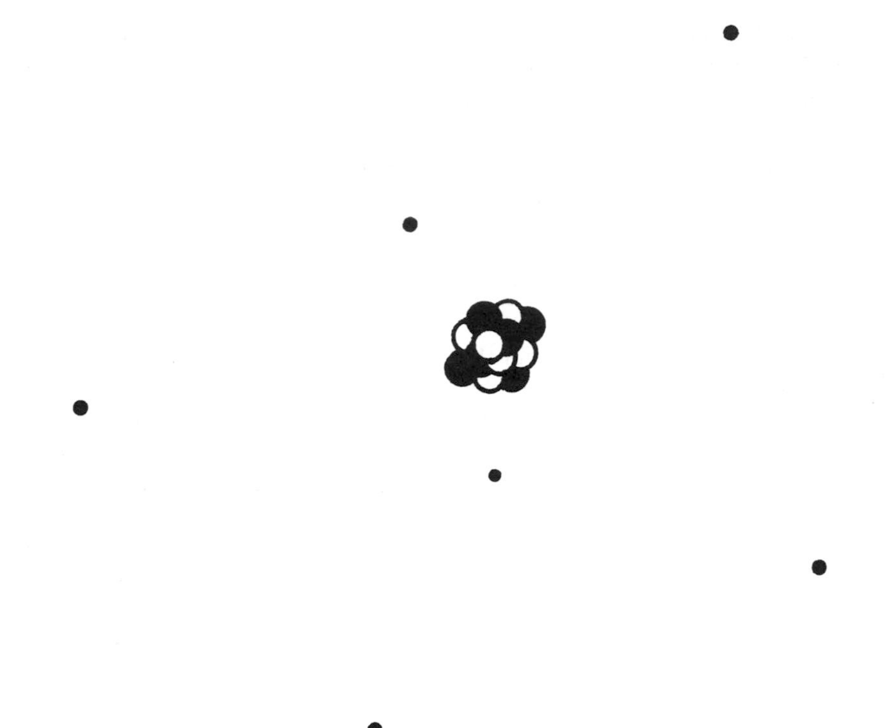

Figure 1.2. Illustration of an atom of carbon showing the nucleus containing protons, solid circles, and protons with the electrons about the nucleus.

In Figure 1.2 notice all of the open, unoccupied space within the atom. Over 99% of the space in an atom is not occupied, yet it appears to be solid. The wall, the floor, and your chair are over 99% empty space, yet they appear to be solid.

1.9 MODERN ATOMIC STRUCTURE

The concept of principal quantum levels or shells is still accepted, and these levels are designated, in the order of increasing energy, from 1 to 7, and so on, or by the letters K, L, M, and so on. The electron exhibits some of the characteristics of a particle, like a bullet, and some of the characteristics of a wave, like a wave in the ocean.

Werner Heisenberg used the term uncertainty principle to describe the inability to locate the position of a specific electron precisely. In general, this lack of precision is related to the energy used in viewing, which causes the particle to move in accordance with the energy used by the viewer. Because of the presence of the viewer and compiler of data, sociological observations are also uncertain.

Erwin Schrödinger, working independently of Heisenberg, used wave mechanics, which can also be used for the study of waves generated in a pool of water, to describe the patterns of an electron surrounding a nucleus. His approach, which led to the description of the movement and location of electrons, has been refined and is called quantum mechanics.

The position of electrons is now described in the general terms of probability pathways called orbitals. Thus, in considering the location of an electron, it is proper to describe it in general terms of probability. This probability pathway is called an orbital, and the maximum number of electrons that can occupy a single orbital is two.

1.10 PERIODICITY

All 110 or so elements are arranged in the order of their increasing atomic numbers in a periodic table. This table is a slight modification of the one devised by Dmitry Mendeleyev in the last part of the nineteenth century. Mendeleyev arranged the elements in order of their increasing atomic weights and successfully used this periodic table to predict physical and chemical properties of all known and some undiscovered elements. In the modern periodic chart, the elements are arranged vertically in groups or families according to their atomic numbers instead of their mass numbers. All members of a group have the same number of electrons in the atoms of their outer or valence shells. The number of electrons in the valence shell increases as one goes from left to right in the horizontal rows or periods. We will be concerned only with the electrons in the outermost or valence shell. Valence, which is derived from the Latin word *valentia*, meaning capacity, is equal to the combining power of an element with other elements. For example, the valence of hydrogen is one and that of carbon is four.

The periodic table is shown in Figure 1.3. It is called "periodic" because there is a recurring similarity in the chemical properties of certain elements. Thus, lithium, sodium, potassium, rubidium, cesium, and francium all react similarly. In the periodic table these elements are arranged in the same vertical column called a group or family. For the main group elements, those designated with the letter "A," the group also corresponds to the number of electrons in the outer or valence shell. Thus, all 1A elements have a single outer, valence electron, 2A elements have two valence electrons, 3A elements have three outer electrons, and so on.

Knowing the number of outer, valence electrons is important because these electrons are responsible for the existence of all compounds through formation of bonds. The elements designated by the letter "B" are called transition elements.

Some of the families have special names. The 1A family is known as the alkali metals, the 2A family is known as the alkaline earth metals, and the Group 7A elements are known as the halogens. Hydrogen has features of both Group 1A and Group 7A elements and yet has properties quite different from these

Atomic number / Symbol / Atomic weight: 92 [S] U Uranium 238.0289

State: [S] Solid; [L] Liquid; [G] Gas; [X] Not found in nature

Main Group metals; Transition metals, lanthanide series, actinide series; Metalloids; Nonmetals, noble gases

Period	1A	2A	3B	4B	5B	6B	7B	8B	8B	8B	1B	2B	3A	4A	5A	6A	7A	8A
1	1 [G] H Hydrogen 1.0079																	2 [G] He Helium 4.0026
2	3 [S] Li Lithium 6.941	4 [S] Be Beryllium 9.0122											5 [S] B Boron 10.811	6 [S] C Carbon 12.011	7 [G] N Nitrogen 14.0067	8 [G] O Oxygen 15.9994	9 [G] F Fluorine 18.9984	10 [G] Ne Neon 20.1797
3	11 [S] Na Sodium 22.9898	12 [S] Mg Magnesium 24.3050											13 [S] Al Aluminum 26.9815	14 [S] Si Silicon 28.0855	15 [S] P Phosphorus 30.9738	16 [S] S Sulfur 32.066	17 [G] Cl Chlorine 35.4527	18 [G] Ar Argon 39.945
4	19 [S] K Potassium 39.0983	20 [S] Ca Calcium 40.078	21 [S] Sc Scandium 44.9559	22 [S] Ti Titanium 47.88	23 [S] V Vanadium 50.9415	24 [S] Cr Chromium 51.9961	25 [S] Mn Manganese 54.9380	26 [S] Fe Iron 55.847	27 [S] Co Cobalt 58.9332	28 [S] Ni Nickel 58.693	29 [S] Cu Copper 63.546	30 [S] Zn Zinc 65.39	31 [S] Ga Gallium 69.723	32 [S] Ge Germanium 72.61	33 [S] As Arsenic 74.9216	34 [S] Se Selenium 78.96	35 [L] Br Bromine 79.904	36 [G] Kr Krypton 83.80
5	37 [S] Rb Rubidium 85.4678	38 [S] Sr Strontium 87.62	39 [S] Y Yttrium 88.9059	40 [S] Zr Zirconium 91.224	41 [S] Nb Niobium 92.9064	42 [S] Mo Molybdenum 95.94	43 [X] Tc Technetium (98)	44 [S] Ru Ruthenium 101.07	45 [S] Rh Rhodium 102.9055	46 [S] Pd Palladium 106.42	47 [S] Ag Silver 107.8682	48 [S] Cd Cadmium 112.411	49 [S] In Indium 114.82	50 [S] Sn Tin 118.710	51 [S] Sb Antimony 121.757	52 [S] Te Tellurium 127.60	53 [S] I Iodine 126.9045	54 [G] Xe Xenon 131.29
6	55 [S] Cs Cesium 132.9054	56 [S] Ba Barium 137.327	57 [S] La Lanthanum 138.9055	72 [S] Hf Hafnium 178.49	73 [S] Ta Tantalum 180.9479	74 [S] W Tungsten 183.85	75 [S] Re Rhenium 186.207	76 [S] Os Osmium 190.2	77 [S] Ir Iridium 192.22	78 [S] Pt Platinum 195.08	79 [S] Au Gold 196.9665	80 [L] Hg Mercury 200.59	81 [S] Tl Thallium 204.3833	82 [S] Pb Lead 207.2	83 [S] Bi Bismuth 208.9804	84 [S] Po Polonium (209)	85 [S] At Astatine (210)	86 [G] Rn Radon (222)
7	87 [S] Fr Francium (223)	88 [S] Ra Radium 226.0254	89 [S] Ac Actinium 227.0278	104 [X] Rf Rutherfordium (261)	105 [X] Db Dubnium (262)	106 [X] Sg Seaborgium (263)	107 [X] Bh Bohrium (262)	108 [X] Hs Hassium (265)	109 [X] Mt Meitnerium (266)	110* [X] — — (269)	111 [X] — — (272)	112 [X] — — (277)						

Lanthanides	58 [S] Ce Cerium 140.115	59 [S] Pr Praseodymium 140.9076	60 [S] Nd Neodymium 144.24	61 [X] Pm Promethium (145)	62 [S] Sm Samarium 150.36	63 [S] Eu Europium 151.965	64 [S] Gd Gadolinium 157.25	65 [S] Tb Terbium 158.9253	66 [S] Dy Dysprosium 162.50	67 [S] Ho Holmium 164.9303	68 [S] Er Erbium 167.26	69 [S] Tm Thulium 168.9342	70 [S] Yb Ytterbium 173.04	71 [S] Lu Lutetium 174.967
Actinides	90 [S] Th Thorium 232.0381	91 [S] Pa Protactinium 231.0359	92 [S] U Uranium 238.0289	93 [X] Np Neptunium 237.0482	94 [S] Pu Plutonium (244)	95 [X] Am Americium (243)	96 [X] Cm Curium (247)	97 [X] Bk Berkelium (247)	98 [X] Cf Californium (251)	99 [X] Es Einsteinium (252)	100 [X] Fm Fermium (257)	101 [X] Md Mendelevium (258)	102 [X] No Nobelium (259)	103 [X] Lr Lawrencium (256)

*Elements 110–112 have not yet been name

Figure 1.3. The periodic table.

elements. Thus it is often shown separately or as a member of both Groups IA and VIIA in periodic charts.

In addition to being an orderly presentation of the elements, from which all matter as we know it is composed, the periodic chart also contains a vast abundance of information. Depending on the particular periodic table, it may contain the chemical name, for example, carbon; the chemical symbol, C; the atomic number, which is the number of protons and in a neutral atom also the number of electrons; and the atomic mass or atomic weight in atomic mass units (amu) or daltons (one dalton = one amu), which is the sum of the number of protons and the average number of neutrons that occur naturally.

$$\text{atomic number} = \text{number of protons} = \text{number of electrons in a neutral atom}$$
$$\text{atomic mass} = \text{number of protons} + (\text{average}) \text{ number of neutrons}$$

For carbon, the atomic mass is not 12 but rather 12.011 since carbon exists in nature with two different numbers of neutrons. About 99% of carbon has six protons and six neutrons, and about 1% of carbon has six protons and seven neutrons. Atoms that are of the same element (that is, have the same number of protons in their nucleus) but have different numbers of neutrons are called isotopes. Thus carbon has three naturally occurring isotopes: carbon-12 (99%), carbon-13 (1%), and carbon-14 (trace).

Hydrogen's isotopes are so well known that they even have their own names. Hydrogen with one proton and no neutrons is simply called hydrogen; hydrogen with one proton and one neutron is called deuterium; and hydrogen with one proton (it would not be hydrogen if it had any number other than one proton) and two neutrons is called tritium. The beginning letters for the isotopes of hydrogen can be remembered from Hot, DoT.

The nuclei of many elements are unstable and spontaneously emit, or give off, particles, energy, or both. Such isotopes are called radioactive isotopes or radioisotopes. The three most common forms of natural radiation are shown in Table 1.1. The alpha particle is a package of two neutrons and two protons. This corresponds to the nucleus of helium. It has a positive two charge since each proton is positively charged and there are no electrons present to neutralize the positive charges. They

Table 1.1 Characteristics of three common radioactive emissions

Name	Identity	Charge	Relative Mass (amu)	Penetrating Power
Alpha	Two protons and two neutrons	+2	4.0026	Low
Beta	Electron	−1	0.0005	Low to moderate
Gamma	High-energy radiation similar to x rays	0	0	High

are fast traveling (about 5–10% of the speed of light), but relative to the other two radioactive emissions they are slower. Alpha particles are massive; thus their destructive capability is great. Fortunately, their massiveness also allows them to be stopped by thin sheets of aluminum foil, several sheets of paper, or human skin to prevent internal damage. The beta particle travels up to about 90% of the speed of light, whereas the gamma particle travels at the speed of light. Because of the small mass associated with these two emissions and their great speeds, both have penetrating powers greater than that of the alpha particle.

1.11 MOLECULAR STRUCTURE

As noted in section 1.10, the periodic table lists elements that are composed of a single kind of atom based on the number of protons. Combinations that contain two or more different kinds of elements are called compounds. Thus, CO_2 is a compound because it contains both carbon and oxygen; H_2O is a compound since it contains hydrogen and oxygen; SiO_2, the representative formula for sand, is a compound because it contains silicon and oxygen; and so on.

The formation of compounds from atoms is dependent on the formation of primary chemical bonds either through exchange of electrons (ionic bonding) or through the sharing of electrons, (covalent bonding). Our emphasis with giant molecules will be on covalent bonds. Thus, giant molecules are largely, but not totally, based on nonmetal elements.

Properties of compounds are dependent on the particular arrangement of the atoms within the compound and the arrangement of the atoms is dependent on the atoms that are in the compound.

G. N. Lewis represented valence or outer electrons as dots. Thus, hydrogen with one valence electron, oxygen with six valence electrons, and nitrogen with five valence electrons may be represented as ·H, :Ö:, and :Ṅ:. We use the Lewis representations or structures to show the valence electrons of the hydrogen, oxygen, and nitrogen molecules as follows: H:H, Ö::Ö, and :N⋮⋮N:. The shared bonds between the atoms are usually represented by single, double (two bonds), and triple (three bonds) bonds as follows: H–H, O=O, and N≡N.

The goal in predicting chemical structures is to look for stable electronic structures that allow for preferred (where possible) bonding arrangements. In general, hydrogen forms one bond sharing its single electron with another atom. Carbon forms four bonds with four, three, or two different atoms; oxygen forms two bonds either with one other atom as in Cl_2CO (phosgene) shown below or with two different atoms as in the case of water, H_2O, below. Nitrogen typically forms three bonds such as above in molecular nitrogen and in ammonia, NH_3, below. Most of the second row elements, lithium through neon, attempt to get eight valence electrons about them. This is the so-called rule of eight. Notice the Lewis dot formulas for water, methane, phosgene, and ammonia where each "dot" represent an outer or valence electron and two "dots" represents a pare of shared electrons, that is a covalent bond, or an unbonded electron pair (in water and ammonia). Each of the central elements has eight valence electrons surrounding it. But other non-second-row elements, such as sulfur and phosphorus, routinely have more than eight electrons about them as they form compounds.

The Lewis dot formulas for water, methane, phosgene, and ammonia are as follows, where each "dot" represents an outer or valence electron and two "dots" represent a pair of shared electrons, that is, a covalent bond.

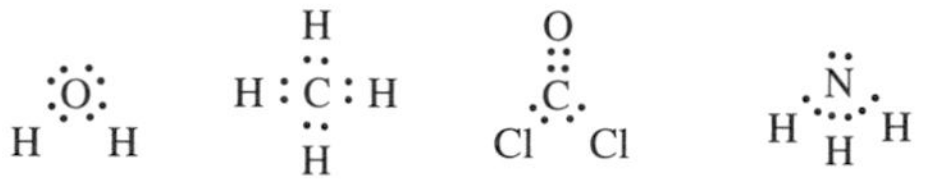

It is not customary to show the presence of unbonded electrons but to use simple structural representations such as

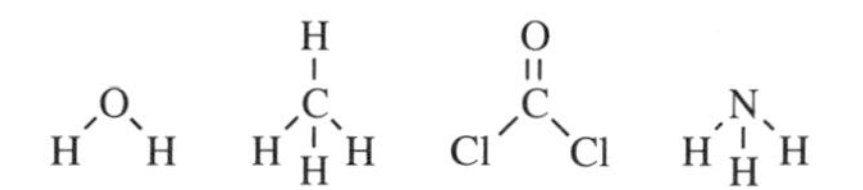

In chemical formulas, one simply notes the atoms present and their relative abundance as shown by

$$H_2O \qquad CH_4 \qquad COCl_2 \qquad NH_3$$

It is important to note that both H_2O and OH_2 are correct, but it is customary to write the formula for a molecule of water as H_2O. We will not be concerned with such rules in this book, but it is critical that you remember that the water molecule contains two atoms of hydrogen bonded to one atom of oxygen by covalent bonds.

As noted before, the valence or outer electrons can be easily remembered for many of the main group elements by simply looking at the family or group number. Thus, sodium, Na, is a 1A element, meaning it has one valence electron. Calcium is a 2A element, meaning it has two valence or outer electrons. Oxygen is a 6A element and has six outer electrons.

Figure 1.4 contains a representation for methane, CH_4. Notice the nucleus of the carbon with six protons and six neutrons and two inner or nonbonding electrons. Also notice the four single protons that represent the nuclei of the four hydrogen atoms. Finally note the four sets of electron pairs with each pair shared between carbon and a single hydrogen. Again, notice the unoccupied space.

The bonding for these nonmetallic molecules generally occurs so that the nuclei of the other surrounding atoms and the nonbonded electron pairs are as far away from one another but they are attached through the sharing of electrons. This is because the positively charged nuclei repeal one another and the nonbonded

Table 1.2 Geometric models for simple molecules

Molecule	Geometry	Bond Angle	Structure
CO_2	Linear	180°	
H–C(=O)–H	Trigonal planar	120°	
CH_4	Tetrahedral	109°	
NH_3	Tetrahedral/trigonal pyramid	107°	
H_2O	Tetrahedral/bent or "V"	105°	

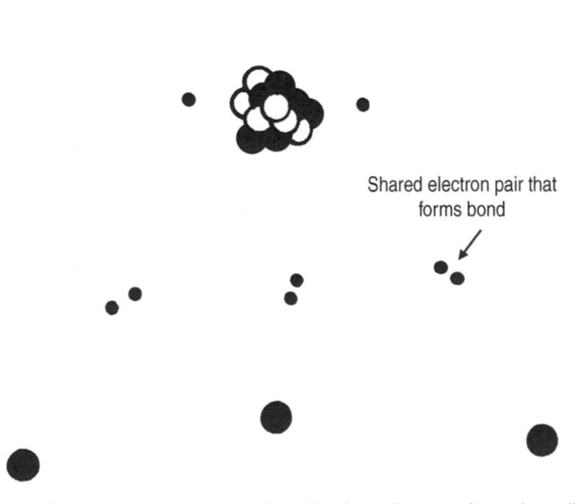

Figure 1.4. Representation of methane showing the bonding and nonbonding electrons.

valence electron pairs also repeal other electrons. Bonding occurs because of the attraction between the negatively charged nucleus and the positively charged electrons. Table 1.2 shows some common geometrical arrangements found in giant molecules.

1.12 CHEMICAL EQUATIONS

In the same manner that unstable atoms, like hydrogen, oxygen, and nitrogen, combine to form stable diatomic molecules with a complete electron duet for hydrogen and a complete electron octet for oxygen and nitrogen, dissimilar atoms also enter into combinations to produce more complex molecules. Many of these reactions release energy in the form of heat and are said to be exothermic. In contrast, those in which energy must be added to the reactants to cause a chemical reaction are called endothermic.

The equation for the exothermic reaction between hydrogen and oxygen molecules for the formation of water molecules is shown as

$$H_2 + O_2 \rightarrow H_2O$$

According to the law of conservation of mass, the weight or mass of reactants (H_2 and O_2) must equal the mass of the product (H_2O). Hence, we must balance the equation by placing small integers before the symbols for the molecules; that is, we must also ascertain that the same number of atoms of each element is on each side of the arrow. In this example, we obtain a balanced equation by placing the number 2 before both H_2 and H_2O:

$$2H_2 + O_2 \rightarrow 2H_2O$$

A balanced equation is very important. For the production of water, it states that two molecules of H_2O will be produced by the combination of two molecules of H_2 and one molecule of O_2. The prefixes tell us the number of whatever follows. Thus, the 2 in front of H_2 means two hydrogen molecules or four hydrogen atoms; the 2 in front of H_2O means two water molecules or H_2O, H_2O or a total of four hydrogen atoms and two oxygen atoms. Whenever there is 1 in front of a unit, it is omitted with the understanding that there is only one of this unit. Thus, there is an understood 1 in front of O_2 meaning there is one oxygen. Now let us again look at the number of atoms of each element on the left side of the arrow and do the same for the right side. We see that the left side has four hydrogen atoms and two oxygen atoms while the right side has four hydrogen atoms and two oxygen atoms. For balanced equations the number of each kind of atom is the same on both sides of the reaction arrow.

Figure 1.5 illustrates this reaction beginning in the center where we have 5 oxygen molecules (open intersected circles; for a total of 10 oxygen atoms) and 10 hydrogen molecules (solid intersected circles; total of 20 hydrogen atoms). To the right is the formation of 10 water molecules containing a total of 10 oxygen atoms and 20 hydrogen atoms so that the number of oxygens and hydrogens are the same on both sides of the reaction arrow. Furthermore, the ratio of hydrogen

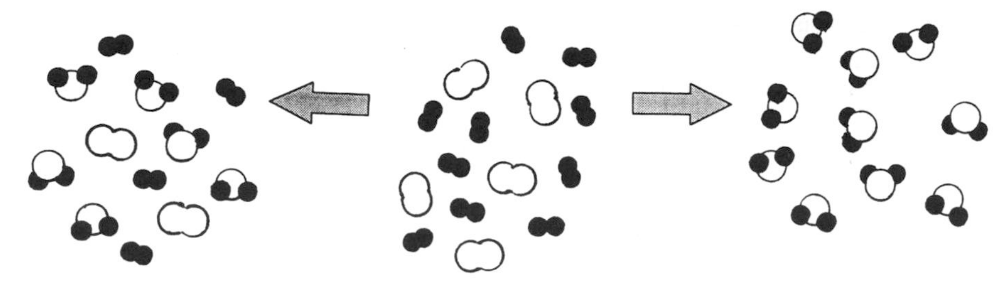

Figure 1.5. Oxygen and hydrogen (solid) molecules (*middle*) reacting to completely form water molecules (*right*) or incompletely forming water molecules (*left*).

molecules to oxygen molecules to water molecules is 2 to 1 to 2, the ratio of the prefixes on the balanced equation. The ratio of each reactant and product corresponds to the prefix numbers. Thus, 100 hydrogen molecules will react with 50 oxygen molecules to give 50 water molecules—a 2 to 1 to 2 ratio.

Figure 1.5 also shows a situation were all the hydrogen and oxygen molecules did not form water molecules. The ratio of reacted hydrogen molecules to oxygen molecules to water molecules is still 2 to 1 to 2, but instead of forming the maximum number of water molecules—namely 10—only 6 were formed. This often occurs with reactions where less than 100% of the possible product is formed. The percentage yield is calculated by dividing the actually formed product by the possible product and multiplying this fraction by 100. The maximum possible yield is also called the theoretical yield. For the present situation the percentage yield is then

$$\begin{aligned} \text{Percentage yield} &= (\text{actual yield}/\text{theoretical yield}) \times 100 \\ &= (6/10) \times 100 \\ &= 60\% \text{ yield} \end{aligned}$$

If only four molecules of water were formed, then the percentage yield would be $(4/10) \times 100 = 40\%$ yield.

While less than nearly 100% product yields are permissible for laboratory-scale reactions, industrial-scale reactions are generally run under conditions where the overall yield is nearly 100%. This is necessary since even a 99% yield for an industrial-scale reaction where 100 million pounds of product is synthesized leaves a million pounds of material to be discarded or otherwise taken care of. These high yields are accomplished through years of determining just the right conditions for the reaction. Also, in most cases solvent and unreacted materials are recycled.

For some reactions there is an excess one of the reactants. Figure 1.6 shows a reaction where there is an excess of hydrogen molecules. The number of water molecules is limited by the number of oxygen molecules, so oxygen is called the limiting reactant or limiting reagent while hydrogen is called the reactant in excess. In this situation, only 10 water molecules could be formed because there were only 5 oxygen molecules as the limiting reactant. It does not matter that there

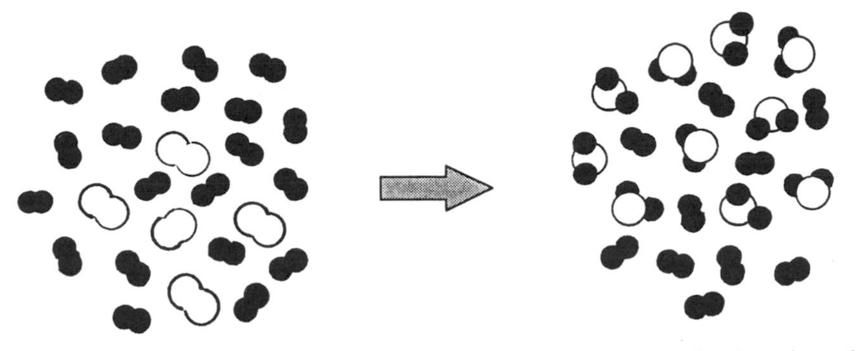

Figure 1.6. Hydrogen (solid) and oxygen molecules forming water molecules where the hydrogen molecules are in excess.

were 10 extra hydrogen molecules; the maximum number of water molecules formed is 10, based on the limiting reactant.

Reaction systems often also have a number of other molecules present that are not reactive under the reaction conditions. Thus, if the reaction forming water were carried out in the air, there would also be helium, nitrogen, carbon dioxide, and so on, molecules present that are not involved in the reaction so these molecules are ignored and not present in the balanced equation.

Ammonium, NH_3, is an important compound. It is a form of so-called fixed nitrogen. "Fixed" means to be in a usable form for plants. Before World War I, (WWI), nitrogen compounds essential for fertilizers and explosives were obtained from the nitrate deposits of northern Chile. During WWI, Germany was cut off from this source and turned to a new process discovered by Fritz Haber that involved combining hydrogen and nitrogen from the atmosphere using high temperature and pressure and special catalysts. This process remains an important process and is described by equation below in a balanced equation:

$$N_2 + 3H_2 \rightarrow 2NH_3$$

Notice that the balanced equation has 2 nitrogen atoms and 6 hydrogen atoms on both sides of the arrow. Furthermore, that the coefficients are 1, 3, 2 so that 1 nitrogen molecule reacts with 3 hydrogen molecules to give 2 ammonium molecules or that 10 nitrogen molecules will react with 30 hydrogen molecules to give 20 nitrogen molecules, and so on.

We know that molecules and atoms are very small. In fact a single drop of water holds about 2,000,000,000,000,000,000,000 or 2×10^{21} molecules of water. The concept of the mole is used when dealing with such large numbers. Essentially every mole of a material contains the same number of units. That number is called Avogadro's number and is 6×10^{23}. This number also corresponds to the atomic weights found in the periodic table. So, 23 grams of sodium metal contains one mole and 6×10^{23} atoms of sodium; 32 grams of molecular oxygen (remember that oxygen is diatomic, so we take the atomic weight of oxygen times two) contains one mole and 6×10^{23} molecules of oxygen; 18 grams of water, H_2O, contains one mole (2 hydrogen atoms with an atomic weight of 1 each plus 1 oxygen with an atomic weight of 16); 58.5 grams of sodium chloride contains one mole (atomic weight of sodium is 23 plus the atomic weight of chlorine is $35.5 = 58.5$); and so on. The weight of a mole of CO_2 is 44; and that of methane, CH_4, is 16. In other words, a mole is simply the summation of the atomic weights given in the formula for the compound. These values are referred to as gram formula weights and gram moles.

We do not always have one mole of a material. Thus, we often calculate the number of moles of a material by simply dividing the weight in grams of the material by the formula weight. Thus, the number of moles in 10 grams of water is 10 grams divided by 18 grams in a mole of water $= 0.56$ moles. The number of moles in 22 grams of CO_2 is 22 grams/44 grams in a mole of $CO_2 = 0.5$ moles.

The weight of a mole of material is called the molecular weight. Thus, the molecular weight of CO_2 is 44, the molecular weight of H_2O is 18, and so on. The unit of molecular weight is generally atomic mass unit (amu or simply "u"), or Daltons (or daltons). Thus, the molecular weight of CO_2 is 44 amu or 44 Daltons or 44 u, and so on. Often the molecular weight is given without units so that the molecular weight of CO_2 is simply 44.

1.13 CHEMICAL BONDING

There are a number of periodic properties related to the periodic table. Figure 1.7 shows the relative atomic sizes for the main group, those with an "A" at the top of the vertical rows. As we move within any horizontal period atomic size generally decreases as we move from left to right. Thus, Na > Mg >

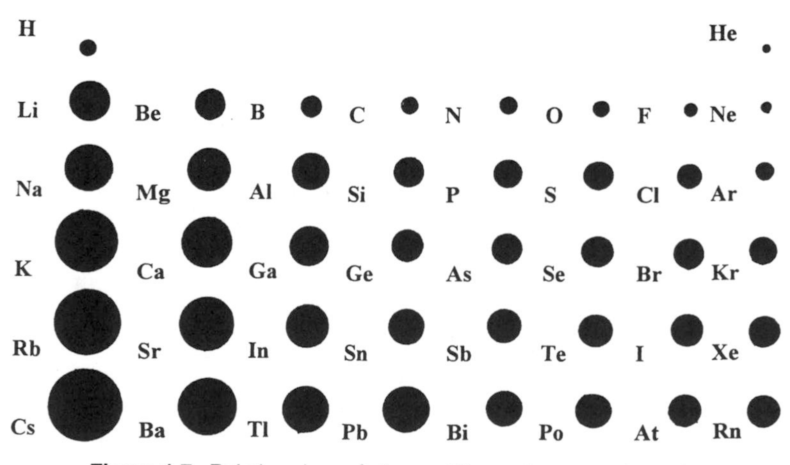

Figure 1.7. Relative sizes of atoms of the main group elements.

Al > Si > P > S > Cl > Ar. Next we see that as we go from bottom to top in any given family (i.e., vertical column), we get smaller. Thus, for the halides, Group VII A or Group 7A, we have At > I > Br > Cl > F. Along with the relationship to size there is an inverse relationship to tendency to attract and hold electrons within a bond. In general, the smaller an atom (not including the rare gases, Group 8A, which generally do not readily form compounds), the greater the tendency for it to attract and hold electrons. We find that elements that are in the upper right-hand corner have the greatest ability to attract and hold on to electrons while those atoms that are to the lower left have the least ability to hold and attract electrons. Since bonding within compounds involves electrons, this general trend allows us to predict the type of bonding between two atoms.

This tendency to attract/hold on to electrons can be expressed in terms of a relative scale developed by Linus Pauling and is consistent with the trend of atomic size such that smaller elements in the upper right have the greatest electronegative values, the greatest tendency to attract and hold on to electrons, while those in the lower left have the lowest electronegative values and thus the lower tendency to attract or to hold on to electrons.

We can divide the periodic table into metals and nonmetals by drawing an imaginary stair-step that goes between boron and aluminum, silicon and germanium, arsenic and antimony, and tellurium and polonium. The elements that are to the left of this are nonmetals, while those to the right are called metals. The elements touching this stair-step are called metalloids and can be either metals or nonmetals. The further away elements are within the periodic table, the greater the difference in the ability of these two atoms to hold on to or attract electrons and the greater the tendency for them to form ionic bonds because the atom to the right will attract the electrons in the bond while the atom to the left will give them up.

Compounds are formed from combinations of different elements. The binding together to these different elements is referred to as primary bonds and are stronger than secondary bonding, which will be considered elsewhere.

We can divide the type of primary bonds formed between atoms in a compound into ionic and covalent bonding. Ionic bonds are formed by exchange of electrons where one atom receives an electron(s) and the other gives up an electron(s). Since electrons are negatively charged and the nucleus holds positively charged protons, the net charge on an atom is calculated by assessing the net charge. Thus, when a compound is formed between sodium and chloride, it is an ionic bond because sodium and chloride are far away from one another in the periodic table. The sodium atom gives up one electron, taking on a net positive one charge; we write as Na^{+1} or simply Na^{+} because it now has 11 positive protons and only 10 negatively charged electrons (because it lost or gave up one electron), giving a net positive one charge. We call such positively charged ionic atoms (or groups of atoms) cations. Chloride takes on an electron to become negatively charged because it now has a net negative one charge. The chloride atom now has 17 positively charged protons but 18 negatively charged electrons or a net negative one, written as Cl^{-1} or simply Cl^{-}. Such negatively charged ionic atoms (or groups of atoms) are called anions. While we write the formula for sodium chloride as NaCl, it is really Na^+, Cl^-. The ionic compound formed from sodium and chlorine is called sodium chloride and it is what we call common table salt.

The most ionic bonds are formed between ions from atoms having large differences in electronegativity.

While ionic bonding is very important in inorganic chemistry, it is less important in the science of the giant molecule, which mainly contains atoms bonded together with covalent bonds.

The bonds between similar atoms or between atoms with similar electronegativity values are formed by sharing of electrons and are called covalent bonds. The electronegativities of hydrogen and carbon are similar and hence covalent bonds between carbon and hydrogen atoms are present in hydrocarbons, such as methane:

H
|
H−C−H   or CH_4
|
H

Polar covalent bonds are formed when the difference in the electronegativity values is greater than that in molecules, such as hydrogen and methane. Thus, methyl chloride,

H
|
H−C−Cl   or H_3CCl
|
H

is a polar molecule because of the relatively large difference between the electronegativity values of carbon and chlorine.

The combination between carbon and oxygen is found in a number of small and giant molecules, forming what is called a carbonyl. It is present in the small molecules carbon dioxide, CO_2, carbon monoxide, CO, and formaldehyde, $H_2C{=}O$ as below.

Focusing on the C=O combination, carbon is to the left of oxygen so it has a lesser tendency to attract electrons; thus, on the average the electrons will act as though they are more associated with the oxygen, making the C=O bond like a little dipole where the carbon is a little positively charged and the oxygen, because it has more of the electron, a little negatively charged. Such dipoles attract one another, forming secondary bonds called dipole bonds.

H δ−   H δ−
C=O δ+ ······C=O δ+
H   H

While it is hard to tell from the periodic table, hydrogen has an electronegativity value similar to that of carbon. Thus, the combination of nitrogen and hydrogen, such as in dimethyl amine as below, has the nitrogen with a partial negative charge and the nitrogen with a partial positive charge, giving another example of a secondary dipole bond.

```
H3C  δ-       H3C  δ-
   \             \
    N—H δ+ ······ N—H δ+
   /             /
H3C           H3C
```

There is a special kind of dipole bond called the hydrogen bond, where a hydrogen atom bonded to an electronegative atom such as nitrogen, oxygen, chlorine, fluorine, and phosphorus is "caught" between another electronegative bond. The dipole bond in dimethyl amine above is an example of this. Another would be the bond formed between formaldehyde and dimethyl amine.

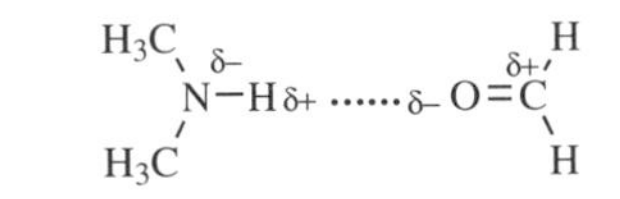

Still another example, and the most important, is the bonding that exists within water. Water is a liquid, rather than a gas, at room temperature because of this hydrogen bonding that makes water appear to act not as single H_2O molecules but rather it acts as if it were lots of water molecules because of the hydrogen bonding.

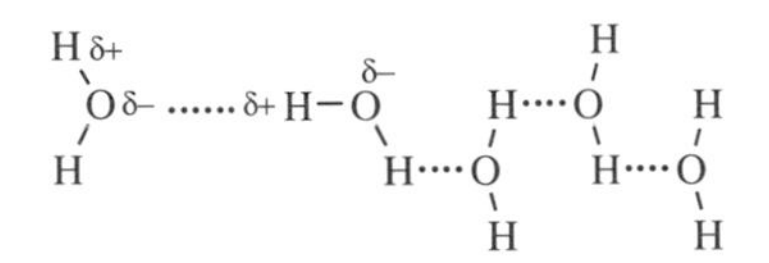

Hydrogen bonding is important in nylons, proteins, and nucleic acids. The polar bond is also important in most polymers containing atoms such as Cl, N, and O, so look for them. More about secondary bonds in Section 1.14.

In general, single bonds, such as those present in ethane ($H_3C–CH_3$), are called sigma bonds. Additional bonds such as those present in ethylene ($H_2C=CH_2$) are called pi bonds. The pi bonds are located above and below the bonding axis of the sigma bond. The bonds in ethylene, which are called double bonds, are not twice as strong as single (sigma) bonds. Actually, because of the presence of the pi bonds, double bonds are much more reactive than single sigma bonds.

Throughout the text, different types of formulas and models will be employed to emphasize various aspects of the chemical structures (Figure 1.8). General molecular formulas are employed for brevity, whereas skeletal formulas are used to emphasize main-chain or other desired characteristics such as branching and to show structural features related to bond angles. Generalized line drawings convey more extensive generalizations, in expanded structural formulas which emphasize the bonding among the different atoms. Ball-and-stick models (Table 1.2) are used to convey bonding, bonding angles, possible relative positions of the various atoms,

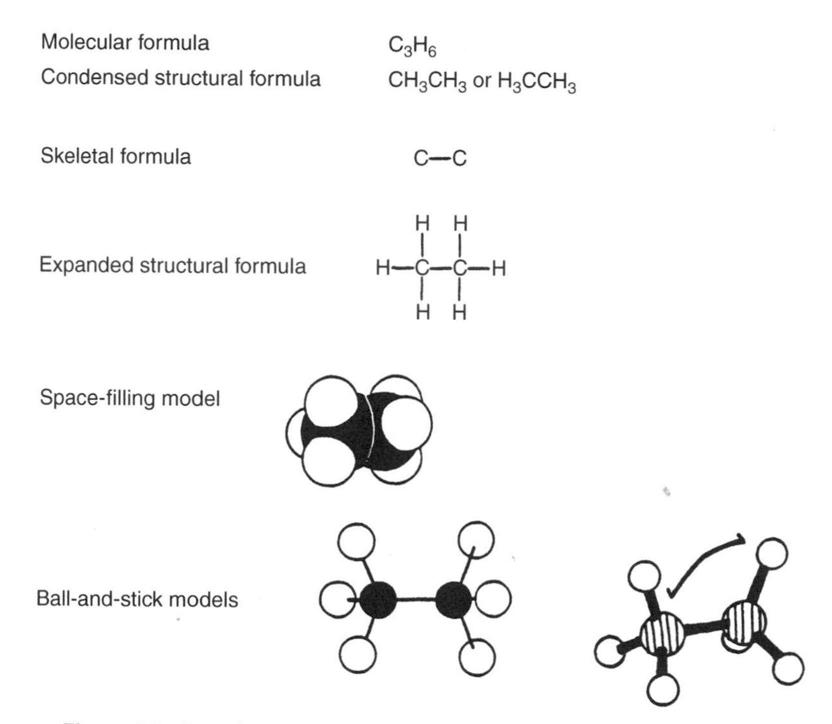

Figure 1.8. Sample models for depicting the molecular structure of ethane.

and associated geometric properties of the atoms. Space-filling models are constructed from atomic models whose relative size is related to the actual volumes occupied by the particular atoms. Still other pictorial models convey further aspects of the overall geometry and shape of molecules.

Again, carbon forms four primary bonds, oxygen forms two primary bonds, nitrogen forms two bonds, and hydrogen forms a single bond. (The lone main exception to this is carbon monoxide.) Look for this as you move through the book. More about this bonding can be found in Chapters 2 and 3.

1.14 INTERMOLECULAR FORCES

Ionic bonds between atoms with large differences in electronegativity values and covalent bonds between atoms with small differences in electronegativity values are called primary covalent bonds. The length of primary covalent bonds varies from 0.09 to 0.2 nm, and that of the carbon–carbon single bond is 0.15–0.16 nm. These primary bonds are strong bonds with energies usually greater than 90 kcal/mol.

There are also attractive forces between molecules, called secondary forces. These forces operate over long distances of 25–50 nm and have lower energy values (1–10 kcal/mol) than primary bonds. These secondary forces are called van der

Waals forces. Intermolecular forces increase cumulatively as one goes from methane (CH_4) to ethane (C_2H_6) to propane (C_3H_6), and so on, in a homologous series. A homologous series here is one in which each member differs by a methylene group (CH_2).

These secondary forces may be classified as weak London or dispersion forces (about 2 kcal/mol), dipole–dipole interactions (2–6 kcal/mol), and hydrogen bonds (about 10 kcal/mol). Since these forces are cumulative, the secondary bond energies and boiling points increase as one goes from methane (CH_4) to ethane (CH_3CH_3) to propane ($CH_3CH_2CH_3$) to butane ($CH_3CH_2CH_2CH_3$), and so on.

1.15 UNITS OF MEASUREMENT

Scientists and citizens of most other nations use the meter–gram–second (mgs) or metric system for measuring distance, weight, and time. The metric system will be used occasionally in this book. However, since Americans are moving very slowly, inch by inch (25.4 mm), from the outmoded foot–pound–second (fps) system to the metric (mgs) system, we will use the English system throughout this book. A conversion table for changing fps units to mgs units is given in Table 1.3.

We will use the Celsius (centigradc) temperature scale in which water freezes at 0°C and boils at 100°C, as well as the Fahrenheit temperature scale, in which water freezes and boils at 32°F and 212°F, respectively. We will also use the Kelvin (K) temperature scale (absolute temperature scale), in which water frezes and boils at 273 K and 373 K, respectively.

Table 1.3 Useful conversions to metric measures

Symbol	When You Know (fps)	Multiply by	To Obtain (mgs)	Symbol
in.	Inch	2.5	Centimeter	cm
yd	Yard	0.9	Meter	m
mi	Mile	1.6	Kilometer	km
oz	Ounce	28	Gram	g
lb	Pound	0.45	Kilogram	kg
tsp	Teaspoon	5	Milliliter	mL
Tbsp	Tablespoon	15	Milliliter	mL
fl. oz	Fluid ounce	30	Milliliter	mL
c	Cup	0.24	Liter	L
qt	Quart	0.95	Liter	L
gal	Gallon	3.8	Liter	L
yd^3	Cubic yard	0.76	Cubic meter	m^3
°F	Fahrenheit	5/9 (after subtracting 32)	Celsius	°C
°C	Celsius (centigrade)	Add 273	Kelvin	K

Table 1.4 Prefixes for multiples and submultiples

Multiple or Submultiple		Prefix	SI Symbol
10^{12}	1000 000 000 000	tera	T
10^{9}	1000 000 000	giga	G
10^{6}	1000 000	mega	M
10^{3}	1000	kilo	k
10^{2}	100	hecto	h
10^{1}	10	deka	da
10^{0}	1		
10^{-1}	0.1	deci	d
10^{-2}	0.01	centi	c
10^{-3}	0.001	milli	m
10^{-6}	0.000 001	micro	μ
10^{-9}	0.000 000 001	nano	n
10^{-12}	0.000 000 000 001	pico	p
10^{-15}	0.000 000 000 000 001	femto	f
10^{-18}	0.000 000 000 000 000 001	atto	a

As shown in Table 1.4, multiples or submultiples of 10 are used as prefixes to the mgs units in the metric system. The prefixes kilo (k), mega (M), and giga (G) represent multiples of one thousand (10^3), one million (10^6), and one billion (10^9). (The exponent denotes the number of integers after the first integer, as illustrated in Table 1.4.) Other common prefixes are centi (c), milli (m), micro (μ), and nano (n) for submultiples of one hundredth (10^{-2}), one thousandth (10^{-3}), one millionth (10^{-6}), and one billionth (10^{-9}). (The negative exponent denotes the number of decimal places that precede the first integer.) It should be pointed out that 1 billion in the United States is 10^9 but is 10^{12} in the United Kingdom and many other countries.

GLOSSARY

Anion: A negatively charged ion.

Atomic number: A number that is equal to the number of protons in a specific atom.

Atom: The building blocks of the universe. An atom is the smallest stable part of an element.

Avogadro's number: 6.023×10^{23} particles in a mole.

Cation: A positively charged ion.

Celsius: Temperature scale in which water freezes at 0°C and boils at 100°C.

Covalent bond: Bonds formed by sharing of electrons.

Dipole–dipole interaction: Moderately strong van der Waals forces.

Electron: The negatively charged building unit for all atoms.

Electronegativity: A measure of the tendency of an atom to attract electrons.

Element: A substance, such as carbon, consisting of identical atoms.

Endothermic reaction: A reaction in whcih energy is absorbed.

Exothermic reaction: A reaction in which energy, in the form of heat, is released.

Gram mole: Mass of 6.023×10^{23} particles in grams.

Homologous series: A series of related organic compounds, such as each differing by a methylene group (CH_2).

Hydrogen bond: Strong secondary forces resulting from the attraction of the hydrogen atom to an oxygen or nitrogen atom.

Ion: A charged atom.

Ionic bond: Bonds formed by an exchange of electrons.

kcal: Kilocalorie (1000 calories).

Kelvin: An absolute temperature scale in which water freezes at 273 K and boils at 373 K.

Law of Conservation of Matter: Principle stating that the total amount of mass remains unchanged in chemical reactions.

Lewis representation: Designation of outer or valence electrons as dots.

London dispersion force: Weak van der Waals dispersion forces.

Mass: A quantity of matter that is independent of gravity.

Mass number: A number that is equal to the number of protons plus the number of neutrons in a specific atom.

Matter: Anything that has mass and occupies space.

Metric system: A decimal system of units for length in meters (m), mass in grams (g), and time in seconds (s).

Mole: 6.023×10^{23} particles.

Molecule: A combination of atoms capable of independent existence—for example, hydrogen (H_2), which is a diatomic molecule.

Neutron: An uncharged building unit for all atoms except hydrogen. The mass of the neutron is approximately 1 amu.

nm: Nanometer, 10^{-9} m.

Nucleon: Nuclear particles, that is, protons and neutrons.

Octet Rule: Rule of 8; that is, a stable compound has 8 outer electrons in the outer shell of the atom.

Orbital: The probable pathway of an electron in an atom.

Periodic Law: The arrangement of elements in order of increasing atomic numbers, which shows the periodic variation in many chemical and physical properties.

Periodic table: A systematic arrangement of atoms in the order of their increasing atomic numbers.

Periodic table families: Groups of elements arranged in vertical columns, all having the same number of valence electrons.

Periodicity: The position of an element in the periodic table.

Pi bonds: The bonds above and below the sigma bonds in double-bonded atoms, such as ethylene.

Primary chemical bond: Bonds between atoms in molecules in which the electrons are shared or exchanged.

Principal quantum number: Numbers used to describe the gross distance of electrons from the nucleus in an atom.

Proton: The positively charged building unit for all atoms. The mass of the proton is approximately 1 amu. This mass is approximately 1840 times that of the electron.

Quantum level: A specific energy level for an electron in the atomic shell.

Quantum mechanics: A description of the movement and location of an electron in an atom.

Quantum number: Numbers used to describe the average position and possible pathway of an electron.

Secondary quantum number: Numbers used to describe the shapes of the probable path of an electron. The letters *s*, *p*, *d*, and *f* are used to describe the subshells.

Sigma bond: Single covalent bonds between atoms.

Valence electron: Electrons in the outer shell of an atom.

van der Waals force: Attractive force between the nonpolar atoms. Also called London dispersion force.

Wavelength: The distance between waves.

x ray: Electromagnetic radiation of extremely short wavelengths.

REVIEW QUESTIONS

1. Will the mass of an astronaut be greater or less in outer space than on the earth's surface?

2. What is 0 kelvin (K) on the Celsius scale?

3. What is 0 kelvin (K) on the Fahrenheit scale?

4. Which of the following are actually chemical elements: water, fire, carbon, hydrogen?

5. How many atoms are there in a molecule of methane (CH_4)?

6. How does a proton differ from a hydrogen atom?

7. Which has the greater mass: an electron or a proton?
8. Which has the longer wavelength: visible light or an x ray?
9. What is the atomic number of hydrogen, carbon, nitrogen, and oxygen?
10. What is the atomic weight of hydrogen, carbon, nitrogen, and oxygen?
11. What element has the same number of electrons as carbon in its outer shell? (*Hint*: Use the periodic table, Figure 1.3.)
12. Which of the following have covalent bonds: CH_4, H_2O, C_2H_6?
13. Which of the following have ionic bonds: NaCl, LiF, HCl?
14. Show the Lewis dot representation for methane.
15. Which of the following is an exothermic reaction: boiling eggs or a burning candle?
16. If the total weight of reactants in a chemical reaction is 18 g, which is the weight of the products of this reaction? (Assume 100% reaction.)
17. How many particles are there in 0.1 mol?
18. What is the sign of the charge of an anion?
19. How many nanometers (nm) are there in 1 meter (m)?
20. If CH_4 and C_3H_8 are members of a homologous series, what is the formula for the homologue with two carbon atoms?
21. Which is stronger, a dipole–dipole interaction or a London dispersion force?
22. Which is stronger, a hydrogen bond or a covalent bond in CH_4?

BIBLIOGRAPHY

Chandrasekhar, P. (1999). *Conducting Polymers: Fundamentals and Applications—A Practical Approach*, Kluwer, New York, 1999.

Chung, T. (2001). *Advances in Therotropic Liquid Crystal Polymers*, Technomic, Lancaster, PA.

Collings, P. J., and Hird, M. (1997). *Introduction to Liquid Crystals, Chemistry and Physics*, Taylor and Francis, London.

Frechet, J., and Tomalia, D. (2002). *Dendrimers and Other Dendritic Polymers*, Wiley, New York.

Gebelein, C., and Carraher, C. (1995). *Industrial Biotechnological Polymers*, Technomic, Lancaster, PA.

Kawazoe, X., Ohno, K., and Kondow, T. (2001). *Clusters and Nanomaterials*, Springer, New York.

Mishra, M., and Kobayashi, S. (1999). *Star and Hyperbranched Polymers*, Marcel Dekker, New York.

Newkome, G., Moorefield, C., and Vogtle, F. (2001). *Dendrimers and Dendrons*, Wiley, New York.

McCormick, C. (2000). *Stimuli-Responsive Water-Soluble Polymers*, ACS, Washington, D.C.

Rupprecht, L. (1999). *Conductive Polymers and Plastics*, ChemTec, Toronto.

Wallace, G., and Spinks, G. (1996). *Conductive Electroactive Polymers: Intelligent Materials Systems*, Technomic, Lancaster, PA.

ANSWERS TO REVIEW QUESTIONS

1. The weight of the astronaut will be less in outer space but the mass will be unchanged.
2. −273°C.
3. −459°F.
4. Carbon and hydrogen.
5. 5.
6. The proton (H^+) has one less electron than the hydrogen atom (H).
7. The proton has a mass 1840 times that of the electron.
8. Visible light has a much longer wavelength.
9. 1, 6, 7, 8.
10. 1, 12, 14, 16.
11. Silicon; and other 4A members.
12. All three.
13. NaCl and LiF.
14.
```
   H
   ..
H : C : H
   ..
   H
```
15. A burning candle.
16. 18 g.
17. 6.023×10^{22}.
18. Negative.
19. 1×10^9 or 1 billion.
20. C_2H_6.
21. Dipole–dipole interaction.
22. The covalent bond.

2

SMALL ORGANIC MOLECULES

2.1 INTRODUCTION

Because they serve as good examples, organic molecules were used to illustrate covalent bonding and intermolecular forces in Chapter 1. Some additional information on organic chemistry that should be useful in the study of giant molecules is presented in this chapter.

Giant Molecules: Essential Materials for Everyday Living and Problem Solving, Second Edition, by Charles E. Carraher, Jr.
ISBN 0-471-27399-6

2.2 EARLY DEVELOPMENTS IN ORGANIC CHEMISTRY

In 1685, N. Lemery classified all matter as being animal, vegetable, or mineral. The latter class included inorganic compounds, such as salts (sodium chloride, NaCl), acids (hydrochloric acid, HCl), and alkalies (sodium hydroxide, NaOH). The former classes—that is, animal and vegetable—consisted almost entirely of organic or carbon-containing compounds.

Although Solomon referred to the reaction of vinegar with chalk in some of his proverbs several millenia ago (*Proverbs* 10:26, 25:20), the fact that all organic compounds contain carbon was not recognized until Johann Gmelin made this observation in 1848. Jöns Berzelius used the term inorganic to describe chalk and other minerals in the nineteenth century. However, both he and his contemporaries insisted that although inorganic compounds could be synthesized, it was not possible to synthesize organic compounds in the laboratory, since they believed that a "vital force" was essential for such a synthesis.

In 1828, Friedrich Wöhler demonstrated that the so-called vital force was not absolute when he produced an organic compound, urea (H_2NCONH_2), by heating an inorganic compound, ammonium cyanate, as shown by the equation

$$NH_4NCO \xrightarrow{\Delta} H_2NCONH_2$$

Polymer science might have been limited to natural polymers, such as proteins, nucleic acids, starch, and cellulose, if Wöhler had not demonstrated that the vital force was not essential in the synthesis of organic chemicals. In general, organic compounds are classified as aliphatic—that is, molecules with linear chains of atoms—or as aromatic—that is, molecules with unsaturated cyclic structures. Some organic molecules occur as saturated cyclic structures, like cyclohexane (C_6H_{12}), or heterocyclic structures, like ethylene oxide,

$$\begin{array}{c} H_2C\!-\!\!-\!CH_2 \\ \diagdown\ \diagup \\ O \end{array}$$

2.3 ALKANES

Methane (CH_4), which is the major component of natural gas and one of the decomposition products of organic matter, is the simplest and one of the most abundant organic compounds. Methane is the first member of a homologous series, called the alkane or paraffin hydrocarbon series. All alkanes have the empirical formula $H{+}CH_2{+}_nH$, where n is equal to 1 for methane and is equal to 500 or more for the giant molecule polyethylene. In spite of its name, polyethylene is a member of the alkane and not the ethylene (ethene) homologous series.

Because of different intermolecular forces, homologues have different physical properties, such as boiling points, melting points, and densities. The prefixes used for the names of the low-molecular-weight members of a homologous series are related to the number of carbon atoms present in these compounds. Thus, $H(\!-\!CH_2\!-\!)_3H$ is called propane and $H(\!-\!CH_2\!-\!)_4H$ is called butane after butyric acid, which is also a four-carbon compound and is responsible for the odor of rancid butter.

The prefixes for homologues with five or more carbon atoms are similar to those used for geometrical figures. Thus, $H(\!-\!CH_2\!-\!)_5H$ is the formula for pentane, $H(\!-\!CH_2\!-\!)_6H$ is the formula for hexane, $H(\!-\!CH_2\!-\!)_7H$ stands for heptane, and $H(\!-\!CH_2\!-\!)_8H$ is the formula for octane. The residue, after the removal of a hydrogen atom ($H^{\bullet}$) from an alkane, is called an alkyl radical. It has the general formula $H(\!-\!CH_2)^{\bullet}_n$ and is represented by the symbol $R^{\bullet}$. Specific radicals related to the alkane homologues are called methyl ($CH_3^{\bullet}$), ethyl ($C_2H_5^{\bullet}$), propyl ($C_3H_7^{\bullet}$), and so on.

Structural formulas for alkanes are simply attempts to represent models of these molecules on paper. Experimental evidence is available to show that all the carbon–hydrogen bonds in methane are of equal length and directed toward the corners of a tetrahedron. Accordingly, the bond angles between the carbon and hydrogen atoms are 109.5° and this angle is characteristic for every carbon–hydrogen bond and every carbon–carbon bond in all alkane hydrocarbons.

We may also use the Lewis representation to show the structural formulas for methane, ethane, and propane:

```
    H           H  H            H  H  H
    ..          .. ..           .. .. ..
H : C : H   H : C : C : H   H : C : C : C : H
    ..          .. ..           .. .. ..
    H           H  H            H  H  H

 Methane       Ethane          Propane
```

The names and structural formulas for several alkanes are shown in Table 2.1.

For reasons of simplicity, we shall represent the covalent bonds by short lines called single bonds. Thus, we may write the formula for methane as

```
    H
    |
H — C — H
    |
    H
```

It should be understood that each carbon atom will be surrounded by four pairs of dots representing eight electrons, and these four electron pairs represent four covalent bonds joined to carbon or hydrogen atoms. Accordingly, we may simplify these structural formulas and use less detailed skeletal formulas in which the presence of the hydrogen atoms on the carbon atom is understood. Thus, methane, ethane, and propane can be represented by the skeletal formulas

C, C—C, C—C—C

Table 2.1 Names of unbranched alkanes (normal alkanes)

Name	Number of Carbons	Geometrical Formulas	Molecular Formulas
Methane	1	CH_4	CH_4
Ethane	2	H_3CCH_3	C_2H_6
Propane	3	$H_3CCH_2CH_3$	C_3H_8
Butane	4	$H_3C(CH_2)_2CH_3$	C_4H_{10}
Pentane	5	$H_3C(CH_2)_3CH_3$	C_5H_{12}
Hexane	6	$H_3C(CH_2)_4CH_3$	C_6H_{14}
Heptane	7	$H_3C(CH_2)_5CH_3$	C_7H_{16}
Octane	8	$H_3C(CH_2)_6CH_3$	C_8H_{18}
Nonane	9	$H_3C(CH_2)_7CH_3$	C_9H_{20}
Decane	10	$H_3C(CH_2)_8CH_3$	$C_{10}H_{22}$
Undecane	11	$H_3C(CH_2)_9CH_3$	$C_{11}H_{24}$
Dodecane	12	$H_3C(CH_2)_{10}CH_3$	$C_{12}H_{26}$
Tridecane	13	$H_3C(CH_2)_{11}CH_3$	$C_{13}H_{28}$
Tetradecane	14	$H_3C(CH_2)_{12}CH_3$	$C_{14}H_{30}$
Pentadecane	15	$H_3C(CH_2)_{13}CH_3$	$C_{15}H_{32}$

If we proceed to write these simplified skeletal structural formulas for higher homologues, we will observe that two skeletal formulas can be written for butane and that three structures can be shown for pentane:

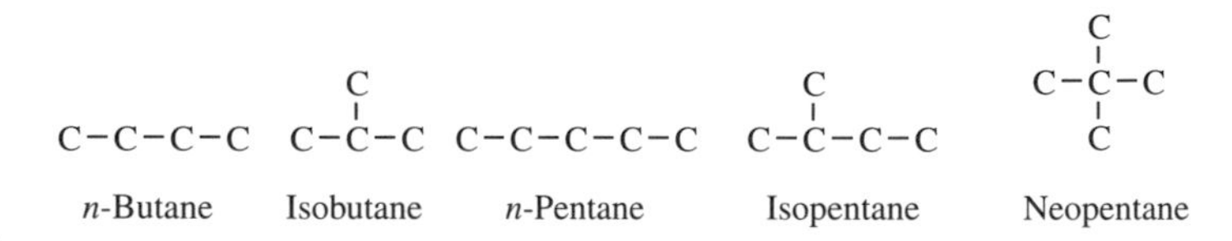

These structures represent actual structural isomers that have the common names shown. It is important to note that the number of structural isomers increases as the number of carbon atoms in the alkane molecules increases. Physical constants for isomers of pentane and hexane are shown in Table 2.2.

Systematic nomenclature has been developed by the International Union of Pure and Applied Chemistry (IUPAC), but the trivial (common) names are used almost universally for the simple alkanes. In the IUPAC system, the compound is named as a derivative of the longest chain, and the positions of the substituents are designated by appropriate numbers. Thus, the pentanes are named as follows:

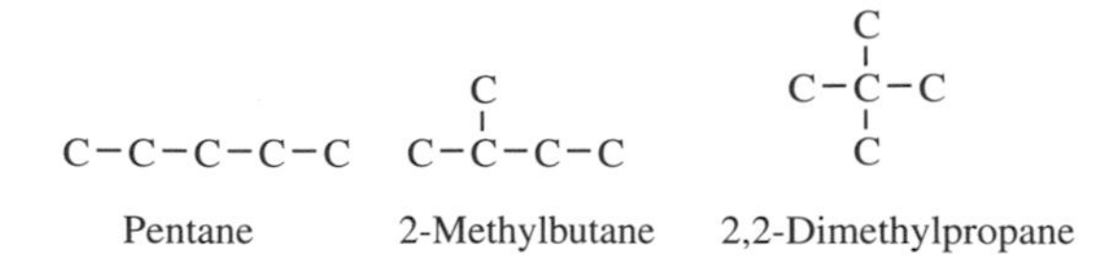

When a linear hydrocarborn, such as pentane, is burned, it produces carbon dioxide, water, and thermal energy, but it sputters during the combustion process.

Table 2.2 Geometrical isomers and physical constants for pentane and hexane

Molecular Formulas	Structural Formulas	Melting Point (°C)	Density (g/mL)
C_5H_{12}	$H_3CCH_2CH_2CH_2CH_3$	−130	0.626
C_5H_{12}	$H_3C-CH-CH_2CH_3$ $\quad\;\; \vert$ $\quad CH_3$	−160	0.620
C_5H_{12}	$\quad CH_3$ $\quad\;\; \vert$ $H_3C-C-CH_3$ $\quad\;\; \vert$ $\quad CH_3$	−20	0.613
C_6H_{14}	$H_3CCH_2CH_2CH_2CH_2CH_3$	−95	0.660
C_6H_{14}	$H_3CCHCH_2CH_2CH_3$ $\quad\;\; \vert$ $\quad CH_3$	−154	0.653
C_6H_{14}	$H_3CCH-CHCH_3$ $\quad\;\; \vert$ $\quad CH_2CH_3$	−129	0.662
C_6H_{14}	$\quad CH_3$ $\quad\;\; \vert$ $H_3C-C-CH_2CH_3$ $\quad\;\; \vert$ $\quad CH_3$	−98	0.649

However, less sputtering (more complete combustion) is observed when branched hydrocarbons, such as 2-methylbutane and 2,2-dimethylpropane, are burned. The antiknock properties of hydrocarbons in unleaded gasoline are related to the extent of branching in the molecules. Ball-and-stick models for some aliphatic hydrocarbons are shown in Figure 2.1.

2.4 UNSATURATED HYDROCARBONS (ALKENES)

As mentioned in Chapter 1, the carbon–carbon bond in alkanes is called a sigma (σ) bond and the bond angle is 109.5°. Thus, ethane ($H(CH_2)_2H$) contains seven bonds, six of which are carbon–hydrogen bonds. When ethane is heated in the presence of an appropriate catalyst, it loses two atoms of hydrogen, and the product is called by the trivial name of ethylene. The Greek symbol for fire (Δ) is used to show that heat is added to the reactants. The equation for this dehydrogenation is

$$\begin{array}{c} H\;\;H \\ |\;\;\;| \\ H-C-C-H \\ |\;\;\;| \\ H\;\;H \end{array} \xrightarrow{\Delta} H_2 + \begin{array}{c} H \qquad H \\ \diagdown\;\;\diagup \\ C=C \\ \diagup\;\;\diagdown \\ H \qquad H \end{array}$$

The reverse reaction is called catalytic hydrogenation.

Ethylene is one of the principal starting materials for the petrochemical industry. This hydrocarbon, which is called ethylene in IUPAC systematic nomenclature, is the first member of the alkene homologous series and may be represented by the empirical formulas of $H(CH_2)_nCH{=}CH_2$ or C_2H_{2n} or $R{-}CH{=}CH_2$, where R is

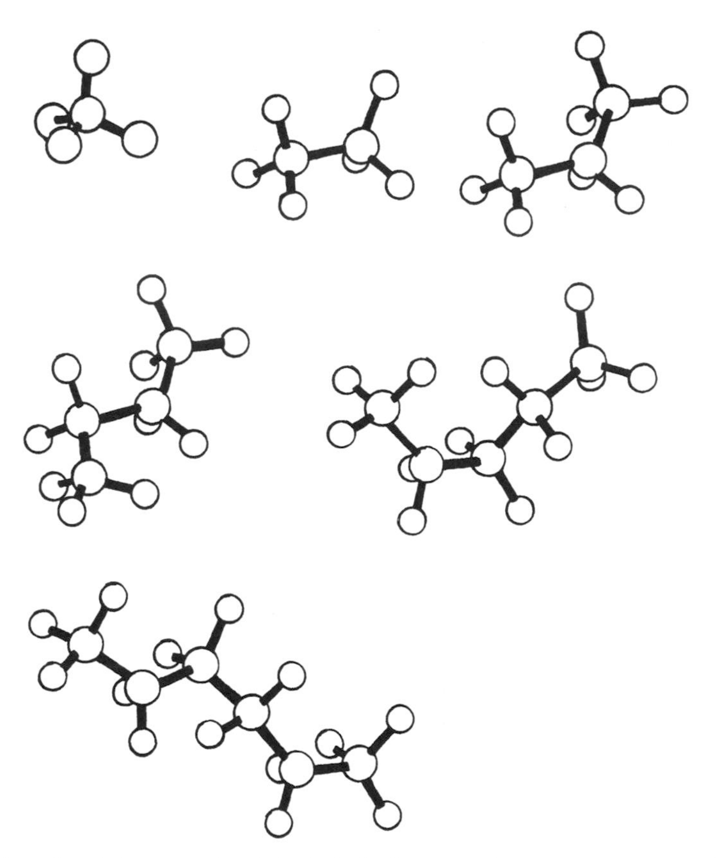

Figure 2.1. Ball-and-stick models of the first six linear members of the alkane hydrocarbon family (from left to right, top to bottom): methane, ethane, propane, butane, pentane, and hexane.

an alkyl radical. Since these homologues have fewer hydrogen atoms than the alkanes, they are sometimes called unsaturated hydrocarbons and, accordingly, the alkanes are called saturated hydrocarbons. The alkenes are also called by the common name of olefins.

The names of the alkene homologues containing three, four, and five carbon atoms are propene or propylene, butene or butylene, and pentene. As shown by the following skeletal formulas, there are three butenes that have systematic IUPAC names:

C=C−C−C 1 2 3 4	C−C=C−C 1 2 3 4	C | C=C−C 1 2 3
1-Butene (butylene)	2-Butene	2-Methylpropene (isobutylene)

We could make simple models of 2-butene using two toothpicks to hold two gumdrops together for carbons 2 and 3. If single toothpicks were used to bond gumdrops representing carbons 1 and 4 at an angle of 109.5°, we would find that these two carbon atoms could be either on one side of the plane of the double-bonded carbon atoms or on opposite sides of this plane. These models represent actual geometrical isomers that are called cis ("on this side") and trans ("across") isomers, as shown in the skeletal structures

```
                  C
C   C             |
|   |           C=C
C=C               |
                  C
```

cis-2-Butene   *trans*-2-Butene

The word isomer is derived from the Greek word *iso*, meaning equal, and *mer*, meaning parts. Isomers have the same chemical composition and empirical formulas (in this case C_4H_8) but have different structural formulas.

As shown in Table 2.3, cis and trans isomers may be distinguished from each other by physical properties, such as melting and boiling points. It is of interest to note that naturally occurring rubber is an olefinic hydrocarbon made up of repetitive cis linkages. In contrast, another naturally occurring hydrocarbon with a similar empirical (simplest) formula, $(C_5H_8)_n$, is a nonelastic, rigid material. The latter, which is called gutta-percha, consists of repetitive trans linkages.

The empirical formula $(C_5H_8)_n$ is the general formula for many naturally occurring materials called terpenes. The value for n may range from 1 to over 1000. When there are many repeating units, such as in rubber and gutta-percha, the product is called a giant molecule, macromolecule, or polymer. The prefix macro is from the Greek word *makrus*, meaning big one.

The U.S. production for these polymers of ethylene and propylene in 2000 was 17 million and 7.7 million tons, respectively. The nonpolymeric olefins, such as ethylene, which are called monomers, are also used for the production of many other organic compounds (e.g., petrochemicals).

The cycloalkanes—for example, cyclopropane $(CH_2)_3$, cyclobutane $(CH_2)_4$, and cyclopentane $(CH_2)_5$—represent another homologous hydrocarbon series. Asphalt, which is used on roofs and for road surfaces, contains high-molecular-weight cycloalkanes. As shown in Table 2.4, in contrast to the linear structures of the alkenes, the skeletal structures of the lower cycloalkene homologues are simple geometrical figures, such as triangles, squares, pentagons, and hexagons.

Table 2.3 Properties of selected cis–trans isomers

	Melting Point (°C)	Boiling Point (°C)	Density (g/mL)
cis-2-Butene	−139	3.7	0.621
trans-2-Butene	−106	0.9	0.604
cis-1-Chloro-1-butene	—	63	0.915
trans-1-Chloro-1-butene	—	68	0.921
cis-2-Chloro-2-butene	−117	71	0.924
trans-2-Chloro-2-butene	−106	63	0.914

Table 2.4 Structures of some common cycloalkanes

Name	Molecular Formula	Structural Formula
Cyclopropane	C_3H_6	H_2C ring: CH_2 / H_2C-CH_2
Cyclobutane	C_4H_8	H_2C-CH_2 / H_2C-CH_2
Cyclopentane	C_5H_{10}	CH_2 / H_2C, CH_2 / H_2C-CH_2
Cyclohexane	C_6H_{12}	CH_2 / H_2C, CH_2 / H_2C, CH_2 / CH_2

The alkynes constitute another homologous series. Acetylene (HC≡CH), which is the first and most important member of this series, may be produced by the catalytic cracking or decomposition of saturated hydrocarbons. However, as shown by the following equation, acetylene may also be readily produced by the addition of water to calcium carbide (CaC_2):

$$\underset{\text{Calcium carbide}}{CaC_2} + \underset{\text{Water}}{2H_2O} \longrightarrow \underset{\text{Acetylene}}{HC{\equiv}CH} + \underset{\text{Calcium hydroxide}}{Ca(OH)_2}$$

Acetylene is burned in oxyacetylene welding torches. It is also the starting material for many other organic chemicals.

Some typical organic chemical compounds are discussed in Section 2.5.

The term configuration is used to describe different orientations in molecules that are locked in because of the bonding that is present. The molecule ethylene has two bonds connecting the two carbon atoms. We call such two bond combinations "double bonds." Such double bonds restrict the rotation of the two connected carbon atoms so that we have two compounds, as noted above, called the cis and the trans compounds.

By comparison, there is rotation about single bonds such as is present in ethane. Long chains of such carbon atoms along with the appropriate number of hydrogen atoms are similar to a rope in that they can exist in a number of different orientations ranging from circular, to straight, to any combination of such geometries. We call such different shapes conformations. In solution, molecules are in constant local movement so that at one moment a chain of hydrocarbons like *n*-pentane may be linear and then the next moment it may vacillate to a somewhat curved

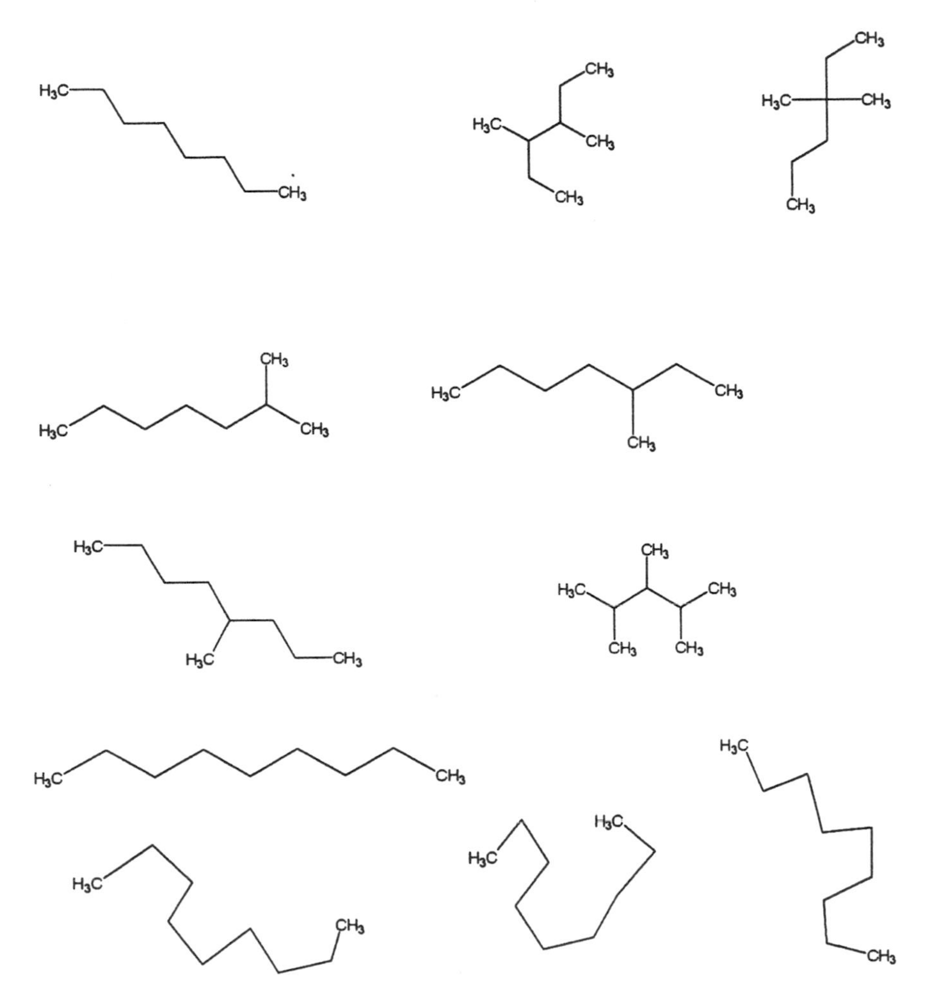

Figure 2.2. The top seven structures are geometric isomers of *n*-octane. The bottom four structures are different conformational structures of *n*-nonane.

structure and in another moment to a more circular conformation and finally in another moment back to the original linear conformation. Such changes in conformation result from the simple rotation about single bonds. Figure 2.2 contains both conformational and configurational structures for alkane structures containing eight carbon atoms (top) and nine carbon atoms (below).

2.5 ALIPHATIC COMPOUNDS

Alkanes, which are saturated hydrocarbons, form compounds through substitution of a hydrogen atom by another moiety. For simple substituted alkanes, the particular radical formed is named by replacing the -ane suffix by -yl. Thus we have

Hydrocarbon	Radical
Methane (CH_4)	Methyl (CH_3-)
Ethane (CH_3CH_3)	Ethyl (CH_3CH_2-)
Propane ($CH_3CH_2CH_3$)	Propyl ($CH_3CH_2CH_2-$)

A functional group, such as a hydroxyl group (OH), may replace a hydrogen atom in an alkane in the same manner that the hydrogen atom in propane is replaced by a methyl radical in isobutane (C—C(C)—C). Organic compounds containing hydroxyl groups are called by the trivial name of alcohols or by the systematic name of alkanols.

The alcohols and organic compounds, with other functional groups, belong to characteristic homologous series having nomenclature related to that used for the alkanes. Thus, the names of the lower alcohol homologues are methanol (H_3COH), ethanol ($H(CH_2)_2OH$), and propanol ($H(CH_2)_3OH$). As shown by the following skeletal formulas, there are two propanol isomers:

```
                         OH
                         |
C-C-C-OH               C-C-C
```

1-Propanol

2-Propanol
(isopropyl alcohol)

Since most nonscientists are not aware of the many homologues in the alkanol series, they may erroneously assume that any alcohol may be used as a beverage. Hence, to avoid mistaken identity, one should use the systematic names and not call an alkanol, such as methanol, by the name methyl alcohol. The latter and all other alkanols, except ethanol, are extremely toxic. For example, the poisonous fusel oils present in improperly distilled liquor are pentanols, which are also called by the trivial name of amyl alcohols. Ethanol, of course, is toxic and actually lethal in large quantities. Alcohols are designated as primary (RCH_2OH), secondary (R_2CHOH), and tertiary (R_3COH) based on the decreasing number of hydrogen atoms present on the carbon atom in addition to the hydroxyl group.

When two alkyl radicals are joined by an oxygen atom, such as in $H_5C_2OC_2H_5$, these compounds are called alkyl ethers. The following formula is for ethyl ether or diethyl ether, which was used for over a century as an anesthetic:

$$CH_3\text{—}CH_2\text{—}O\text{—}CH_2\text{—}CH_3$$

When a hydrogen atom or an alkyl radical is joined to a carbonyl group (C=O), the resulting compound, such as $H_2C{=}O$, is called an aldehyde. The systematic name for this molecule is methanal, but the trivial name formaldehyde is used universally. When two alkyl groups are joined to a carbonyl group, the compound, such as $H_3C(CO)CH_3$, is called a ketone. The preceding formula is for acetone.

Acetic acid (H_3CCOOH), which is the major constituent of vinegar, consists of the methyl radical (CH_3) and the carboxyl group (COOH). The first member of this homologous series is called formic acid (HCOOH) after the Latin word *formica*, meaning ant. The blister that forms as a result of an ant bite is caused by formic acid.

When the hydrogen atom of the carbonyl group is replaced by an alkyl radical, a neutral ester (RCOOR′) is obtained. Esters, such as amyl acetate ($H_3CCOO(CH_2)_5H$), which is also called banana oil, are responsible for many characteristic fruit odors.

Alcohols and ethers may be considered to be derivatives of water (HOH) in which either one or both hydrogen atoms are replaced by alkyl groups. Likewise, amines, which are organic bases, may be considered as derivatives of ammonia (NH_3) in which one or more of the hydrogen atoms are replaced by alkyl radicals, as shown by the following formulas for methylamines. These compounds are classified as primary, secondary, and tertiary amines in accordance with the number of hydrogen atoms displaced by alkyl groups:

H_3C-NH_2	$H_3C-N(H)-CH_3$	$H_3C-N(CH_3)-CH_3$
Methylamine (primary)	Dimethylamine (secondary)	Trimethylamine (tertiary)

Alkyl groups may also replace the hydrogen atom in hydrogen halides (such as hydrogen chloride, HCl), in hydrocyanic acid (HCN), and in hydrogen sulfide (H_2S) and may replace the hydroxyl radical in nitric acid ($HONO_2$) and sulfuric acid ($HOSO_2OH$). These replacements would produce compounds such as ethyl chloride (C_2H_5Cl), methyl cyanide or acetonitrile (CH_3CN), ethyl mercaptan (C_2H_5SH), nitromethane (CH_3NO_2), and ethyl sulfonic acid ($C_2H_5SO_2OH$). A list of typical functional groups is shown in Table 2.5.

It is important to note that more than one substituent may be present in an organic compound and that different substituents may be present in the same molecule. Ethylene glycol, which is used as an antifreeze, has the formula $HO(CH_2)_2OH$; and D-glucose, which is a simple sugar or carbohydrate, contains five alcohol groups and one ether group; and alanine, an amino acid, that contains one amine, NH_2, group and one acid, COOH, group.

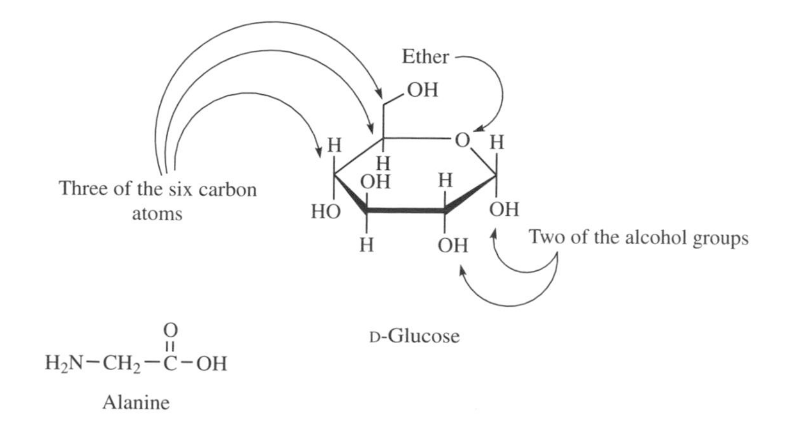

Table 2.5 Typical functional groupings

Acid	$R-\overset{O}{\overset{\|}{C}}OH$	Ether	$R-O-R$
Acid chloride	$R-\overset{O}{\overset{\|}{C}}Cl$	Isocyanate	$R-NCO$
Alcohol	$R-OH$	Ketone	$R-\overset{O}{\overset{\|}{O}}-R$
Aldehyde	$R-\overset{O}{\overset{\|}{C}}H$	Nitrile	$R-CN$
Amide	$R-\overset{O}{\overset{\|}{C}}\ \overset{H}{\overset{\mid}{N}}-R$	Sulfide	$R-S-R$
Amine	$R-NH_2$	Thiol	$R-SH$
Anhydride	$R-\overset{O}{\overset{\|}{C}}O\overset{O}{\overset{\|}{C}}-R$	Urea	$R-\overset{H}{N}-\overset{O}{\overset{\|}{C}}-\overset{H}{N}-R$
Carbonate	$R-O\overset{O}{\overset{\|}{C}}O-R$	Urethane	$R-\overset{H}{\overset{\mid}{N}}\ \overset{O}{\overset{\|}{C}}O-R$
Ester	$R-\overset{O}{\overset{\|}{C}}O-R$		

Polymer Type	Interunit Linkage	Polymer Type	Interunit Linkage
Polyester	$-\overset{O}{\overset{\|}{C}}-O-$	Polyamide	$-\overset{O}{\overset{\|}{C}}-NH-$
Polyanhydride	$-\overset{O}{\overset{\|}{C}}-O-\overset{O}{\overset{\|}{C}}-$	Polyurethane	$-\overset{O}{\overset{\|}{C}}-NH-$
Polyether	$-\overset{H}{\underset{R}{C}}-O-$	Polyurea	$-NH-\overset{O}{\overset{\|}{C}}-NH-$

2.6 UNSATURATED COMPOUNDS

The term saturated simply means that atoms cannot be added without removal of other atoms. As noted in Section 2.3, alkanes are saturated hydrocarbon (meaning containing only hydrogen and carbon) compounds. Compounds that contain double (two) and triple (three) bonds can add atoms without loss of atoms. Thus, ethylene is an unsaturated compound since we can add atoms to it without loss of any atoms. As an illustration, molecular hydrogen and hydrogen chloride are added to ethylene:

$$H_2C{=}CH_2 + H_2 \longrightarrow H_3C{-}CH_3$$

$$H_2C{=}CH_2 + H{-}Cl \longrightarrow H_3C{-}\underset{\displaystyle Cl}{\underset{|}{CH_2}}$$

As noted in Section 2.4, hydrocarbon compounds containing a double bond are called alkenes. Ethylene above is an alkene. Notice the "ene" ending that indicates that ethylene is an alkene. The term vinyl is used to describe the $H_2C{=}CH{-}$ grouping. Most vinyl polymers are derived from substituted vinyl compounds where the double bond "adds" to another double bond, eventually forming giant molecules.

Hydrocarbons that contain a triple bond are called alkynes. The best-known alkyne is ethyne (note the "yne" ending), which is better known by its common name of acetylene. Acetylene is the fuel for acetylene torches used to cut metal.

$$HC{\equiv}CH$$

Benzene, considered in the next section, is an unsaturated compound because of the presence of a series (three to be precise) of double bonds.

Many compounds contain more than one functional group. Methyl methacrylate contains both a carbon–carbon double bond and an ester group.

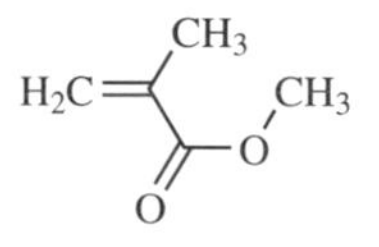

2.7 BENZENE AND ITS DERIVATIVES (AROMATIC COMPOUNDS)

Benzene, which has the formula C_6H_6, and its derivatives are called aromatic compounds. The symbol for benzene is

The circle within the hexagon indicates that this is a resonance hybrid in which all bonds have equal angles and length and, as in ethylene, there is a region of high electron density above and below the flat hexagonal ring. The two contributing

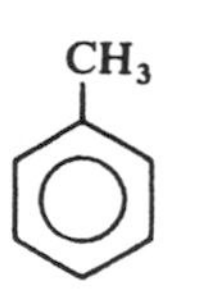

CH_2CH_3

Ethylbenzene,
$C_6H_5CH_2CH_3$

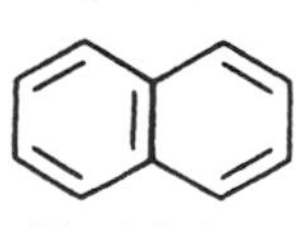

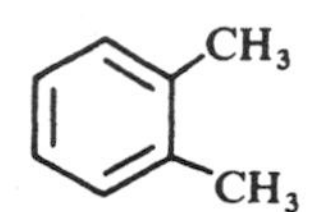

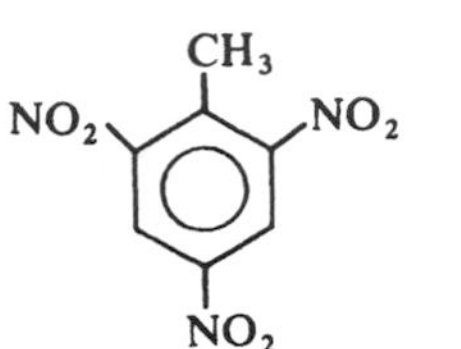

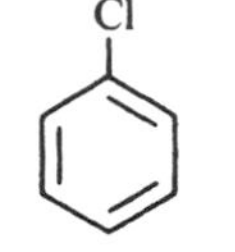

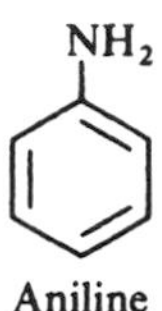

Figure 2.3. Selected aromatic compounds.

forms for this hybrid are

The true structure is actually a combination of these two forms.

The six hydrogen atoms that are not shown in the skeletal formula for benzene may be replaced by alkyl groups (R), aromatic aryl (Ar) groups, and any of the functional groups cited in the preceding discussion of aliphatic chemistry. Some typical aromatic compounds are shown in Figure 2.3.

2.8 HETEROCYCLIC COMPOUNDS

In addition to linear and cyclic aliphatic compounds, such as hexane and cyclohexane, and aromatic compounds, such as benzene, there are also heterocompounds that have other atoms besides carbon in their molecules. The heterocompounds may be linear, such as ethyl ether, or heterocyclic, such as ethylene oxide (oxirane).

Structures of these typical compounds are

$H(CH_2)_2O(CH_2)_2H$ — Ethyl ether

H_2C—CH_2 bridged by O, or triangle with O — Ethylene oxide (oxirane)

Some of the more important cyclic and heterocyclic compounds are shown in Table 2.6.

Table 2.6 Structures of selected simple cyclic and heterocyclic compounds

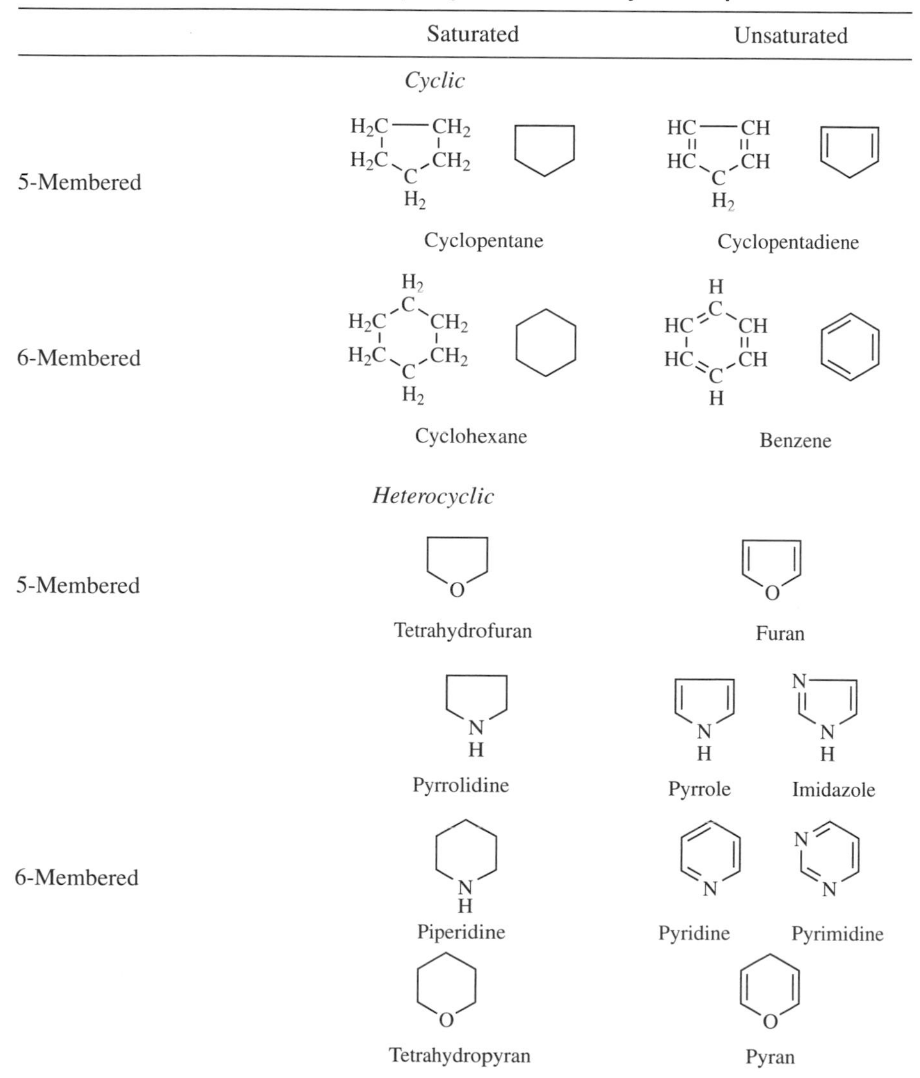

	Saturated	Unsaturated
	Cyclic	
5-Membered	Cyclopentane	Cyclopentadiene
6-Membered	Cyclohexane	Benzene
	Heterocyclic	
5-Membered	Tetrahydrofuran	Furan
	Pyrrolidine	Pyrrole, Imidazole
6-Membered	Piperidine	Pyridine, Pyrimidine
	Tetrahydropyran	Pyran

Table 2.6 (*Continued*)

	Saturated	Unsaturated
	Fused Rings	
6-Membered plus 5-membered	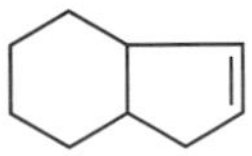 Hydrindane	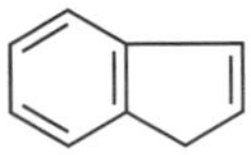 Indene
Three 6-membered plus one 5-membered	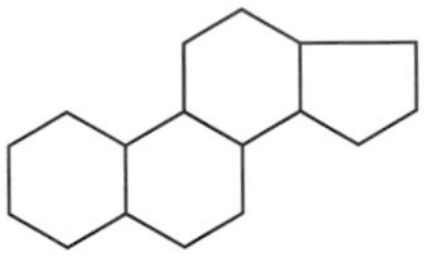 Sterane	
	Fused Heterocyclic	
6-Membered plus 5-membered	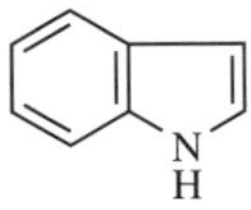 N H	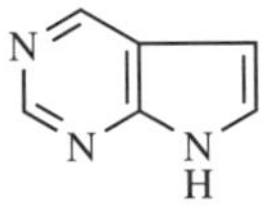 N N N H Purine
6-Membered plus 6-membered	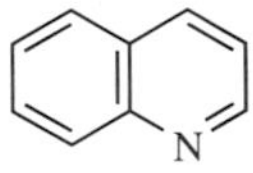 N Quinoline	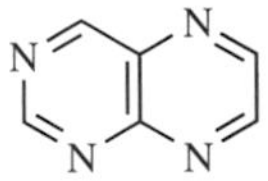 N N N N Pteridine

2.9 POLYMERIC STRUCTURE

As will be described in Chapter 3, the structure of organic polymers is similar to that of small organic compounds. The principal difference is that polymers are made up of long sequences of the smaller molecules, which are called repeating units. Thus, as shown by the following structural formulas, the principal difference between the small molecule decane and a selected polyethylene molecule is the number of repeating units:

$$H\text{-}(CH_2)_{10}\text{-}H \qquad H\text{-}(CH_2CH_2)_{500}\text{-}H$$

Decane Polyethylene

Obviously, there is much more to organic chemistry than we have discussed in this chapter. However, the brief discussions in these first two chapters should provide sufficient background for an appreciation of giant molecules, which are discussed in subsequent chapters in this book.

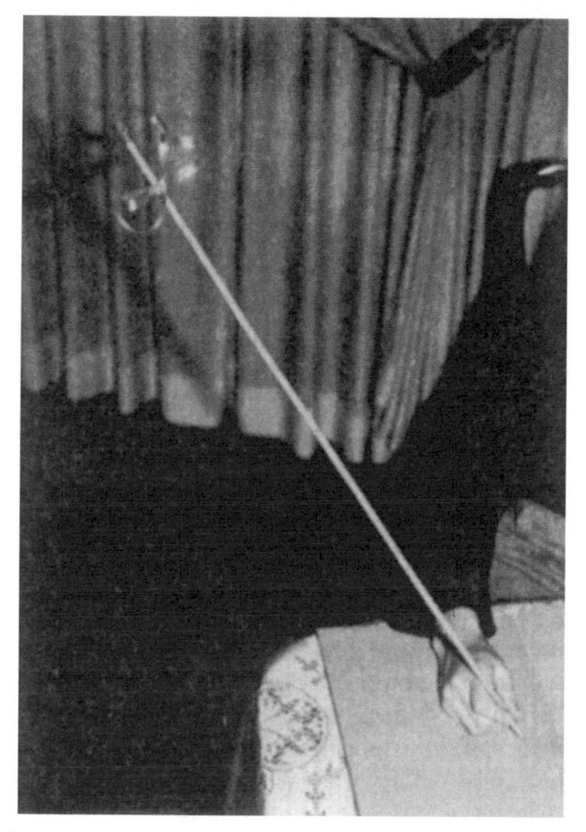

Figure 2.4. Giant polymer pencil. Courtesy of Kenrich Petrochemicals, Inc.

You may also gain some appreciation of the relative size of giant molecules, as compared to ordinary molecules, by observing the giant polymer pencil shown in Figure 2.4. This pencil was made by forcing a mixture of wood flour and polystyrene through a hot circular die in a process called extrusion. For the convenience of the photographer, a 3-ft section was cut from the continuous extrudate. Had a 200-ft section been photographed, it would be about 300 times the length of an ordinary pencil and in the range of the ratio of the length of a giant molecule to the length of an ordinary molecule.

2.10 STRUCTURES

We will be using a number of different vehicles to convey information at the molecular level. We will use two reactions to illustrate these vehicles. The first is the reaction between hydrogen gas and oxygen gas to form water. The second is the reaction of ethylene to form polyethylene. One vehicle is the chemical equation and chemical formula. Here, the broad details of the chemical nature are given. Thus, for the reaction between molecular diatomic hydrogen and molecular diatomic oxygen to form water we have

Hydrogen	Plus	Oxygen	React to from	Water
H_2	$+$	O_2	$\rightarrow$	H_2O

As we noted before, we generally want the equation to be balanced with respect to having the same number of each element on each side of the reaction arrow, the pointed arrow. To do this we use coefficients in front of each element that tells us how many times that unit is needed. Here we have

$$2H_2 + 1O_2 \rightarrow 2H_2O$$

When the coefficient is a "1" we omit it, giving us

$$2H_2 + O_2 \rightarrow 2H_2O$$

In terms of single atoms, this means that we have $2H_2$ or four total hydrogen atoms, and O_2 means that we have 2 total oxygen atoms, giving us $2H_2O$ or two water molecules that contain a total of 2 water molecules or 4 hydrogens and 2 oxygens so that the number of hydrogen atoms and oxygen atoms are the same on both sides of the equation. This equation is then referred to as a balanced equation.

Now for the reaction between ethylene to form polyethylene. There are a number of equivalent chemical formulas for the molecule ethylene. You will notice that the common or same feature for these formulas is that each ethylene molecule has 2 carbon atoms and 4 hydrogen atoms. These can be bunched together, giving

$$C_2H_4$$

or it can be divided to show that each carbon is connected to two hydrogen and one carbon atoms, giving

$$H_2CCH_2 \quad \text{or} \quad CH_2CH_2$$

or it may be given to emphasize that there are two chemical bonds between the two carbons such as

$$H_2C{=}CH_2 \quad \text{or} \quad CH_2{=}CH_2$$

Thus, the overall reaction can be given as

Ethylene	Reacts to give	Polyethylene
$CH_2{=}CH_2$	$\longrightarrow$	$\left(CH_2{-}CH_2\right)$

A single polyethylene chain may have a variable number of ethylene units ranging from several hundred to many thousands of ethylene units, with the number of ethylene units being different for different polyethylene chains so that only an average is often given. This average is generally identified with the use of a lowercase letter, often the letters "n" and "m". Thus, we can rewrite the equation as

$$n\,CH_2{=}CH_2 \longrightarrow -\!\!(CH_2-CH_2)\!\!-_n$$

The symbols "( )" and "[ ]" are generally used to describe the repeat of the common unit, here the ethylene unit. Thus a polyethylene chain a hundred units long would have one hundred ethylene or $-CH_2-CH_2-$ units arranged in a chain.

These chemical equations can be further detailed to show the actual arrangement of atoms. One equation of this variety for the formation of water would be

$$O_2 + H_2 \longrightarrow \underset{H\quad H}{\overset{O}{/\backslash}}$$

where the arrangement of hydrogens attached to oxygen are described in greater detail.

The physical and chemical properties of polymers are often easily conveyed in pictures so that this book also contains a number of drawings that depict actual polymer structures on the molecular level. Here ethylene can be represented as below where the presence of the carbon atoms are specifically given (left structure) or implied (right structure).

H H H H
C=C
H H H H

Other representations are commonly used for a particular emphasis. Where the actual relative volume or space of a molecule is being emphasized, the so-called space-filling representations are useful (below left structure). When the geometrical arrangement is being emphasized, the so-called ball and stick representations can be used (below right structure).

Various skeletal models are often used in describing the general volume occupied (below left) or where the general geometry is being emphasized (below right). Here, only the atoms that depict the particular emphasis are given. Hydrogen atoms are often understood to be present but omitted from the structure. Below are skeletal representations of ethylene using space-filling (left structure) and ball-and-stick (right structure) representations.

Basic polymer behaviors lend themselves particularly to structural illustrations because they often behave as you would predict from viewing the structural illustrations. Each of the above representations will be employed to convey important structural concepts.

A number of computer generated structures are also used. Here, often only the skeletal backbone is given. Thus, for a short chained polyethylene the representation can be

$CH_3\text{-}CH_2\text{-}CH_2\text{-}CH_2\text{-}CH_2\text{-}CH_2\text{-}CH_2\text{-}CH_2\text{-}CH_2\text{-}CH_2CH_2\text{-}CH_2\text{-}CH_2\text{-}CH_2\text{-}CH_2\text{-}CH_2\text{-}CH_2\text{-}CH_3$

or $CH_3\text{-}(CH_2\text{-}CH_2)_8\text{-}CH_3$ or

H_3C (zigzag chain) CH_3

GLOSSARY

Acetaldehyde: H_3CCHO.

Acetic acid: H_3CCOOH.

Acetone: H_3CCOCH_3.

Acetylene: $HC{\equiv}CH$.

Alcohol: Compounds with hydroxyl (OH) substituents.

Aldehyde: $H(CH_2)_nCHO$ or

$$R-\overset{\overset{\displaystyle O}{\|}}{C}-H$$

Aliphatic: Open chains of atoms like C_2H_6 and C_4H_{10}.

Alkane: Belonging to the series having the empirical formula $H(CH_2)_nH$.

Alkene homologous series: Unsaturated hydrocarbons having the formula $H(CH_2)_nCH{=}CH_2$.

Alkyl: $H(CH_2)_n$.

Alkyne: Compounds having the empirical formula $H(CH_2)_nC{\equiv}CH$.

Amine: RNH_2, for example, CH_3NH_2, R_2NH, or R_3N.

Amine, primary: An amine with two hydrogen atoms on the nitrogen atom, that is, RNH_2.

Amine, secondary: An amine with one hydrogen atom on the nitrogen atom, that is, R_2NH.

Amine, tertiary: An amine with no hydrogen atom on the nitrogen atom, that is, R_3N.

Aniline: Aminobenzene ($C_6H_5NH_2$).

Aromatic: Cyclic unsaturated molecules, like benzene (C_6H_6).

Bond, single: Bond formed by sharing two electrons, represented by a single bar.

Branch: Substituents or chain extensions on the main chain of an organic compound.

Butane: C_4H_{10}.

Carboxyl: —COOH.

Catalyst: A substance that accelerates the attainment of equilibrium in a chemical reaction. Only a small amount of catalyst is required, and this substance can be recovered unchanged.

Cellulose: A naturally occurring carbohydrate made up of repeating units of D-glucose ($C_6H_{12}O_6$).

Chalk: Calcium carbonate ($CaCO_3$).

cis: An unsaturated organic compound with substituents on the same side of the plane of the double bond.

Ester: RCOOR′, for example, amyl acetate ($H(CH_2)_5OOCCH_3$).

Ethanol: C_2H_5OH.

Ether: ROR; also used as a trivial name for ethyl ether ($(C_2H_5)_2O$).

Ethylene glycol: $CH_2(OH)CH_2(OH)$.

Formic acid: HCOOH.

Formula, skeletal: A structural formula in which the hydrogen atoms have been omitted, for example, ethane (C—C).

Formula, structural: Two-dimensional representation of molecules on paper, for example,

```
    H
    |
H — C — H
    |
    H
```

Functional group: A group capable of reacting further.

Fusel oil: Pentanols.

Gutta-percha: A rigid, naturally occurring hydrocarbon polymer with trans arrangement around the double bonds.

H$^\bullet$: Hydrogen atom.

$H(CH_2)_n^\bullet$: Alkyl radical, for example, CH_3.

Heterocyclic compound: Cyclic compounds with other atoms in addition to carbon atoms in the ring.

Homologous series: A series of related compounds with formulas differing by a constant unit like CH_2.

Hydroxyl group: –OH.

Inorganic chemistry: The chemistry of minerals and related compounds.

iso: An organic compound with a substituent on carbon number 2.

Isomer, geometrical: Unsaturated compounds with substituents on each of the double-bonded carbon atoms. Because of lack of free rotation, part of these substituents may be on one side of the plane of the double bond or on alternate sides.

IUPAC: International Union of Pure and Applied Chemistry.

IUPAC System: A preferred systematic nomenclature for organic compounds.

Ketone: $H[(CH_2)_n]_2CO$ or

$$R-\overset{\overset{\displaystyle O}{\|}}{C}-H$$

Monomer: An organic compound capable of forming a giant molecule.

neo: An organic compound with two substituents on the same carbon atom, for example, neopentane ($CH_3C(CH_3)_2CH_3$).

Nitric acid: $HONO_2$ or HNO_3.

Normal: A straight or continuous (linear) chain structure.

Olefin: Alkenes or unsaturated hydrocarbons,

Organic chemistry: The chemistry of carbon-containing compounds.

Pentane: C_5H_{12}.

Petrochemical: Compounds derived from petroleum.

Phenol: Hydroxybenzene (C_6H_5OH).

Polyethylene: A giant molecule belonging to the alkane homologous series.

Polymer: A giant molecule or macromolecule.

Propane: C_3H_8.

Propene: Propylene ($H_3CCH{=}CH_2$).

R$^\bullet$: Alkyl radical, that is, $H(CH_2)_n^\bullet$.

Resonance hybrid: A molecule that can be represented by two or more structures that differ only in the disposition of electrons. The true formula (hybrid) is one that is in between the two contributing forms and is unusually stable.

Rubber, natural: An elastic, naturally occuring hydrocarbon with cis arrangement around the double bonds.

Starch: A naturally occurring carbohydrate made up of repeat units of D-glucose ($C_6H_{12}O_6$).

Sulfuric acid: $HOSO_2OH$.

Terpene: Compounds with the empirical formula $(C_5H_8)_n$.

Toluene: Methylbenzene ($C_6H_5CH_3$).

trans: An unsaturated organic compound with substituents on opposite sides of the plane of the double bond.

Vital force: An essential force formerly believed to be associated with living organisms.

REVIEW QUESTIONS

1. Which of the following are organic chemicals: $CaCO_3$, CH_4, C_6H_6 (benzene), $C_6H_{12}O_6$ (glucose)?
2. Which of the following are polymers: ethylene, protein, cellulose, polyethylene?
3. What is the empirical (simplest) formula for hexane?
4. How many electrons are present in the hydrogen atom?
5. What is the formula for the propyl radical?
6. What is the structural formula for ethylene?
7. What is the IUPAC name for isobutane?
8. What is the structural formula for propylene (propene)?
9. What is the difference between an alkene, an olefin, and an unsaturated hydrocarbon?
10. What is the structural formula for *trans*-2-butene?
11. What is the difference in the structure of elastic natural rubber and rigid gutta-percha?
12. What is the structural formula for acetylene?
13. What functional group is always present in an alcohol?
14. What is the formula for ethyl ether?
15. What is the general formula for an aliphatic aldehyde?
16. What is the formula for propionic acid?
17. What is the formula for ethyl acetate?
18. Is diethylamine ($(C_2H_5)_2NH$) a primary or secondary amine?
19. How many hydroxyl groups are there in ethylene glycol?

20. The hybrid benzene can be represented by two different structures:

21. What is the formula for ethylbenzene?

22. Is aniline a primary or secondary amine?

BIBLIOGRAPHY

Connell, N., and Baker, E. (1999). *Surfaces of Nanoparticles and Porous Materials*, Marcel Dekker, New York.

Craver, C., and Carraher, C. (2000). *Applied Polymer Science*, Elsevier, New York.

Datta, S., and Lohse, D. (1996). *Polymeric Compatibilizers*, Hanser-Gardner, Cincinnati.

Lutz, J., and Grossman, R. (2000). *Polymer Modifiers and Additives*, Marcel Dekker, New York.

Wypych, G. (2000). *Handbook of Fillers*, Chem Tech, Toronto.

Zweifel, H. (2001). *Plastics Additives Handbook*, Hanser-Gardner, Cincinnati.

ANSWERS TO REVIEW QUESTIONS

1. CH_4, C_6H_6, $C_6H_{12}O_6$.

2. Protein, cellulose, polyethylene.

3. $C_6H_{14}(H(CH_2)_6H)$.

4. One.

5. C_3H_7.

6.
```
H     H
 \   /
  C=C
 /   \
H     H
```

7. 2-Methylpropane.

8.
```
    H   H   H
    |   |   |
H - C - C = CH
    |
    H
```

9. They are identical.

10.
```
    CH3
    |
H - C = C - H
        |
        CH3
```

11. Gutta-percha is a *trans*-polyisoprene; rubber is a *cis*-polyisoprene.

12. $HC\equiv CH$.

13. The hydroxyl group (OH).

14. $C_2H_5OC_2H_5$.

15. RCHO.

16. H_3CCH_2COOH.

17. $H_3CCOOC_2H_5$.

18. Secondary.

19. Two.

20. Both.

21. $C_6H_5{-}C_2H_5$.

22. Primary.

3

INTRODUCTION TO THE SCIENCE OF GIANT MOLECULES

Giant Molecules: Essential Materials for Everyday Living and Problem Solving, Second Edition, by Charles E. Carraher, Jr.
ISBN 0-471-27399-6

3.1 A BRIEF HISTORY OF CHEMICAL SCIENCE AND TECHNOLOGY

The science of giant molecules is relatively new, and many living polymer scientists have spent their entire lifetimes in the development of our present knowledge. Many of the developments in polymer science have taken place in the twentieth century, and most of these have occurred during the last half of the 20th century.

Of course, humans have always been dependent on giant molecules (i.e., starch, protein, and cellulose) for food, shelter, and clothing, but little was known about these essential products until recently. Organic chemistry was poorly understood until 1828, when Friedrich Wöhler demonstrated that it was possible to synthesize organic molecules.

Progress in organic chemistry was extremely slow until the 1850s and 1860s, when Friedrich August Kekulé discovered a new way to write the structural formulas for organic compounds. Many breakthroughs in organic chemistry occurred in the last years of the nineteenth century, when chemists recognized the practicability of synthesis and were able to write meaningful structural formulas for organic compounds.

Most giant molecules are organic polymers, but little progress was made in polymer science until the 1930s because few organic chemists accepted the concepts of polymer molecules giant molecules as formulated by Hermann Staudinger; he did not receive the Nobel prize for his elucidation of the molecular structure of polymers until 1953. Many of his contemporaries maintained that polymers were simply aggregates of smaller molecules held together by physical rather than chemical forces.

Nevertheless, in spite of the delays in the development of polymer science, there were several important empirical discoveries in the technology of giant molecules in the nineteenth century. Charles Goodyear and his brother Nelson separately transformed natural rubber (*Hevea braziliensis ulei*) from a sticky thermoplastic to a useful elastomer (vulcanized rubber, Vulcanite) and a hard thermoset plastic (Ebonite or Vulcacite), respectively, by heating natural rubber with controlled amounts of sulfur in the late 1830s. Thomas Hancock, who discovered the process of curing natural rubber via reverse research—that is, by an examination of the Goodyears' product—coined the term vulcanization after the Roman god Vulcanos (Vulcan).

Likewise, Christian F. Schönbein produced cellulose nitrate by the reaction of cellulose with nitric acid, and J. P. Maynard made collodion by dissolving the cellulose nitrate in a mixture of ethanol and ethyl ether in 1847. Collodion, which was used as a liquid court plaster (Nuskin), also served in the 1860s as Parkes and Hyatt's reactant for making celluloid (the first man-made thermoplastic) and Chardonnet's reactant in 1884 for making artificial silk (the first man-made fiber). This "Chardonnet silk" was featured at the World Exposition in Paris in 1889.

Although most of these early discoveries were empirical, they may be used to explain some terminology and theory in modern polymer science. It is important to note that, like the ancient artisans, all of these inventors converted naturally occurring polymers to more useful products. Thus, in the transformation of heat-softenable thermoplastic castilla rubber to a less heat-sensitive product, Charles Goodyear introduced a relatively small number of sulfur cross-links between the long individual chainlike molecules of natural rubber (polyisoprene).

Nelson Goodyear used sulfur to introduce many cross-links between the polyisoprene chains so that the product was no longer a heat-softenable thermoplastic but rather a heat-resistant thermoset plastic. Thermoplastics are two-dimensional (linear) molecules that may be softened by heat and returned to their original states by cooling, whereas thermoset plastics are three-dimensional network polymers that cannot be softened and reshaped by heating. The prefix thermo is derived from the Greek word *thermos*, meaning warm, and *plasticos* means to shape or form. Since these pioneers did not know what a polymer was, they had no idea of the complex changes that had taken place in the pioneer production of these useful man-made rubber, plastic, and fibrous products.

It was generally recognized by the leading organic chemists of the nineteenth century that phenol would condense with formaldehyde. Since they did not recognize the essential concept of functionality—that is, the number of available reactive sites in a molecule—Baeyer, Michael, Kleeburg, and other eminent organic chemists produced worthless cross-linked goos, gunks, and messes and then returned to their classical research on reactions of monofunctional reactants. However, by the use of a large excess of phenol, Smith, Luft, and Blumer were able to obtain useful thermoplastic condensation products.

Although there is no evidence that Leo Baekeland recognized the existence of macromolecules, he did understand functionality, and by the use of controlled amounts of trifunctional phenol and difunctional formaldehyde he produced thermoplastic resins that could be converted to thermoset plastics (Bakelite). Other polymers had been produced in the laboratory before 1910, but Bakelite was the first truly synthetic plastic. The fact that the processes used today are essentially the same as those described in the original Baekeland patents demonstrates this inventor's ingenuity and knowledge of the chemistry of the condensation of trifunctional phenol with difunctional formaldehyde.

Prior to World War I, celluloid, shellac, Galalith (casein), Bakelite, cellulose acetate, natural rubber, wool, silk, cotton, rayon, and glyptal polyester coatings, as well as bitumen/asphalt, coumarone/indene, and petroleum resins, were all commercially available. However, as shown chronologically in Table 3.1, because of the lack of knowledge of polymer science, there were few additional significant developments in polymer technology prior to World War II.

The following advice was given to Dr. Staudinger by his colleagues in the 1920s: "Dear Colleague: Leave the concept of large molecules well alone.... There can be no such thing as a macromolecule." Fortunately, this future Nobel laureate disregarded their unsolicited advice and laid the groundwork for modern polymer science in the 1920s when he demonstrated that natural and synthetic polymers were not aggregates, like colloids, or cyclic compounds, like cyclohexane, but instead were long, chainlike molecules with characteristic end groups. In 1928,

Table 3.1 Chronological development of commercial polymers

Date	Material (Brand/Trade Name and/or Investor)	Typical Application
Before 1800	Cotton, flax, wool and silk fibers; bitumen caulking materials; glass and hydraulic cements, leather, cellulose sheet (paper); balata, shellac, guttapercha, *Hevea braziliensis*	
1839	Vulcanization of rubber (Charles Goodyear)	Tires
1846	Nitration of cellulose (Schönbein)	Coatings
1851	Ebonite (hard rubber; Nelson Goodyear)	Electrical insulation
1860	Molding of shellac and gutta-percha	Electrical insulation
1868	Celluloid (CN: Hyatt)	Combs, mirror, frames
1889	Regenerated cellulosic fibers (Chardonnet)	Fabric
	Cellulose nitrate photographic films (Reichenbach)	Pictures
1890	Cuprammonia rayon fibers (Despeisses)	Fabric
1892	Viscose rayon fibers (Cross, Bevan, and Beadle)	Fabric
1893	Cellulose recognized as a polymer (E. Fischer)	
1907	Phenol–formaldehyde resins (PF: Bakelite; Baekeland)	Electrical
1908	Cellulose acetate photographic films (CA)	
1912	Regenerated cellulose sheet (cellophane)	Sheets, wrappings
1923	Cellulose nitrate automobile lacquers (Duco)	Coatings
1924	Cellulose acetate fibers	
	Concept of macromolecules (H. Staudinger)	
1926	Alkyd polyesters (Kienle)	Electrical insulators
1927	Polyvinyl chloride (PVC; Semon; Koroseal)	Wall covering
1927	Cellulose acetate sheet and rods	Packaging Films
1929	Polysulfide synthetic elastomer (Thiokol; Patrick)	Solvent-resistant rubber
1929	Urea–formaldehyde resins (UF)	Electrical switches and parts
1931	Polymethyl methacrylate plastics (PMMA; Plexiglas; Rohm)	Display signs
1931	Polychloroprene elastomer (Neoprene; Carothers)	Wire coatings
1933	Polyethylene (LDPE; Fawcett and Gibson)	Cable coating, packaging, squeeze bottles
1935	Ethylcellulose	Moldings
1936	Polyvinyl acetate (PVAc)	Adhesives
1936	Polyvinyl butyral (PVB)	Safety glass
1937	Polystyrene (PS)	Kitchenware, toys, foam
1937	Styrene–butadiene (Buna-S; SBR), acrylonitrile (Buna-N), copolymer elastomers (NBR)	Tire treads
1938	Nylon 6,6 fibers (Carothers)	Fibers
1938	Fluorocarbon polymers (Teflon; Plunkett)	Gaskets, grease-repellent coatings
1939	Melamine–formaldehyde resins (MF)	Tableware
1938	Copolymers of vinyl chloride and vinylidene chloride (Pliovic)	Films, coatings
1939	Polyvinylidene chloride (PVDC; Saran)	Films, coatings

Table 3.1 (*Continued*)

Date	Material (Brand/Trade Name and/or Investor)	Typical Application
1940	Isobutylene–isoprene elastomer (butyl rubber; Thomas and Sparks)	Adhesives, coatings, caulkings
1941	Polyester fibers (PET; Whinfield and Dickson)	Fabric
1942	Unsaturated polyesters (Foster and Ellis)	Boat hulls
1942	Acrylic fibers (Orlon; Acrylan)	Fabrics
1943	Silicones (Rochow)	Gaskets, caulkings
1943	Polyurethanes (Baeyer)	Foams, elastomers
1944	Styrene–acrylonitrile–maleic anhydride, engineering plastic (Cadon)	Moldings, extrusions
1947	Epoxy resins (Schlack)	Coatings
1948	Copolymers of acrylonitrile butadiene and styrene (ABS)	Luggage, electrical devices
1955	Polyethylene (HDPE; Hogan, Banks, and Ziegler)	Bottles, film
1956	Polyoxymethylenes (acetals)	Moldings
1956	Polypropylene oxide (Hay; Noryl)	Moldings
1957	Polypropylene (Hogan, Banks, and Natta)	Moldings, carpet fiber
1957	Polycarbonate (Schnell and Fox)	Appliance parts
1959	*cis*-Polybutadiene and *cis*-polyisoprene elastomers	Rubber
1960	Ethylene–propylene copolymer elastomers (EPDM)	Sheets, gaskets
1962	Polyimide resins	High-temperature films and coatings
1965	Polybutene	Films, pipe
1965	Polyarylsulfones	High-temperature thermoplastics
1965	Poly-4-methyl-1-pentene (TPX)	Clear, low-density (0.83 g/L) moldings
1965	Styrene–butadiene block copolymers (Kraton)	Shoe soles
1970	Polybutylene terephthalate (PBT)	Engineering plastic
1970	Ethylene–tetrafluoroethylene copolymers	Wire insulation
1971	Polyphenylene sulfide (Ryton; Hill and Edmonds)	Engineering plastic
1971	Hydrogels, hydroxyacrylates	Contact lenses
1972	Acrylonitrile barrier copolymers (BAREX)	Packaging
1974	Aromatic nylons (Aramids; Kwolek and Morgan)	Tire cord
1980	Polyether ether ketone (PEEK; Rose)	High-temperature service
1982	Polyether imide (Ultem)	High-temperature service

Kurt H. Meyer and Herman F. Mark reinforced Staudinger's concepts by using x-ray techniques to determine the dimensions of the crystalline areas of macromolecules in cellulose and natural rubber.

While Staudinger was arguing the case for his concepts of macromolecules in Germany, a Harvard professor working for DuPont was actually producing giant molecules in accord with Staudinger's concepts. In the mid-1930s Wallace Carothers, along with Julian Hill, synthesized a polyamide that they called nylon 6,6. In contrast to Chardonnet's fiber, which was made by the regeneration of naturally occurring cellulose, nylon fiber was a completely synthetic polymer.

Nylon was produced by the condensation of two difunctional reactants, namely, a dicarboxylic acid and a diamine. As shown by the following empirical equation, each product produced in the stepwise reactions was capable of further reaction to produce a linear giant molecule:

$$\text{H-A-R-A-H} + \text{H-B-R}'\text{-B-H} \rightarrow \text{H-A-R-A-B-R}'\text{-B-H} + H_2O$$
$$\text{H-A-R-A-B-R}'\text{-B-H} + \text{H-A-R-A-B-R}'\text{-B-H} \rightarrow \text{H-A-R-A-B-R}'\text{-B-A-R-A-B-R}'\text{-B-H} + H_2O$$
$$\text{H-A-R-A-B-R}'\text{-B-A-R-A-B-R}'\text{-B-H} + \text{H-A-R-A-B-R}'\text{-B-A-R-A-B-R}'\text{-B-H} \rightarrow$$
$$\text{H-A-R-A-B-R}'\text{-B-A-R-A-B-R}'\text{-B-A-R-A-B-R}'\text{-B-A-R-A-B-R}'\text{-B-H} + H_2O \rightarrow\rightarrow\rightarrow\rightarrow$$

where $AH = \text{–}COOH$ and $BH = \text{–}NH_2$

As a result of Carothers' contributions and subsequent discoveries, polymerization (that is, the production of giant molecules from small molecules) has been recognized as one of the greatest discoveries of all time. As was true in the nineteenth century, the art usually preceded the science, but many developments in the mid-twentieth century were based on macromolecular concepts championed by Staudinger, Mark, and Carothers.

Many discoveries in polymer technology were serendipitous or by chance, but in many cases scientists applied polymer science concepts to these accidental discoveries to produce useful commercial products. Among these accidental discoveries are the following: J. C. Patrick obtained a rubberlike product (Thiokol) when he was attempting to synthesize an antifreeze in 1929. Fawcett and Gibson heated ethylene under very high pressure, in the presence of traces of oxygen, and obtained polyethylene (LDPE) in 1933. When the gaseous tetrafluoroethylene did not escape through the open valve in a pressure cylinder, Roy J. Plunkett cut open the cylinder and found a solid product that was polytetrafluoroethylene (Teflon) in 1938.

The leading polymer scientists of the 1930s agreed that all polymers were chainlike molecules and that the viscosities of solutions of these macromolecules were dependent on the size and shape of the molecules in these solutions. Although the large-scale production of many synthetic polymers was accelerated by World War II, it must be recognized that the production of these essential products was also dependent on the concepts developed by Staudinger, Carothers, Mark, and other polymer scientists prior to World War II.

Giant molecules are all about us. The soil we grow our foods from are largely giant molecules as are the foods we eat. The plants about us are largely giant molecules. The buildings we live in are mostly composed of giant molecules. We are walking exhibits as to the widespread nature of giant molecules: These are found in our hair and fingernails, our skin, bones, tendons, and muscles; our clothing (socks, shoes, glasses, undergarments); the morning newspaper; major amounts of our automobiles, airplanes, trucks, boats, spacecraft; our chairs, wastepaper baskets, pencils, tables, pictures, coaches, curtains, glass windows; the roads we drive on, the houses we live in, and the buildings we work in; the tapes and CDs we listen to music on; and packaging—all are either totally polymeric or contain a large amount of polymeric materials. Table 3.2 lists some general groupings of important giant molecules. Welcome to the wonderful world of giant molecules.

You will see that we use essentially interchangeably two other terms to describe giant molecules. These other terms are polymers and macromolecules. More about this in Section 3.5.

The science of giant molecules has common themes that drives their behavior and uses. Look for them. Giant molecules are interesting in that they behave the way you think they should. You should see this as you move along in the book.

An additional reason why both nature and industry have chosen to "major in polymers" is the abundance of the building blocks of polymers readily found in nature, making polymers inexpensive and readily constructible. It is interesting to note that carbon is one of the few elements that readily undergoes catenation (forming long chains) and that both natural and synthetic polymers have high carbon content. Furthermore, this catenation of carbon atoms can be both controlled and varied, permitting both synthesis of materials with reproducible properties and polymers with quite divergent properties.

Table 3.2 Polymer classes—natural and synthetic

Polymeric Materials				
Inorganic			Organic	
Natural	Synthetic	Organic/Inorganic	Natural	Synthetic
Clays	Fibrous glass	Siloxanes	Proteins	Polyethylene
Cement	Poly(sulfur nitride)	Polyphosphazenes	Nucleic acids	Polystyrene
Pottery	Poly(boron nitride)	Polyphosphate esters	Lignins	Nylons
Bricks	Silicon carbide	Polysilanes	Polysaccharides	Polyesters
Sands		Sol–Gel networks	Melanins	Polyurethanes
Glasses			Polyisoprenes	Poly(methyl methacrylate)
Rocklike				Polytetrafluoroethylene
Agate				Polyurethane
Talc				Poly(vinyl chloride)
Zirconia				Polycarbonate
Mica				Polypropylene
Asbestos				Poly(vinyl alcohol)
Quartz				
Ceramics				
Graphite/diamond				
Silicas				

3.2 POLYMERIZATION

In addition to the step reaction polymerization described in Section 3.1, synthetic polymers may also be prepared by chain reactions—that is, addition polymerization reactions. In step reaction polymerization, difunctional reactants, such as ethylene glycol and terephthalic acid, react to produce products with reactive end groups that are capable of further reaction:

$$\mathrm{HO{+}CH_2{+}_2OH} + \mathrm{HO-\overset{O}{\overset{\|}{C}}-C_6H_4-\overset{O}{\overset{\|}{C}}-OH} \rightleftharpoons$$

$$\mathrm{H_2O} + \mathrm{HO{+}CH_2{+}_2O-\overset{O}{\overset{\|}{C}}-C_6H_4-\overset{O}{\overset{\|}{C}}-OH}$$

Polyesters, nylons (polyamides), polyurethanes, epoxy resins, phenolic resins, and melamine resins are produced by step reaction polymerization.

As seen above, there are two arrows, each pointing in the opposite direction, signaling that the reaction is an equilibrium reaction. This means that not only does ethylene glycol react with terephthalic acid, giving the ester and water, but that also the ester and water can react, giving ethylene glycol and terephthalic acid. Condensation reactions are generally equilibrium reactions. The trick to forming polymer is to cause the reaction to favor moving toward the left or toward formation of polymer. This is done by removing the water, H_2O.

Most elastomers (rubbers), some fibers (polyacrylonitrile), and many plastics are produced by chain reaction polymerization. These reactions include three steps: initiation, propagation, and termination. Polymerization chain reactions may be initiated by anions, such as butyl anions ($C_4H_9{:}^-$), by cations, such as protons (H^+), or by free radicals, such as the benzoyl free radicals ($C_6H_5COO^{\bullet}$). As shown in the following equations, the initiator, such as a free radical ($R^{\bullet}$), adds to a vinyl monomer, such as vinyl chloride, to produce a new free radical.

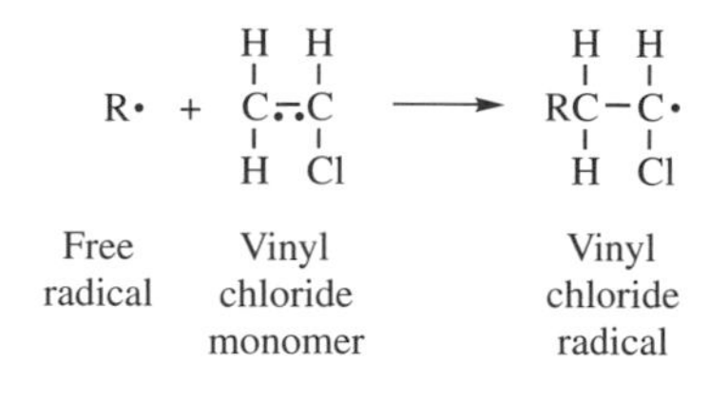

Then, as shown by the following equation, the new free radical adds to another vinyl chloride monomer molecule to produce a dimer radical, and this reaction continues rapidly and sequentially to produce larger and larger macroradicals (n = number of repeating units).

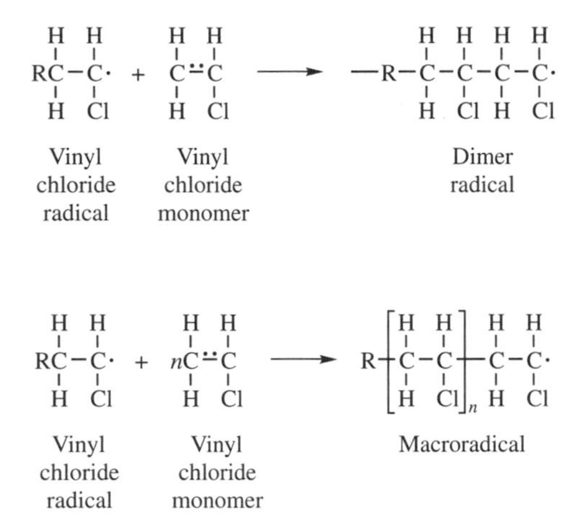

or

The reaction may be terminated by the collision of two macroradicals to produce a dead polymer (inactive polymer) in a coupling reaction or the macroradical may abstract a hydrogen atom from another molecule, called a telogen, to produce a dead polymer and a new radical.

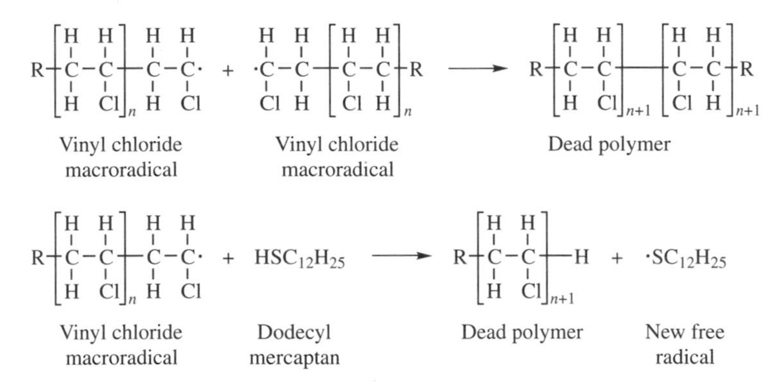

Now let us move to some particulars about the two main types of polymerization: chain and stepwise processes. As noted before, the preparation of nylon and polyesters occurs through what is called a condensation reaction or condensation polymerization. These polymers are called condensation polymers and can generally be identified because the backbone of the polymer chain has elements in addition to carbon in them. Thus, polyamides or nylons, with a repeat unit as shown below, have a nitrogen atom in the backbone.

$$\left(\overset{\overset{\displaystyle O}{\|}}{C}-R-\overset{\overset{\displaystyle O}{\|}}{C}-\overset{\overset{\displaystyle H}{|}}{N}-R'-\overset{\overset{\displaystyle H}{|}}{N}-C\right)_{n}$$

Nylon (polyamide)

while polyesters such as those shown in Section 7.3 have an oxygen in their backbone.

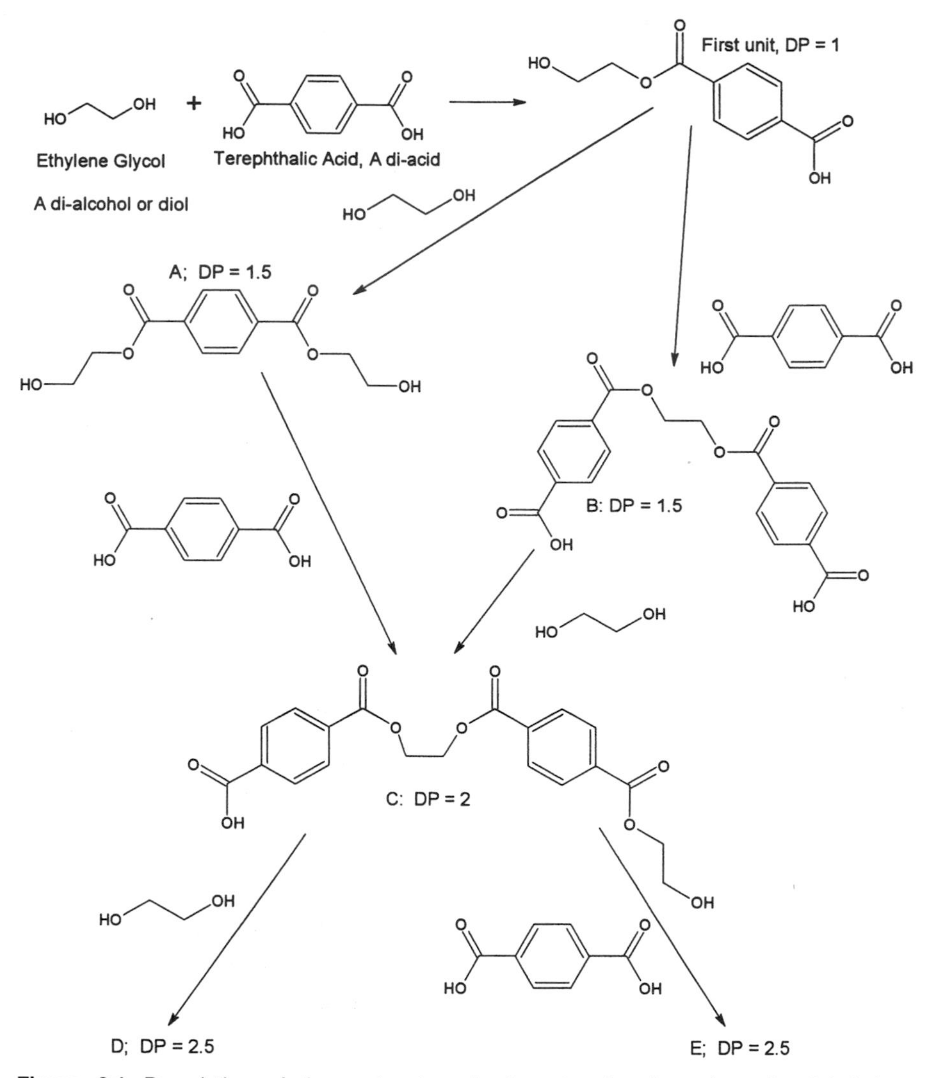

Figure 3.1. Description of the early steps in the stepwise formation of poly(ethylene terephthalate), PET.

Kinetics is the name given to the study of how fast a reaction is and the precise steps involved in the formation of the product. Condensation polymerizations generally are formed though a stepwise kinetic process or a stepwise condensation process. The series of reactions given in Figure 3.1 describe the first steps toward the formation of the polyester poly(ethylene terephthalate) or PET used in making bottles and other common objects. The first step produces a product that contains one part derived from the ethylene glycol and one part derived from terephthalic acid and is actually the beginning of the polyester chain with a degree of polymerization, DP, of 1. It has an acid group at one end and an alcohol group at the other end. The next step involves reaction with either ethylene glycol (with two alcohol groups, a diol) or terephthalic acid (with two acid groups). Reaction with ethylene glycol gives a product with two alcohol end groups (product A). Reaction with terephthalic acid gives a product with two acid end groups (product B). The product with two alcohol end groups then can react with only terephthalic acid giving again a product with one alcohol and one acid end group. The product with two acid end groups can act with only ethylene glycol, giving a product with one alcohol and one acid group (product C), the same product formed from reaction of the two alcohol end group product with the acid (product C). This stepwise sequence continues until the polyester is formed. For each step, water is formed and must be removed to "drive" the reaction toward polymer formation.

Such reactions generally take hours to occur, with products formed in high yield because the steps toward formation of long chains require that the incorporation of the other growing chains and long-chained polymer only occur near the end of the reaction.

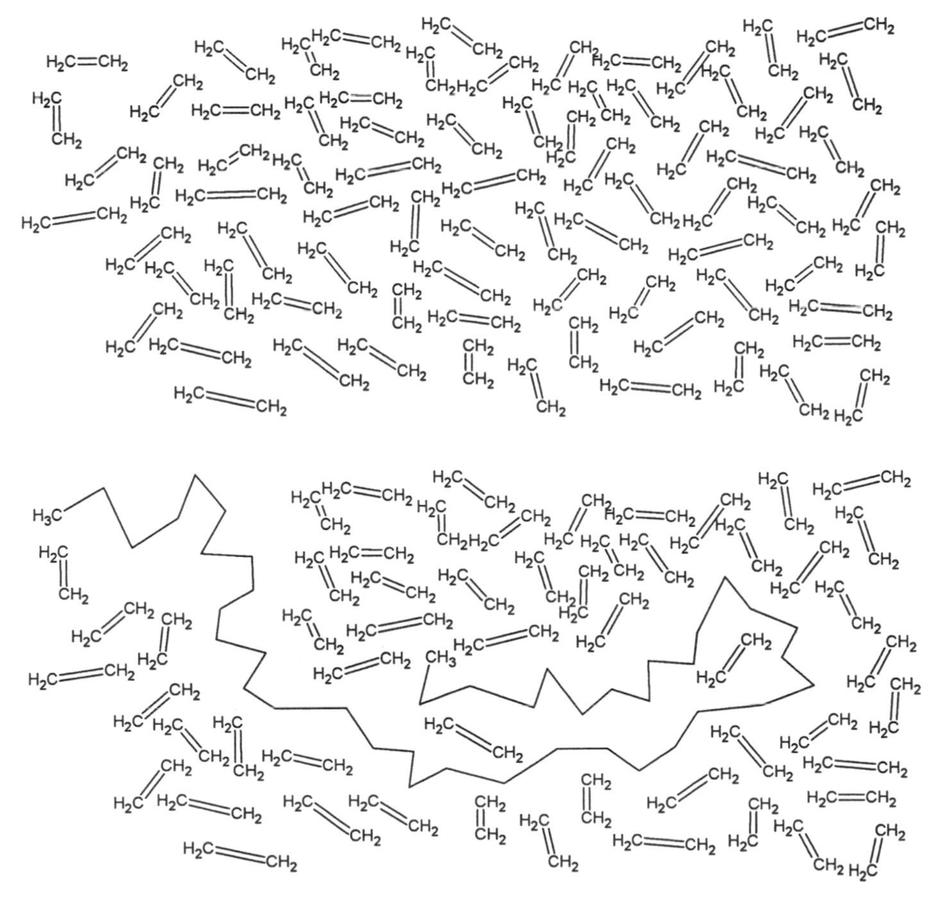

Figure 3.2. Seventy-five vinyl monomers at the beginning of the reaction (*top*) and after one chain of 20 units has been formed (*bottom*).

Vinyl polymers generally have only carbon in their backbone. They are formed from the other main process referred to as a chainwise kinetic process or as simply a chain reaction. Polymers derived from vinyl reactants such as ethylene, styrene, and vinyl chloride are formed from a chain process. Here, an active form of the monomer is created and this active form reacts with another monomer, giving an active end that in turn reacts or adds another monomer giving an active end, and so on, forming a "growing" polymer chain until termination occurs (Figure 3.2). This procedure occurs in three general steps, as noted above, called initiation or initial formation of an active monomer, propagation where monomer units are added, thereby extending the polymer chain, and finally termination, where the growing chain is inactivated. Such single polymer chain growth occurs within parts of a second. Here, polymer yield can be low to high because long-chained giant molecules are grown throughout the process.

Step processes generally require energy, heat, to encourage the reactants to combine and to help drive off the water. Chain processes produce heat (exothermic) and this heat must be controlled by removal of the produced heat. While many free radical processes produce polymer at and above room temperature, anion- and cation-associated polymerizations typically occur below room temperature.

Thus, the terms condensation and stepwise are often used to describe the same polymers such as polyesters and nylons, while the terms vinyl and chainwise are also used to describe the same polymers such as polyethylene and polystyrene.

3.3 IMPORTANCE OF GIANT MOLECULES

There are numerous ways to measure the importance of a specific discipline. One way is to consider its pervasiveness. Polymer science and technology are essential for our housing, clothing, and food and health needs, because polymeric materials are common and integral in our everyday lives. We are concerned with natural polymers, such as (a) proteins in meats and dairy products and (b) starches in our vegetables, and we use them as building blocks and agents of life. Synthetic polymers serve as floor coverings, laminated plastics, clothing, gasoline hoses, tires, upholstery, records, dinnerware, and many other uses.

Another way to measure the importance of a specific discipline is to consider the associated work force. The U.S. polymer industry employs more than 1 million people indirectly and directly. This corresponds favorably to the employment in the entire metal-based industry. Furthermore, about one-half of all professional chemists and chemical engineers are engaged in polymer science and technology, including monomer and polymer synthesis and polymer characterization, and this need will increase as the industry is predicted to continue to increase.

Still another way to measure the importance of an industry is to study its growth. The number of new opportunities in polymer science and technology is on a par with those in the fastest growth areas. A fourth possible consideration is the marketplace influence. After food-related materials, synthetic polymers comprise the largest American export market, both bulkwise and moneywise (Section 3.13).

A fifth consideration is the influence of this science with respect to other disciplines. The basic concepts and applications of polymer science apply equally to natural and synthetic polymers, and thus are important in medical, health, nutrition, engineering, biology, physics, mathematics, computer, space, and ecological sciences and technology.

3.4 POLYMER PROPERTIES

There is a basic question that needs to be answered. Why has polymer science and technology grown into such a large industry, and why has nature chosen the macromolecule to be the very fabric of life and material construction? The obvious answer, and only the tip of the iceberg, is molecular size. Other answers relate to physical and chemical properties exhibited by polymers. We will briefly describe two of these properties.

A. Memory

We use the terms "memory" and "to remember" in similar but different ways when describing the behavior of giant molecules. The first use of the terms "memory" and "to remember" involves reversible changes in the polymer structure usually associated with the bending of rubbery materials where only segments move as the material is deformed–stretched or bent or twisted, but the entire chain does not move with cross-links acting to return the rubbery material to its original shape when the distortion is removed. Thus, the polymer "remembers" its initial segmental arrangement and returns to it through the guiding of the cross-links.

The second use involves nonreversible changes of polymer segments and whole-chain movements also brought about through application of some distortion. These changes include any chain and segmental orientations that have occurred either prior to, during, or after synthesis of the polymer including fabrications effects. These changes involve "permanent" changes in chain and segmental orientation, and in some ways these changes represent the total history of the polymer materials from inception (synthesis) through the moment when a particular property or behavior is measured. These irreversible or nonreversible changes occur with both cross-linked and non-cross-linked materials and are largely responsible for the change in polymer property as the material moves from being synthesized, processed, fabricated, and used in whatever capacity it finds itself. Thus, the polymeric material "remembers" its history with respect to changes and forces that influence chain and segmental chain movements.

The ability of polymers to "remember" and have a "memory" are a direct consequence of their size. Some polymers, such as rubber, return to their original shape and dimensions after being distorted. This "memory" is related to physical and/or chemical bonds (cross-links) between polymer chains for large distortions and to the high cumulative secondary bonding forces present between chains (intermolecular forces) for small distortions. The degree of cross-linking affects many physical properties of polymers. Thus, many elastomers, including natural rubber, change from soft to hard as the amount of cross-linking increases from 1 to 1000 units in the polymer chain.

In Nature, this "memory" is utilized to restrict flow of materials and to transmit information. Memory is also exhibited by the ability of certain macromolecules to pass on impulses (nerve transmissions and electrical conductivity).

B. Solubility and Flexibility

The large size of polymer molecules contributes to their relatively poorer solubility compared to smaller molecules. In general, compared to smaller molecules, polymers are less soluble in a given solvent, soluble in fewer solvents, and more difficult to dissolve. The solubility behavior of polymers (and in fact any solubility) is dependent on both kinetic (how fast) and thermodynamic (energy and order/disorder) factors.

There are two thermodynamic driving forces to be considered when different materials are mixed; these forces determine if (not when) the two materials will mix, or in this case they determine if the solvent molecules will dissolve the polymer chains. These two factors are energy and order/disorder. Let us first look at the energy factor. There is an axiom that says that "like likes like best of all." This axiom applies to solubility. A material is infinitely soluble in itself. It also means that liquids that are similar in general structure to the polymer will be more apt to be a solvent for that polymer.

Thus, amorphous polypropylene is composed of nonpolar units and is soluble in nonpolar liquids like hexane, while poly(vinyl alcohol) contains polar hydroxyl, –OH, groups and is soluble in polar liquids like water.

The other driving force is order/disorder. Nature generally moves from ordered to disordered arrangements. A good example of this is the tendency of our rooms to get messy if we do not expend effort (energy/work) to prevent or correct this situation. The number of geometric arrangements of connected polymer segments in a chain is much less than if the segments were free to act as individual units. Thus, for polymers, there is a decreased tendency, in comparison to small molecules, to achieve random orientations, thereby decreasing the tendency for a polymer to dissolve. In fact, for all mixing, including dissolving, the energy factor is against the mixing to occur because the forces holding together the pure materials are more alike than the forces that hold together unlike molecules. Thus, the driving force for mixing is the increase in disorder that occurs when mixing occurs. The attempt to match polar liquids with polar polymers and to match nonpolar liquids with nonpolar polymers is an attempt to minimize the energy factor that works against mixing.

The kinetic factors are related to how fast something occurs, in this case how fast the polymers are dissolved. Solvent molecules are not able to readily penetrate to the interior of a group of polymer chains with undissolved polymer segments preventing the continuous "moving away" of the dissolved segments.

Many linear polymers undergo solubility through several stages. Initially, the polymer appears to lack solubility. After some time, which may be hours, days, or even months, the polymer appears to become a gel that is swollen because of the presence of solvent molecules. Finally, solubility occurs. We can get some ideal of what is occurring by remembering that solubility requires that solvent molecules come into contact with the polymer chains. Exposure of the internal polymer chains requires that outer polymer chains have already become exposed to solvent molecules to the extent that the solvent molecules can penetrate and reach the internal polymer chains. In some ways this is like pealing an onion layer by layer. As one layer is peeled away, a new layer is exposed, and as this layer is exposed a new layer is exposed, and so on until all the layers are exposed. The gel state or stage occurs when the polymer chains become exposed to the solvent molecules, with the solvent molecules entrapped within the chains so that there are enough solvent molecules present to dissolve parts of the polymer chains but not enough to entirely dissolve the entire assembly of polymer chains. The entrance of the various solvent molecules occurs in a somewhat random manner with progress into the polymer interior requiring time.

For smaller molecules such as simple table sugar in water, the water molecules solubilize the individual sugar molecules, rapidly removing the sugar molecules exposing new sugar molecules that are solubilized, and so on. For a water-soluble polymer such as poly(vinyl alcohol), individual polymer segments can be exposed to the water molecules that effectively "dissolve" that particular segment, but the chain remains undissolved until all the polymer units are dissolved. The fact that the various segments are tied to one another and may exist within several layers makes it more difficult for an abundance of water molecules to be present to entirely dissolve the polymer chain.

While the "connectiveness" of the polymer units makes solubility more difficult, it is useful in applications where you want the polymer to be resistant. Thus, polymers are good materials for outer space applications since the lack of an atmosphere may cause some segments to leave the solid; other segments will not allow the entire chain to "evaporate" into outer space, thereby preventing removal of the entire chain.

Because of the orderly nature of crystalline polymers, there is no room to allow liquid molecules to penetrate within the crystalline structure, and thus most crystalline polymers are less soluble than the same polymer except in the amorphous state. Often, polymer solubility can be increased by heating the polymer to above its glass transition temperature where segmental mobility allows liquid molecules to come into contact with the various chains. Furthermore, cross-linking inhibits solubility, and even as little as 1 to 5 cross-links per hundred units may be sufficient to prevent the polymer from being soluble. These cross-links prevent liquid molecules from penetrating the polymer.

The resistance of a polymer to be readily dissolved permits pseudosolutions or semisolubility to occur. In animals, the proteins retain flexibility through entrapment of water. Thus, our skin is flexible and organs can stretch and bend. In plants, water permits leaves and grass to "flow in the breeze."

If only a few solvent molecules are allowed to be present, these few solvent molecules may be sufficient to allow portions of the polymer chain to be flexible, thereby creating a polymer–solvent mixture that is flexible. These solubilizing molecules are called plasticizers. For the human body, as noted above, water is often a plasticizer allowing the various polymers such as proteins, enzymes, and nucleic acids to be flexible enough to perform their task and not to be so brittle as to break when bent but not too solubilized so as to disturb the necessary shape of the molecule that allows it to perform its essential duties.

Flexibility for polymers requires that portions of the polymer chains be mobile. If the total polymer chain were mobile, then the polymer would behave as a liquid. We talk about segmental mobility when we are describing that a portion of a polymer chain is free to move. Thus, flexibility requires that a portion of the polymer chain be mobile. This mobility is generally achieved by addition of plasticizers or sufficient heat to allow the movement of segments, but not entire chains, of the polymer chain. The temperature range where segmental chain mobility begins is called the glass transition temperature and is given the symbol T_g. Below the T_g the polymer is brittle since chains are unable to move when the polymer is bent or otherwise distorted. Above the T_g the segments of the polymer chains can move, allowing the polymer to be distorted, within limits, without breaking. Most vinyl-backbone-type polymers such as amorphous polypropylene and amorphous polyethylene have T_g values below room temperature. Polysiloxane polymers such as polydimethylenesiloxane have a T_g that is well below room temperature (about $-200°F$), and thus polysiloxane polymers are suitable for use at low temperatures for refrigeration seals and seals for automobiles that are for use in the far north. Polymers with polar groups within their backbones such as polyesters (poly(ethylene terephthate, PET, $T_g = 158°F$) and nylons (nylon 66, $T_g = 140°F$) often have T_g values above room temperature and thus act as solids or brittle plastics around room temperature.

As noted above, another way to gain segmental mobility is to add a plasticizer to the polymer. Poly(vinyl chloride), PVC, ($T_g = 176°F$) and polystyrene (amorphous $T_g = 212°F$) as pure materials are brittle, yet we know that materials made from them, such as PVC piping, are flexible. This is because plasticizers are added that allow the material to be flexible below their T_g.

Plastics can be flexible or stiff, depending on a number of factors. One of the simplest is thickness. Look at commercial vitamin bottles made from a plastic material. The sides are flexible while the neck is not because the sides of the bottles are thinner than the neck. Now look at plastic bags from the store. Most of these bags are made from polyethylene or polypropylene. They are thin and quite flexible. Layers of these remain flexible in spite of the thickening. This is because the particular layers are able to slide past one another. So that thickness alone is not a guarantee to achieving an inflexible material.

Thus, moderately thick bulk "flexible" polymers can become quite rigid and resistant to bending. They often replace metal in building and other applications because they are resistant to many of the environmental problems such as rusting, easily formed into various shapes, readily available, and inexpensive.

C. Cross-Links

Chains can be connected to one another through physical entanglement similar to what happens when a kitten gets a hold of a ball of yarn. These entanglements are referred to as physical cross-links. Chains can also be connected through formation of chemical linkages that chemically hold one chain to another chain. These chemical connections are called chemical cross-links. These cross-links, physical and chemical, act to bind together the connected chains so that they act in some unison rather individually. Some polymers, such as the traditional rubbers of our automobile tires, are highly interconnected (Section 10.8) through chemical bonds, whereas other polymers have only a small amount of chemical interconnections such as often present in so-called permanent-press dress shirts and proteins (Section 14.3).

As noted above, these two types of interconnections, physical and chemical, are referred to as cross-links and the extent of cross-linking is referred to as cross-link density. Cross-linking helps "lock-in" a particular structure. Thus, the formation of cross-links in our hair can lock in curly or straight hair. The "locked-in" structure can be an ordered structure such as the locking-in of a specific shape for a protein (Section 14.3), or the "locked-in" structure can be a general or average shape such as present in the ebonite rubber head of a hammer (Chapter 10). Furthermore, some structures are composed of a maze of cross-linking, a high cross-link density, forming a complex interlocking structure that offers only an average overall structure such as the melamine-formaldehyde dishes (Section 8.4) and silicon dioxide glass (Section 16.5) while other highly cross-linked structures have ordered structures such as in silicon dioxide quartz (Section 16.6).

3.5 A FEW DEFINITIONS OF POLYMERS (MACROMOLECULES)

Briefly, polymer science is the science that deals with large molecules consisting of atoms connected by covalent chemical bonds. Polymer technology is the practical application of polymer science. The word polymer is derived from the Greek *poly* (many) and *meros* (parts). The word macromolecule—that is, giant molecule—is often utilized synonymously for polymer and vice versa.

Some scientists differentiate between the two terms by using the word macromolecule to describe large molecules such as DNA and proteins, which cannot be derived from a single, simple unit, and using the term polymer to describe a large molecule such as polystyrene, which is composed of repetitive styrene units. This differentiation is not always observed and will not be used in this text. The process of forming a polymer is called polymerization.

The degree of polymerization (DP) or average degree of polymerization ($\overline{DP}$) is the number of repeating units (mers) in a polymer chain. The term chain length is used as a synonym for DP. The DP of a dimer is 2, that of a trimer is 3, and so on. Chains with DPs below 10 to 20 are referred to as oligomers (small units) or telomers. Many polymer properties are dependent on chain length, but the change in polymer properties with changes in DP, for most commercial polymers, is small when the DP is greater than 100.

As will be noted in Section 3.7, polymer chains can come in different lengths. This is particularly true for synthetic giant molecules, but not true for biological molecules that are required to have a specific size to perform their function such as proteins and nucleic acids (Chapter 6). For polymers where the chain length varies, we often give some average of the number of units.

Many of the structures used in this book are called repeat units; thus if we repeated the unit for the appropriate number, we would have an adequate structural representation of the polymer. Thus, the repeat unit for polyethylene is

$$-\!(CH_2-CH_2-)_n$$

A chain 100 units long—that is $n = 100$ or the DP is 100, would have 200 carbon atoms arranged in a string along with the appropriate number of hydrogens. The individual unit is referred to as a "mer" as in "polymer."

At both ends of the polymer chain there are "end groups." These are sometimes given as below for polyethylene where the end groups are both CH_3-CH_2-, but typically they are not given.

$$CH_3-CH_2-\!(CH_2-CH_2)_n CH_2-CH_3$$

The set of carbons that are connected to form the chain in polyethylene is referred to as the polymer backbone or simply the backbone. For polyethylene the backbone is then –C–C–, while for poly(ethylene oxide) the backbone is –C–C–O–.

$$-\!(CH_2-CH_2-O)_n$$

Poly(ethylene oxide)

Most of the synthetic polymers considered in this book are linear; that is, they take on the shape of a rope or string. Some polymers have units that come off the main linear polymer chain. These polymers are called branched polymers, and the units that are coming off the main linear polymer chains are referred to as branches. Polyethylene chains often have various branches coming off the main polymer backbone.

If the polymer can be represented as having only one repeat unit, then it is called a homopolymer. Polyethylene is a homopolymer as is nylon 6,6.

$$-\!\!\left(\overset{O}{\overset{\|}{C}}-CH_2-CH_2-CH_2-CH_2-\overset{O}{\overset{\|}{C}}-\overset{H}{\overset{|}{N}}-CH_2-CH_2-CH_2-CH_2-CH_2-CH_2-\overset{H}{\overset{|}{N}}\right)_n$$

Nylon 6,6

But, sometimes more than one repeat unit is necessary. For instance, the polymer Saran™, from which Saran Wrap™ is made, is composed of two different units and is called a copolymer.

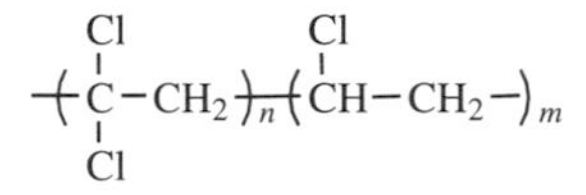

$$-\!\!\left(\begin{matrix}Cl\\|\\C\\|\\Cl\end{matrix}-CH_2\right)_n\!\!\left(\begin{matrix}Cl\\|\\CH\end{matrix}-CH_2-\right)_m$$

Poly(vinylidene chloride-co-vinyl chloride)

Functionality means the number of possible reaction sites. Thus, ethylene has two functional sites, one at each carbon, allowing it react with other ethylene units growing to become a long chain composed of ethylene units. Glycerol has three reactive sites, the three alcohol or OH groups, and thus it has a functionality of three.

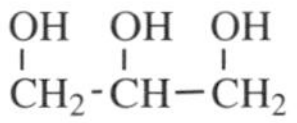

$$\begin{matrix}OH & & OH & & OH\\| & & | & & |\\CH_2 & - & CH & - & CH_2\end{matrix}$$

Glycerol

Linear polymers are formed when the functionality of the reactants is two. If the functionality of any reactant is greater than two, such as with glycerol, the resulting giant molecule will be cross-linked, forming a three-dimensional matrix or network.

3.6 POLYMER STRUCTURE

The terms configuration and conformation are often confused. Configuration refers to arrangements fixed by chemical bonding, which cannot be altered except through primary bond breakage. Terms such as head to tail, D and L isomers, and cis and trans isomers refer to configurations of isomers in a chemical species. Conformation, on the other hand, refers to arrangements around single primary bonds. Polymers in solutions or in melts continuously undergo conformational changes—that is, changes in shape. The principal difference between a hard-boiled egg and a raw egg is an irreversible conformational change.

Monomer units in a growing vinyl chain usually form what is referred to as a head-to-tail arrangement in which the repeating polymer unit ($-CH_2-CHX-$) in the polymer chain can be shown simply as

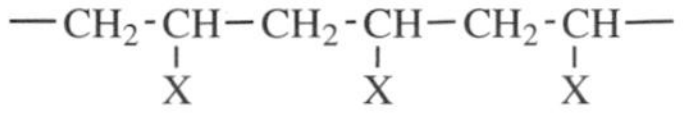

$$\begin{matrix}-CH_2- & CH & -CH_2- & CH & -CH_2- & CH & -\\ & | & & | & & | & \\ & X & & X & & X & \end{matrix}$$

Even with head-to-tail configuration, a variety of structures are possible. For illustrative purposes, we will consider possible combinations derived from the homopolymerization of monomer A and the copolymerization of A with another monomer B. Homopolymerization involves one repetitive monomeric unit in the chain.

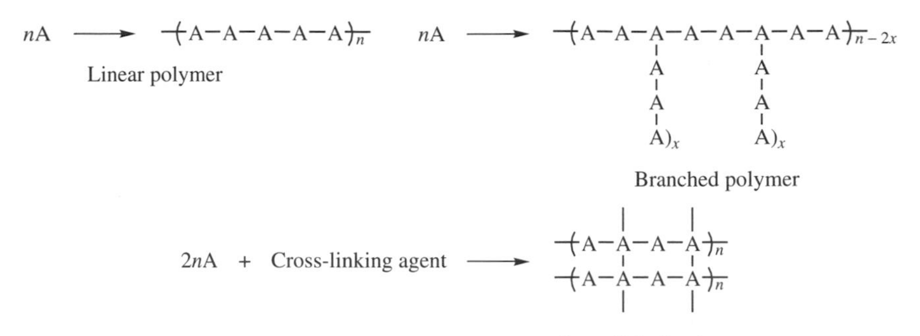

Cross-linked polymer

Copolymerization involves more than one monomeric unit in the chain, and the copolymer structure may differ:

—A—A—B—A—B—B—A—A—B—A—B—

Linear random copolymer

—A—B—A—B—A—B—A—B—A—B—

Linear alternating copolymer

—A—A—A—A—B—B—B—B—B—B—A—A—A—A—B—B—

Linear block copolymer

—A—A—A—A—A—A—A—A—A—A—
(B—B—B—B—B— side chains grafted on the second and ninth A units)

Graft copolymer

It is currently possible to tailor-make polymers of these structures to obtain almost any desired property by utilizing combinations of many of the common monomers.

The term configuration refers to structural regularity with respect to the substituted carbon atoms in the polymer chains. For linear homopolymers derived from monomers of the form $H_2C{=}CHX$, configurations from monomeric unit to monomeric unit can vary randomly (atactic) with respect to the geometry (configurations) about the carbon atom to which the pendant group X is attached or can vary alternately (syndiotactic), or be alike in having all the pendant X groups placed on the same side of a backbone plane (isotactic). These configurations are shown in next page.

Another type of stereogeometry is illustrated by polymers of 1,4-dienes, such as 1,4-butadiene, in which rotation in the polymer is restricted by the presence of

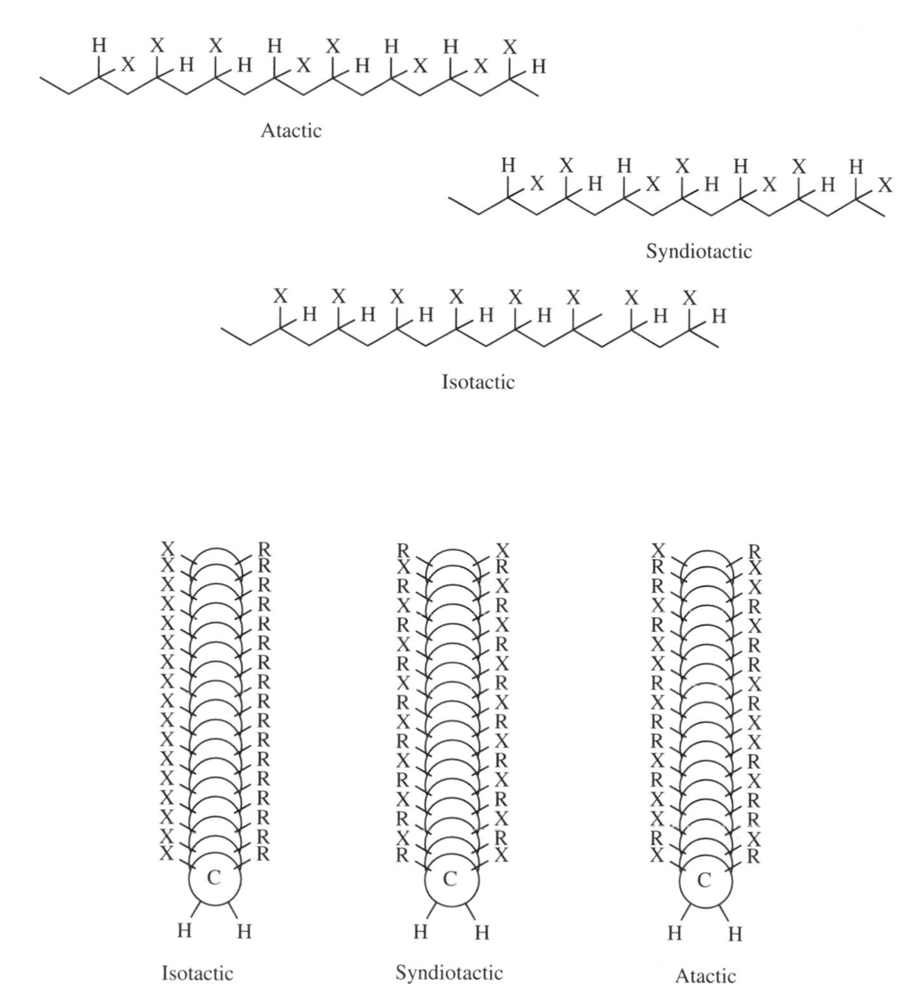

the double bond. Polymerization can occur through a single static double bond to produce 1,2 molecules that can exist in the stereoregular forms of isotactic and syndiotactic and irregular, atactic forms. The stereoregular forms are rigid, crystalline materials, whereas the atactic forms are soft, amorphous elastomers.

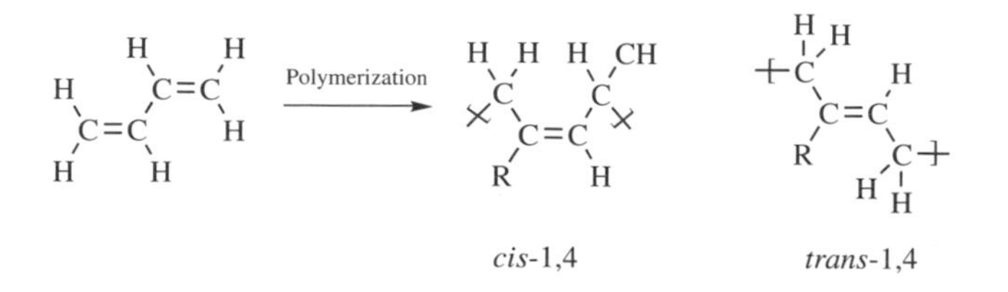

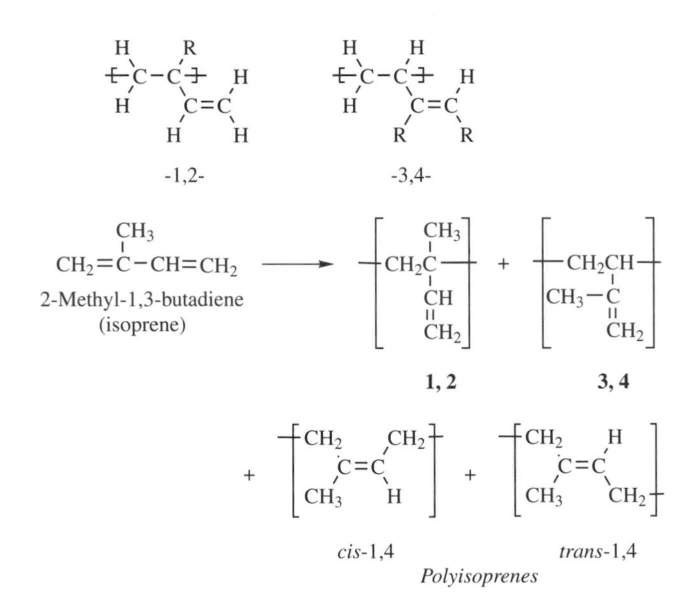

Polymerization of dienes can also produce polymers in which carbon moieties are on the same side of the newly formed double bond (cis) or on the opposite side (trans). The cis isomer of poly-1,4-butadiene is a soft elastomer with a glass transition temperature (T_g) of $-108°C$. The glass transition temperature of the isomer of poly-1,4-butadiene is $-83°C$. The glass transition temperature is the temperature at which a glassy polymer becomes flexible when heated. T_g is a characteristic value for amorphous (noncrystalline) polymers.

3.7 MOLECULAR WEIGHTS OF POLYMERS

Polymerization reactions may produce polymer chains with different numbers of repeating units or degrees of polymerization (DP). Most synthetic polymers and many naturally occurring polymers consist of molecules with different molecular weights and are said to be polydisperse. In contrast, specific proteins and nucleic acids consist of molecules with a specific molecular weight and are said to be monodisperse.

Since typical molecules with DPs less than the critical value required for chain entanglement are weak, it is apparent that certain properties are related to molecular weight. The melt viscosity of amorphous polymers is dependent on the molecular weight distribution. In contrast, density, specific heat capacity, and refractive index are essentially independent of the molecular weight at molecular weight values above the critical molecular weight, which is typically a DP of about 100.

Viscosity is the resistance of a substance to flow when subjected to a shear stress. When applied to solutions of polymers and melts, viscosity is measured by a device called a viscometer. Shear or tangential stress is a force that is applied parallel to the surface, like spreading butter on a piece of toast.

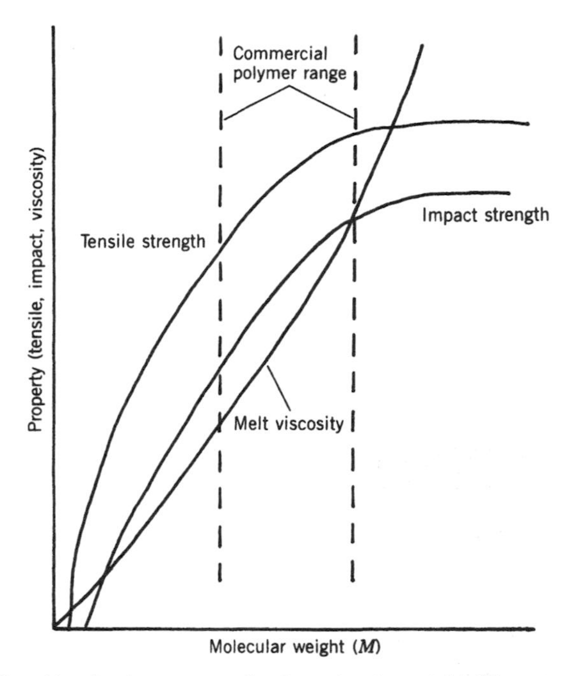

Figure 3.3. Relationship of polymer properties to molecular weight. (From *Introduction to Polymer Chemistry* by R. B. Seymour, McGraw-Hill, New York, 1971. Used with permission of McGraw-Hill Book Company.)

The melt viscosity (η) is usually proportional to the 3.4 power of the average molecular weight at values above the critical molecular weight required for chain entanglement, that is, $\eta = \bar{M}^{3.4}$. ($\bar{M}$ or $\overline{\mathrm{DP}}$ represents an average value for polydisperse macromolecules.) The melt viscosity increases rapidly as the molecular weight increases, and hence more energy is required for the processing and fabrication of these large molecules. However, as shown in Figure 3.3, the strength of a polymer increases as its molecular weight increases, then tends to level off.

Thus, although a value above the threshold molecular weight value (TMWV) is essential for most practical applications, the additional cost for energy required for processing higher-molecular-weight polymers is seldom justified. Accordingly, it is customary to establish a commercial polymer range above the TMWV but below the extremely high molecular weight range. However, it should be noted that since toughness increases with molecular weight, polymers such as ultrahigh-molecular-weight polyethylene (UHMWPE) are used for the production of strong items such as trash barrels.

The value of TMWV is dependent on the glass transition temperature, the intermolecular forces, expressed as cohesive energy density (CED) of amorphous polymers, the extent of crystallinity in crystalline polymers, and the extent of reinforcement present in polymer composites. Although a low-molecular-weight amorphous polymer may be satisfactory for use as a coating or adhesive, a much higher $\overline{\mathrm{DP}}$ value may be required if the polymer is used as an elastomer or plastic. With the exception of polymers with highly regular structures, such as isotactic polypropylene, strong hydrogen intermolecular bonds are required for fibers. Because of the higher CED values resulting from stronger intermolecular forces, lower $\overline{DP}$ values are usually satisfactory for polar polymers used as fibers.

3.8 POLYMERIC TRANSITIONS

Polymers can exhibit a number of different conformational changes, each change accompanied by differences in polymer properties. Two major transitions are the glass transition temperature (T_g), which is dependent on local, segmental chain mobility in the amorphous regions of a polymer, and the melting point (T_m), which is dependent on large-scale chain mobility. The T_m is called a first-order transition temperature, whereas T_g is often referred to as a second-order transition temperature. The values for T_m are usually 33–60% greater than those for T_g, with T_g values being low for typical elastomers and flexible polymers and higher for hard amorphous plastics. The T_g for silicones is −190°F and that for E-glass is 1544°F. The T_g values for most other polymers are in between these extremes.

3.9 TESTING OF POLYMERS

Public acceptance of polymers is usually associated with an assurance of quality based on a knowledge of successful, long-term, and reliable tests. In contrast, much of the dissatisfaction with synthetic polymers is related to failures that possibly could have been prevented by proper testing, design, and quality control. The American Society for Testing and Materials (ASTM), through its committees D-1 on paint and D-20 on plastics, for example, has developed many standard tests that are available to all producers and large-scale consumers of finished polymeric materials. There are also testing and standards groups in many other technical societies throughout the world.

Much of the testing performed by the industry is done to satisfy product specifications using standardized tests for stress–strain relationships, flex life, tensile strength, abrasion resistance, moisture retention, dielectric constant, hardness, thermal conductivity, and so on. New tests are continually being developed, submitted to ASTM, and, after adequate verification through "round-robin" testing, finally accepted as standard tests.

Each standardized ASTM test is specified by a unique combination of letters and numbers, along with exacting specifications regarding data gathering, instrument design, and test conditions, thus making it possible for laboratories throughout the world to compare data with confidence. The Izod test, a popular impact test, has the ASTM number D256-56 (1961), the latter number being the year it was first accepted. The ASTM instructions for the Izod test specify test material shape and size, exact specifications for the test equipment, detailed description of the test procedure, and how results should be reported. More complete information on testing and characteristics of polymers is provided in Chapter 8.

3.10 CHEMICAL NAMES OF POLYMERS

The International Union of Pure and Applied Chemistry (IUPAC) formed a subcommission on Nomenclature of Macromolecules in early 1952 and has continued to periodically study the various topics related to polymer nomenclature. Many of the names that scientists employed for giant molecules are *source-based*; that is, they are named according to the common name of the repeating units in the giant molecule, preceded by the prefix poly. Thus, the name polystyrene (PS) is derived from the common name of its repeating unit, and the name poly(methyl methacrylate) (PMMA) is derived from the name of its repeating unit:

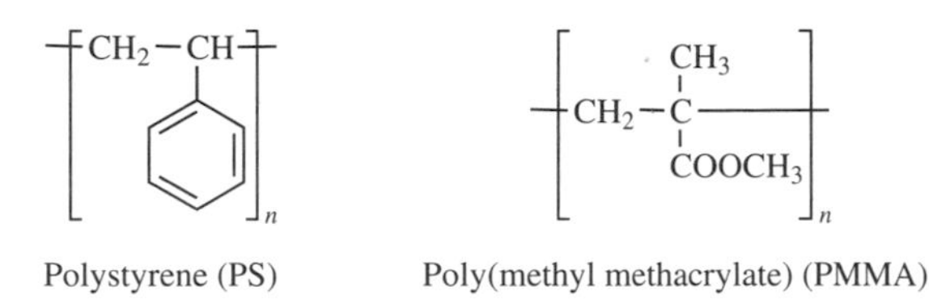

Polystyrene (PS) Poly(methyl methacrylate) (PMMA)

Little rhyme or reason is associated with *common-based* names. Some common names are derived from the "discoverer"; for example, Bakelite was commercialized by Leo Baekeland in 1905. Others are based on the place of origin, such as *Hevea braziliensis*, literally "rubber from Brazil," the name given for natural rubber (NR).

For some important groups of polymers, special names and systems of nomenclature were invented. For example, the nylons were named according to the number of carbons in the diamine and carboxylic acid reactants (monomers) used in their synthesis. The nylon produced by the condensation of 1,6-hexamethylenediamine (6 carbons) and sebacic acid (10 carbons) is called nylon 6,10. Industrially, nylon 6,10 has been designated nylon 6,10, nylon 6 10, or 6-10 nylon.

$$\left[HN-(CH_2)_6NH-\overset{\overset{O}{\|}}{C}-(CH_2)_8-\overset{\overset{O}{\|}}{C} \right]$$

Polyhexamethylenesebacamide (nylon 6,10)

$$\left[HN-(CH_2)_6-NH-\overset{\overset{O}{\|}}{C}-(CH_2)_4-\overset{\overset{O}{\|}}{C} \right]_n$$

Polyhexamethyleneadipamide (nylon 6,6)

$$\left[CH_2-CH_2-\overset{\overset{O}{\|}}{C}-NH \right]_n$$

Polyalanine (nylon-3)

$$\left[\overset{\overset{O}{\|}}{C}-(CH_2)_4-\overset{\overset{O}{\|}}{C}-NH-CH_2-NH \right]_n$$

Polymethyleneadipamide (nylon 1,6)

Similarly, the polymer produced from the single reactant caprolactam (6 carbons) is called nylon-6. The structure-based name for nylons is polyamide because of the presence of the amide grouping. Thus, scientists are talking about the same family of polymers if they are talking about nylons or polyamides.

Abbreviations are also widely employed. Thus PS represents polystyrene and PVC represents poly(vinyl chloride). The media have given abbreviations to some common monomers such as vinyl chloride (VCM) and styrene (SM).

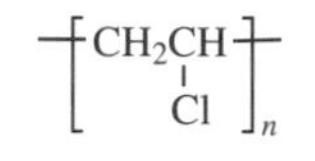

Poly(vinyl chloride) (PVC)

3.11 TRADE NAMES OF POLYMERS

Many firms use trade names to identify specific polymeric products of their manufacture. However, generic names, such as rayon, cellophane, polyesters, and polyurethane, are used more universally. For example, Fortrel polyester is a poly(ethylene terephthalate) (PET) fiber produced by Fiber Industries, Inc. The generic term polyester indicates that the composition of this fiber is based on a condensation product of a dihydric alcohol (glycol, $R(OH)_2$) and terephthalic acid (an aromatic dicarboxylic acid, $Ar(COOH)_2$. Many generic names for fibers, such as polyester, are defined by the Textile Fiber Products Identification Act. This act also controls the composition of fibers such as rayon and polyurethane.

3.12 IMPORTANCE OF DESCRIPTIVE NOMENCLATURE

Unfortunately, there are also many trivial names that tend to cause some confusion. For example, when a nonscientist says alcohol, he or she means ethanol, which is just one of hundreds of alcohols. Likewise, the nonscientist uses the term sugar to indicate a specific sugar (sucrose), salt to indicate a specific salt (sodium chloride), and vinyl to indicate PVC.

The uninformed consumer may not recognize that there are numerous alcohols, sugars, salts, vinyl polymers, synthetic fibers, and plastics. After reading subsequent chapters, you will be aware of the many different polymers whose properties cover the entire spectrum, from insulators to conductors, from liquids to solids, from water-soluble to water-insoluble, and from those that soften at room temperature to those that can be used in combustion engines. Additional structural information on plastics, fibers, and elastomers is given in Table 3.3.

3.13 MARKETPLACE

Giant molecules account for most of what we are [proteins, nucleic acids (DNA and RNA), enzymes], what we eat, and the society in which we live (plants, buildings, roads, animals, clothing, tires, coatings, rugs, newspaper, etc.).

Table 3.3 Structures of industrially important addition polymers

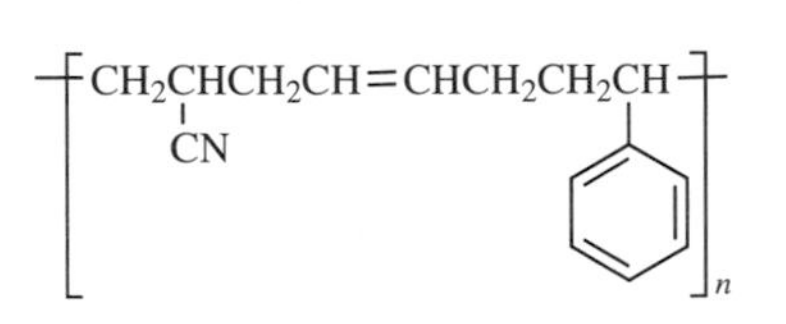

Acrylonitrile–butadiene–styrene terpolymer (ABS)

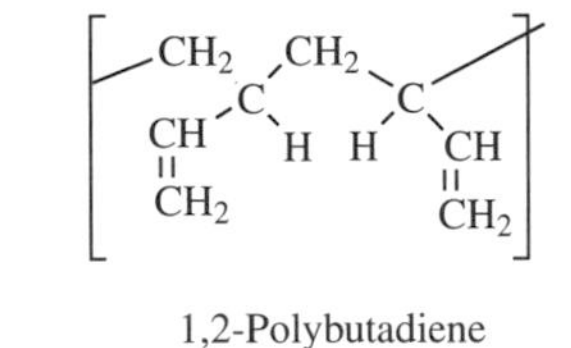

1,2-Polybutadiene

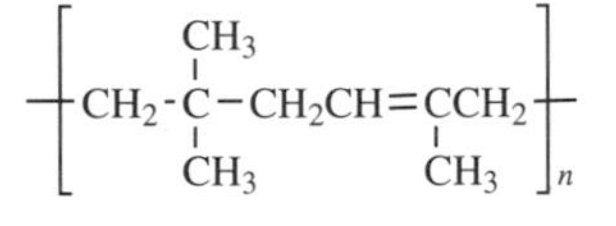

Butyl rubber

$$\left[CH_3-CH=CH-CH_2 \right]_n$$

trans-1,4-Polybutadiene

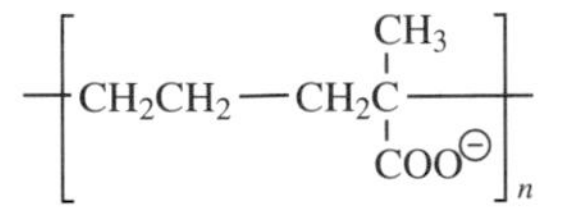

Ethylene-methacrylic acid copolymers (ionomers)

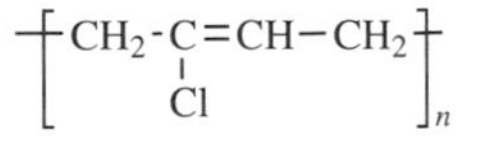

Polychloroprene

$$\left[CH_2\underset{CN}{C}H \right]_n \left[CH_2CH=CHCH_2 \right]_n$$

Nitrile rubber (NBR)

$$\left[CH_2CH_2 \right]_n$$

Polyethylene (PE)

$$\left[CH_2-\underset{CN}{C}H \right]_n$$

Polyacrylonitrile

$$\left[OCH_2CH_2 \right]_n$$

Poly(ethylene glycol) (PEG)

$$\left[(CH_2)_6-S \right]_n$$

Poly(hexamethylene thioether)

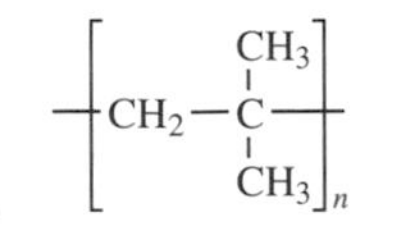

Polyisobutylene (PIB)

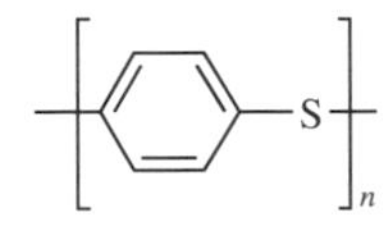

Poly(phenylene sulfide) (PPS)

$$\left[CH_2\underset{Cl}{C}H \right]_n$$

Poly(vinyl chloride) (PVC)

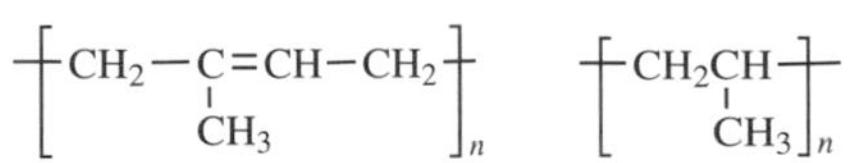

Polyisoprene

$$\left[CH_2\underset{CH_3}{C}H \right]_n$$

Polypropylene (PP)

$$\left[CH_2CCl_2 \right]_n$$

Poly(vinylidene chloride)

3,4-Polyisoprene

Poly(propylene glycol) (PP)

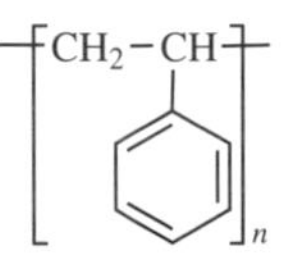

Polyvinylpyridene

Table 3.3 (*Continued*)

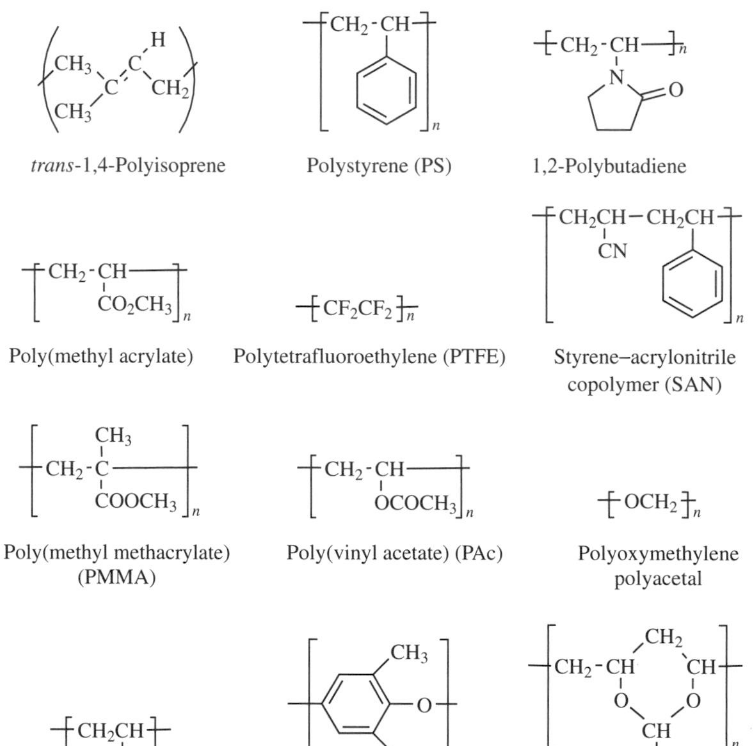

Table 3.4 Summation production amounts for the United States in 2000 in millions of pounds

Grouping	Production
Thermoplastics (Chapter 6) and Engineering plastics (Chapter 7)	79,000
Thermosets (Chapter 8)	10,000
Fibers (Chapter 9)	12,500
Synthetic rubber (Chapter 10)	5,000
Paper and paper products (Chapter 13)	~160,000
Portland cement (Chapter 16)	~200,000

Table 3.5 U.S. chemical industrial employment for 2000 (in thousands)

Sector	Employment
Agricultural	53
Drugs	305
Industrial inorganics	98
Industrial organics	121
Soaps, cleaners, etc.	158
Synthetic polymers	1206

Source: U.S. Department of Labor.

The annual U.S. production of various groupings of giant molecules is given in Table 3.4. Tar and concrete are also principal items of construction and both composed of giant molecules. Portland cement is utilized at an annual rate of greater than 160,000 million pounds annually. All told, this represents an annual production of about 1500 pounds or three-quarters of a ton for each of us including only the items listed in Table 3.4. This does not include such important giant molecules as wood, cellulose, starch, proteins, and tar. Wood products pervade our society as construction materials, and tar is extensively employed in the building of our roads.

On a manufacturing level, the number of persons employed in the synthetic polymer industry alone is greater than those employed in all the metal-based industries combined. More than 60% of all chemical industrial employment in the United States involves synthetic polymers (Tables 3.5 and 3.6).

Polymeric materials, along with the majority of the chemical industrial products, contribute positively to the balance of world trade (Table 3.7). In fact, plastics and resins show the greatest value increase of exports minus imports.

Table 3.6 U.S. production workers for 2000 (in thousands)

Sector	Employment
Agricultural	32
Drugs	140
Industrial inorganics	55
Industrial organics	73
Soaps, cleaners, etc.	97
Synthetic polymers	909

Source: U.S. Department of Labor.

Table 3.7 U.S. chemical trade-important and exports, 2000 (millions of dollars)

Chemical	Exports	Imports
Organic chemicals	18,900	28,600
Inorganic chemicals	5,500	6,100
Oils and perfumes	5,000	3,200
Dyes and colorants	4,200	2,700
Medicinals and pharmaceuticals	13,100	14,700
Fertilizers	2,500	1,700
Plastics and resins	20,100	10,600
Others	12,700	5,700
Total chemicals (includes nonlisted)	82,500	73,600
Total	780,400	1,024,800

GLOSSARY

ABS: A terpolymer of acrylonitrile, butadiene, and styrene.

Alkyd: Polyesters produced by the condensation of a dicarboxylic acid (phthalic acid), a dihydric alcohol (ethylene glycol), and an unsaturated oil, such as linseed oil.

Amorphous: Shapeless.

Anion: ($A:^-$) A negatively charged atom or molecule.

ASTM: American Society for Testing and Materials.

Atactic: A random arrangement of pendant groups in a polymer chain.

Baekeland, Leo: Inventor of phenol–formaldehyde plastics (Bakelite), the first truly synthetic plastic (1910).

Balata: A rigid, naturally occurring *trans*-polyisoprene.

Block copolymer: A polymer made up of a sequence of one repeating unit followed by a sequence of another repeating unit.

Branched copolymer: One with branches on the main chain.

Butadiene:

$$H_2C{=}\overset{\displaystyle H}{\overset{|}{C}}{-}\overset{\displaystyle H}{\overset{|}{C}}{=}CH_2$$

Carothers, W. H.: Inventor of nylon 6,6.

Catenation: Chain formation.

Cation (C^+): A positively charged atom or molecule.

Cellophane: Regenerated cellulose film.

Celluloid: Plasticized cellulose nitrate.

Chloroprene:

$$H_2C{=}\underset{}{\overset{Cl}{\overset{|}{C}}}{-}\overset{H}{\overset{|}{C}}{=}CH_2$$

cis: A geometrical isomer with both constituents on the same side of the plane of the double bond.

Cohesive energy density (CED): Internal pressure of a molecule, which is related to the strength of the intermolecular forces of the molecules.

Configuration: Arrangement of bonds in a molecule. Changes in configurations require breaking and making of covalent bonds.

Conformation: Arrangement of groups about a single bond—that is, shape that changes rapidly without bond breakage as a result of the mobility of the molecule.

Copolymer: A polymer made up of more than one repeating unit.

Coupling: The joining of two macromolecules to produce a dead polymer.

Critical molecular weight: Minimum molecular weight required for chain entanglement.

Cross-links: Chemical bonds between polymer chains—for example, bonds between *Hevea* rubber molecules produced by heating natural rubber with sulfur.

Degree of polymerization (DP): Number of repeating units (mers) in a polymer chain.

Dicarboxylic acid: An organic compound with two carboxylic acid groups.

Dimer: A combination of two smaller molecules.

DNA: Deoxynucleic acid.

Dope: Solution of cellulose acetate.

Elastomer: A rubbery polymer.

Functionality: The number of reactive groups in a molecule.

Glass transition temperature (T_g): Temperature at which segmental motion occurs when a polymer is heated, for example, glassy polymers become flexible.

Glyptal: Polyester protective coating.

Goodyear, Charles: Vulcanized *Hevea* rubber by heating it with small amounts of sulfur (1839).

Graft copolymer: A copolymer in which polymeric branches have been grafted onto the main polymer chain.

HDPE: Linear polyethylene, of higher density than LDPE.

***Hevea braziliensis*:** Natural rubber.

Homopolymer: A polymer made up of similar repeating units.

Impact strength: Resistance to breakage, degree of lack of brittleness.

Initiation: The first step in a chain reaction.

Isoprene:

$$\begin{array}{c} \quad CH_3 \;\; H \\ \quad | \qquad | \\ H_2C{=}C{-}C{=}CH_2 \end{array}$$

Isotactic: An arrangement in which the pendant groups are all on one side of the polymer chain.

IUPAC: International Union of Pure and Applied Chemistry.

Kekulé, Friedrich A.: Developed methods for writing structural formulas of organic compounds (1850s and 1860s).

Kinetic: Related to motion of molecules.

LDPE: Low-density polyethylene, a highly branched polymer.

Linear low-density polyethylene (LLDPE): Low-density polyethylene consisting of copolymers of ethylene and 1-butene or 1-hexene.

Linear polymer: A polymer consisting of a continuous straight chain.

$\bar{M}$**:** Average molecular weight.

Macro: Large.

Macroradical: An electron-deficient macromolecule.

mer: Repeating unit.

Monodisperse: A macromolecule in which all molecules have identical molecular weights.

η **(eta):** Viscosity.

Neoprene: Polychloroprene.

Nylon 6,6: A polymer produced by heating the salt from the reaction of hexamethylenediamine ($H_2N(CH_2)_6NH$) and adipic acid ($HOOC(CH_2)_4COOH$).

Oligomer: Polymer consisting of 10 to 20 repeating units.

Patrick, J. C.: Inventor of America's first synthetic elastomer (rubber).

Phenol: Hydroxybenzene (C_6H_5OH).

Plasticizer: An additive that enhances the flexibility of plastics.

Polyacetals (POM): Polymers of formaldehyde with the repeating unit $\left(O{-}CH_2\right)$.

Polyamide (PA): A polymer with repeating amide units, such as nylon 6,6.

Polyamide–imide (PAI): A high-temperature-resistant polymer with alternating amide and imide groups.

Polyarylate: A high-temperature-resistant polymer produced by the condensation of bisphenol A and an equimolar mixture of iso and terephthalic acids.

Poly(butylene terephthalate) (PBT): High-performance polymer produced by the condensation of terephthalic acid and 1,4-dihydroxybutane.

Polycarbonate (PC): Tough, high-performance polymer produced by the condensation of bisphenol A and phosgene.

Polychloroprene: An elastomer with the repeating units

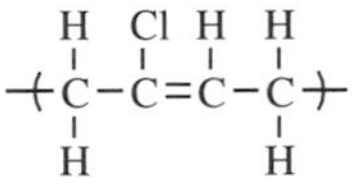

$$\left(\begin{array}{cccc} H & Cl & H & H \\ | & | & | & | \\ C- & C= & C- & C \\ | & & & | \\ H & & & H \end{array} \right) \quad \text{(neoprene)}$$

Polydisperse: A mixture of macromolecules with different molecular weights.

Polyether imide (PEI): A high-performance polymer with alternating ether and imide groups.

Polyether ketone (PEEK): A high-performance polymer containing the carbonyl (C=O) stiffening group in the polymer chain.

Polyethylene: A polymer with the repeating unit (CH_2-CH_2).

Poly(ethylene terephthalate) (PET): High-performance polymer produced by the condensation of terephthalic acid and ethylene glycol.

Polyimide (PI): High-temperature-resistant polymer produced by the condensation of an aliphatic diamine and an aromatic dianhydride.

Polymerization: A process in which large molecules (giant molecules or macromolecules) are produced by a combination of smaller molecules.

Poly(methyl methacrylate) (PMMA): A polymer with the repeating unit

$$\left(\begin{array}{cc} H & CH_3 \\ | & | \\ C- & C \\ | & | \\ H & C-OCH_3 \\ & \| \\ & O \end{array} \right)$$

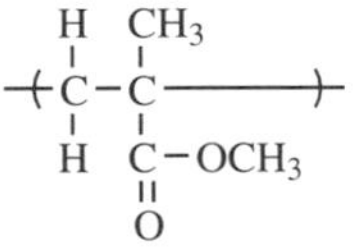

Polymethylpentene (TPX): A polyolefin with the repeating unit

$$\left(\begin{array}{cc} & H_3C \diagdown \diagup CH_3 \\ & CH \\ & | \\ H & CH_2 \\ | & | \\ C- & C \\ | & | \\ H & H \end{array} \right)$$

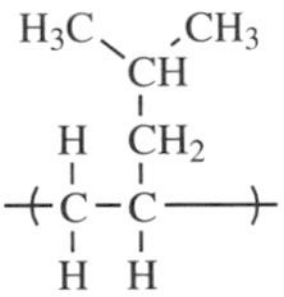

Poly(phenylene oxide) (PPO): A high-temperature-resistant polymer with phenylene and oxygen units in the chain.

Poly(phenylene sulfide) (PPS): A high-temperature-resistant polymer with the repeating unit (C_6H_4-S).

Polyphosphazene: Inorganic polymer with the repeating unit $(N=P(OR)_2)$.

Polypropylene (PP): A polymer with the repeating unit

$$\left(\begin{array}{cc} CH_3 & H \\ | & | \\ C- & C \\ | & | \\ H & H \end{array} \right)$$

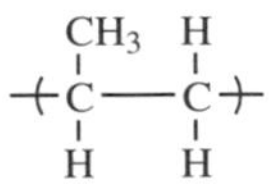

Polystyrene (PS): A polymer with the repeating unit

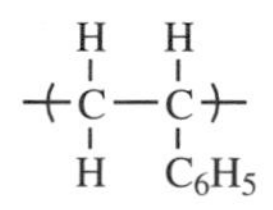

Polysulfone (PES): A high-performance polymer with the repeating unit $-\!(C_6H_4SO_2C_6H_4)\!-$ produced by the condensation of bisphenol A and a dichlorodiphenyl sulfone.

Polyurethane (PUR): A polymer produced by the reaction of a diisocyanate ($Ar(CNO)_2$) and a dihydric alcohol ($R(OH)_2$).

Poly(vinyl chloride) (PVC): A polymer with the repeating unit $-\!(CH_2CHCl)\!-$.

Propagation: The growth steps in a chain reaction.

Radical (R·): An electron-deficient molecule.

Rayon, cupraammonia: Cellulose fibers regenerated from a solution of cellulose in cupraammonium hydroxide.

Rayon, viscose: Cellulose fibers regenerated from cellulose xanthate.

Round-robin testing: Independent testing by different individuals.

Saran: Trade name for polymers of vinylidene chloride (PVDC).

Schönbein, Christian F.: Produced cellulose nitrate by the nitration of cellulose (1846).

Silicones: Inorganic polymers produced by the hydrolysis of dialkyldimethoxysilanes ($R_2Si(OCH_3)_2$), the repeating unit of which is

$$-\!\left(O-\underset{\displaystyle R}{\overset{\displaystyle R}{\overset{|}{\underset{|}{Si}}}}\right)\!-$$

SMA: Copolymers of styrene and maleic anhydride.

Staudinger, Hermann: Developed modern concepts of polymer macromolecular science (1920s).

Step reaction polymerization: Polymerization that occurs by a stepwise condensation of reactants.

Syndiotactic: An alternate arrangement of pendant groups on a polymer chain.

T_g: Glass transition temperature.

T_m: Melting point.

Technology: Applied science.

Teflon: Trade name for polytetrafluoroethylene (PTFE).

Telomer: A low-molecular-weight polymer.

Termination: The final step in a chain reaction.

Thermoplastic: A linear polymer that can be softened by heat and cooled to reform the solid.

Thermoset plastic: A cross-linked (three-dimensional) polymer that does not soften when heated.

Thiokol: Trade name for polyethylene sulfide rubber.

trans: A geometrical isomer with substituents on alternate sides of the double bond.

Transition: Change.

UHMWPE: Ultrahigh-molecular-weight polyethylene.

Wöhler, Friedrich: First chemist to synthesize an organic molecule from an inorganic compound (1828).

REVIEW QUESTIONS

1. How many functional groups are present in glycerol?

```
 H    H  H
 |    |  |
(HC — C - CH)
 |    |  |
 OH  OH OH
```

2. What is *Hevea braziliensis*?
3. Which has more cross-links: flexible vulcanized rubber or hard rubber?
4. Which of the following are thermoplastics: hard rubber, Bakelite, PVC, polystyrene, polyethylene?
5. Which of the following are thermoset plastics: Melamine dishware, Bakelite, hard rubber?
6. What is the functionality of phenol?

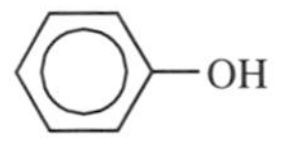

7. How does rayon differ from cotton from a chemical viewpoint?
8. What is the principal structural difference between LDPE and HDPE?
9. Why was former President Reagan called the Teflon President?
10. Which is the faster reaction: step reaction or chain reaction polymerization?
11. What is the propagating species in cationic polymerization?
12. What is the molecular weight of polyethylene with a DP of 1000?
13. Which has the higher value for a specific polymer with both amorphous and crystalline regions: T_g or T_m?
14. Is a protein a polydisperse or monodisperse polymer?

15. Why should the molecular weight of structural polymers be greater than the critical molecular weight required for chain entanglement?

BIBLIOGRAPHY

Allcock, H. R., and Lampe, F. W. (2003). *Contemporary Polymer Chemistry*, 3rd ed., Wiley, New York.

Callister, W. (2000). *Materials Science and Engineering*, 5th ed., Wiley, New York.

Campbell, I. (2000). *Introduction to Synthetic Polymers*, Oxford, New York.

Carraher, C. (2003). *Polymer Chemistry*, Marcel Dekker, New York.

Craver, C., and Carraher, C. (2000). *Applied Polymer Science*, Elsevier, New York.

Ehrenstein, G. (2001). *Polymeric Materials*, Hanser-Gardner, Cincinnati.

Elias, H. G. (1997). *An Introduction to Polymers*, Wiley, New York.

Fried, J. R. (2002). *Polymer Science and Technology*, 2nd ed., Prentice-Hall, Upper Saddle River, NJ.

Grosberg, A. and Khokhlov, A. R. (1997). *Giant Molecules*, Academic Press, Orlando, FL.

Hummel, R. E. (1998). *Understanding Materials Science: History, Properties, Applications*, Springer-Verlag, New York.

Nicholson, J. W. (1997). *The Chemistry of Polymers*, Royal Society of Chemistry, London.

Ravve, A. (2000). *Principles of Polymer Chemistry*, Kluwer, New York.

Rodriguez, F. (1996). *Principles of Polymer Systems*, 4th ed., Taylor and Francis, Philadelphia.

Salamone, J. C. (1998). *Concise Polymeric Materials Encyclopedia*, CRC Press, Boca Raton, FL.

Sandler, S., Karo, W., Bonesteel, J., and Pearce, E. M. (1998). *Polymer Synthesis and Characterization*, Academic Press, Orlando, FL.

Seymour, R., Carraher, C. (1997). *Introduccion a la Quimica de los Polymeros*, Editorial Reverte, S. A., Barcelona, Spain.

Sperling, L. (2001). *Introduction to Physical Polymer Science*, 2nd ed., Wiley, New York.

Thrower, P. (1996). *Materials in Today's World*, 2nd ed., McGraw-Hill, New York.

Tonelli, A. (2001). *Polymers Inside Out*, Wiley, New York.

Walton, D. (2001). *Polymers*, Oxford University Press, New York.

ANSWERS TO REVIEW QUESTIONS

1. Three.

2. Natural rubber.

3. Hard rubber.

4. PVC, polystyrene, polyethylene.

5. All are thermosets.

6. Depends on the kind of reaction. Generally 3.
7. No difference; rayon is regenerated cellulose.
8. LDPE is highly branched and therefore has a lower density (higher volume) than linear HDPE.
9. Teflon (polytetrafluoroethylene) is slippery because of the four fluorine pendant groups on each repeating unit. Hence, few things will stick to PTFE.
10. Chain reaction polymerization.
11. A macrocation.
12. 28,000 (1000×28).
13. T_m.
14. Monodisperse.
15. In order to achieve strength through entanglement.

4

RELATIONSHIPS BETWEEN THE PROPERTIES AND STRUCTURE OF GIANT MOLECULES

Giant Molecules: Essential Materials for Everyday Living and Problem Solving, Second Edition,
by Charles E. Carraher, Jr.
ISBN 0-471-27399-6

4.1 GENERAL

Plastic bags, our skin, hair, foam picnic plates, plastic spoons, nylons, rubber bands, tire treads, curtains, skirts, paper, glass, cement, diamonds, wood, paint, rugs, tape, potatoes, dandelions, fabrics, shower curtains, raincoats, shoes, ... all are composed of giant molecules. What makes some giant molecules suitable for long-term memory such as in rubber bands and our DNA while other giant molecules are strong, rigid, and tough, allowing their use in bullet-resistant vests while others have properties intermediate such as the flexible automobile dashboards, still others act as good adhesives such as glues and paints, while others are strong and flexible such as fabrics, ...? This chapter lays the groundwork for answering these questions.

Many of the properties of giant molecules (polymers) are unique and not characteristic of other materials, such as metals and salts. Polymer properties are related not only to the chemical nature of the polymer, but also to such factors as extent and distribution of crystallinity, distribution of polymer chain lengths, and nature and amount of additives. These factors influence polymeric properties, such as hardness, biological response, comfort, chemical resistance, flammability, weatherability, tear strength, dyeability, stiffness, flex life, and electrical properties.

We can get an idea of the influence of size in looking at the series of methylene hydrocarbons as the number of carbon atoms increases. For low numbers of carbons (methane, ethane, propane, butane), the materials are gases at room temperature (Table 4.1). For the next groupings (Table 4.1, gasoline, kerosine, light gas oil) the materials are liquids. The individual hydrocarbon chains are held together by

Table 4.1 Typical properties of straight-chain hydrocarbons

Average Number of Carbon Atoms	Boiling Range (°C)	Name	Physical State at Room Temperature	Typical Uses
1–4	<30	Gas	Gas	Heating
5–10	30–180	Gasoline	Liquid	Automotive fuel
11–12	180–230	Kerosene	Liquid	Jet fuel, heating
13–17	230–300	Light gas oil	Liquid	Diesel fuel, heating
18–25	305–400	Heavy gas oil	Viscous liquid	Heating
26–50	Decomposes	Wax	Waxy	Wax candles
50–1000	Decomposes		Tough waxy to solid	Wax coatings of food containers
1000–5000	Decomposes	Polyethylene	Solid	Bottles, containers, films
>5000	Decomposes	Polyethylene	Solid	Waste bags, ballistic wear, fibers, automotive parts, truck liners

dispersion forces that are a sum of the individual methylene and end group forces. There is a gradual increase in boiling point and total dispersion forces for the individual chains as hydrocarbon units are added until the materials become a waxy solid such as found in bees waxes and finally where the total dispersion forces are sufficient to be greater than individual carbon–carbon bond strengths so that the chains decompose prior to their evaporation. As the chain length increases, we get to the point where the chain lengths are sufficient to give tough and brittle solids we call polyethylene. It is interesting to note that these long-chain straight-chain hydrocarbons, without any branching, become very strong but they are brittle. They are crystalline and as with most other crystalline materials, such as quartz and diamonds, they are strong but brittle. Fortunately, synthetic polyethylene contains both (a) crystalline regions where the polymer chains are arranged in ordered lines and (b) regions where the chains are not arranged in orderly lines. These latter arrangements are often imposed on the polyethylene because of the presence of branching off of the linear polymer backbone. These amorphous regions are responsible for allowing the polyethylene to have some flexibility. Thus, many polymers contain both amorphous and crystalline regions that provide both flexibility and strength.

In this chapter we briefly describe the chemical and physical nature of polymeric materials that permits their classification into broad "use" divisions, such as elastomers or rubbers, fibers, plastics, adhesives, and coatings. Descriptions relating chemical and physical parameters to general polymer properties and structure are included.

4.2 ELASTOMERS

Elastomers are giant molecules possessing chemical and/or physical cross-linking. For industrial applications, the "use" temperature of an elastomer must be above the T_g (to allow for segmental "chain" mobility), and the polymer must be amorphous in its normal (unextended) state. The restoring force, after elongation, is largely due to entropy effects. As the elastomer is elongated, the random chains are forced to occupy more ordered positions; but on release of the applied force, the chains tend to return to a more random state. Entropy is a measure of the degree of randomness or lack of order in a material.

Elastomers possess what is referred to as memory; that is, they can be deformed, misshaped, and stretched, and after the stressing (applied) force is removed, they return to their original, prestressed shape.

The actual mobility of polymer chains in elastomers must be low. The cohesive energies density forces (CED) between chains should be low enough to permit rapid and easy extension of the random-oriented chain. In its extended (stretched) state, an elastomeric polymer chain should have a high tensile strength, whereas at low extensions it should have a low tensile strength. Polymers with low cross-linked density usually meet the desired property requirements. After deformation, the material should return to its original shape because of the presence of the cross-links, which limit chain slippage to the chain sections between the cross-links (principal sections).

South American Indians use the names "hhevo" and "Cauchuc," which mean "weeping wood," to describe the native rubber tree. The French continue to use the word "caoutchouc," but when it was found to be more effective than bread crumbs in removing pencil marks, E. Nairne and J. Priestley called it rubber. The term elastomer is now used to describe both natural and synthetic rubbers.

4.3 FIBERS

Characteristic fiber properties include high tensile strength and high modulus (high stress for small strains, i.e., stiffness). These properties are related to high molecular symmetry and high cohesive energy density forces between chains. Both of these properties are related to a relatively high degree of crystallinity present in fiber molecules.

Fibers are normally linear and drawn (oriented) in one direction to enhance mechanical properties in the direction of the draw. Typical condensation polymers, such as polyesters and nylons, often exhibit these properties.

If the fiber is to be ironed, its T_g should be above 350°F, and if it is to be drawn from the melt, its T_g should be below 570°F. Branching and cross-linking in fibers are undesirable since they disrupt crystalline formation, but a small amount of cross-linking may increase some physical properties if introduced after the material has been drawn and processed. In fact, a small amount of cross-linking is introduced for permanent press fabrics to help hold in a desired shape.

Cotton, linen, wool, and silk were used for over 2000 years before cellulose nitrate filaments were spun by H. Chardonnet. Regenerated cellulose produced by spinning cellulose xanthate was introduced in 1892 by C. Cross, E. Bevan, and C. Beadle. Cellulose xanthate is produced by the reaction of cellulose and carbon disulfide (CS_2) in the presence of alkali. The term rayon is now used to describe all regenerated cellulose, including derivatives such as acetate rayon. Nylon, which was the first synthetic fiber, was produced by W. Carothers and J. Hill in the 1930s.

4.4 PLASTICS

Materials with properties that are intermediate between those of elastomers and fibers are grouped together under the general term "plastics." Thus, plastics exhibit some flexibility and hardness and varying degrees of crystallinity. The molecular requirements for a thermoplastic are that it have little or no cross-linking and that it be used below its glass transition temperature, if amorphous, and/or below its melting point, if crystalline. Thermoset plastics must be sufficiently cross-linked to severely restrict molecular motion. The term cross-linked density is used to describe the extent of cross-linking in a material.

4.5 ADHESIVES

Adhesives can be considered to be coatings sandwiched between two surfaces. Early adhesives were water-susceptible and biodegradable animal and vegetable glues obtained from hides, blood, and starch. Adhesion may be defined as the process that occurs when a solid and a movable material (usually in a liquid or solid form) are brought together to form an interface and the surface energies of the two substances are transformed into the energy of the interface.

Starch was used to glue sheets of papyrus by the Egyptians 6000 years ago, and hydrolyzed collagen from bones, hides, and hooves (carpenter's glue) was used as an adhesive in 1500 B.C. Starch, which was partially degraded by vinegar, was used as an adhesive for paper in 120 B.C. These early adhesives continue to be used but have been largely displaced by solutions and hot melts of synthetic polar polymers.

A unified science of adhesion has yet to be developed. Adhesion can result from mechanical bonding and chemical and/or physical forces between the adhesive and adherend. Contributions through chemical and physical bonding are often more important and illustrate why nonpolar polymeric materials, such as polyethylene, are difficult to bond, whereas polar polycyanoacrylates, such as butyl-2-cyanoacrylate, are excellent adhesives. There are numerous types of adhesives, including solvent-based, latex, pressure-sensitive, reactive, and hot-melt adhesives.

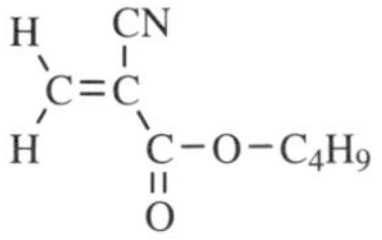

$$H_2C{=}C(CN){-}C({=}O){-}O{-}C_4H_9$$

Butyl-α-cyanoacrylate

The combination of an adhesive and adherend is a laminate. Commercial laminates are produced on a large scale with wood as the adherend and phenolic, urea, epoxy, resorcinol, or polyester resins as the adhesives. Some wood laminates are called plywood. Laminates of paper or textile include items with the trade names Formica and Micarta. Laminates of phenolic, nylon, or silicone resins with cotton, asbestos, paper, or glass textiles are used as mechanical, electrical, and general-purpose structural materials. Plastic composites of mat or sheet fibrous glass and epoxy or polyester resins are widely employed as fiber-reinforced plastic (FRP) structures.

4.6 COATINGS

The annual cost of corrosion is over $100 billion in the United States. With the exception of metal and ceramic types, nearly all surface coatings are based on polymeric films. The surface-coating industry originated in prehistoric times. By 1000 B.C., naturally occurring resins and beeswax were used as constituents of paints. The coatings industry used drying oils, such as linseed oil, and natural resins, such as rosin, shellac, and copals, prior to the early 1900s.

Linseed oil, which is obtained from the seeds of flax (*Lininum usitatissium*), was the first vegetable oil binder used for coatings. This unsaturated (drying) oil hardens (polymerizes) in air when a heavy metal salt (drier, siccative) is present. Presumably, some free oleic acid reacts with the white lead pigment to produce a drier (catalyst) when the linseed oil and pigment are heated. Subsequently, ethanolic solutions of shellac were displaced by collodion (a solution of cellulose nitrate), but oleoresinous paints continue to be used.

Phenolic, alkyd, and urea resins were used as coatings in the 1920s. Interior paints based on lattices of poly(vinyl acetate), poly(methyl methacrylate), and styrene–butadiene copolymers were introduced after World War II. Latex paints for exterior use were marketed in the late 1950s.

The fundamental purposes of coatings as being decorative and protective are giving way to more complex uses in energy collection devices and burglar alarm systems. Even so, the problems of the coating's adhesion, weatherability, permeability, corrosion inhibition, flexural strength, endurance, application, preparation, and application procedures continue to be the major issues. Effective coatings generally yield tough, flexible films with moderate to good adhesion to metal or wood surfaces.

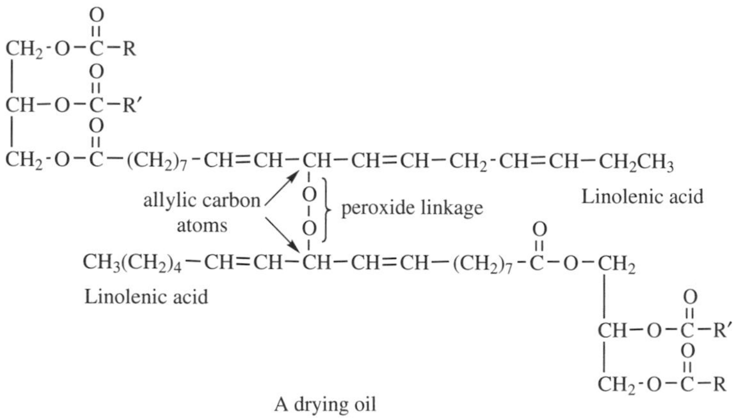

$CH_2{-}O{-}C({=}O){-}R$

$CH{-}O{-}C({=}O){-}R'$

$CH_2{-}O{-}C({=}O){-}(CH_2)_7{-}CH{=}CH{-}CH{-}CH{=}CH{-}CH_2{-}CH{=}CH{-}CH_2CH_3$

Linolenic acid

allylic carbon atoms | $O{-}O$ peroxide linkage

$CH_3(CH_2)_4{-}CH{=}CH{-}CH{-}CH{=}CH{-}(CH_2)_7{-}C({=}O){-}O{-}CH_2$

Linolenic acid

$CH{-}O{-}C({=}O){-}R'$

$CH_2{-}O{-}C({=}O){-}R$

A drying oil

4.7 POLYBLENDS AND COMPOSITES

Polyblends are made by mixing components together in extruders or intensive mixers or on mill rolls. Most heterogeneous systems consist of a polymeric matrix in which another polymer is embedded. Whereas the repeating units of copolymers are connected through primary bonds, the components of polyblends are connected through secondary bonding forces. In contrast to polyblends, which are blends of polymers, composites consist of a polymeric matrix in which a polymeric material is dispersed. Composites typically contain fillers, such as carbon black, wood flour, and talc, or reinforcing materials, such as glass fibers, hollow spheres, and glass mats.

4.8 CRYSTALLINE–AMORPHOUS STRUCTURES

Polymers typically contain a combination of ordered (often called crystalline) and disordered (called amorphous) areas, regions, or domains (all three terms are used to describe essentially the same thing). In general, the crystalline regions are stiffer and stronger and contribute to the materials strength and inflexibility. The amorphous regions contribute to a materials flexibility particularly when the material is above its T_g. Highly crystalline giant molecules generally exhibit higher melting points, higher glass transition temperatures, and higher densities, are less soluble, have lowered permeabilities, and are stiffer relative to polymers with less crystallinity. This is a consequence of a tighter, more compact structure that has fewer open spaces and where the closeness and ordered structure allow for the secondary forces to be more effective. Compare a ball of yarn from the store to the ball of yarn after a kitten gets in it. The "pre-kitten" ball of yarn is more tightly packed and is similar to the crystalline portions, whereas the "after-kitten" ball of yarn illustrates the amorphous regions of a giant molecule.

Figure 4.1 contains a polyethylene chain where both crystalline and amorphous are present. Note the presence of side chains or arms that inhibit the chains containing these side arms or branching from coming close together, thus discouraging crystalline formation. For comparison, linear polyethylene, also called high-density polyethylene, is largely crystalline with a density of about 0.96 g/mL and a melting point of about 130°C whereas branched polyethylene, also called low-density polyethylene, has a density of about 0.91 g/mL, a melting temperature of about 100°C. The crystalline polyethylene is stronger, tougher, and less attacked by chemicals; it is also less permeable, meaning fewer molecules can get through.

The particular structure and combinations of amorphous and crystalline portions vary with the structure of the polymer chains and the conditions that are imposed on the polymer. For instance, rapid cooling generally decreases the amount of crystallinity because there is not enough time to allow the long chains to organize themselves into more ordered structures. Polymers with large bulky groups are less apt to form high degrees of crystallinity.

In general, linear polymers form a variety of single crystals when crystalized from very dilute solutions. For instance, highly linear polyethylene can form diamond-shaped single crystals with a thickness on the order of 20 ethylene units when crystallized from dilute solution. The surface consists of "hairpin-turned" methylene units as depicted in Figure 4.2 The polymer chain axes is perpendicular

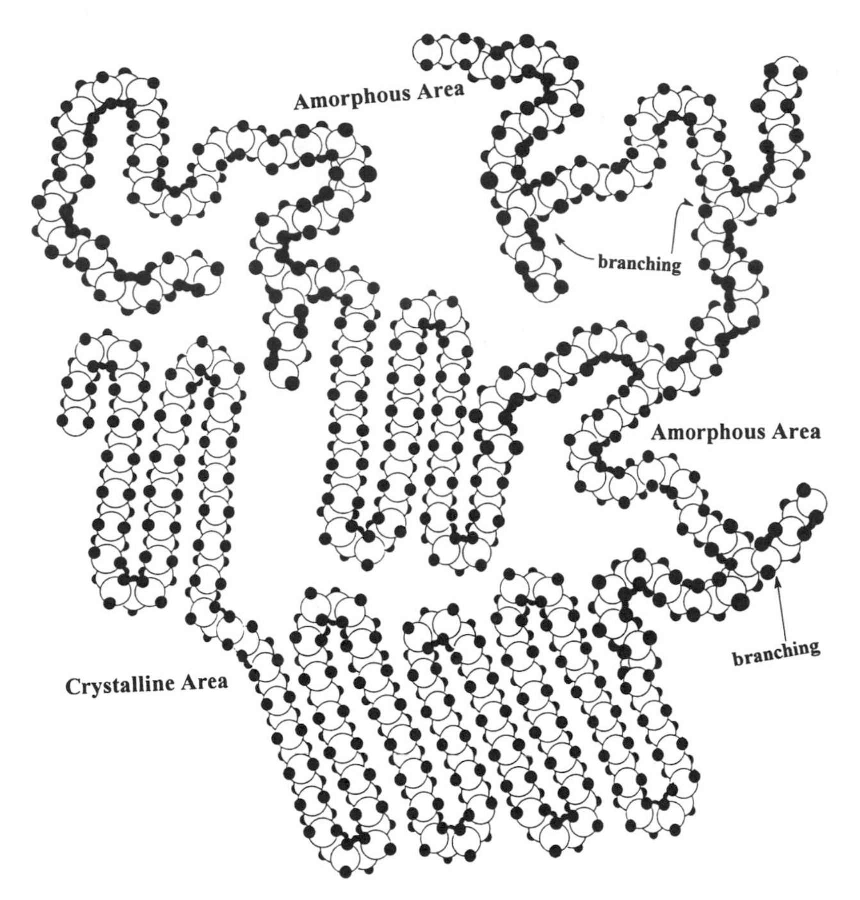

Figure 4.1. Polyethylene chain containing about 250 ethylene (or 500 methylene) units arranged into crystalline and amorphous areas and containing some branching.

to the large flat crystal faces. A single polymer chain with 1000 ethylene (2000 methylene) units might undergo on the order of 50 of these hairpin turns on the top surface and another 50 turns on the bottom face with about 20 ethylene units between the two surfaces.

Many polymers form more complex single crystals when crystallized from dilute solution including hollow pyramids that often collapse on drying. As the polymer concentration increases, other structures occur including twins, spirals, and multilayer dendritic structures, with the main structure being spherulites.

When a polymer is heated, it can form a fluid mixture called a melt where both segmental and whole chain movement readily occurs. On cooling, mixtures of amorphous and crystalline regions are formed and locked in. These mixtures of amorphous and minicrystalline structures or regions may consist of somewhat random chains containing some chains that are parallel to one another, forming

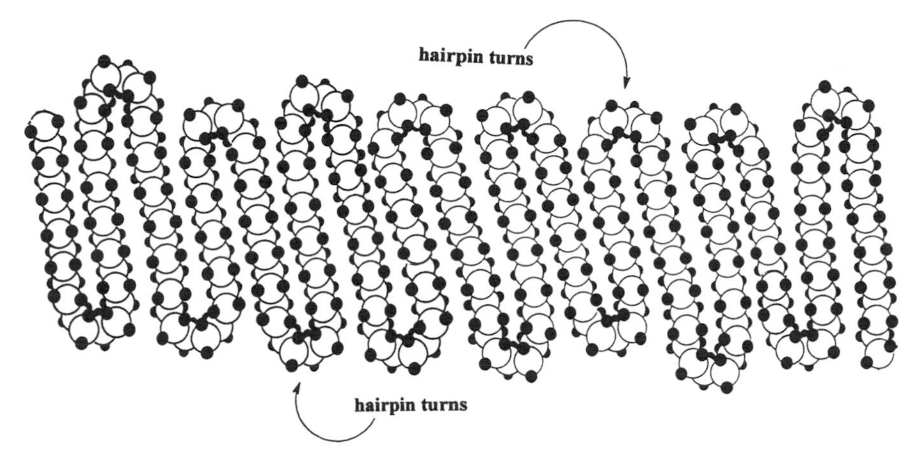

Figure 4.2. Linear polyethylene chain (about 125 ethylene or 250 methylene units) illustrating hairpin turns and linear inner structural arrangement.

short-range minicrystalline regions. Crystalline regions may be formed from large-range ordered plateletlike structures including polymer single crystals or they may form even larger organizations such as spherulites (Figure 4.3). Short- and longer-range ordered structures can act as physical cross-links.

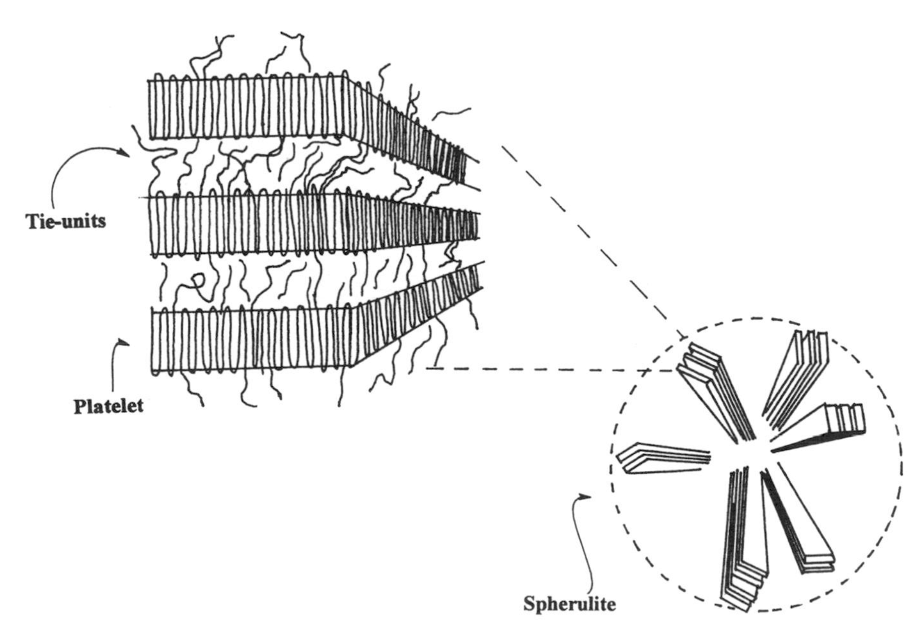

Figure 4.3. Spherulite structure showing the molecular-level lamellar chain-folded platelets and tie and frayed chain arrangements.

When polymers are produced from their melt, the most common structures are these spherulites. For linear polyethylene the initial structure formed is a single crystal with folded-chain lamellae. These quickly lead to the formation of sheaflike structures. As growth proceeds, the lamellae develop on either side of a central reference plane. The lamellae continue to fan out, occupying increasing volume sections through the formation of additional lamellae at appropriate branch points. The result is the formation of spherulites as pictured in Figure 4.3.

While the lamellar structures present in spherulites are similar to those present in polymer single crystals, the folding of chains in spherulites is less organized. Furthermore, the structures that exist between these lamellar structures are generally occupied by amorphous structures.

The individual spherulite lamellae are bound together by "tie" molecules that are present in several lamellae within the spherulite (Figure 4.4). Sometimes these tie segments form intercrystalline links between different spherulites. These tie segments are threadlike structures that are important in developing the characteristic good toughness found in semicrystalline polymers since they connect or tie together the strong inflexible spherulites with the more flexible threadlike tie segments. They then act to tie together the entire assembly of spherulites into a more or less coherent "package."

But, if the polymer is caused to flow through a pipe as the melted polymer is transported so it can be turned into a pipe or sheet, crystallization with repeated

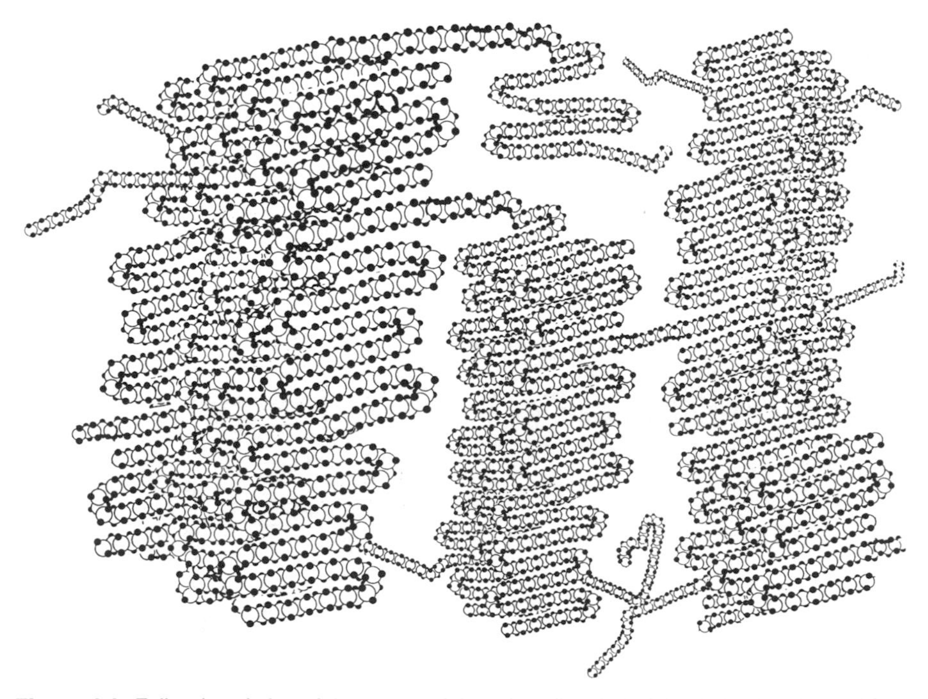

Figure 4.4. Fuller description of three sets of three lamellar chain-folded platelets formed from polyethylene. Each of the bottom two platelets contains about 850 ethylene units while the upper on contains about 1500 ethylene units. Notice the tie lines between the platelets.

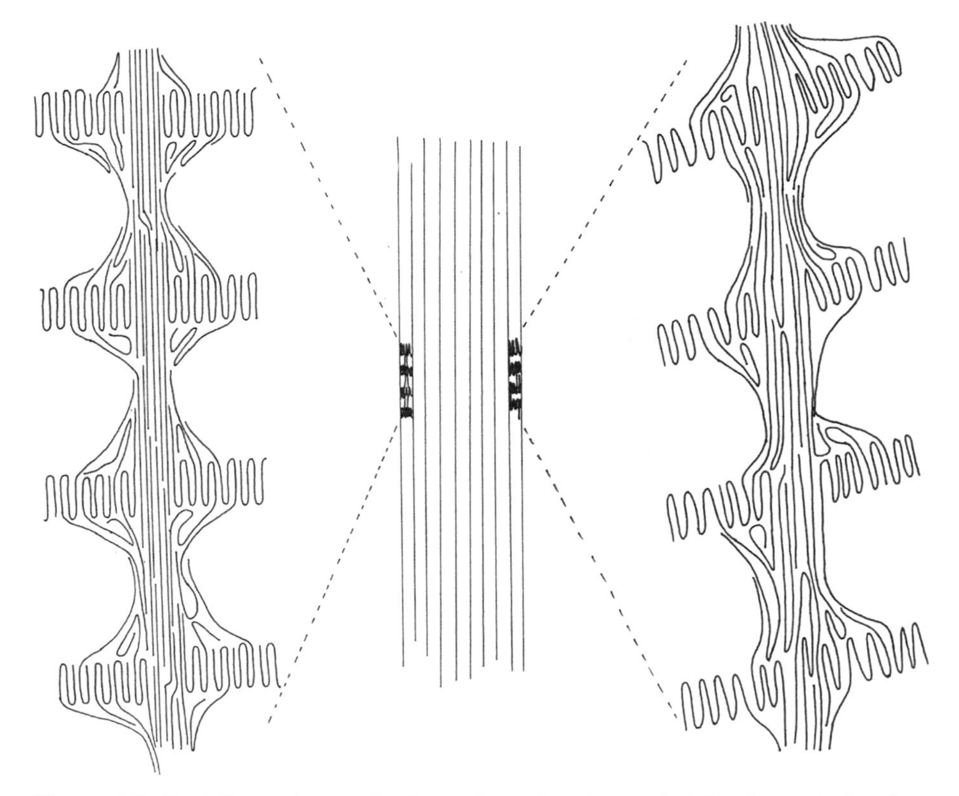

Figure 4.5. Crystalline polymer structures formed under applied tension including flow conditions. Left shows the tertiary monofibrillar structure including platelets, right shows tilted arms caused by an increased flow rate, and the center shows these monofibrillar structures bundled together forming a quaternary structure fibril.

back-and-forward folding such as present in the spherulite form about an inner shaft (Figure 4.5, left) with more linear-chain crystallization occurring within the shaft, forming a shish-kebab arrangement. The center part of Figure 4.5 is a bundle of polymer shafts. If the flow becomes faster, then the outside chains are pulled relative to the inner chains; the result is a tilt to the crystals forming about the inner shaft giving an upward shift to the arm crystalline portions (Figure 4.5, right). Both crystalline and amorphous regions exist in these shish-kebab structures. These shish-kebab structures often organize into quaternary structures consisting of bundles of shish-kebab single-strand filaments forming fibrils as shown in the center of Figure 4.5.

These structures are "locked in" when the giant molecules cool. This illustrates another common theme of giant molecules. Materials, particularly giant molecules, "remember" what has occurred to them. Thus, if they are cooled when they are largely in a crystalline form, then the resultant material will be largely crystalline and the material will behave as a largely crystalline material.

As noted before, the amorphous regions within the spherulite confer onto the material some flexibility while the crystalline platelets give the material strength,

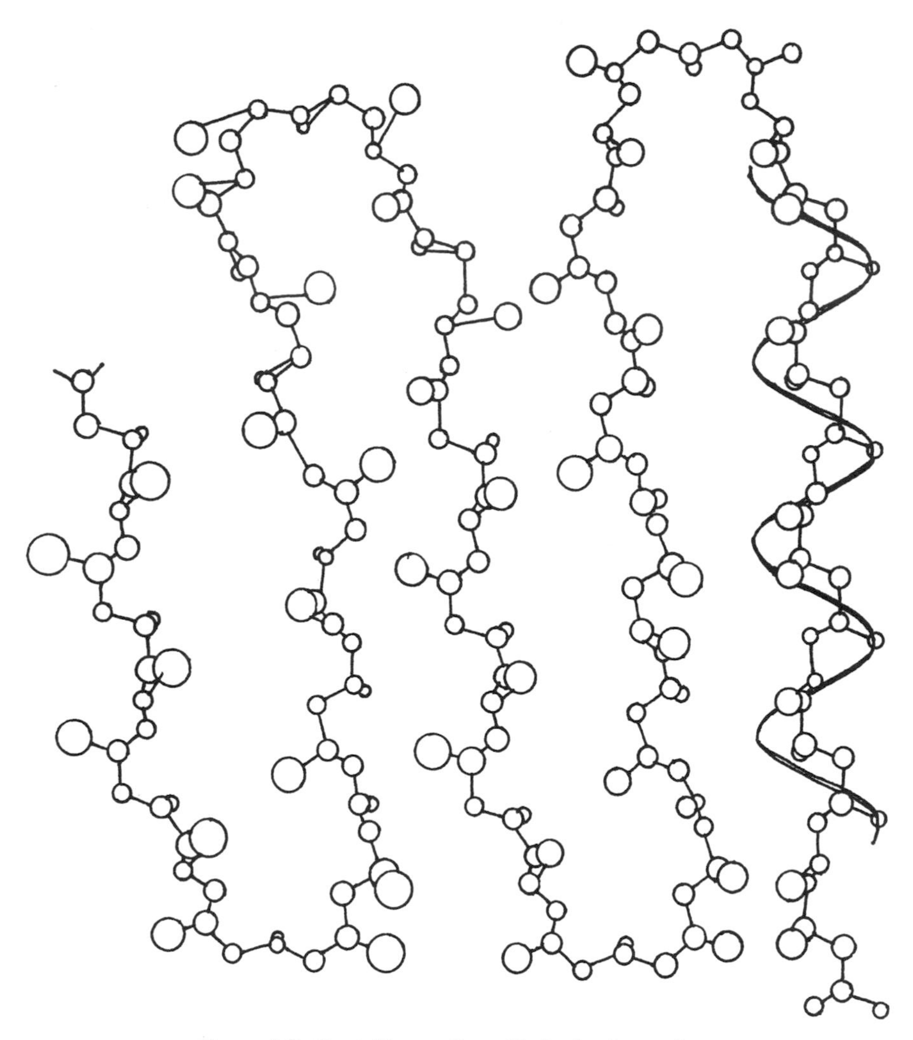

Figure 4.6. Crystalline portion of helical polypropylene.

just as in the case with largely amorphous materials. This theme of amorphous flexibility and crystalline strength (and brittleness) is a central idea in polymer structure–property relationships.

It must be remembered that the secondary structure of both the amorphous and crystalline regions typically tend toward a helical arrangement of the backbone as illustrated in Figure 4.6.

The kind, amount, and distribution of polymer chain order/disorder (amorphous/crystalline) is driven by the processing (including pre- and post-) conditions, and thus it is possible to vary the polymer properties through a knowledge of and ability to control the molecular-level structures. Factors that contribute to the inherent crystalline–amorphous-forming tendencies of polymers are discussed next.

A. Chain Flexibility

The tendency toward crystallinity in some polymers increases as flexibility is increased. Polymers containing regularly spaced single C–C and C–O bonds allow rapid conformational changes that contribute to the flexibility of a polymer chain and the tendency toward crystal formation. This is also true in the case of linear polyethylene, polypropylene, and poly(vinyl chloride), whose structures are shown in Figure 4.7.

Chain stiffness may also enhance crystalline formation by permitting only certain "well-ordered" conformations to occur within the polymer chains. Thus,

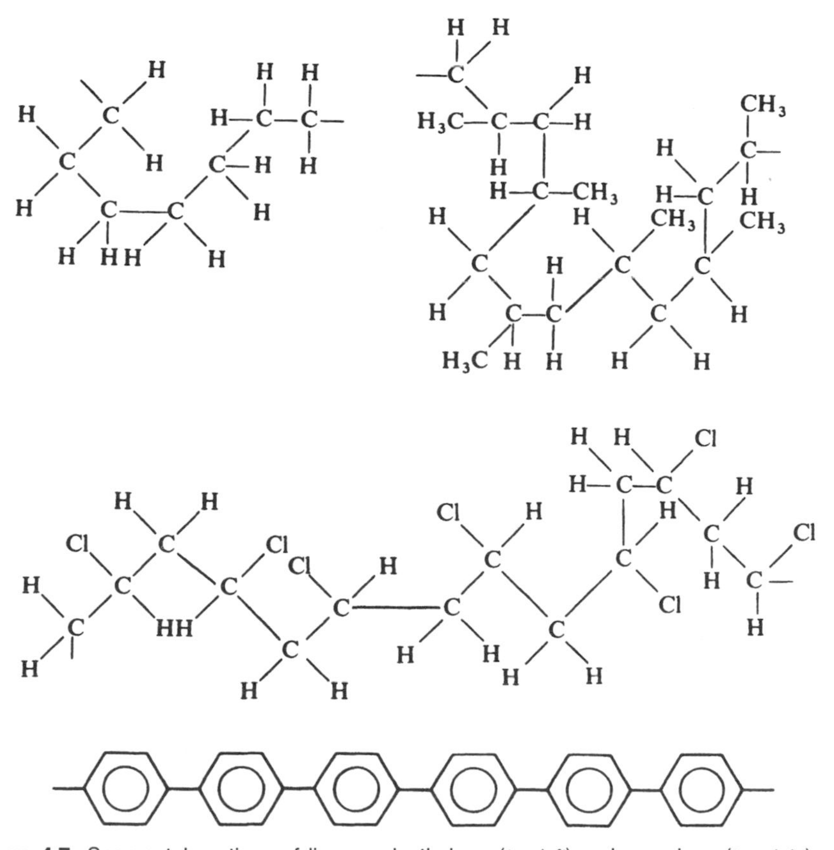

Figure 4.7. Segmental portions of linear polyethylene (*top left*), polypropylene (*top right*), and polyvinyl chloride (middle), illustrating chain flexibility, and poly-*p*-phenylene (*bottom*), illustrating chain stiffness.

poly-*p*-phenylene is a linear chain that cannot "fold over" at high temperatures. Hence, such species are crystalline, high-melting, rigid, and insoluble.

B. Intermolecular Forces

Crystallization is favored by the presence of regularly spaced units that permit strong intermolecular interchain associations. The presence of moieties that carry dipoles or are highly polarizable promotes strong interchain exchanges. This is particularly true for interchain hydrogen bond formation. Thus, the presence of regularly spaced carbonyl (C=O), amine (NH_2), amide ($CONH_2$), sulfoxide (SO_2), and alcohol (OH) moieties promotes crystallization.

C. Structural Regularity

Structural regularity also enhances the tendency for crystallization. Thus, it is difficult to obtain linear polyethylene (HDPE) in any form other than a highly crystalline one. Low-density, branched polyethylene (LDPE) is typically largely amorphous. The linear polyethylene chains are nonpolar, and the crystallization tendency is mainly based on its flexibility, which permits it to achieve a regular, tightly packed conformation, which takes advantage of the special restrictions

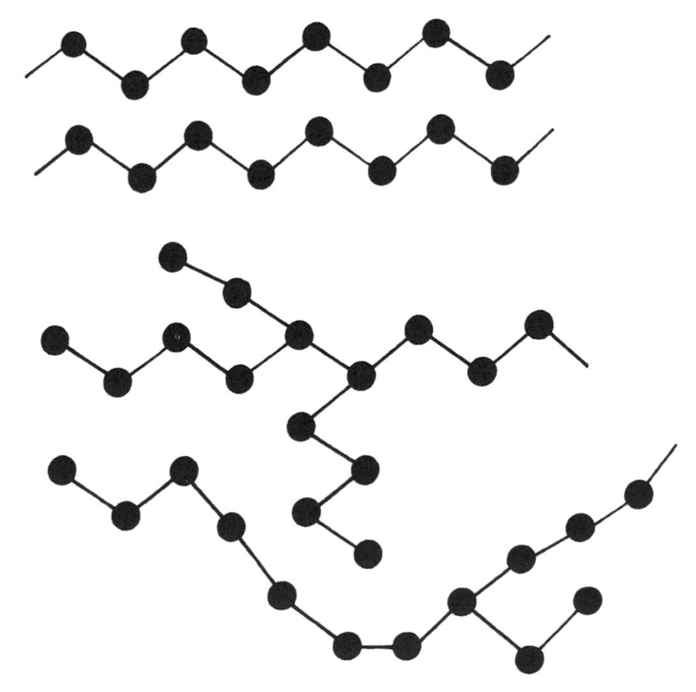

Figure 4.8. (*Top*) Simulated structure of high-density, linear polyethylene, emphasizing the tendency toward intrachain regularity. (*Bottom*) Low-density, branched polyethylene, illustrating the inability for intrachain regularity.

inherent in the dispersion forces. Simulated structures of HDPE and LDPE are shown in Figure 4.8.

Monosubstituted vinyl monomers ($CH_2{=}CHX$) can produce polymers with different configurations, that is, two regular structures (isotactic and syndiotactic) and a random, atactic form. Polymers with regular structures exhibit greater rigidity and are higher-melting and less soluble than the atactic form.

Extensive work with condensation polymers and copolymers confirms the importance of structural regularity on crystallization tendency and associated properties. Thus, copolymers containing regular alternation of each copolymer unit, either ABABAB type or block type, show a distinct tendency to crystallize, whereas corresponding copolymers with random distributions of the two monomers are intrinsically amorphous, less rigid, and lower melting and have greater solubility. The concept of micelles in natural fibers was first expressed by C. Nagele in 1858, but since most nineteenth-century scientists preferred to consider polymers as aggregates of molecules (colloids) rather than as individual giant molecules, the existence of polymer crystals was deemphasized until 1917 when H. Ambronn suggested that cellulose nitrate fiber had crystalline characteristics. In 1920, R. Herzog and M. Polanyi used x-ray diffraction techniques to show the presence of crystallites in flax fiber.

D. Steric Effects

The effect of substituents on polymer properties depends on the location, size, shape, and mutual interactions of the substituents. Methyl and phenyl substituents (pendant groups) tend to lower chain mobility but prevent good packing of chains. These substituents produce unit dipoles, which contribute to the crystallization tendency.

Aromatic substituents contribute to intrachain and interchain attraction tendency through the mutual interactions of the aromatic substituents. Their bulky size retards crystallization and promotes rigidity because of increased interchain distances. Thus, polymers containing bulky aromatic substituents tend to be rigid, high-melting, less soluble, and amorphous.

Substituents from ethyl to hexyl tend to lower the tendency for crystallization, since they increase the average distance between chains and decrease the contributions of secondary bonding forces. Thus, LLDPE is an amorphous polymer. If these linear substituents are larger (12 to 18 carbon atoms), these side chains form crystalline domains on their own (side-chain crystallization).

4.9 SUMMARY

Polymer properties are directly dependent on both the inherent shape of the polymer and its treatment. Contributions of polymer shape to polymer properties are often complex and interrelated but can be broadly divided into terms related to chain regularity, interchain forces, and steric effects.

GLOSSARY

Additive: Substances added to polymers to improve properties, such as strength, ductility, stability, and resistance to flame.

Adherend: A substance whose surface is adhered by an adhesive.

Adhesive: A substance that bonds two surfaces together.

Amorphous: Shapeless, noncrystalline.

Cohesive energy density (CED): A measure of intermolecular forces between molecules.

Composite: A mixture of a polymer and an additive, usually a reinforcing fiber or filler.

Density, cross-linked: A measure of the extent of cross-linking in a polymer network.

Elastomer: Amorphous, flexible polymers that are usually cross-linked to a small extent.

Entropy: A measure of the degree of disorder or randomness in a polymer.

Fiber: A threadlike substance in which the ratio of the length to diameter is at least 100:1. Fibers are characterized by strong intermolecular forces.

Fringed micelle concept: A diagrammatic representation of aligned polymer chains (crystalline) separated by regions of nonaligned or amorphous areas.

FRP: Fiberglass-reinforced plastic.

HDPE: High-density (linear) polyethylene.

Laminate: A composite resulting from adhering two surfaces together.

Latex: A stable dispersion of a polymer in water.

LDPE: Low-density (branched) polyethylene.

LLDPE: A low-density linear polyethylene, usually a copolymer of ethylene and 1-butene or 1-hexene.

Modulus: The ratio of strength to elongation, a measure of stiffness.

Paint: A mixture of a pigment, unsaturated oil, resin, and drier (catalyst).

Plastic: Substances with properties in between those of elastomers and fibers.

Plywood: A laminate of thin sheets of wood and adhesives.

Principal section: Portion of a polymer chain between cross-links.

Steric: Arrangement in space.

REVIEW QUESTIONS

1. What is the function of additives in polymers?
2. What are the characteristics of an elastomer?
3. Which has the higher entropy: stretched or unstretched rubber?

4. Which has the higher cohesive energy density (CED): an elastomer or a fiber?
5. Which has the higher cross-linked density: soft vulcanized rubber or hard rubber?
6. Which has the longer principal sections: soft vulcanized rubber or hard rubber?
7. Which has the higher modulus: soft vulcanized rubber or hard rubber?
8. Which has a higher degree of crystallinity: HDPE or LDPE?
9. What are the adhesive and adherend widely used in reinforced plastics?
10. What is the trade name of a laminate used for kitchen countertops?
11. What is the difference between a paint and a protective coating?
12. Why are latex-based coatings popular?

BIBLIOGRAPHY

Bicerano, J. (2002). *Predicting of Polymer Properties*, 2nd ed. Marcel Dekker, New York.

Blau, W., Lianos, P., and Schubert, U. (2001). *Molecular Materials and Functional Polymers*, Springer-Verlag, New York.

Brown, W (1996). *Light Scattering: Principles and Development*, Springer-Verlag, New York.

Carraher, C., Swift, G., and Bowman, C. (1997). *Polymer Modification*, Plenum, New York.

Hansen, C. (200O). *Hanson Solubility Parameters*, CRC, Boca Raton, FL.

Higgins, J., and Benoit, H. C. (1997). *Polymers and Neutron Scattering*, Oxford University Press, Cary, NC.

Roe, R. (2000). *Methods of X-Ray and Neutron Scattering in Polymer Science*, Oxford University Press, Cary, NC.

Seymour, R., Carraher, C. (1984). *Structure–Property Relationships in Polymers*, Plenum, New York.

Schultz, J. (2001). *Polymer Crystallization*, Oxford University Press, Cary, NC.

Tsujii, K. (1998). *Surface Activity*, Academic Press, Orlando, FL.

Woodward, A. (1995). *Understanding Polymer Morphology*, Hanser Gardner, Cincinnati, OH.

Wypych, G. (2001). *Handbook of Solvents*, ChemTec, Toronto, Can.

Yagci, Y., Mishra, M., Nuyken, O., Ito, K., Wnek, G. (2000). *Tailored Polymers and Applications*, VSP, Leiden, Netherlands.

ANSWERS TO REVIEW QUESTIONS

1. They improve properties.
2. It is amorphous when unstretched, has weak intermolecular forces, and usually has a low cross-linked density.

3. Unstretched rubbers have a greater degree of randomness or disorder.
4. A fiber usually contains intermolecular hydrogen bonds.
5. Hard rubber.
6. Soft vulcanized rubber.
7. Hard rubber.
8. HDPE has a more ordered structure.
9. The resin (polyester, epoxy) is the adhesive and the fiberglass or graphite is the adherend.
10. Micarta or Formica.
11. Paint is a protective coating, but there are many other types of protective coatings.
12. They are easy to produce and do not affect the environment adversely as do solvent-based coatings. They have a low volatile organic concentration (VOC).

5

PHYSICAL AND CHEMICAL TESTING OF POLYMERS

Giant Molecules: Essential Materials for Everyday Living and Problem Solving, Second Edition,
by Charles E. Carraher, Jr.
ISBN 0-471-27399-6

5.1 TESTING ORGANIZATIONS

Giant molecules are asked to perform many tasks in today's society. Often they are required to perform these tasks again and again and again . . . A plastic hinge must be able to work thousands of times, yet some giant molecules are asked to perform repeated tasks many more times. Our hearts, composed of complex protein muscles, provide about 2,500,000,000 (2.5 billion) beats within a lifetime, moving oxygen throughout the approximately 144,000 km or 90,000 miles of the circulatory system with some blood vessels the thickness of hair and delivering about 8000 L or 2100 gallons of blood every day with little deterioration of the cell walls. Nerve impulses travel within the body largely though the use of giant molecules at a speed of about 300 m/min or 12,000 in./min. Our bones, largely composed of giant molecules, have a strength about five times that of steel on a weight basis. Genes, again composed of giant molecules, appear to be about 99.9% the same, with only 0.1% acting to produce individuals with a variety of likes, dislikes, strength, abilities, and so on, thereby making each of us unique.

Public acceptance of materials containing giant molecules is associated with an assurance of quality based on a knowledge of successful long-term and reliable tests. In contrast, dissatisfaction is often related to failures that might have been prevented by proper testing, design, and quality control.

The selection of general-purpose polymers has sometimes been the result of trial and error, misuse of case history data, or questionable guesswork. However, since polymeric materials must be functional, it is essential that they be tested using meaningful use-oriented procedures. Both the designer and the user should have an understanding of the testing procedure used in the selection of a polymeric material for a specific end use. They should know both the advantages and the disadvantages of the testing procedure, and designers should continue to develop additional empirical tests.

Fortunately, there are many standards and testing organizations whose sole purpose is to ensure the satisfactory performance of materials. The largest standards organization is the International Standards Organization (ISO), which consists of members from about 90 countries and many cooperative technical committees. There is also the American National Standards Institute (ANSI) and the American Society for Testing and Materials (ASTM), which publishes its tests on an annual basis. Other important reports on tests and standards are published by the National Electrical Manufacturing Association (NEMA), Deutsches Institut fur Normenausschuss (DIN), and the British Standards Institute (BSI).

The ASTM tests have a listing after them. For instance, the coefficient of linear expansion test has a number ASTM D696-79 meaning that the test has successfully completed the "round robin" testing and been accepted in 1979. The particular test, 696, has specifications that include exact specifications regarding data gathering, instrument design, sample specifications, and test conditions that allow laboratories throughout the world to reproduce the test and test results if given the same test material. Most tests developed by one testing society have analogous tests or more often use the same tests so that they may have both ASTM, ISO, ANIS, and so on, designations. The coefficient of linear expansion test is actually ANIS/ASTM D696-79, so it is accepted by both ASTM and ANIS.

Many tests are based on whether the tested material is chemically changed or is left unchanged. Nondestructive tests are those that involve no (detectable) chemical change. Destructive tests involve a change in the chemical structure of at least a portion of the tested material.

There often occurs a difference in "mind-set" between the nucleic acid and protein biopolymers covered in this chapter and other biopolymers and synthetic polymers covered in other chapters. Nucleic acids and proteins are site-specific with one conformation. Generally, if it differs from the specific macromolecule called for, it is discarded. Nucleic acids and proteins are not a statistical average, but rather a specific material with a specific chain length and conformation. By comparison, synthetic and many other biopolymers are statistical averages of chain lengths and conformations. The distributions are often kinetic/thermodynamic-driven.

This difference between the two divisions of biologically important polymers is also reflected in the likelihood that there are two molecules with the exact same structure. For molecules such as polysaccharides and those based on terpenelike structures, the precise structures of individual molecules vary, but for proteins and nucleic acids the structures are identical from molecule to molecule. This can be considered a consequence of the general function of the macromolecule. For polysaccharides the major, though not the sole, functions are energy and structural. For proteins and nucleic acids, main functions include memory and replication, in addition to proteins sometimes also serving a structural function.

Another difference between proteins and nucleic acids and other biopolymers and synthetic polymers involves the influence of stress/strain activities on the materials properties. Thus, application of stress on many synthetic polymers and some biopolymers encourages realignment of polymer chains and regions, often resulting in a material with greater order and strength. However, application of stress to certain biopolymers, such as proteins and nucleic acids, causes a decrease in performance (through denaturation, etc.) and strength. For these biopolymers, this is a result of the biopolymer already existing in a compact and "energy favored" form and already existing in the "appropriate" form for the desired performance. The performance requirements for the two classifications of polymers is different. For one set, including most synthetic and some biopolymers, performance behavior involves response to stress/strain application with respect to certain responses such as chemical resistance, absorption enhancement, and other physical properties. By comparison, the most cited performances for nucleic acids and proteins involves selected biological responses requiring specific interactions occurring within a highly structured environment that demands a highly structured environment with specific shape and electronic requirements.

For special-use giant molecules, specific tests are performed that are related to the end use of the material. In all testing, the end use of the giant molecule should guide in the testing and the evaluation of the results.

A brief listing of some important physical and chemical tests follows.

Electrical
- Bulk resistivity
- Dissipation factor (ASTM D-150)
- Power factor
- Electrical resistance (ASTM D-257)
- Dielectric constant (ASTM-150-74)
- Dielectric strength (ASTM D-149)
- Arc resistance
- Dielectric strength

Optical Properties
- Index of refraction (such as ASTM D-542)
- Optical clarity
- Adsorption and reflectance (such as ASTM E-308)
- Index of refraction (ASTM D-542-50 (1970)

Spectral

Density

Thermal
- Glass transition temperature (such as ASTM D-3418)
- Thermal conductivity (ASTM C-177-71)
- Thermal expansion (such as ASTM D696-79)
- Heat capacity
- Melting point
- Softening point (such as ASTM D-1525)
- Heat deflection temperature (ASTM D-648)

Flammability (such as ASTM D-635)

Surface characterization

Particle size

Mechanical
- Tensile strength (ASTM D-638-72)
- Creep
- Shear strength
- Elongation
- Compression strength (such as ASTM D-695)
- Impact strength (such as Izod-ASTM D-256; Charpy-ASTM D-256)
- Hardness (such as Rockwell-ASTM D-785-65 (1970); Pencil tests-ASTM D-3363); Tabor-ASTM D-1044; Deformation underload; Indentation tests-ASTM D2240, D-2583-67; D-674; D-671)

- Brittleness (such as ASTM D-746 and D-1790-62)
- Failure
- Flexural strength (ASTM D790-71/78)

Chemical resistance (such as ASTM D-543)

Weatherability

- Outdoors (ASTM D-1345)
- Accelerated (ASTM G-S23)
- Accelerated-light (ASTM-625 and 645)
- Water absorption (ASTM D570-63 (1972))

Following is a cross section of important tests routinely applied to bulk materials. Do not worry about the particular conditions and specifications included in describing many of the following tests. The particular conditions are given to remind ourselves of the nature of the tests and the importance to have such standardized conditions. When you need to carry out a particular test, the specifications are given in the ASTM book that deals with that particular test.

5.2 EVALUATION OF TEST DATA

Unlike the physical data compiled for metals and ceramics, the data for polymers are dependent on the life span of the test, the rate of loading, temperature, preparation of the test specimen, and so on. Some of these factors, but not all, have been taken into account in obtaining the data listed in tables in subsequent chapters of this book. Published data may vary for the same polymer fabricated on different equipment or produced by different firms and for different formulations of the same polymer or composite. Hence, the values cited in the tables are usually labeled "Properties of Typical Polymers."

Many tests used by the polymer industry are adaptations of those developed previously for metals and ceramics. None is so precise that it can be used with 100% reliability. In most instances, the physical, thermal, and chemical data are supplied by the producers, who are expected to promote their products in the marketplace. Hence, in the absence of other reliable information, positive data should be considered as upper limits of average test data and an allowance should be assumed by the user or designer.

5.3 STRESS/STRAIN RELATIONSHIPS

Mechanical testing involves a complex of measurements including creep, tensile and shear strength, impact strengths, and so on. Tensile strength is one of a grouping of tests that rely on application of a force and looking at what happens. Thus, when force is applied to a flexible plastic spoon, it bends with the extent of the bend dependent on the amount of force applied and the flexibility of the spoon. The force that is applied is given the name stress and the extent of bending referred to as strain. Stress/strain measurements are employed to help evaluation the usefulness of many giant molecules. Stiff materials have a high stress/strain ratio, whereas flexible materials have relatively low stress/strain ratios. A high stress/strain ratio simply means that it take lots of force to distort or bend the material a little.

Polymers are viscoelastic materials, meaning they can act as liquids (the "visco" portion) and as solids (the "elastic" portion). Descriptions of the viscoelastic properties of materials generally falls within the area called rheology. Determination of the viscoelastic behavior of materials generally occurs through stress/strain and related measurements. Whether a material behaves as a "viscous" or "elastic" material depends on temperature, the particular polymer and its prior treatment, polymer structure, and the particular measurement or conditions applied to the material. The particular property demonstrated by a material under given conditions allows polymers to act as solid or viscous liquids, as plastics, elastomers, or fibers.

As noted above, stress/strain results are related to a number of factors. Two important factors are the rate at which the force is applied, also called the interaction time, and temperature. If the rate of applying the stress exceeds the ability of the chain segments to move, then the material will act as a brittle solid.

For most plastics to be flexible, the temperature must be above the T_g or sufficient plasticizer is present to allow the chain segments to be mobile. On a cold day in South Dakota the temperatures get to $-30°C$ and a plastic spoon made of polypropylene, with a T_g of $-20°C$, is brittle and the stress/strain ratio is high. By comparison when it is brought indoors, where it warms up to above the T_g, the plastic spoon is now flexible and the stress/strain ratio is less.

Stress/strain testing is typically carried out using holders where one member is movable and contained within a load frame. Studies typically vary with either the stress or strain fixed and the result response measured. In a variable stress experiment a sample of given geometry is connected to the grips. Stress, load, is applied, generally by movement of the grip heads either toward one another (compression) or away from one another (elongation). This causes deformation, strain, of the sample. The deformation is recorded as is the force necessary to achieve this deformation.

Results of stress/strain tests are often modeled to look at the relative importance of chain segment movement, bond flexing, and other molecular motions. In general terms, a spring is used to represents bond flexing while a piston within a cylinder filled with a viscous liquid (called a dashpot) is used to represent chain and local segmental movement. Stress/strain behavior is related to combinations of dashpots and springs as indicators of the relative importance of bond flexing and segmental movement.

In general terms, below their T_g, polymers can be modeled as having a behavior where the spring portion is more important. Above their T_g, where segmental mobility occurs, the dashpot portion is more important.

The relative importance of these two modeling parts, the spring and the dashpot, is also dependent on the rate at which an experiment is carried out. Rapid interaction, such as striking a polymer with a hammer, is more apt to result in a behavior where bond flexibility is more important, while slow interactions are more apt to allow for segmental mobility to occur.

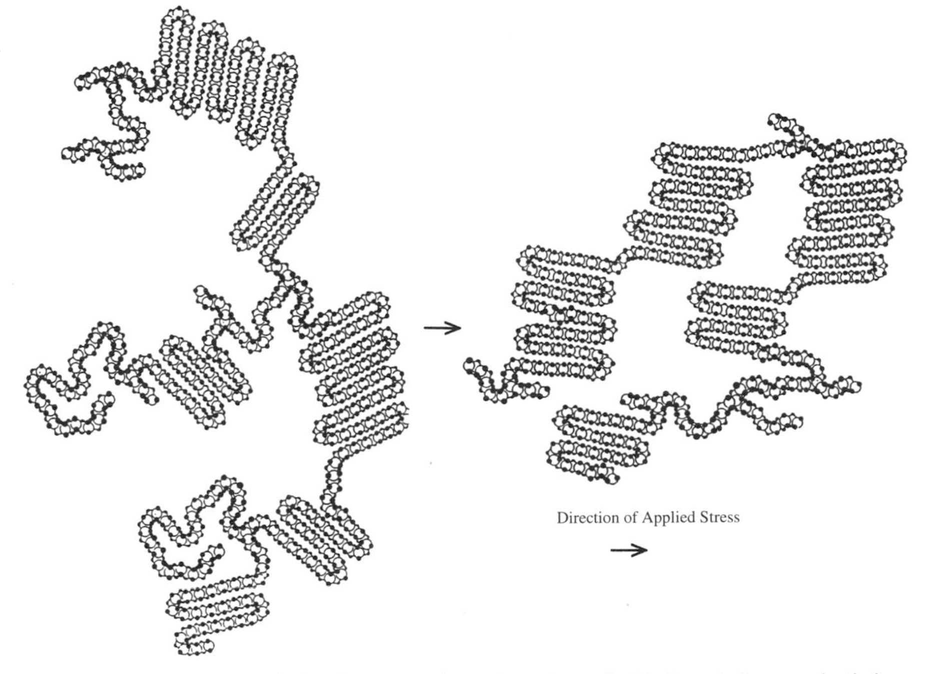

Figure 5.1. Visualization of what happens when stress is applied to largely linear polyethylene that contains both crystalline and amorphous regions.

Figure 5.1 gives a typical stress/strain experiment looking at what happens on a molecular level. As stress, pulling, occurs the molecules align themselves along the direction of the pull. The crystalline portions remain intact and the amorphous regions will align themselves, often forming crystalline regions themselves (not shown here). There is less "free volume" or unoccupied space in the stressed sample. In a sheet of stressed material, this results in the material being stronger in the direction of the pull and the sheet itself being less permeable; that is, gases and liquids are less apt to get through the film. Thus, such thin stressed films of polyethylene should be more suitable to being used as a strong, tough barrier to maintain fruit and vegetable freshness in comparison to nonstressed films.

Based on stress/strain behavior, Carswell and Nason assigned five classifications to polymers (Figure 5.2). Under normal room conditions an example of the soft weak class, A, is polyisobutylene (Chapter 10); polystyrene (6.11) is an example of a hard and brittle, B, material; plasticized poly(vinyl chloride) (6.13) behaves as a soft and tough, C, material; rigid poly(vinyl chloride) (6.13) is an example of a hard and strong, D, material; while ABS copolymers (10.6) behave as hard and tough, E, materials. As you go through the various chapters, think about which classification the particular material covered in that chapter might be in.

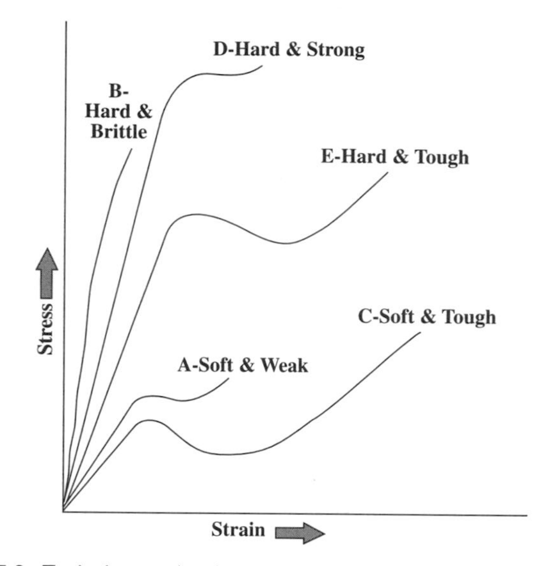

Figure 5.2. Typical stress/strain curves for plastics under room conditions.

5.4 HEAT DEFLECTION TEST

The heat deflection standard, which is now called Deflection Temperature of Plastics under Flexural Load (DTUL) (ANSI/ASTM D648-72/78), is a result of "round-robin" testing by all interested members of the ASTM Committee D20. This standard was accepted several decades ago. As shown by the numbers after D648 in the test designation, it was revised and reapproved in 1972 and reapproved in 1978, respectively.

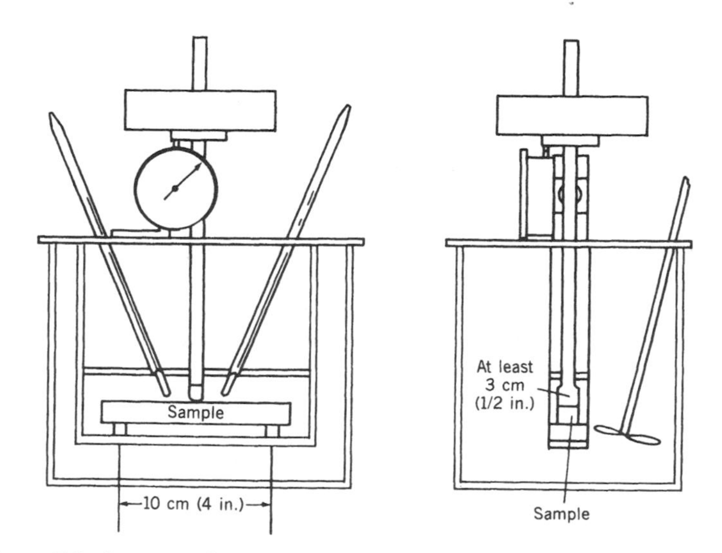

Figure 5.3. Apparatus for heat deflection under load (1.820 or 0.460 MPa) test.

The DTUL test measures the temperature at which an arbitrary deformation occurs when plastic specimens are subjected to an arbitrary set of testing conditions. The standard molded test span measures 127 mm in length, 13 mm in thickness, and 3–13 mm in width. The specimen is placed in an oil bath under a 0.455- or 1.820-MPa load in the apparatus shown in Figure 5.3, and the temperature is recorded when the specimen deflects by 0.25 mm.

The results of this test must be used with caution. The established deflection is extremely small and in some instances may be, in part, a measure of warpage or stress relief. The maximum resistance to continuous heat is an arbitrary value for useful temperatures, which is always below the DTUL value.

5.5 COEEFICIENT OF LINEAR EXPANSION

Since it is not possible to exclude factors such as changes in moisture, plasticizer, or solvent content, or release of stresses with phase changes, ANSI/ASTM D696-79 provides only an approximation of the true thermal expansion. The values for thermal expansion of unfilled polymers are high, relative to that of other materials of construction, but these values are dramatically reduced by the incorporation of fillers and reinforcements.

In this test, the specimen, measuring between 50 and 125 mm in length, is placed at the bottom of an outer dilatometer tube and below the inner dilatometer tube. The outer tube is immersed in a bath, and the temperature is measured. The increase in length (ΔL) of the specimen as measured by the dilatometer is divided by the initial length (L_0) and multiplied by the increase in temperature to obtain the coefficient of linear expansion (α). The formula for calculating this value is

$$\alpha = (\Delta L/L_0)T$$

5.6 COMPRESSIVE STRENGTH

Compressive strength, or the ability of a specimen to resist a crushing force, is measured by crushing a cylindrical specimen in accordance with ASTM-D695.

The test material is mounted in a compression tool as shown in Figure 5.4, and one of the plungers advances at a constant rate. The ultimate compression strength is equal to the load that causes failure divided by the minimum cross-sectional area. Since many materials do not fail in compression, strengths reflective of specified deformation are often reported.

5.7 FLEXURAL STRENGTH

Flexural strength or crossbreaking strength is the maximum stress developed when a bar-shaped test piece, acting as a simple beam, is subjected to a bending force perpendicular to the bar (ANSI/ASTM D790-71/78). An acceptable test specimen

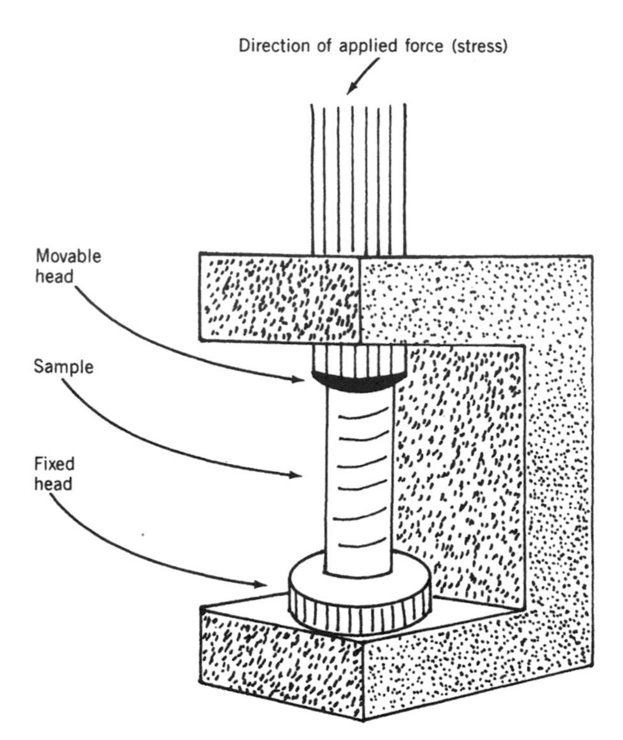

Figure 5.4. Apparatus for measurement of compression-related properties.

is one that is at least 3.2 mm in depth and 12.7 mm in width and long enough to overhang the supports, but the overhang should be less than 6.4 mm on each end.

The load should be applied at a specified crosshead rate, and the test should be terminated when the specimen bends or is deflected by 0.05 mm/min. The flexural strength (S) is calculated from the following expression in which P is the load at a given point on the deflection curve, L is the support span, b is the width of the bar,

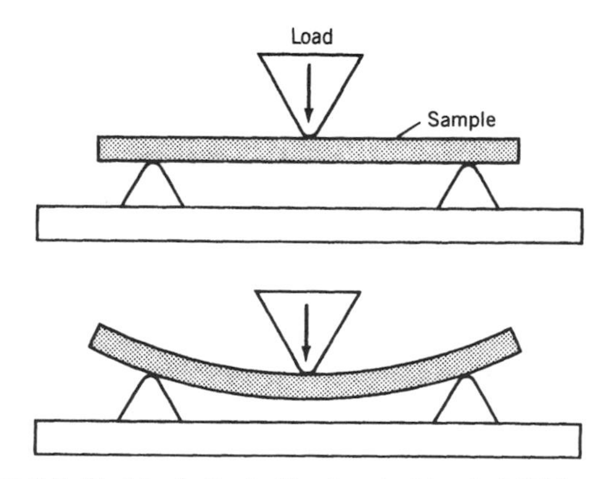

Figure 5.5. Sketch of effect of load on test bar in ASTM test 790.

and d is the depth of the beam. Figure 5.5 shows a sketch of the test, and the expression for calculating flexural strength is $S = PL/bd^2$.

One may use the following expression in which D is the deflection to obtain the maximum strain (r) of the specimen under test:

$$r = 6Dd/L$$

One may also obtain data for flexural modulus, which is a measure of stiffness, by plotting flexural stress (S) versus flexural strain (r) during the test and measuring the slope of the curve obtained.

5.8 IMPACT TEST

Impact strength may be defined as toughness or the capacity of a rigid material to withstand a sharp blow, such as that from a hammer. The information obtained from the most common test (ANSI/ASTM D256-78) on a notched specimen (Figure 5.6) is actually a measure of notch sensitivity of the specimen.

In the Izod test, a pendulum-type hammer, capable of delivering a blow of 2.7 to 21.7 J, strikes a notched specimen (measuring 127 mm × 12.7 mm × 12.7 mm with a 0.25 mm notch), which is held as a cantilever beam. The distance that the pendulum travels after breaking the specimen is inversely related to the energy required to break the test piece, and the impact strength is calculated for a 25.4-mm test specimen.

5.9 TENSILE STRENGTH

Tensile strength or tenacity is the stress at the breaking point of a dumbbell-shaped tensile test specimen (ANSI/ASTM D638-77). The elongation or extension at the

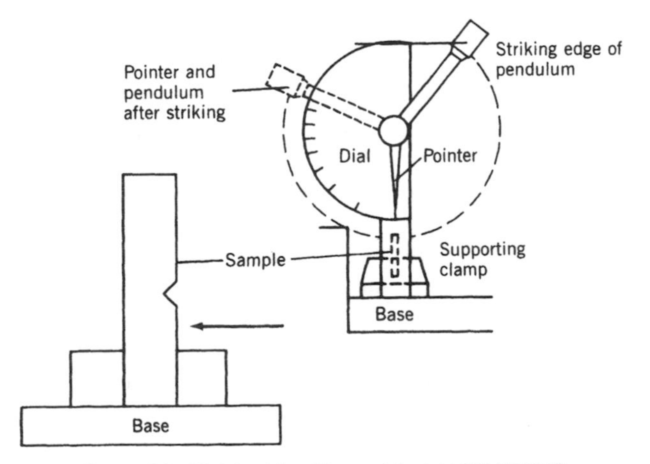

Figure 5.6. Notched Izod impact test (ASTM D256).

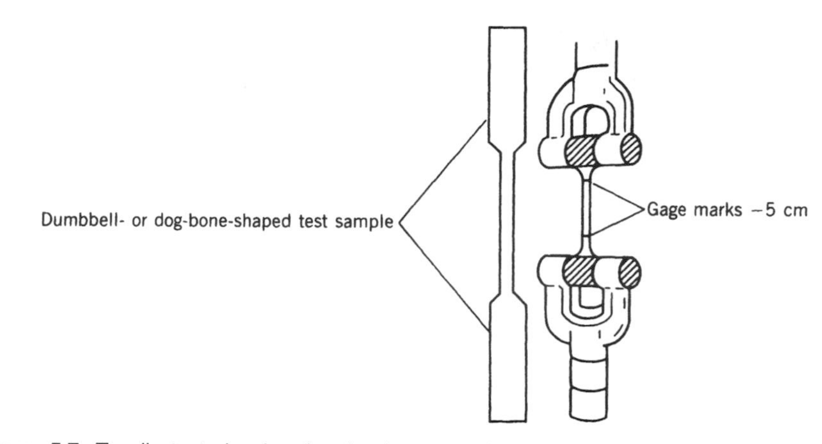

Figure 5.7. Tensile test showing the dog-bone specimen clamped in the jaws of an Instron tester.

breaking point is the tensile strain. As shown in Figure 5.7, the test specimen is 3.2 mm thick and has a cross section of 12.7 mm. The jaws holding the specimen are moved apart at a predetermined rate, and the maximum load and elongation at break are recorded. The tensile strength is the load at break divided by the original cross-sectional area. The elongation is the extension at break divided by the original gauge length multiplied by 100. The tensile modulus is the tensile stress divided by the strain. As an alternative to reporting the tensile strength, one may determine the slope of the tangent to the initial portion of the elongation curve.

5.10 HARDNESS TEST

The term hardness is a relative term. Hardness is the resistance to local deformation that is often measured as the ease or difficulty for a material to be scratched, indented, marred, cut, drilled, or abraded. It involves a number of interrelated properties such as yield strength and elastic modulus. Because polymers present such a range of behavior, they are viscoelastic materials, the test conditions must be carefully described. For instance, elastomeric materials can be easily deformed, but this deformation may be elastic with the indentation disappearing once the force is removed. While many polymeric materials deform in a truly elastic manner returning to the initial state once the load is removed, the range of total elasticity is often small, resulting in limited plastic or permanent deformation. Thus, care must be taken in measuring and in drawing conclusions from results of hardness measurements.

Hardness is related to abrasion resistance-resistance to the process of wearing away the surface of a material. The major test for abrasion resistance involves

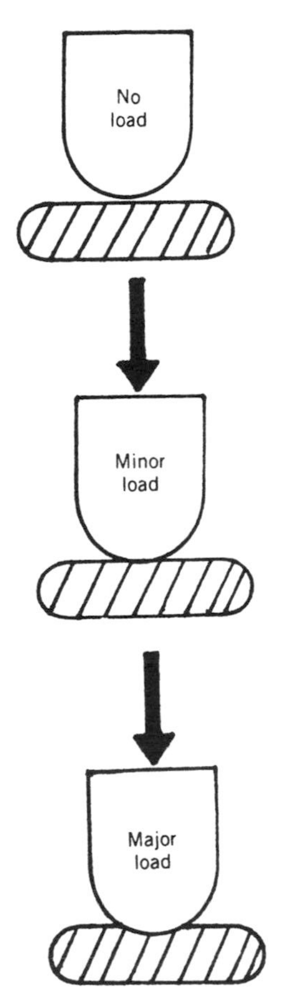

Figure 5.8. Illustration of Rockwell hardness test equipment.

rubbing an abrader against the surface of the material under specified conditions [ASTM D-1044].

Static indentation is most widely employed as a measure of hardness. Here, permanent deformation is measured. One test utilizes an indentor, which may be a sharp-pointed cone in the Shore D Durometer test or ball in the Rockwell test (Figure 5.8).

The indention stresses, while focused within a concentrated area, are generally more widely distributed to surrounding areas. Because of the presence of a combination of elastic and plastic or permanent deformation, the amount of recovery is also often determined. The combination of plastic and elastic deformation is dependent on the size, distribution, and amount of various crystalline and amorphous regions as well as physical and chemical cross-links and polymer structure.

5.11 GLASS TRANSITION TEMPERATURE AND MELTING POINT

Qualitatively, the glass transition temperature corresponds to the onset of short-range (typically one- to five-atom chains) coordinated motion. Actually, many more (often 10 to 100) atoms may attain sufficient thermal energy to move in a coordinated manner at T_g.

The glass transition temperature (ASTM D-3418) is the temperature at which there is an absorption or release of energy as the temperature is raised or lowered. T_g may be determined using any technique that signals an energy gain or loss.

It must be emphasized that the actual T_g of a sample is dependent on many factors, including pretreatment of the sample and the method and conditions of determination. For instance, the T_g for linear polyethylene has been reported to be from about 140 to above 300 K. Calorimetric values for polyethylene centralize about two values, 145 and 240 K; thermal expansion values are quite variable within the range of 140 to 270 K; NMR values occur between 220 and 270 K; and mechanical determinations range from 150 to above 280 K. The method of determination and the end property use should be related. Thus, if the area of concern is electrical, then determinations involving dielectric loss are appropriate.

Whether a material is above or below its T_g is important in describing the material's properties and potential end use. Fibers are composed of generally crystalline polymers that contain polar groups. The polymers composing the fibers are usually near their T_g to allow flexibility. Cross-links are often added to prevent gross chain movement. An elastomer is cross-linked and composed of essentially nonpolar chains; the use temperature is above its T_g. Largely crystalline plastics may be used above or below their T_g. Coatings or paints must be used near their T_g so that they have some flexibility but are not rubbery. Adhesives are generally mixtures in which the polymeric portion is above its T_g. Thus the T_g is one of the most important physical properties of an amorphous polymer.

As is the case with the glass transition temperature, melting will be observed to occur over a temperature range since it takes time for the chains to unfold. If the temperature is raised very slowly, a one- to two-degree range will be observed. The determination of the melting point requires only visual observation of when melting occurs as the sample is heated.

5.12 DENSITY (SPECIFIC GRAVITY)

Specific gravity is simply the density (mass per unit volume) of a material divided by the density of water. In cgs units the density of water is about 1.00 g/cc at room temperature. Thus, at room temperature the density and specific gravity values are essentially the same. Specific gravity is often used because it is unitless, whereas a density, although commonly given in cgs units, can be given in other weight per volume units such as pounds per quart.

Table 5.1 Stability of various polymers to various conditions

Polymer	Nonoxidizing Acid 20% Sulfuric	Oxidizing Acid 10% Nitric	Aqueous Salt Solution NaCl	Aqueous Base NaOH	Polar Liquids—Ethanol	Nonpolar Liquids—Benzene	Water
Nylon 6,6	U	U	S	S	M	S	S
Polytetrafluoroethylene	S	S	S	S	S	S	S
Polycarbonate	M	U	S	M	S	U	S
Polyester	M	M	S	M	M	U	S
Polyetheretherketone	S	S	S	S	S	S	S
LDPE	S	M	S	—	S	M	S
HDPE	S	S	S	—	S	S	S
Poly(phenylene oxide)	S	M	S	S	S	U	S
Polypropylene	S	M	S	S	S	M	S
Polystyrene	S	M	S	S	S	U	S
Polyurethane	M	U	S	M	U	M	S
Epoxy	S	U	S	S	S	S	S
Silicone	M	U	S	S	S	M	S

S, satisfactory; M, moderately to poor; U, unsatisfactory.

5.13 RESISTANCE TO CHEMICALS

The resistance of polymers to chemical reagents has been measured as described in ANSI/ASTM D543-67/78, which covers 50 different reagents. In the past, the change in weight and appearance of the immersed test sample have been reported. However, this test has been updated to include changes in physical properties as a result of immersion in test solutions.

Most high-performance polymers are not adversely affected by exposure to nonoxidizing acids and alkalies. Some are adversely affected by exposure to oxidizing acids, such as concentrated nitric acid, and all amorphous linear polymers will be attacked by solvents with solubility parameters similar to those of the polymer. Relatively complete tables showing resistance of polymers to specific corrosives have been published.

Tables 5.1 and 5.2 contain a summary of typical stability values for a number of polymers and elastomers against typical chemical agents. As expected, condensation polymers (Section 3.2) generally exhibit good stability to nonpolar liquids while they are generally only (relatively) moderately or unstable toward polar agents and acids and bases. This is because of the polarity of the connective "condensation" linkages within the polymer backbone. By comparison, vinyl type of polymers (Section 3.2) exhibit moderate to good stability toward both polar and nonpolar liquids and acids and bases. This is because the carbon–carbon backbone is not particularly susceptible to attack by polar agents and because nonpolar liquids, at best, will simply solubilize the polymer. All of the materials show good stability to water alone because all of the polymers have sufficient hydrophobic character to repeal the water.

Table 5.2 Stability to various elemental conditions of selected elastomeric materials

Polymers	Weather—Sunlight	Oxidation	Ozone Cracking Aging	NaOH-Dil/Con	Acid-Dil/Con	Degreasers Chlorinated Hydrocarbons	Aliphatic Hydrocarbons
Butadiene	P	G	B	F/F	F/F	P	P
Neoprene	G	G	G	G/G	G/G	P	F
Nitrile	P	G	F	G/G	G/G	G	G
Polyisoprene (Natural)	P	G	B	G/F	G/F	B	B
Polyisoprene (Synthetic)	B	G	B	F/F	F/F	B	B
Styrene-Butadiene	P	G	B	F/F	F/F	B	B
Silicone	G	G	G	G/G	G/F	B	F-P

G, good; F, fair; P, poor; B, bad.

5.14 WATER ABSORPTION

Water absorption can be determined through weight increase when a dried sample is placed in a chamber of specified humidity and temperature (ANSI/ASTM D570-63(1972)).

GLOSSARY

ASTM—American Society for Testing and Materials: USA society responsible for codifying and approving standard tests that help in ensuring satisfactory performance of materials.

Coefficient of linear expansion: Measure of the change in length of a standard sized material as the temperature is changed.

Compression strength: Measure of the ability of a material to resist a crushing force.

Dashpot: Cylinder filled with a viscous liquid that is used to represent the liquid behavior of a viscoelastic material. Used to represent chain and chain segment movement.

Density: Mass of a material per volume.

Destructive testing: Tests that involve the chemical structural change of at least a portion of the tested material.

Flexural strength: Measure of the ability of a material to resist breaking when a bending force is applied.

Free volume: Unoccupied space in a material.

Glass transition temperature, Tg: Temperature where segments of a giant molecule have enough thermal (heat) energy to move.

Hardness: Resistance of a material to local deformation, marring, and scratching.

Heat deflection test: Measures the deformation of a giant molecule material under a specified "load" or applied force.

Impact strength: Ability of a material to withstand a sharp blow such as being hit by a hammer.

ISO—International Standards Organization: International organization responsible for codifying and approving standard tests and procedures.

Melting point, Tm: Temperature where there is sufficient thermal (heat) energy to allow entire giant molecule chains to move.

Nondestructive tests: Tests that involve no detectable chemical change in the material tested.

Specific gravity: The density of a material where the mass is measured in grams and the volume in cubic centimeters, cc or cm^3, divided by the density of water that is about 1.00 gram/cc.

Spring: Used to represent the elastic or solid behavior of a viscoelastic material. Used to represent bond flexing.

Strain: Deformation of a material brought about because of application of a stress or force.

Stress: Force applied to a material.

Tensile strength: Measure of the resistance of a material to pulling stresses.

Viscoelastic materials: Materials, such as giant molecules, that act as a liquid and solid depending on factors such as temperature. Dashpots and springs are used to model viscoelastic behavior.

REVIEW QUESTIONS

1. Why is the coefficient of linear expansion important to know for materials used in aircraft?
2. Why is it important to know the rate of addition of load in the compressive strength test?
3. Why is it important to establish standard test conditions?
4. If the heating rate of a sample was low, would you expect the melting point obtained to be lower or higher than a melting point obtained when heating the sample faster?
5. Compare the impact test with the test for flexural strength.
6. What is the density of a piece of plastic that weighs 30 g and that occupies a volume of 20 cc?
7. Arrange the following in order of increased density: wood that floats on water, a piece of heavy plastic that sinks when placed in water, and a paper clip made of metal.

BIBLIOGRAPHY

Adamson, A., and Gast, A. (1997). *Physical Chemistry of Surfaces*, 6th ed., Wiley, New York.

Ando, I., and Askakura, T. (1998). *Solid State NMR of Polymers*, Elsevier, New York.

Brandolini, A., and Haney, D. (2000). *NMR Spectra of Plastics*, Marcel Dekker, New York.

Brandrup, J., Immergut, E. H., and Grulke, E., (1999). *Polymer Handbook*, 4th ed., Wiley.

Brostow, W. (2000). *Performance of Plastics*, Hanser-Gardner, Cincinnati.

Brostow, W., D'Souza, N., Menesses, V., and Hess, M. (2000). *Polymer Characterization*, Wiley, New York.

Calleja, F., and Fakirov, S. (2000). *Microhardness of Polymers*, Cambridge University Press, New York.

Cohen, S., and Lightbody, M. (1999). *Atomic Force Microscopy/Scanning Tunneling Microscopy*, Kluwer, New York.

Craver, C., and Carraher, C. (2000). *Applied Polymer Science*, Elsevier, New York.

Fawcett, A. H. (1996). *Polymer Spectroscopy*, Wiley, New York.

Friebolin, H. (1998). *Basic One- and Two-Dimensional NMR Spectroscopy*, Wiley, 1998.

Hatakeyama, T., and Quinn, F. (1999). *Thermal Analysis: Fundamentals and Application to Polymer Science*, Wiley, New York.

Hilado, C. (1998). *Flammability Handbook for Plastics*, Technomic, Lancaster, PA.

Hoffman, H., Schwoereer, M., and Vogtmann, T. (2000). *Macromolecular Systems: Microscopic Interactions and Macroscopic Properties*, Wiley, New York.

Hummel, D. (2000). *Hummel Infrared Industrial Polymers*, Wiley, New York.

Koenig, J. (1999). *Spectroscopy of Polymers*, Elsevier, New York.

Kosmulski, M. (2001). *Chemical Properties of Material Surfaces*, Marcel Dekker, New York.

Lee, T. (1998). *A Beginners Guide to Mass Spectral Interpretations*, Wiley, New York.

Macomber, R. (1997). *A Complete Introduction to Modern NMR*, Wiley, New York.

Mark, J. E., ed. (1999). *Polymer Data Handbook*, Oxford University Press, New York.

Mathot, V. (1994). *Calorimetry and Thermal Analysis of Polymers*, Hanser-Gardner, Cincinnati.

McCreery, R. (2000). *Raman Spectroscopy for Chemical Analysis*, Wiley, New York.

Moalli, J. (2001). *Plastics Failure Analysis and Prevention*, ChemTec, Toronto.

Moore, D., Pavan, A. and Williams, J. (2001): *Fracture Mechanics Testing Methods for Polymers, Adhesives and Composites*, Elsevier, New York.

Myer, V. (1999). *Practical High-Performance Liquid Chromatography*, 3rd ed., Wiley, New York.

Oliver, R. (1998). *HPLC of Macromolecules: A Practical Approach*, 2nd ed., Oxford University Press, Cary, NC.

Pollack, T. C. (1995). *Properties of Matter*, 5th ed., McGraw-Hill, New York.

Pourdeyhimi, B. (1999). *Imaging and Image Analysis Applications for Plastics*, ChemTec, Toronto.

Riande, E. (1999). *Polymer Viscoelasticity*, Marcel Dekker, New York.

Roberts, G. C. K. (1993). *NMR of Macromolecules: A Practical Approach*, Oxford University Press, Cary, NC.

Roe, R. (2000). *Methods of X-Ray and Neutron Scattering in Polymer Science,* Oxford University Press, New York.

Rohn, C. (1995). *Analytical Polymer Rheology*, Hanser-Gardner, Cincinnati.

Seymour, R., and Carraher, C. (1984). *Structure–Property Relationships in Polymers*, Plenum, New York.

Schmida, M., Antonietti, M., Coelfen, H., Koehler, W., and Schaefer, R. (2000). *New Developments in Polymer Analytics*, Vols I and II, Springer-Verlag, New York.

Shah, V. (1998). *Handbook of Plastics Testing Technology*, Wiley, New York.

Sibilia, J. P. (1996). *A Guide to Materials Characterization and Chemical Analysis*, 2nd ed., Wiley, New York.

Silverstein, R., and Webster, F. X. (1997). *Spectrometric Identification of Organic Compounds*, 6th ed., Wiley, New York.

Smith, C. (2002). *Pocket Handbook of Polymers and Other Macromolecules Instrumental Techniques for Analytical Chemistry*, Prentice-Hall, Upper Saddle River, NJ.

Smith, M., Busch, K. L. (1999). *Understanding Basic Mass Spectra: A Basic Approach*, Wiley, New York.

Solymar, L., and Walsh, D. (1998). *Electrical Properties of Materials*, Oxford University Press, New York.

Turi, E., ed. (1997). *Thermal Characterization of Polymeric Materials*, 2nd ed., Academic Press, Orlando, FL.

Urban, M. (1996). *Attenuated Total Reflectance Spectroscopy of Polymers: Theory and Practice*, Oxford University Press, Cary, NC.

Weber, U. (1998). *NMR-Spectroscopy*, Wiley, New York.

White, M. A. (1999). *Properties of Materials*, Oxford University Press, Cary, NC.

Wilkins, C. (2002). *Pocket Handbook of Mass Spectroscopy: Instrumentation Techniques for Analytical Chemistry*, Prentice-Hall, Upper Saddle River, NJ.

Willaims, J., and Pavan, A. (2000). *Fracture of Polymers, Composites and Adhesives*, Elsevier, New York.

Wineman, A., and Rajagopal, K. (2000). *Mechanical Response of Polymers: An Introduction*, Cambridge University Press, New York.

Workman, J. (2000). *Handbook of Organic Compounds: NIR, IR, Raman, UV & VIS Spectra Featuring Polymers and Surfactants*, Academic Press, San Diego.

Zerbi, G. (1999). *Modern Polymer Spectroscopy*, Wiley, New York.

ANSWERS TO REVIEW QUESTIONS

1. Temperatures for aircraft operation can vary greatly, thus it is important that a good match exists between bonded materials in the aircraft.
2. Some materials will act differently dependent on the rate of load application.
3. So that comparison of test results are more reliable.
4. Less—since the slower heating rate will allow the chains a longer time to unfold.
5. See Section 5.6 and 5.9. In the impact test the load is more rapidly applied.
6. $D = 30$ g/20 cc $= 1.5$ g/cc.
7. Wood, plastic, clip.

6

THERMOPLASTICS

Giant Molecules: Essential Materials for Everyday Living and Problem Solving, Second Edition,
by Charles E. Carraher, Jr.
ISBN 0-471-27399-6

6.1 INTRODUCTION

All polymers are classified as either thermoplastics, that is, linear or branched polymers that can be reversibly softened by heating and solidified by cooling, or thermosets, that is, cross-linked polymers that cannot be softened by heating without degradation. The word plastic is derived from the Greek word *plastikos*, meaning able to be molded. Both thermoplastics and thermoset prepolymers can be molded into desirable shapes.

It is of interest to note that the human desire to produce shaped articles was satisfied in the early cultures by chipping stone, chiseling wood, casting bronze, and shaping warmed tortoise shell and horn. Artisans who shape ivory are still called horners. Since these products could be molded by heat, the advent of ebonite (hard rubber) and celluloid in the nineteenth century provided a new outlet for shaping.

Shellac, gutta-percha, balata, casein, and bitumens are naturally occurring thermoplastics. Derivatives of natural rubber and cellulose—that is, cyclized rubber, cellulose nitrate, and cellulose acetate—are also thermoplastics. However, the first synthetic moldable plastic was a thermoset—that is, the reaction product of phenol and formaldehyde—which was produced commercially by Leo Baekeland in the early 1900s. The thermosets, which were the principal plastics prior to World War II and now account for less than 10% of all moldable plastics, are described in Chapter 8.

Each year the United States consumes about 79,000 million pounds of plastic and synthetic resins, or about 260 pounds for every citizen. The use of lightweight plastics has helped increase gas mileage in automobiles (Table 6.1), and this trend will increase with plastic car bodies being more widely used in the near future.

It is important to recognize that of the total yearly U.S. oil and gas consumption, 60% is used as stationery fuels in home heating and fuel to run power plants, 33% is

Table 6.1 Major plastics applications in automobiles

Area	Application	Material (Usual)
Interior	Crash pad	Urethane, ABS, PVC
	Headrest pad	Urethane, PVC
	Trim, glove box	Polypropylene, PVC, ABS
	Seat	Urethane
	Upholstery, carpet	PVC, nylon
Exterior	Fender apron	Polypropylene
	Front end	Unsaturated polyester
	Wheel covers	ABS, polyphenylene oxide
	Fender extension	Unsaturated polyester, nylon
	Grille	ABS, polyphenylene oxide
	Lamp housing (rear)	Polypropylene
	Styled roof	PVC
	Bumper sight shield	EPDM rubber, urethane
	Window louvers	Poly(butylene terephthalate)
Under the hood	Ducts	Polypropylene
	Battery case	Polypropylene
	Fan shroud	Polypropylene
	Heater and air conditioning	Unsaturated polyester
	Electrical housing and wiring	Phenolic, PVC, silicone
	Electronic ignition components	Poly(butylene terephthalate)

used as transportation fuels, and 7% is used for the manufacture of petrochemicals, including fertilizer, rubber, paints, fibers, solvents, and medicines. Only about 2.5% is employed for polymer applications, yet this 2.5% makes possible the production of many useful products.

The difference between thermoplastics and engineering plastics is often a thin line and depends on whether the material is to be cut, drilled, and so on. Some plastics like Teflon™ can be drilled and cut but with some difficulty so that it is between a thermoplastic and an engineering plastic. Here Teflon will be covered as a thermoplastic. Both thermoplastics and engineering plastics are thermoplastics. Those vinyl polymers that do not contain highly polar (like carbonyls) and hydrogen-bonding units (like amides) have relatively weaker secondary forces between chains and are thus included in this chapter.

The production of thermoplastics and engineering plastics for the year 2000 is given in Table 6.2. Most of the thermoplastics are derived from vinyl reactants. The formation of the vinyl polymers occurs as described in Section 3.2 whereby the pi bond of the carbon–carbon double bond, C=C, is broken and new sigma bonds are formed that link the monomer units together, forming the polymer chain. The formation of the additional sigma bonds creates energy. Thus, these vinyl polymer forming reactions are exothermic; that is, they create energy in the form of heat. Large-scale vinyl reactions need to be cooled because of the formation of this heat.

Table 6.2 U.S. production of thermoplastics and engineering plastics, 2000

Plastic	Production (Millions of Pounds)
Acrylonitrile–butadiene–styrene, ABS	3,100
Polyamides, nylons	1,400
Polyesters	4,400
Polyethylene, high density	15,400
Polyethylene, low density	17,900
Polystyrene	6,600
Styrene–acrylonitrile	124
Polypropylene	15,400
Poly(vinyl chloride) and copolymers	14,300

The vinyl polymerization for polyethylene is represented as follows;

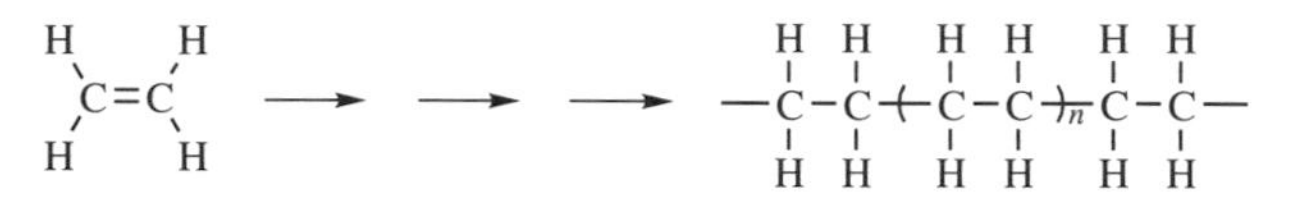

In the ethylene monomer each carbon is surrounded by another carbon atom and two hydrogen atom, whereas in polyethylene each carbon atom is surrounded by two carbon atoms and two hydrogen atoms.

Vinyl polymerizations are started, initiated, by something that breaks the pi bond like heat or high energy light. In the dentist's office an ultraviolet lamp is often used to break this double bond, resulting in the dental cement or filling to be cured or cross-linked through breakage of the pi bonds and formation of new sigma bonds that cross-link or set the dental material. For many reactions, a compound is added to the monomer mixture that has chemical bonds that are easily broken to give products that start the reaction process. These compounds are called initiators because they initiate or start the polymer forming reaction. Ring-opening reactions such as used to form epoxy resins also generally give off energy.

By comparison, engineering thermoplastics are generally formed through condensation reactions (Section 3.2) that require heating with the overall reaction begin endothermic or heat consuming.

By bulk, almost all vinyl polymers are made by four processes: free radical (>50%), complex coordinate (12–15%), anionic (10–15%), and cationic (8–12%) (Table 6.3).

6.2 POLYETHYLENES—HISTORY

Polyethylene for commercial attention was probably initially synthesized by M. E. P. Friedrich while he was a graduate student working for Carl S. Marvel in 1930 when it was an unwanted byproduct from the reaction of ethylene and a

Table 6.3 Major technique used in the production of important vinyl polymers

Free Radical	
Low-density polyethylene, LDPE	Poly(vinyl chloride)
Poly(vinyl acetate)	Polyacrylonitrile and acrylic fibers
Poly(methyl methacrylate)	Polyacrylamide
Polychloroprene	Styrene–acrylonitrile copolymers, SAN
Polytetrafluoroethylene	Poly(vinylene fluoride)
Acrylonitrile–butadiene–styrene copolymers, ABS	
Ethylene–methacrylic acid copolymers	Styrene–Butadiene copolymers, SBR
Nitrile rubber, NBR	Polystyrene
Cationic	
Polyisobutylene	Butyl rubber
Polyacetals	
Anionic	
Thermoplastic olefin elastomers (copolymers of butadiene, isoprene, and styrene)	
Polyacetals	
Complex, Organometallic Catalysis	
High-density polyethylene, HDPE	Ethylene–propylene elastomers
Polybutadiene	Polypropylene
Polyisoprene	

lithium alkyl compound. In 1932, British scientists at the Imperial Chemical Industries (ICI) accidently made polyethylene while they were looking for products that could be produced from the high-pressure reaction of ethylene with various compounds. In March 1933, they found the formation of a white solid when they combined ethylene and benzaldehyde under high pressure. They correctly identified the solid as polyethylene. They attempted the reaction again, but with ethylene alone. Instead of again getting the waxy white solid, they got a violent reaction and the decomposition of the ethylene. They delayed their work until December 1935 when they had better high-pressure equipment. At 350°F, the pressure inside of the reaction vessel containing the ethylene decreased, consistent with the formation of a solid. Because they wanted to retain the high pressure, they pumped in more ethylene. The observed pressure drop could not be totally due to the formation of polyethylene, but something else was contributing to the pressure loss. Eventually they found that the pressure loss was also due to the presence of a small leak that allowed small amounts of oxygen to enter into the reaction vessel. The small amounts of oxygen turned out to be the right amount needed to catalyze the reaction of the additional ethylene that was pumped in after the initial pressure loss—another major discovery by accident. Even so, the ICI scientists saw no real use for the new material.

By chance, J. N. Dean of the British Telegraph Construction and Maintenance Company heard about the new polymer. He had needed a material to coat under-water cables. He reasoned that polyethylene would be water-resistant and suitable to coat the wire, thus protecting it from the corrosion caused by the saltwater in the ocean. In July of 1939, enough polyethylene was made to coat one nautical mile of cable. Before it could be widely used, Germany invaded Poland and polyethylene production was diverted to making flexible high-frequency insulated cable for ground and airborne radar equipment. Polyethylene was produced, at this time, by ICI in the United Kingdom and by DuPont and Union Carbide for the United States.

Polyethylene did not receive much commercial use until after the war when it was used in the manufacture of film and molded objects. Polyethylene film displaced cellophane in many applications being used for packaging produce, textiles, frozen and perishable foods, and so on. This polyethylene was branched and had a relatively low softening temperature (below 212°F), preventing its use for materials where boiling water was needed for sterilization.

Karl Ziegler, director of the Max Planck Institute for Coal Research in Muelheim, Germany was extending early work on polyethylene attempting to get ethylene to form polyethylene at lower pressures and temperatures. They found compounds that allowed the formation of polyethylene under much lower pressures and temperatures. Furthermore, these compounds produced a polyethylene that had fewer branches and a higher softening temperature (above 212°F), allowing the material to be sterilized when needed.

The branched polyethylene is called low-density, high-pressure polyethylene because of the high pressures usually employed for its production; and because of the presence of the branches, the chains are not able to closely pack, leaving voids and subsequently producing a material that has a lower density in comparison to low-branched polyethylene.

Giulio Natta, a consultant for the Montecatini company of Milan, Italy applied the Zeigler catalysts to other vinyl monomers such as propylene and found that the polymers had higher densities, exhibited higher melting points, and were more linear than those produced by the then classical techniques such as free radical initiated polymerization.

Ziegler and Natta shared the Nobel prize in 1963 for their efforts in the production of vinyl polymers using what we know today as solid-state stereoregulating catalysts.

Today there exist a number of polyethylenes that differ mainly in the amount and length of branching as well as in chain length and chain length variation. The next several sections will look at some of the most important of these polyethylenes.

6.3 HIGH-DENSITY POLYETHYLENE

The alkanes, such as paraffin wax, high-density polyethylene (HDPE), and other polyolefins, have the empirical formula $H{-}(CH_2){_n}{-}H$. Nevertheless, the degree of polymerization (DP) of paraffin wax and lower-molecular-weight alkanes is too low to permit entanglement of the polymer chains. Another highly branched polyolefin, called elasterite, occurs naturally in the fossil *Fungus subterraneus* but was never used as a commercial plastic.

The first synthetic polyethylene was produced by von Peckman in the 1890s by the catalytic decomposition of diazomethane (CH_2N_2). W. Carothers, the coinventor of nylon, produced a low-molecular-weight linear polyethylene in the early 1930s by the coupling of decamethylene dibromide ($Br(CH_2)_{10}Br$) in the presence of sodium metal. C. Marvel also produced HDPE in the early 1930s by the polymerization of ethylene in the presence of lithiumalkyl (LiR) and an arsonium compound. This polymer was investigated by duPont, but that company failed to recognize the potential use of HDPE at that time.

HDPE is now a commercial plastic, with over 7.7 million tons being produced annually in the United States.

The first commercial HDPE was produced independently by J. Hogan and R. Banks in the United States and later by Ziegler in Germany in the early 1950s. Ziegler's synthesis was related to that used by Marvel. K. Ziegler and coworkers used aluminumtriethyl ($Al(C_2H_5)_3$) and titanium trichloride ($TiCl_3$) for their polymerization catalyst in what they called the "aufbau" or building-up reaction. Ziegler was awarded the Nobel prize in 1963.

Hogan and Banks used chromic oxide (CrO_3), supported on silica (SiO_2), as their catalyst system for making HDPE. The polymer obtained by the German and American chemists was a linear crystalline polymer. The regularity in the HDPE chain favored the formation of crystals, and this crystalline structure contributed

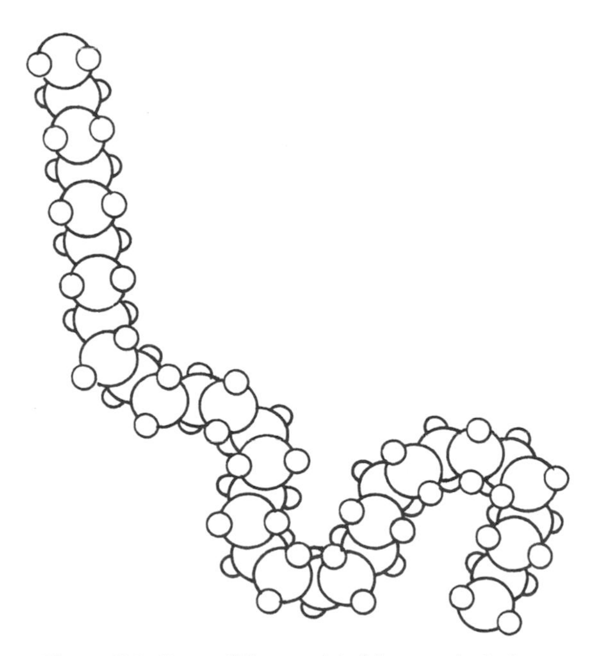

Figure 6.1. Space-filling model of linear polyethylene.

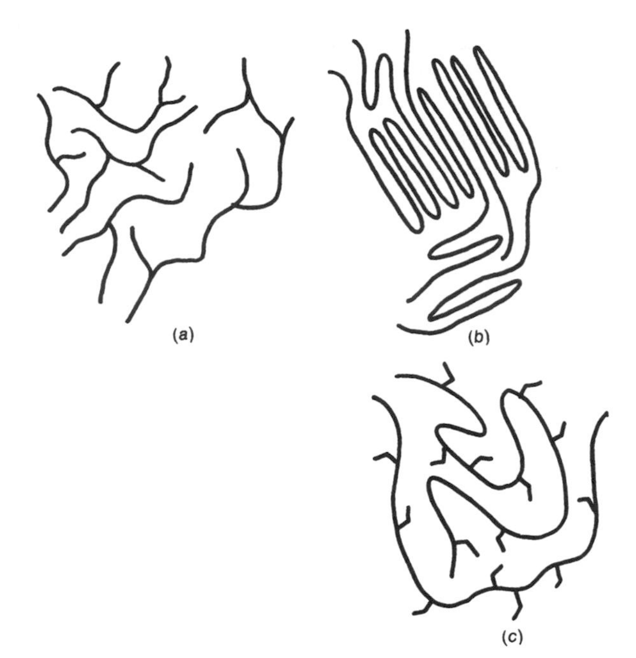

Figure 6.2. Representations of different forms of polyethylene to illustrate branching and nonbranching aspects. (a) Low-density polyethylene (LDPE). (b) High-density polyethylene (HDPE). (c) Linear low-density polyethylene copolymer with 1-butene (LLDPE). The 1-butene discourages crystallization.

to the higher specific gravity of HDPE (0.96). In contrast, the specific gravity of the highly amorphous branched polyethylene (LDPE) is about 0.91. A space-filling model and simulated portions of HDPE and LDPE chains are shown in Figures 6.1 and 6.2. Physical properties of commercial polyethylene are listed in Table 6.4.

The terms high-density and low-density polyethylene are derived from their densities. Density is a measure of the weight of material contained within a given volume. Thus the air at sea level has a density about 1.2×10^{-3} grams per cubic centimeter (g/cc) of volume. Wood has a density of about 0.1 to 1.4 g/cc. Most plastics and organic liquids have densities of about 0.7 to 1.0 g/cc, whereas water has a density of 1.0 g/cc at ordinary temperatures. Denser materials include the metals, such as mercury and tungsten.

Density is related to how tightly material can be packed. Thus linear polyethylene (HDPE) can be tightly packed since linear chains can be efficiently folded as noted in Figure 6.2. Conversely, branched polyethylene (LDPE) packs less firmly as a result of the presence of the branches, which prohibit close, regular folding. Thus, more ethylene units can be packed within a specific volume for linear polyethylene than for branched polyethylene, resulting in a higher weight per volume and consequently higher density for linear polyethylene.

Specific gravity is simply the ratio of the density of the material compared to the density of water. Employing the units of g/cc, the density and specific gravity of a

Table 6.4 Properties of typical polyethylenes

Property	HDPE	LDPE	LLDPE	PP	PP (40% talc)
Melting point (T_m, °F)	275	200	122	170	165
Glass transition temp. (T_g, °F)	—	20	—	—	—
Processing temp.(°F)	450	400	450	450	450
Molding pressure (10^3 psi)[a]	15	10	10	15	15
Mold shrinkage (10^{-3} in./in.)	25	30	20	20	12
Heat deflection temp. under flexural load of 264 psi (°F)	190	110	—	130	175
Maximum resistance to continuous heat (°F)	175	110	130	125	160
Coefficient of linear expansion (10^{-6} in./in., °F)	40	150	125	40	30
Compressive strength (10^3 psi)	30	—	—	65	75
Impact strength Izod (ft-lb/in. of notch)[b]	2	No break	No break	1.0	0.5
Tensile strength (10^3 psi)	35	30	33	50	45
Flexural strength (10^3 psi)	30	—	—	60	80
% elongation	200	300	400	400	5
Tensile modulus (10^3 psi)	155	35	45	50	45
Flexural modulus (10^3 psi)	150	30	50	200	500
Shore hardness	D70	D50	D55	R90	R100
Specific gravity	0.96	0.91	0.93	0.91	1.25
% water absorption	0.01	0.01	0.01	0.01	0.02
Dielectric constant					
Dielectric strength (V/mil)	500	750	700	500	500
Resistance to chemicals at 750°F[c]					
Nonoxidizing acids (20% H_2SO_4)	S	S	S	S	S
Oxidizing acids (10% HNO_3)	Q	Q	Q	Q	Q
Aqueous salt solutions (NaCl)	S	S	S	S	S
Polar solvents (C_2H_5OH)	S	S	S	S	S
Nonpolar solvents (C_6H_6)	Q	Q	Q	Q	Q
Water	S	S	S	S	S
Aqueous alkaline solutions (NaOH)	S	S	S	S	S

[a] psi/0.145 = kPa (kilopascals).
[b] ft-lb/in. of notch/0.0187 = cm · N/cm of notch.
[c] S, satisfactory; Q, questionable; U, unsatisfatisfactory.

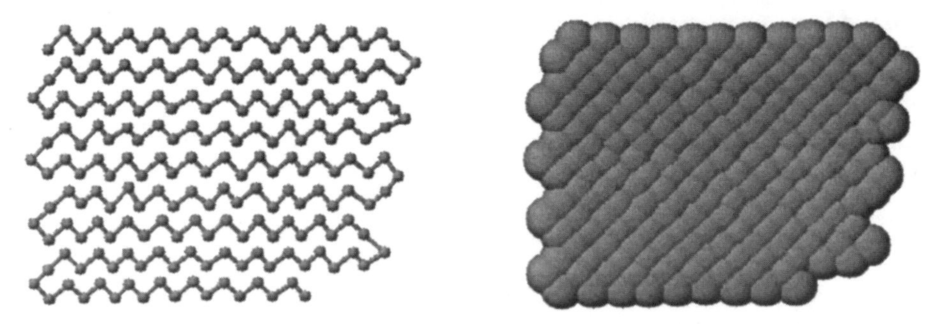

Figure 6.3. Two-hundred-carbon crystalline linear polyethylene chain.

material are essentially the same since the density of water is 1.0 g/cc and a number divided by one is the number itself.

We can get a better ideal of what linear polyethylene looks like on a molecular level in Figure 6.3. It is usually crystalline (see Section 4.8), forming a close-packed molecular package. This package is strong and brittle and relative to amorphous structures does not allow molecules to pass through. Furthermore, this tightly packed molecular bundle is not easily penetrated by unwanted chemicals such as acids and bases. Also, because these acids and bases, like acetic acid in vinegar and citric acid from lemons, are water-loving (hydrophillic) and the ethylene chain is nonpolar and does not love water (hydrophobic) these acids and bases are not attracted. Thus, polyethylene is useful in our homes, where small amounts of acids and bases are a part of life. By comparison, oils like motor oils are themselves composed of hydrocarbon chains so that they are polyethylene-like, but because of the close packing the linear crystalline polyethylene does not allow the motor oil to pass though or to do substantial damage.

While linear polyethylene has a great tendency to form the closely packed crystalline package, it can be made to take an amorphous structure through mixing the melted polymer and then quick-cooling the material locking in the amorphous structure (Figure 6.4).

The amorphous chain pictured above has about 50% empty space. Even so, it is largely impervious to most chemicals such as oxygen. The thickness of a page of this book is about 0.05 mm. A sheet of polyethylene with the same thickness (0.05 mm) would have about 10^{11} layers of methylene units. Thus, while it looks like there are large holes, the sum total presents a large maze for the oxygen molecules to go through. Even so, some oxygen gets through. Crystalline polyethylene (discussed below) has many fewer empty spaces and allows less oxygen to get through.

The thickness of an ordinary envelope is also on the order of 0.05 mm in thickness. Yet the anthrax bacteria, a much larger array of atoms than the simple oxygen, is able to get through the pores of the envelope and contaminate the machinery and personal that comes in contact with the anthrax-containing envelope. The difference is that the polyethylene film is a continuous layer upon layer of polymer chains while the envelope is composed of paper fibers which, even when matted together,

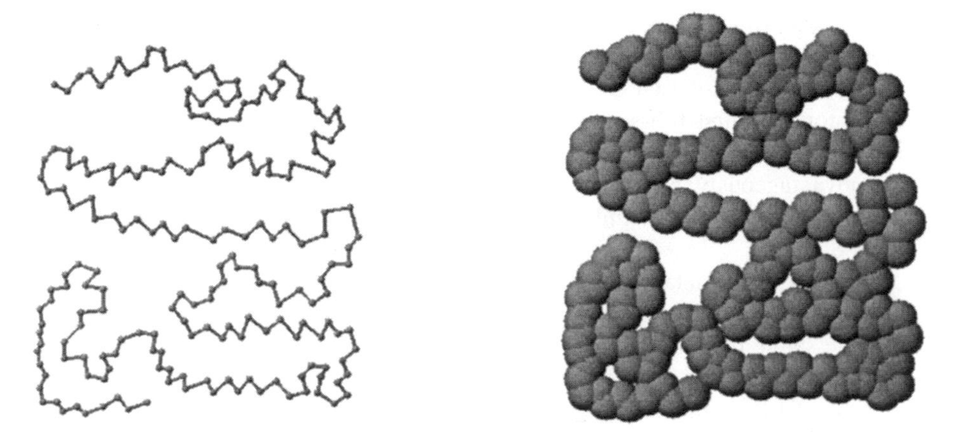

Figure 6.4. Two-hundred-carbon polyethylene chain in an amorphous orientation.

have, on a molecular level, holes large enough to allow the anthrax to get through. The matted paper fibers are not continuous on a molecular level and even though it appears to the naked eye to be continuous, it is not.

Thus, both crystalline and amorphous polyethylene are good barriers to most materials.

HDPE is produced using catalysts and has normally less than 15 (usually 1 to 6) short branches per 1000 ethylene units. It has a melting point of about 130°C, so it can be made into objects that need to be sterilized through the use of boiling water. Typical products are bottles, trays, drums, tanks, cans, pails, housewares, toys, food containers, conduit, wire and cable coating, foam, insulation for coaxial and communication cables, pipes, bags, films, and crates.

6.4 LOW-DENSITY POLYETHYLENE

Tupperware was the idea of Earl Silas Tupper, a New Hampshire tree surgeon and plastics innovator. He began experimenting with polyethylene during the early part of WW II. In 1947 he designed and patented the famous "Tupper seal" that "sealed in" freshness. In order to close the container it had to be "burped" to remove air. Tupperware was also bug-proof and spill-proof, did not rot or rust, and did not break when dropped. Even with all of these advantages, few were sold. Enter Brownie Wise, a divorced single mother from Detroit who desperately needed to supplement her income as a secretary. She had an idea, namely, "Tupperware Parties." By 1951 Tupper had withdrawn all of the Tupperware from the stores and turned over their sales to Brownie Wise with the only source of the ware being through the Tupperware Parties.

The development of low-density polyethylene (LDPE) was based on less than 1 g of a residue that was accidentally produced by E. Fawcett and R. Gibson in 1933 in their unsuccessful attempt to condense ethylene and benzaldehyde at 340°F and at extremely high pressure. However, they did produce a trace of polyethylene.

Larger amounts of LDPE were obtained when a trace of oxygen was used as an initiator. The first full-scale LDPE plant went "on stream" on the day of the outbreak of World War II. LDPE, which is a highly branched polymer, was used advantageously as an insulator for coaxial cable in radio detecting and ranging (radar). About 9 million tons of LDPE are produced annually in the United States.

As shown by the data in Table 6.4, LDPE has a lower modulus (is more flexible) and has a lower melting point than HDPE. It is also less stable than HDPE, though still quite stable in comparison to most other giant molecules, to acids, bases, oils, and so on; and gases such as carbon dioxide and oxygen are able to penetrate sheets and films of LDPE more readily because these invading molecules have a greater opportunity to penetrate the less tightly packed structure (Figure 6.5). LDPE has between 50 and 150 short alkyl branches for every 1000 ethylene units. This branching is sufficient to discourage crystalline formation, resulting in a material that is about 50% amorphous.

LDPE films are nearly clear even though they contain a mixture of crystalline and amorphous regions. This is because the crystalline portions are space-filling and not isolated spherulites, allowing a largely homogeneous structure with respect to refractive index that results in the material being transparent. In fact, the major reason why LDPE films appear hazy or not completely transparent is because of the roughness of the surface and is not due to the light scattering of the interior material.

LDPE is used for bags like the ones you get at the checkout counters in stores; it is also used for packaging products, films, sheeting, piping, industrial containers, and household items.

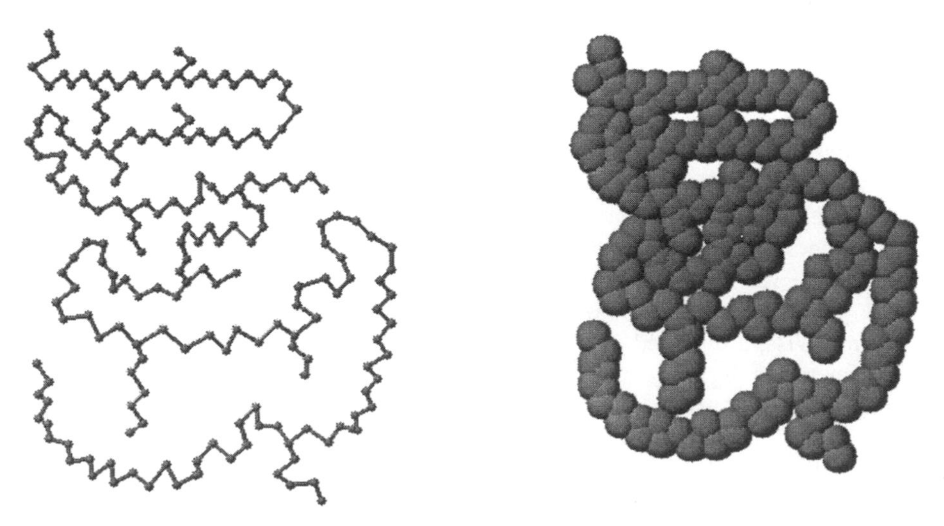

Figure 6.5. Two-hundred-carbon polyethylene chain with branching occurring about every 20 carbons. (*Left*) Ball and stick (*Right*) Space filling.

6.5 ULTRAHIGH-MOLECULAR-WEIGHT POLYETHYLENE

A minimum or threshold molecular weight (about 100 $\overline{\text{DP}}$) is required for entanglement of HDPE. Since high-molecular-weight polymers are difficult to process, polymers with molecular weights slightly above the threshold molecular weight are usually produced commercially. However, the toughness of HDPE and other polymers increases with molecular weight. Hence, ultrahigh-molecular-weight polyethylene (UHMWPE) ($\overline{\text{DP}} = 1$ million) is produced commercially for use where unusual toughness is essential, such as in trash cans and liners for coal freighters.

Polymers, because of their "connectiveness," are able to readily share energy and distortions, thereby redistributing the energy and distortions throughout the chain and between chains. This acts to dissipate impact forces. This is illustrated when looking at the fabrication of protective armor such as vests, vehicle exteriors, body armor, football helmets, riot gear, bomb blankets, explosive containment boxes, bus and taxi shields, and so on. Let us look at body armor.

Many of the so-called bullet-proof vests (really bullet-resistant) have been made of very tough and strong polycarbonates such as Kevlar™. More recently, layers of ultrahigh-molecular-weight polyethylene were found to a have similar "stopping power" for a lesser weight, and today many of the vests are made using polyethylene sheets. Rapid dissipation of energy is critical, allowing the impact energy to be spread over a wide area. Body armor material should be strong enough so as not to immediately break when the bullet hits but have enough contact with other parts of the body armor to allow ready transfer of some of the impact energy. If the material can adsorb some of the energy though bond breakage or heating, then additional energy can be absorbed at the impact site. Along with high strength, the material should have some ability to stretch, allowing the material to transfer some of the energy to surrounding material. If connective forces between components are too strong, total energy dissipation is reduced because strong bonding between the various parts of the body armor discourages another form of energy sharing, that of having the parts slide past one another, thereby redistributing the impact energy. Thus, a balance is needed between material strength, strength holding the various components of the body together, and the ability to readily dissipate the impact energy. Certain sequences of layered material have been found to be more effective at dissipation of the impact energy than others. One of the employed combinations contains sheets of strong and more rigid aramid fiber adjacent to sheets of ultrahigh-molecular-weight polyethylene that is less strong but flexible and stretchable.

Another factor is the breakup of the projectile. Again, superstrong giant molecules composed of composites containing boron carbide ceramics and aramids, ultrahigh-molecular-weight polyethylene, or fibrous glass are effective at breaking up the projectile.

6.6 LINEAR LOW-DENSITY POLYETHYLENE

Commercial copolymers in which both ethylene and 1-butene ($H_2C{=}CH{-}CH_2CH_3$) are present as repeating units in the polymer chain are linear, but because of the bulky pendant (C_2H_5) groups they occupy greater volume and have a lower specific gravity than HDPE. Linear low-density polyethylene (LLDPE) may be produced at low pressure in the gaseous phase or in solution. Higher homologues such as 1-octene ($H_2C{=}CH(CH_2)_5CH_3$) may also be used as the comonomers in LLDPE. New coordination catalysts, which are related to those used for making HDPE, are also used in the production of LLDPE.

LDPE is characterized by good flexibility and hence can be used as a film and in squeeze bottles. HDPE is stiffer and more heat-resistant and is used as rigid pipe. LLDPE is stronger than LDPE and can be used as thinner films for making bags, for example.

Plastomers is the name given to copolymers of ethylene that have a little crystallinity but are largely amorphous. They are also called very low density polyethylene (VLDPE). They are more elastic than LLDPE but less stiff. They are used as a sealing layer in film applications and controlled permeation packaging for vegetables and fruits.

6.7 CROSS-LINKED POLYETHYLENE

Figure 6.6 shows a slightly cross-linked polyethylene chain. Notice how the cross-link prevents wholesale movement and discourages molecules from invading it.

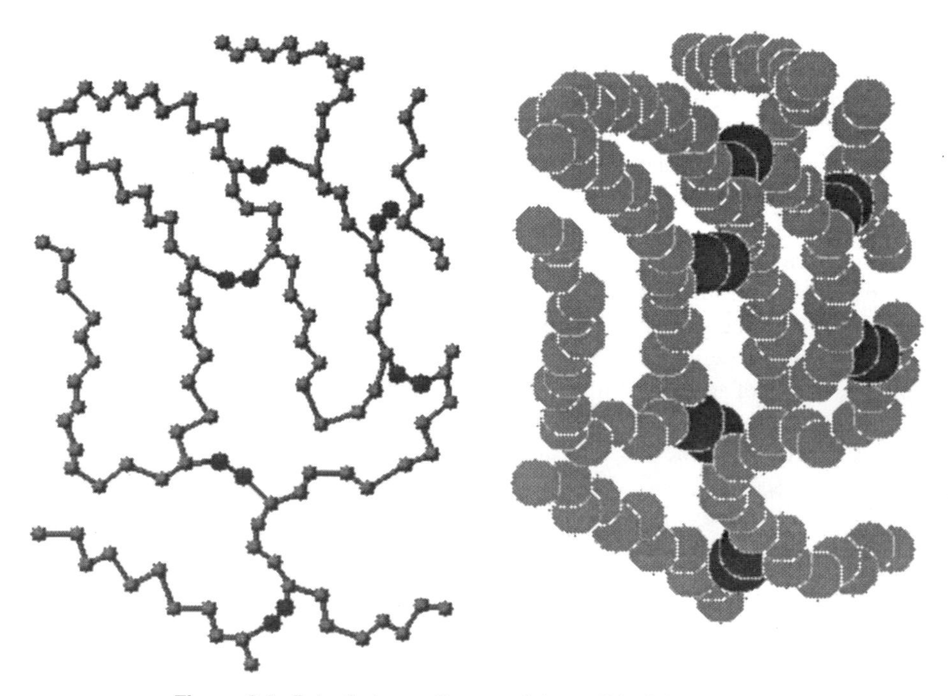

Figure 6.6. Polyethylene with cross-links at 5% of the carbons.

LDPE cross-links when exposed to high-energy radiation. The cross-linked product, which is insoluble in solvents even at elevated temperatures, is used as heat-shrinkable tubing. The stretched, cross-linked product has "elastic memory" and returns to its original dimensions when heated.

6.8 OTHER COPOLYMERS OF ETHYLENE

In addition to copolymers of 1-olefins, such as LLDPE, there are several other commercial copolymers of ethylene. The copolymer of ethylene and vinyl acetate is an amorphous copolymer that may be cast as a clear film or used as a melt coating. The copolymer of ethylene and methacrylic acid ($CH_2{=}C(CH_3)COOH$) is also a moldable thermoplastic. This copolymer, when partially neutralized to form monovalent and divalent metal-containing materials, is called an ionomer. These ionomer salts have a stable cross-linked structure at ordinary temperatures but can be injection-molded. These tough copolymers are used as golf ball covers in place of balata.

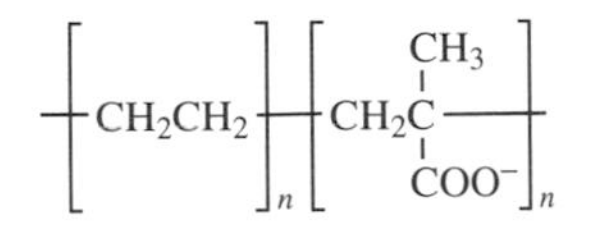

Ethylene–methacrylic acid copolymers (ionomers)

Both HDPE and polypropylene are high-melting crystalline polymers. However, the random copolymer of these two comonomers is an amorphous, low-melting elastomer. It is customary to add a cross-linking monomer, such as dicyclopentadiene, to the comonomers to produce a vulcanizable elastomer (EPDM). EPDM is used as the white sidewalls of tires and as single-ply roofing material.

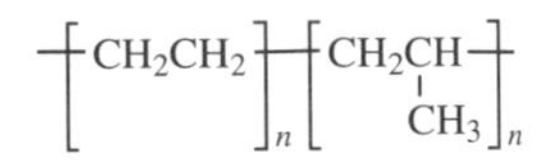

Ethylene–propylene copolymer

These ethylene–propylene copolymers are also employed in other automotive applications such as radiator and heater hoses, seals, mats, weather strips, bumpers, and body parts. Nonautomotive applications include coated fabrics, gaskets and seals, hoses and wire, and cable insulators.

The block copolymer of ethylene and propylene, which contains long sequences of ethylene and propylene repeating units, is a clear, moldable copolymer and is used in place of HDPE in many applications. Its specific gravity is similar to that of LDPE.

6.9 POLYPROPYLENE

Nobel laureate K. Ziegler patented HDPE but failed to include polypropylene (PP) in his patent application. However, many other chemists used the Ziegler catalyst

($TiCl_3 \cdot Al(C_2H_5)_3$) to produce PP in the early 1950s. Nobel laureate G. Natta of Montedison, W. Baxter of DuPont, and E. Vanderburg of Hercules filed for patents for the production of PP using the Ziegler catalyst. J. Hogan and R. Banks of Phillips and A. Zletz of Amoco filed for patents using supported metal oxide catalysts. In 1973 the U.S. Patent Office granted a patent for PP to Natta, but reversed its decision in favor of Hogan and Banks in 1983.

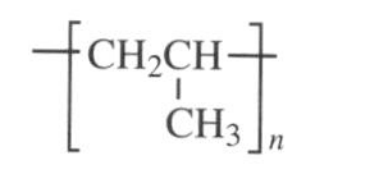

Polypropylene (PP)

Polypropylene (PP) is one of the three most heavily produced synthetic polymers. This abundance of PP is called for because of its variety and versatility being employed today in such diverse applications as a film in disposable diapers and hospital gowns to geotextile liners; plastic applications as disposable food containers and automotive components; and fiber applications such as carpets, furniture fabrics, and twine. Typical properties are given in Table 6.4.

Unlike polyethylene, polypropylene has atoms in addition to hydrogen attached to the polymer backbone. The presence of the methyl group substituting for one of the hydrogens gives rise to several different structural isomers or "tacticities," called (a) isotactic, and (b) syndiotactic isomers, which are both regular or ordered structure, and (c) atactic isomers, which have only a somewhat random arrangement

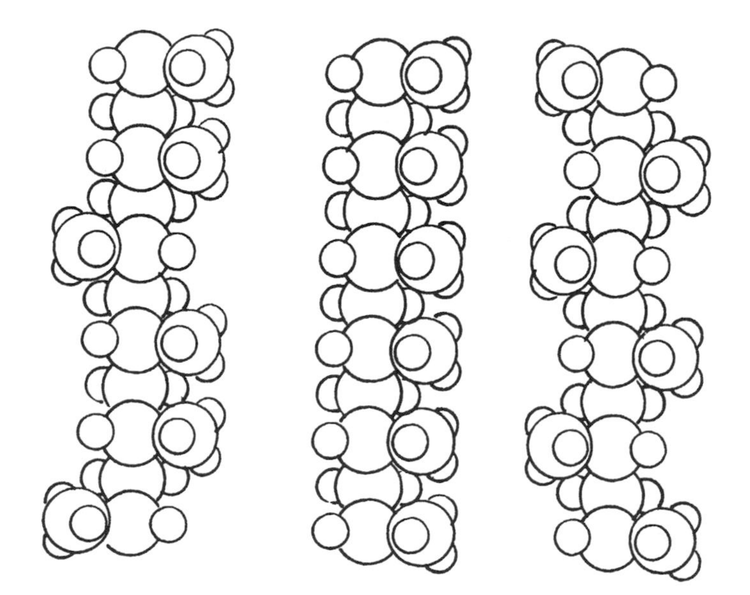

Figure 6.7. Atactic (*left*), isotactic (*middle*), and syndiotactic (*right*) polypropylene. [Some of our youth tell us it is really a model of poly(teddy bears).]

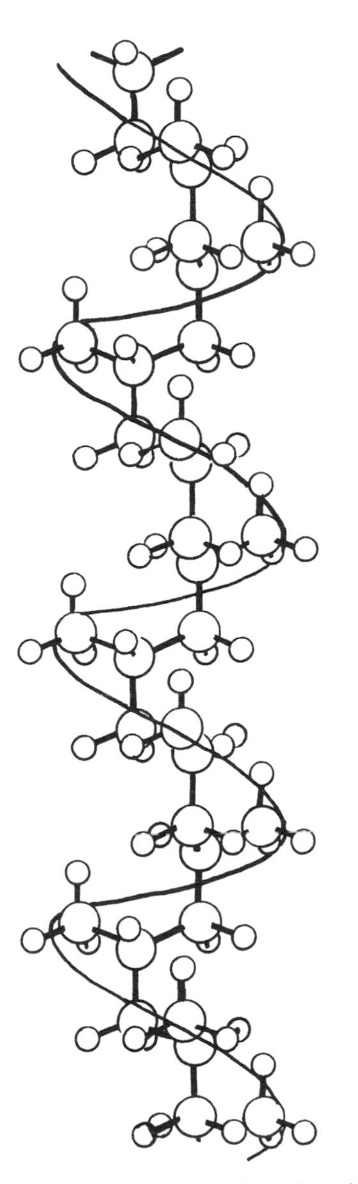

Figure 6.8. Stick-and-ball model of the helical conformation of isotactic polypropylene.

of methyl groups (Figures 6.7–6.9). The regular structures, isotactic and syntiotactic, allow the polymer chains to more readily come closer together, forming crystalline structures. These crystalline structures are reflected in so-called stereoregular PP being stronger, less permeable to acids, oils, and gas molecules, and higher-melting. The disordered random atactic, (where the prefix "a" means having nothing to do with) structure gives a less crystalline product.

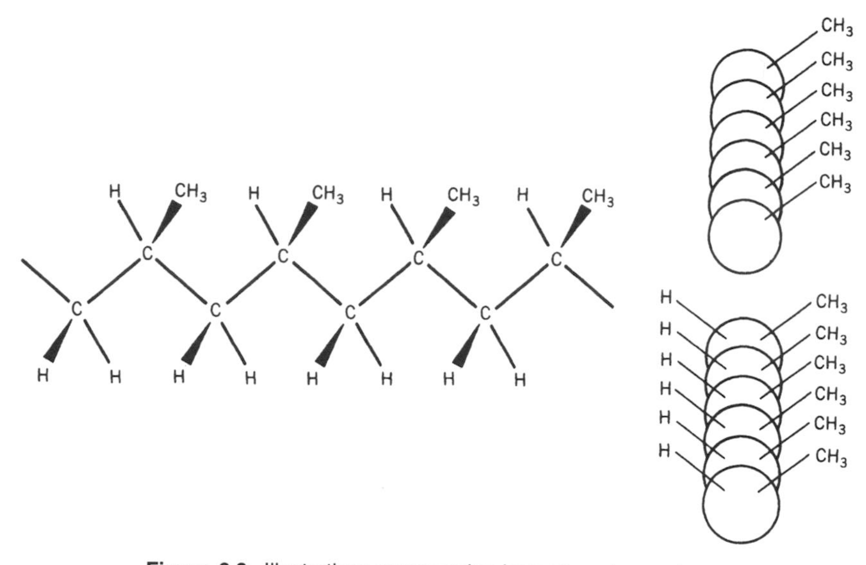

Figure 6.9. Illustrations representing isotactic polypropylene.

While PP was produced for some time, it only became commercially available in the late 1950s with the production of somewhat isotactic PP (iPP) by Natta and co-workers at Phillips. The first PP was not highly crystalline because the tacticity was only approximate; but with the use of the Natta–Zeigler catalysts, iPP was produced that contained greater amounts of stereoregular material with a corresponding increase in crystallinity and associated properties such as an increased stiffness, better clarity, and a higher distortion temperature. Today, with better catalysts, including the soluble metallocene catalysts, the tacticity has been increased so that 99% isotactic material can be readily produced. The more traditional Zeigler–Natta catalysts systems today have activities as high as producing 100,000 pounds iPP per pound of catalysts eliminating the need for catalyst removal.

While most of the PP used today is of the regular or stereoregular form, atactic or amorphous forms of PP are also used. Initially, atactic PP (aPP) was obtained as a byproduct of the production of iPP. As an inexpensive byproduct it is used as a modifier for asphalt for roofing and in adhesives. As the effectiveness of catalyst systems becomes better, less aPP is available so that today some aPP is intentionally made for these applications.

Because iPP is stiff and has a relatively high T_g (about 0°C), some effort has involved the lowering of the T_g and to achieve greater toughness. One approach is to employ a little ethylene in the polymerizing mixture, thus creating a copolymer that is largely i-PP but with enough polyethylene (PE) to effectively lower the amount of crystallinity. Such copolymers are called reactor copolymers (RCPs). Another approach is to blend iPP with rubber. These materials form the important class of ethylene–propylene copolymers.

The use of metallocene catalyst systems allows the formation of an important group of copolymers from the use of alpha-olefins as comonomers with propylene. These catalysts also allow the production of wholly PP block copolymer elastomers that contain blocks of aPP and iPP. The aPP blocks act as the soft and/or non-cross-linked portion, while the iPP blocks act as the hard and/or physically cross-linked portion.

6.10 OTHER POLYOLEFINS

Butlerov produced amorphous, low-melting polyisobutylene in 1873 by the cationic polymerization of isobutylene by using boron trifluoride (BF_3).

$$\left[-CH_2-\underset{CH_3}{\overset{CH_3}{\underset{|}{\overset{|}{C}}}}- \right]_n$$

Polyisobutylene (PIB)

This polymer is used as a chewing gum base and as a caulking material, but when cold it flows much like unvulcanized rubber. This deficiency was overcome by Sparks and Thomas, who produced butyl rubber by copolymerizing isobutylene with small amounts (10%) of isoprene ($H_2C{=}C(CH_3)CH{=}CH_2$). Butyl rubber is resistant to permeation by gases, and this property is enhanced by chlorination.

Polybutene-1, $\left(CH_2{-}CH(C_2H_5)\right)$, is produced by the Ziegler-catalyzed polymerization of butene-1. The gas barrier properties of polybutene-1 are inferior to those of butyl rubber.

Polymethylpentene (TPX) is produced by the Ziegler polymerization of 4-methylpentene ($H_2C{=}CH{-}CH_2{-}CH(CH_3)_2$. Because of the bulky pendant group, TPX has a relatively high volume and low specific gravity (0.83). This high-melting (465°F) transparent polymer is used for laboratory ware.

6.11 POLYSTYRENE

Styrene monomer was discovered by Newman in 1786. The initial formation of polystyrene was by Simon in 1839. While polystyrene was formed almost 175 years ago, the mechanism of formation was not discovered until the early twentieth century. Staudinger, using styrene as the principle model, identified the general free radical polymerization process described in Section 3.2 in 1920. Initially, commercialization of polystyrene, as in many cases, awaited the ready availability of the monomer. While there was available ethyl benzene, it underwent thermal cracking rather than dehydrogenation until the appropriate conditions and catalysts were discovered. Dow first successfully commercialized polystyrene formation in 1938. While most commercial polystyrene (PS) has only a low degree of stereoregularity,

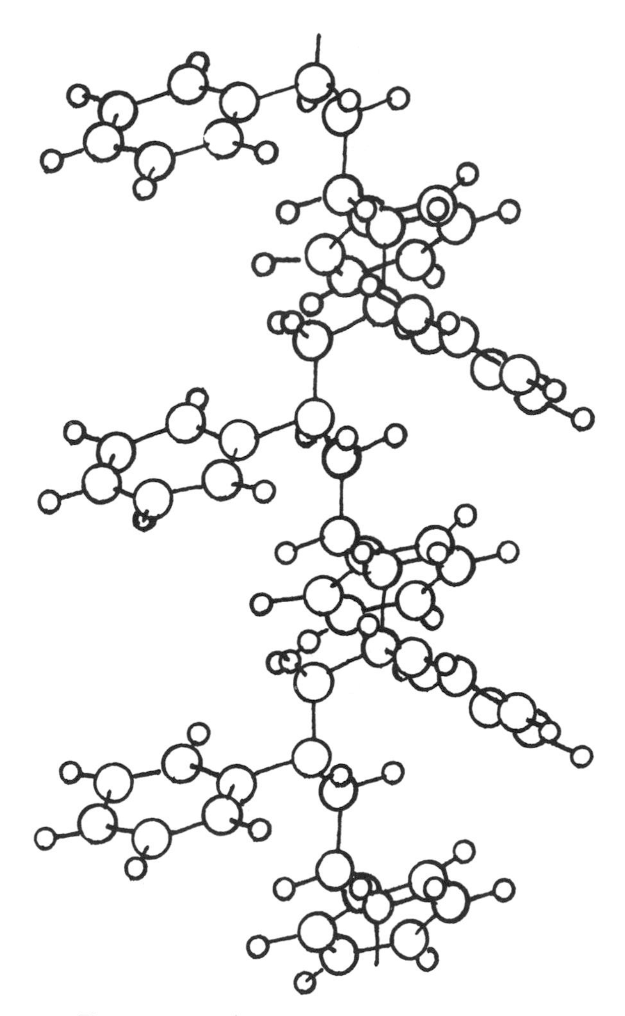

Figure 6.10. Ball-and-stick model of polystyrene.

it is rigid and brittle because of the resistance of easy movement of the more bulky phenyl-containing units in comparison, for example, to the methyl-containing units of polypropylene (Figure 6.10). This is reflected in a relatively high T_g of about 212°F for polystyrene. It is transparent because of the low degree of crystalline formation.

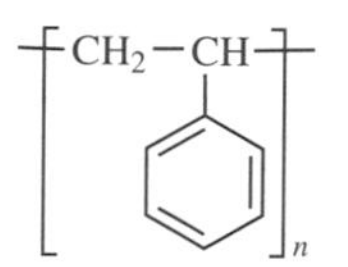

Polystyrene (PS)

While PS is largely commercially produced using free radical polymerization, it can be produced by all four of the major techniques: anionic, cationic, free radical, and coordination-type systems. All of the tactic forms can be formed employing these systems. The most important of the tactic forms is syndiotactic PS (sPS). Metallocene-produced sPS is a semicrystalline material with a T_m of 520°F. It was initially produced by Dow in 1997 under the trade name of Questra. It has good chemical and solvent resistance in contrast with "regular" PS, which has generally poor chemical and solvent resistance because of the presence of voids, due to the presence of the bulky phenyl groups, that are exploited by the solvents and chemicals.

Physical properties of PS are dependent on the molecular weight and presence of additives (Table 6.6). While higher-molecular-weight PS offers better strength and toughness, it also offers poorer processability. Low-molecular-weight PS allows for good processability but poorer strength and toughness. Generally a balance is sought where intermediate chain lengths are used.

Because PS is brittle with little impact resistance under normal operating conditions, early work was done to impart impact resistance. The best-known material from this work is called high-impact polystyrene or HIPS. HIPS is produced by dispersing small particles of butadiene rubber in with the styrene monomer. Polymerization captures the butadiene rubber particles within the polymerizing styrene.

Major uses of PS are in packaging and containers, toys and recreational equipment, insulation, disposable food containers, electrical items and electronics, housewares, and appliance parts. Expandable PS is used to package electronic equipment such as TV's, computers, and stereos.

Legislation was put in place in some states to ensure the recycling of PS. Interestingly some of this legislation was written such that all PS had to be recycled within some period of time such as a year. This legislation was changed to reflect the real concern of fast food containers when it was pointed out that less than 10% PS is used in this manner and that well over twice as much was used as house insulation that should not be recycled every year or so.

6.12 STYRENE COPOLYMERS

In addition to the SBR elastomer described in Chapter 10, a less rubbery copolymer with a lower percentage of butadiene is used as a tough plastic. Styrene–acrylonitrile copolymers (SANs) have relatively high heat deflection temperatures. Because of their thermal stability, SANs are employed in the production of "dishwasher-safe" houseware, such as blender bowls, humidifier parts, detergent dispensers, and refrigerator vegetable and meat drawers. Also, fiberglass-reinforced automotive battery cases and dashboard components are molded from SAN.

Blends of SAN and butadiene–acrylonitrile rubber (NBR) have superior impact resistance. Sheets of this acrylonitrile–butadiene–styrene (ABS) terpolymer are thermoformed for the production of suitcases, crates, and appliance housings.

Table 6.5 Properties of typical styrene polymers

Property	PS	HIPS	SAN	ABS
Melting point (T_m, °F)				
Glass transition temperature (T_g, °F)	100	95	120	115
Processing temperature (°F)	350	400	350	350
Molding pressure (10^3 psi)[a]	15	15	15	20
Mold shrinkage (10^{-3} in./in.)	5	5	4	6
Heat deflection temperature under flexural load of 264 psi (°F)	175	185	210	200
Maximum resistance to continuous heat (°F)	175	175	200	180
Coefficient of linear expansion (10^{-6} in./in., °F)	40	—	40	35
Compressive strength (10^3 psi)	12	—	14	7
Impact strength Izod (ft-lb/in. of notch)[b]	0.4	2.5	0.5	2
Tensile strength (10^3 psi)	12	30	10	5
Flexural strength (10^3 psi)	15	50	12	10
% elongation	2	40	2	20
Tensile modulus (10^3 psi)	400	250	550	350
Flexural modulus (10^3 psi)	425	200	500	400
Rockwell hardness	M65	R65	R83	R110
Specific gravity	1.04	1.04	1.07	1.2
% water absorption	0.02	0.02	0.2	0.4
Dielectric constant	2.5	3.0	2.5	3.0
Dielectric strength (V/mil)	550	—	425	400
Resistance to chemicals at 75°F[c]				
Nonoxidizing acids (20% H_2SO_4)	S	S	S	S
Oxidizing acids (10% HNO_3)	Q	Q	Q	Q
Aqueous salt solutions (NaCl)	S	S	S	S
Polar solvents (C_2H_5OH)	S	S	S	S
Nonpolar solvents (C_6H_6)	U	U	U	U
Water	S	S	S	S
Aqueous alkaline solutions (NaOH)	S	S	S	S

[a] psi/0.145 = kPa (kilopascals).
[b] ft-lb/in. of notch/0.0187 = cm · N/cm of notch.
[c] S, satisfactory; Q, questionable; U, unsatisfatisfactory.

The weather resistance and clarity of ABS is improved by replacing the acrylonitrile by methyl methacrylate. The properties of SAN and ABS are listed in Table 6.5.

ABS terpolymers are actually a family of polymers that can be used as foams, plastics, and elastomers. Acrylonitrile repeating units contribute good strength, heat stability, and chemical resistance. Styrene units contribute rigidity, processability,

Figure 6.11. Blow-molded Cycolac (Borg-Warner Chemicals, Division of General Electric). ABS bumpers were chosen because of processibility, ability to reproduce mold detail, and adequate impact strength.

and good gloss, whereas butadiene repeating units contribute impact strength, toughness, and unsaturation that can form cross-links. As shown in Figure 6.11, ABS plastics and rubbers are used in automotive grills, trim, instrument panels, and bumpers and as appliance housings and cabinets.

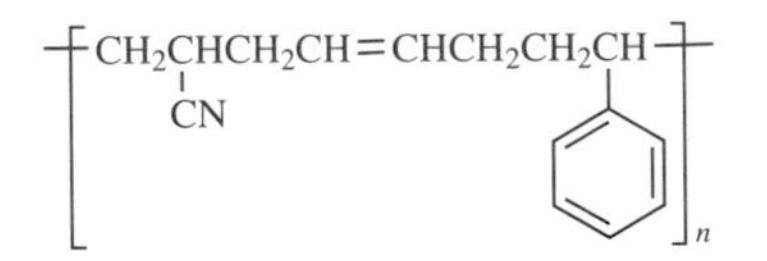

Acrylonitrile–butadiene–styrene terpolymer (ABS)

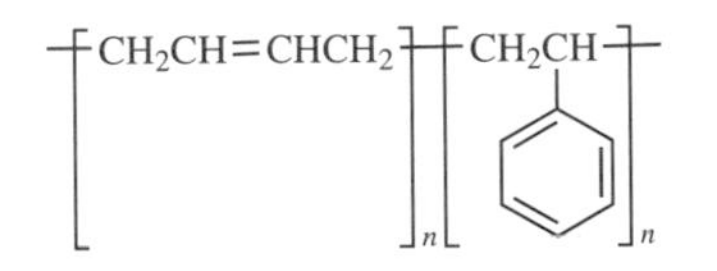

Styrene–butadiene rubber (SBR)

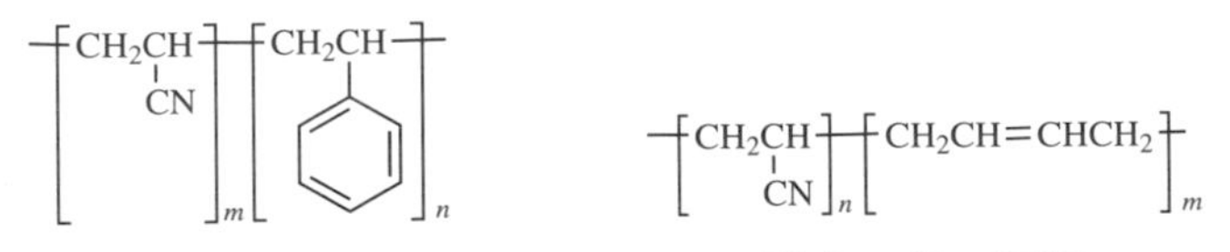

Styrene–acrylonitrile copolymer (SAN)

Nitrile–rubber (NBR)

6.13 POLY(VINYL CHLORIDE) AND COPOLYMERS

PVC is one of the earliest produced polymers. In 1835 Justus von Liebig and his research student Victor Regnault reacted ethylene dichloride with alcoholic potash, forming the monomer vinyl chloride. Later Regnault believed he polymerized vinyl chloride, but later studies showed it to be poly(vinylidene chloride). In 1872 E. Baumann exposed vinyl chloride sealed in a tube to sunlight and produced a solid, PVC. Klasse, in Germany, found that vinyl chloride could be made by addition of hydrogen chloride to acetylene in a system that could be scaled up for commercial production. (Today most of the vinyl chloride is made from the oxychlorination reaction with ethylene.) By World War I, Germany was producing a number of flexible and rigid PVC products. During World War I, Germany used PVC as a replacement for corrosion-resistant metals.

Waldo Semon was responsible for bringing many of the poly(vinyl chloride) products to market. The difficulty in fabricating PVC was associated with its tendency to decompose at temperatures typically used for molding and extrusion. Semon introduced materials called plasticizers, which allowed PVC to be processed at lower temperatures. As a young scientist at BF Goodrich, he worked on ways to synthesize rubber and to bind the rubber to metal. In his spare time he discovered that PVC, when mixed with certain liquids, gave an elastic-like, pliable material that was rainproof and fire-resistant and that did not conduct electricity. Under the trade name Koroseal, the rubbery material came into the marketplace, beginning about 1926, as shower curtains, raincoats, and umbrellas. During World War II it became the material of choice to protect electrical wires for the Air Force and Navy. Another of his inventions was the synthetic rubber patented under the name Ameripol, which was dubbed "liberty rubber" since it replaced natural rubber in the production of tires, gas masks, and other military equipment. Ameripol was a butadiene-type material.

Doolittle and Powell reduced the temperature needed to process PVC by copolymerizing vinyl chloride with vinyl acetate. During World War II, German chemists added heat stabilizers to allow the molding of unplasticized PVC.

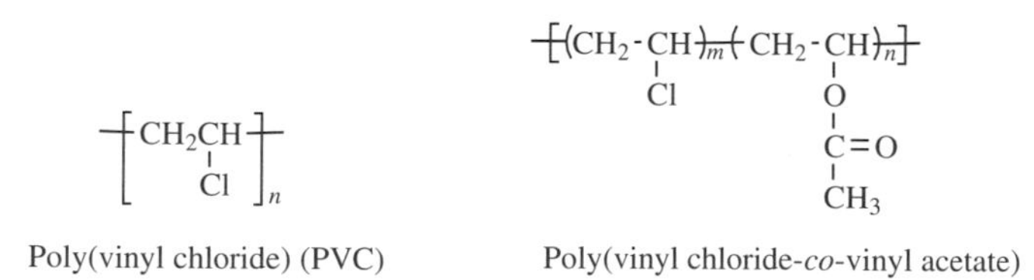

Poly(vinyl chloride) (PVC) Poly(vinyl chloride-*co*-vinyl acetate)

As a side note, there is today a debate concerning the use of chlorine-containing materials and their effect on the atmosphere. This is a real concern and one that is being addressed by industry. PVC and other chloride-containing materials have in the past been simply disposed of through combustion that often created unwanted hydrogen chloride. This practice has largely been stopped, but care should be continued to see that such materials are disposed of properly. Furthermore, simply outlawing of all chloride-containing materials is not possible or practical. For instance we need common table salt for life, and common table salt is sodium chloride. Chlorine is widely used as a water disinfectant both commercially (for our drinking water) and for pools. Furthermore, PVC is an important material that is not easily replaced. Finally, the amounts of chloride-containing residue that is introduced into the atmosphere naturally is large in comparison to that introduced by PVC. Even so, we must exercise care because we want to leave a better world for our children and grandchildren, so a knowledge-based approach must be taken.

Because of its versatility, some unique performance characteristics, ready availability, and low cost, poly(vinyl chloride), PVC, is now the second largest produced synthetic polymer behind polyethylene. PVC materials are often defined to contain 50% or more by weight vinyl chloride units. PVC is generally a mixture of a number of additives and often other units such as ethylene, propylene, vinylidene chloride, and vinyl acetate. In comparison to many other polymers, PVC employs an especially wide variety of additives. For instance, a sample recipe or formulation for common stiff PVC pipe such as used in housing and irrigation applications may contain (in addition to the PVC resin) tin stabilizer, acrylic processing aid, acrylic lubricant-processing aid, acrylic impact modifier, calcium carbonate, titanium dioxide, calcium sterate, and paraffin wax. Such formulations vary according to the intended processing and end use. In such nonflexible PVC materials the weight amount of additive is on the order of 5–10%.

PVC has a built-in advantage over many other polymers in that it is itself flame-resistant. About 50% of PVC is used as rigid pipe. Other uses of rigid PVC are as pipe fittings, electrical outlet boxes, and automotive parts. Uses of flexible PVC include in gasoline-resistant hose, hospital sheeting, shoe soles, electrical tape, stretch film, pool liners, vinyl-coated fabrics, roof coatings, refrigerator gaskets, floor sheeting, and electrical insulation and jacketing. A wide number of vinyl chloride copolymers are commercially used. Many vinyl floor tiles are copolymers of PVC. Typical properties are given in Table 6.6.

Plastisol PVC products are made by heating to 300°F finely divided PVC suspended in a liquid plasticizer that is in a mold.

In addition to its copolymer with vinyl acetate (Vinylite), vinyl chloride is also copolymerized with vinylidene chloride ($H_2C{=}CCl_2$) (Saran, Pliovic).

$$\text{+}CH_2CCl_2\text{+}_n$$

Poly(vinylidene chloride)

6.14 FLUOROCARBON POLYMERS

In spite of their similarity in structure to PVC, the fluorine counterparts were not discovered until the 1930s, and even here the discovery was accidental. Polytetrafluoroethylene, better known as its trade name of Teflon, was accidently discovered by Roy J. Plunkett, a Dupont chemist who had just received his Ph.D. from Ohio State two years before. He was part of a group searching for nontoxic refrigerant gases. On April 6, 1938, he and his assistant, Jack Rebok, had filled a tank with

Table 6.6 Properties of typical vinyl polymers

Property	Rigid PVC	Plasticized PVC
Melting point (T_m, °F)		
Glass transition temperature (T_g, °F)	85	85
Processing temperature (°F)	325	365
Molding pressure (10^3 psi)[a]	25	20
Mold shrinkage (10^{-3} in./in.)	4	20
Heat deflection temperature under flexural load of 264 psi (°F)	150	—
Maximum resistance to continuous heat (°F)	140	125
Coefficient of linear expansion (10^{-6} in./in., °F)	35	65
Compressive strength (10^3 psi)	8	1
Impact strength Izod (ft-lb/in. of notch)[b]	0.5	2
Tensile strength (10^3 psi)	6	2
Flexural strength (10^3 psi)	10	—
% elongation	60	300
Tensile modulus (10^3 psi)	450	—
Flexural modulus (10^3 psi)	400	—
Shore hardness	275	475
Specific gravity	1.4	1.2
% water absorption	0.1	0.2
Dielectric constant	2.5	3.0
Dielectric strength (V/mil)	400	350
Resistance to chemicals at 75°F[c]		
Nonoxidizing acids (20% H_2SO_4)	S	S
Oxidizing acids (10% HNO_3)	S	Q
Aqueous salt solutions (NaCl)	S	S
Polar solvents (C_2H_5OH)	S	S
Nonpolar solvents (C_6H_6)	S	Q
Water	S	S
Aqueous alkaline solutions (NaOH)	S	S

[a] psi/0.145 = kPa (kilopascals).
[b] ft-lb/in. of notch/0.0187 = cm·N/cm of notch.
[c] S, satisfactory; Q, questionable; U, unsatisfatisfactory.

tetrafluoroethylene. After some time, they opened the value but no gas came out. The tank weight indicated that there was no weight loss—so what happened to the tetrafluoroethylene? Using a hacksaw, they cut the cylinder in half and found a waxy white powder. He correctly surmised that the tetrafluoroethylene had polymerized. The waxy white powder had some interesting properties. It was quite inert toward strong acids, bases, and heat and was not soluble in any attempted liquid. It appeared to be quite "slippery."

$$-\!\!\left[CF_2CF_2 \right]\!\!-_n$$

Polytetrafluoroethylene (PTFE)

$$-\!\!\left[CF_2CFCl \right]\!\!-_n$$

Polychlorotrifluoroethylene (PCTFE)

Table 6.7 Properties of typical polyfluorocarbons

Property	PTFE	PCTFE
Melting point (T_m, °F)	325	220
Processing temperature (°F)	—	500
Molding pressure (10^3 psi)[a]	—	5
Mold shrinkage (10^{-3} in./in.)	—	10
Heat deflection temp. under flexural load of 264 psi (°F)		
Maximum resistance to continuous heat (°F)	300	250
Coefficient of linear expansion (10^{-6} in./in., °F)	35	25
Compressive strength (10^3 psi)	2	6
Impact strength Izod (ft-lb/in. of notch)[b]	3	3
Tensile strength (10^3 psi)	4	5
Flexural strength (10^3 psi)	—	10
% elongation	200	150
Tensile modulus (10^3 psi)	65	200
Flexural modulus (10^3 psi)	80	200
Rockwell hardness	D60 (Shore)	R85
Specific gravity	2.2	2.1
% water absorption	0	0
Dielectric constant		
Dielectric strength (V/mil)	500	550
Resistance to chemicals at 750°F[c]		
Nonoxidizing acids (20% H_2SO_4)	S	S
Oxodizing acids (10% HNO_3)	S	S
Aqueous salt solutions (NaCl)	S	S
Polar solvents (C_2H_5OH)	S	S
Nonpolar solvents (C_6H_6)	S	S
Water	S	S
Aqueous alkaline solutions (NaOH)	S	S

[a] psi/0.145 = kPa (kilopascals).
[b] ft-lb/in. of notch/0.0187 = cm · N/cm of notch.
[c] S, satisfactory; Q, questionable; U, unsatisfatisfactory.

Little was done with this new material until the military, working on the atomic bomb, needed a special material for gaskets that would resist the corrosive gas uranium hexafluoride, which was one of the materials being used to make the atomic bomb. General Leslie Groves, responsible for the U.S. Army's part in the atomic bomb project, had learned of Dupont's new inert polymer and had Dupont manufacture it for them.

Teflon was introduced to the public in 1960 when the first Teflon-coated muffin pans and frying pans were sold. Like many new materials, problems were encountered. Bonding to the surfaces was uncertain at best. Eventually the bonding problem was solved. Teflon is now used for many other applications including acting as a biomedical material in artificial corneas, substitute bones for nose, skull, hip, nose, and knees; ear parts, heart valves, tendons, sutures, dentures, and artificial tracheas. It has also been used in the nose cones and heat shield for space vehicles and for their fuel tanks. Because of its resistance to solvents and corrosives, it is used as gaskets and as a coating in cooking ware. Over one-half million vascular graft replacements are performed yearly. Most of these grafts are made of PET and PTFE. These relatively large diameter grafts work when blood flow is rapid, but they generally fail for smaller vessels. Some of its properties are given in Table 6.7.

Polytetrafluoroethylene is produced by the free radical polymerization process. While it has outstanding thermal and corrosive resistance, it is a marginal engineering material because it is not easily machinable. It has low tensile strength, resistance to wear, and low creep resistance. Molding powders are processed by press and sinter methods used in powder metallurgy. It can also be extruded using ram extruder techniques.

The polymers of monochlorotrifluoroethylene ($CClF{=}CF_2$) (PCTFE), vinylidene fluoride ($CH_2{=}CF_2$), PVDF, and vinyl fluoride ($CH_2{=}CHF$) have lower resistance to heat and corrosives and have reduced lubricity in accordance with the reduced fluorine content. The flexibility of polyfluorocarbons is increased by copolymerizing with ethylene. The copolymer of vinylidene fluoride and hexafluoropropylene is a heat-resistant elastomer.

6.15 ACRYLIC POLYMERS

Acrylic acid ($H_2C{=}CHCOOH$) was synthesized in 1843, and ethyl methacrylate ($H_2C{=}C(CH_3)COOC_2H_5$) was synthesized and polymerized in 1865 and 1877, respectively. Otto Rohm produced acrylic plastics in the early 1900s, and a lacquer based on acrylic polymer was marketed by Rohm and Haas in 1927 in Germany and in 1931 in the United States. One of the first uses of acrylic polymers was as an interlining for automobile windshields, but poly(methyl methacrylate) sheet (Plexiglas, Lucite) soon became the principal use of acrylic plastics.

Poly(methyl methacrylate) (PMMA), $+CH_2{-}CH(CH_3)COOCH_3+$, has a light transmittancy of about 92% and has good resistance to weathering. It is widely used in thermoformed signs, aircraft windshields, and bathtubs. The properties of PMMA are summarized in Table 6.8.

Poly(methyle acrylate) Poly(methyl methacrylate) (PMMA)

Poly(methyl methacrylate) is used as an automobile lacquer and polyacrylonitrile, $(CH_2{-}CHCN)_n$, is used as a fiber. Poly(ethyl acrylate), $(CH_2{-}CHCOOC_2H_5)_n$, is

Table 6.8 Properties of typical poly(methyl methacrylate)

Property	
Glass transition temp. (T_g, °F)	100
Processing temperature (°F)	350
Molding pressure (10^3 psi)[a]	15
Mold shrinkage (10^{-3} in./in.)	3
Heat deflection temperature under flexural load of 264 psi (°F)	185
Maximum resistance to continuous heat (°F)	180
Coefficient of linear expansion (10^{-6} in./in., °F)	40
Compressive strength (10^3 psi)	15
Impact strength Izod (ft-lb/in. of notch)[b]	0.5
Tensile strength (10^3 psi)	10
Flexural strength (10^3 psi)	15
% elongation	5
Tensile modulus (10^3 psi)	400
Flexural modulus (10^3 psi)	400
Rockwell hardness	M80
Specific gravity	1.2
% water absorption	0.2
Dielectric constant	3.0
Dielectric strength (V/mil)	450
Resistance to chemicals at 75°F[c]	
Nonoxidizing acids (20% H_2SO_4)	S
Oxidizing acids (10% HNO_3)	Q
Aqueous salt solutions (NaCl)	S
Polar solvents (C_2H_5OH)	S
Nonpolar solvents (C_6H_6)	Q
Water	S
Aqueous alkaline solutions (NaOH)	Q

[a] psi/0.145 = kPa (kilopascals).
[b] ft-lb/in. of notch/0.0187 = cm · N/cm of notch.
[c] S, satisfactory; Q, questionable; U, unsatisfatisfactory.

more flexible and has a lower softening temperature than PMMA. Poly(hydroxyethyl methacrylate), $\text{+}CH_2\text{—}C(CH_3)COOC_2H_4OH\text{+}_n$, is used for contact lenses, and poly(butyl methacrylate) is used as an additive in lubricating oils.

6.16 POLY(VINYL ACETATE)

Vinyl acetate and its polymer were described by Klatte in 1912, and the polymer (PVAc) was produced commercially under the trade name of Elvacet and Gelva in 1920. Because of its low softening point, PVAc is not used as a moldable plastic but is used as an adhesive and in waterborne coatings.

Over 200,000 tons of PVAc are produced annually in the United States. Some of this polymer is hydrolyzed to produce a water-soluble polymer (poly(vinyl alcohol), PVA), and some of the PVA is reacted with butyraldehyde to produce poly(vinyl butyral) (PVB). Poly(vinyl butyral) is used as an inner layer of safety windshield glass.

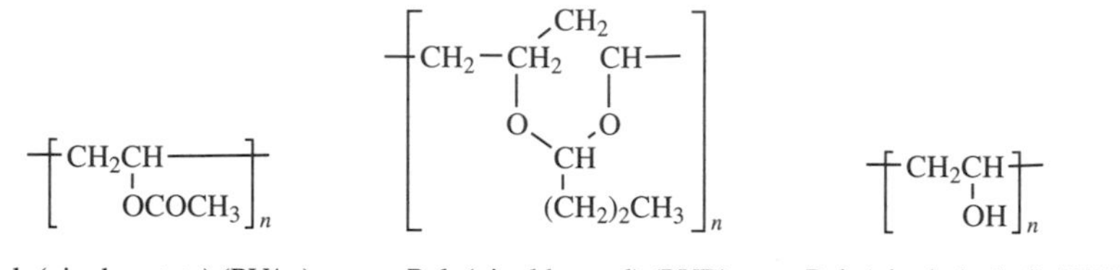

Poly(vinyl acetate) (PVAc) Poly(vinyl butyral) (PVB) Poly(vinyl alcohol) (PVA)

6.17 POLY(VINYL ETHERS)

Vinyl ethers, such as vinyl isobutyl ether, are readily polymerized by Lewis acids, such as boron trifluoride, to produce polymers that have excellent adhesive properties. The copolymer of vinyl isobutyl ether and maleic anhydride (Gantrez) is used as a water-soluble component of floor waxes.

$$\left[CH_2\text{-}CH \right]$$
$$|$$
$$O$$
$$|$$
$$CH_2$$
$$|$$
$$CH(CH_3)_2$$

Poly(vinyl isobutyl ether)

6.18 CELLULOSICS

Cellulose, which is a polymer made up of D-glucose repeating units, occurs widely, but because of strong intermolecular hydrogen bonds between the oxygen atoms in one molecule and the hydrogen atoms in another molecule, it cannot be molded by standard procedures. However, this high-molecular-weight linear polymer can be processed and fabricated when it is converted to derivatives. Thus, A. Parkes in England (1862) and J. and I. Hyatt in the United States (1869) were able to soften cellulose nitrate and shape it into useful articles. Parkes used cottonseed oil, and the Hyatts used camphor to soften (plasticize) cellulose nitrate (erroneously called nitrocellulose). These pioneer plastics were called Parkesine and Celluloid.

From a social and economical viewpoint, it is of interest to note that Leominster, Massachusetts, became the center of the fabrication of celluloid products because of the development of plastic fabrication machinery, and the National Plastics Museum is located in that city. Hyatt's firm (Merchant's Manufacturing Co.) in Newark, New Jersey, became the nation's largest producer of Celluloid. This firm was purchased by E. I. du Pont de Nemours & Co. in order to start that firm's plastics operations.

Other large U.S. firms, such as Celanese, Eastman, Hercules, and Monsanto, also entered the plastics business via Celluloid. Hyatt's incentive for producing Celluloid was a $10,000 award offered by a producer of billiard balls (Pheland and Collendar) for a substitute for ivory. Although Celluloid was widely used for shirt fronts, collars, combs, and brush handles in the nineteenth century, its use today is limited.

Because of its explosive nature, Celluloid could not be extruded or injection-molded. However, cellulose diacetate was produced by the partial saponification of cellulose triacetate by G. Miles in 1905. This product, which is flammable but not explosive, continues to be used as a molding resin and for the production of films and fibers.

A mixed ester of cellulose called cellulose acetate butyrate, which was developed in 1935, is a tough transparent plastic that is widely used for molding steering wheels, ballpoint pens, and typewriter keys. Ethers of cellulose, such as ethylcellulose, have been molded and extruded to produce molded articles and sheets. Ethylcellulose melts have also been used for tool handles and waterproof packaging.

6.19 PLASTICS PROCESSING

A. Introduction

Both natural and synthetic polymers must be processed before use. The seeds must be separated from cotton in the ginning process, pigments and driers must be added to oleoresinous paints, and the latex of *Hevea* rubber or gutta-percha must be coagulated to obtain the solid elastomer plastic. Synthetic polymers must also be compounded and fabricated into useful shapes. Plastics are converted into their final shapes by utilizing a variety of techniques and machinery.

Polymer processing can be defined as the process whereby raw materials are converted into products of desired shape and properties. Thermoplastic resins are generally supplied as pellets, marbles, or chips of varying sizes, and they may contain some or all of the desired additives. When heated above their T_g, thermoplastic materials soften and flow as viscous liquids that can be shaped using a variety of techniques and then cooled to "lock" in the micro- and gross structure.

Thermosetting feedstocks are normally supplied as meltable and/or flowable prepolymer, oligomers, or lightly or non-cross-linked polymers that are subsequently cross-linked, forming the thermoset article.

The processing operation can be divided into three general steps: pre-shaping, shaping, and post-shaping. In pre-shaping, the intent is to produce a material that can be shaped by application of heat and/or pressure. Important considerations include:

- Handling of solids and liquids including mixing, low, compaction, and packing
- Softening through application of heat and/or pressure
- Addition and mixing/dispersion of added materials
- Movement of the resin to the shaping apparatus through application of heat and/or pressure and other flow aiding processes
- Removal (and desired and recycling) of unwanted solvent, unreacted monomer(s), byproducts, and waste (flash)

The shaping step may include any single or combination of the following:

- Die forming (including sheet and film formation, tube and pipe formation, fiber formation, coating, and extrusion)
- Molding and casting
- Secondary shaping (such as film and blow molding, thermoforming)
- Surface treatments (coating and calendering)

Post-shaping processes include welding, bonding, fastening, decorating, cutting, milling, drilling, dying, and gluing.

Polymer processing operations can be divided into five broad categories:

- Spinning (generally for fibers)
- Calendering
- Coating
- Molding
- Injection

Essentially all of the various processing types utilize computer-assisted design (CAD) and computer-assisted manufacture (CAM). CAD allows the design of a part and incorporates operating conditions to predict behavior of the pieces prior to real operation. CAD also transfers particular designs and design specifications to other computer-operated systems (CAMs) that allow the actual construction of the part or total apparatus. CAM systems operate most modern processing systems, many allowing feedback to influence machine operation.

Processing, chemical structure, physical structure (amorphous/crystalline), and performance are interrelated to one another. Understanding these factors and their interrelationships becomes increasingly important as the specific performance requirements become more specific. Performance is related to the chemical and physical structure and to the particular processing performed on the material during its lifetime. The physical structure is a reflection of both the chemical structure and the total history of the synthesis and subsequent exposure of the material to additional force that contributes to the secondary (and greater) structure–stress/strain, light, chemical, and so on.

A single material may be processed using only a single process somewhat unique to that material (such as liquid crystals) or by a variety of processes (such as polyethylene) where the particular technique is dictated by such factors as end us and cost.

Following is a brief description of some of the most widely used techniques employed in the processing of plastics.

B. Casting

One of the simplest and least expensive methods for the production of plastic articles is casting. In this process, which is illustrated in Figure 6.12, a prepolymer, such as a catalyzed epoxy resin, is placed in a mold and allowed to harden, preferably with additional heat. This technique may also be used with urethane reactants (RIM), phenolic resins, unsaturated polyesters, PVC plastisols, and acrylic resins.

With the exception of plastisols, most of these processes are exothermic and thus the articles should be small or the mold must be cooled. Plastisols, which consist of a dispersion of a finely divided polymer, usually PVC, in a liquid plasticizer, must be heated to at least 300°F to fuse the plasticizer–polymer mixture. Polymers, like ethylcellulose and ethylene–vinyl acetate copolymers, which can be melted without decomposition, can be cast as hot melts. Solutions of polymers can be cast as films.

Polymer concrete is produced by a casting process. Simulated marble consists of a filled-peroxy-catalyzed unsaturated polyester prepolymer that polymerizes *in situ*. Comparable mortars consisting of filled-catalyzed phenolic, epoxy, or polyester resins are used for joining brick and tile. Casting is used in manufacturing both thermosetting and thermoplastic resins for making eyeglass lenses, plastic jewelry, and cutlery handles.

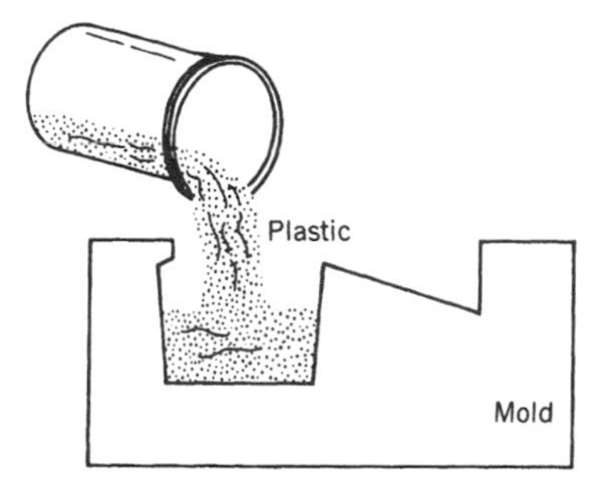

Figure 6.12. Illustration of the casting method of molding plastics.

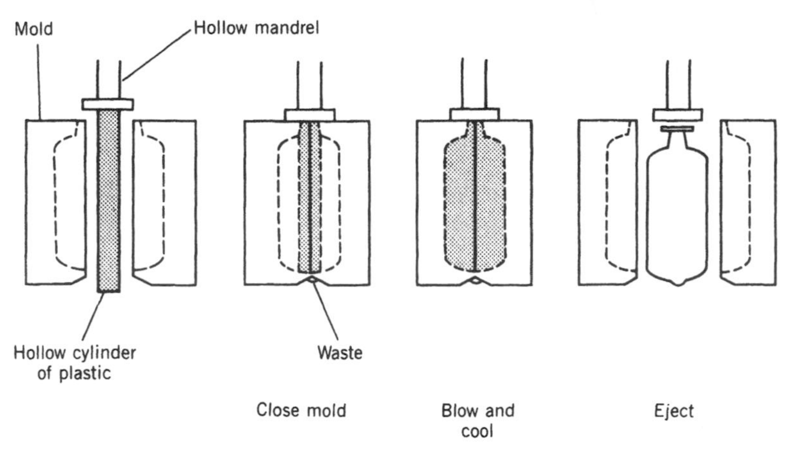

Figure 6.13. Blow molding technique.

C. Blow Molding

Blow molding and plug-assisted vacuum thermoforming are employed to make hollow items such as bottles and many hollow, thin-walled toys and bowls. For blow molding, a plastic parison is placed in the mold and air is applied through the opening of the cylinder-shaped plastic, blowing the plastic toward the mold walls (Figure 6.13). In the plug-assisted molding sequence the plastic resin is present as a sheet.

D. Injection Molding

In injection molding, a large volume of thermoplastics is injection-molded to produce a variety of articles at a rapid rate. As shown in Figure 6.14, the polymer pellets may be heated, softened, and formed or forced (injected) by a ram into a closed, cooled mold. The split mold is opened and closed after the molded article is ejected and the cycle is then repeated. As shown in Figure 6.15, a reciprocating

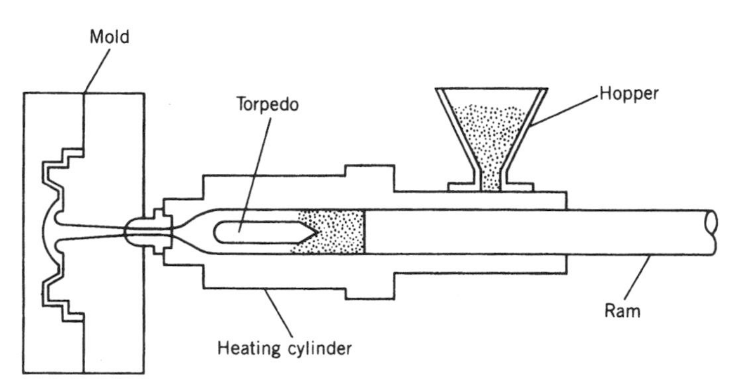

Figure 6.14. Injection molding technique.

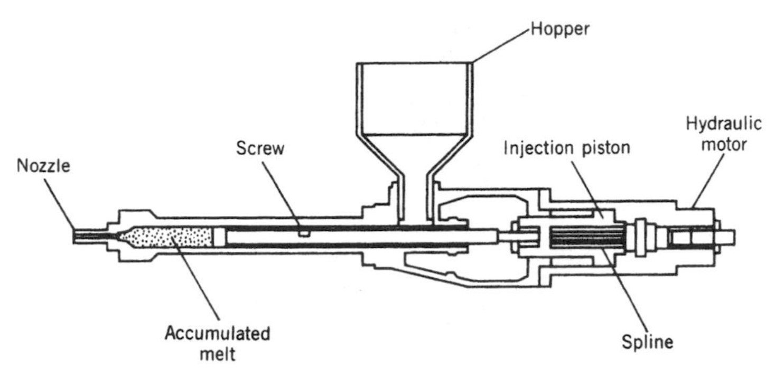

Figure 6.15. Modified injection molding technique.

preplasticating screw that moves forward to eject the softened polymer may be used in place of the ram.

In contrast to slow compressive molding, injection molding is rapid. Complex parts may be produced in a few seconds in multicavity molds. Containers, gears, honeycombs, and trash cans are produced by the injection molding of selected thermoplastics (Figures 6.16 to 6.20).

E. Laminating

In laminating, sheets of metal foil, paper, other plastic, or cloth are treated with a plastic resin. They are then run through rollers that squeeze the sheets together and heat them as shown in Figure 6.21. Paneling and electronic circuits are examples of products produced through this process, which is similar to making sandwiches.

Figure 6.16. A Mobay laboratory technician inspects a compact disc that was molded on the Meiki injection-molding system shown in the background.

Figure 6.17. A Sailor robot automatically removes a compact disc from the mold.

Calendering is similar to laminating except rollers spread melted resin over the sheets to be covered, providing a protective coating as in the case of playing cards and treated wallpapers (Figure 6.22). This is similar to spreading jelly on a slice of bread, with the jelly being the resin.

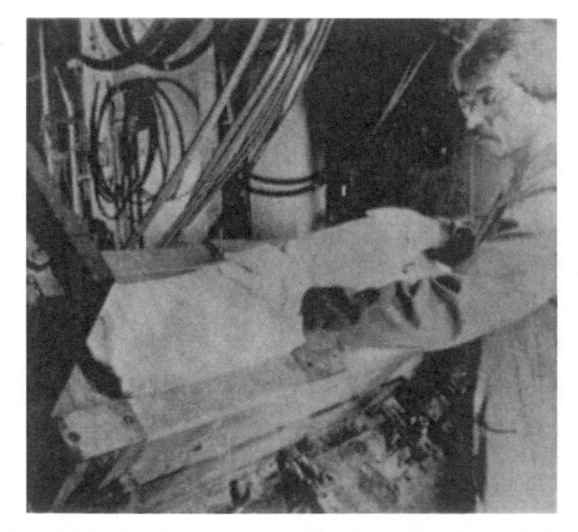

Figure 6.18. A Mobay lab technician places combinations of preformed glass reinforcement for a bumper beam in the RIM tool.

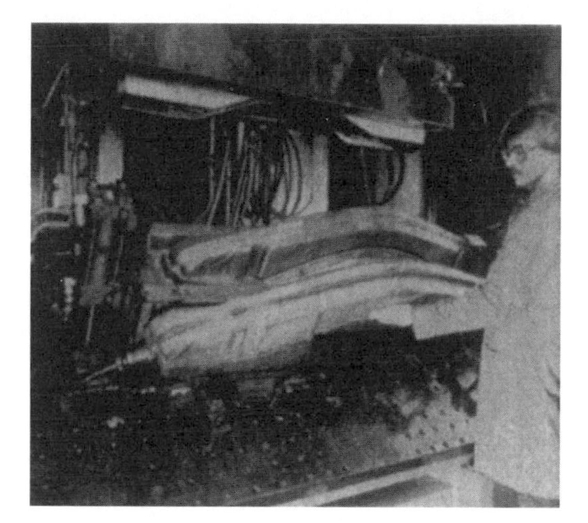

Figure 6.19. A finished bumper beam is removed from the mold less than a minute after the polyurethane mixture is injected into the tool.

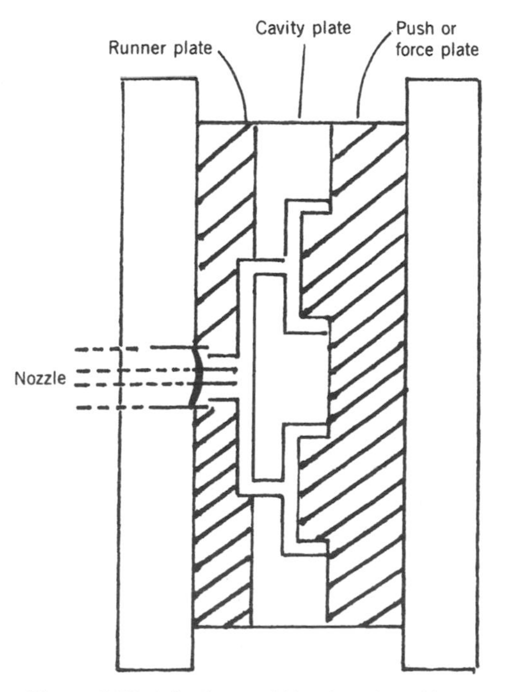

Figure 6.20. Injection mold in closed position.

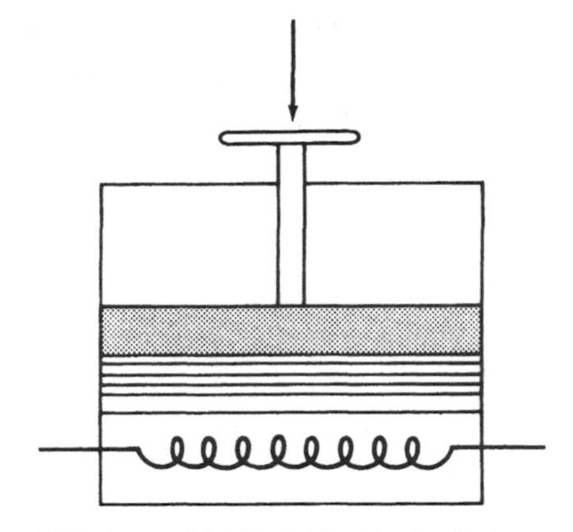

Figure 6.21. Assembly illustrating the laminating process.

F. Compression Molding

There are a wide variety of molding techniques. Simple molding entails squeezing plastic between two halves of a mold. It is similar to making waffles, where the batter is the plastic and the waffle iron is the mold.

One of the simplest molding techniques is compression molding, which is illustrated in Figure 6.23. In this molding process, a heated hydraulic press is used to soften plastic pellets and shape the plastic in a mold. When thermosets are used, the prepolymer is completely polymerized in the closed hot mold and is then ejected when the mold is opened. When thermoplastics are molded by compression molding, the mold cavity must be cooled before ejecting the plastic article.

The labor-intensive compression molding process may be upgraded by preheating a preform of the molding powder in a transfer pot and forcing this softened prepolymer into hot multicavity molds. Transfer molding is illustrated in Figure 6.24.

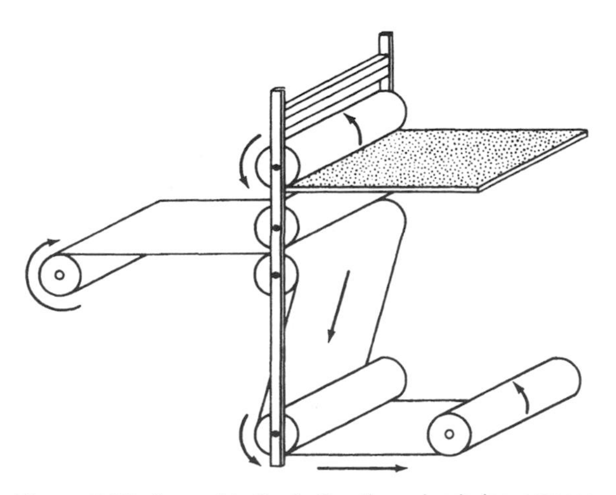

Figure 6.22. Assembly illustrating the calendering process.

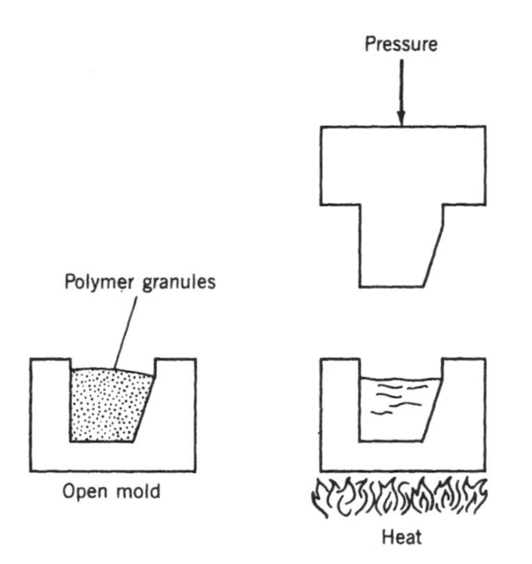

Figure 6.23. Compression molding technique.

G. Rotational Molding

One of the more versatile molding techniques is rotational molding in which a hollow mold containing a resin powder or a liquid plastisol is heated and rotated simultaneously on two perpendicular axes. The mold is then cooled and the hollow object, such as a pipe fitting, is removed.

H. Calendering

One of the most commonly used techniques for making thermoplastic or elastomeric sheet is calendering. As shown in Figure 6.22, the polymer is transported through heated rollers, like those in a rubber mill, to a series of heated wringer-type

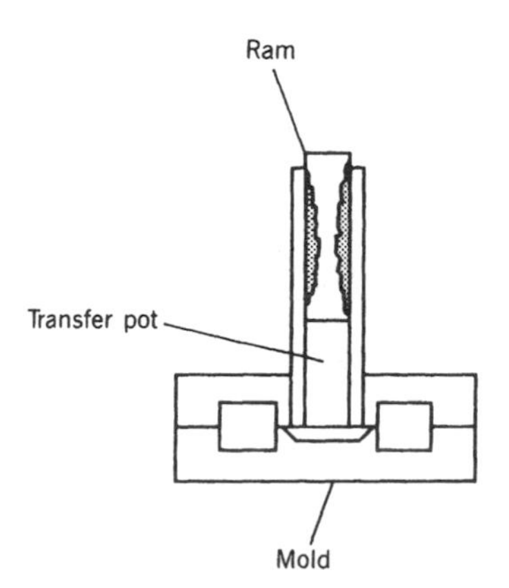

Figure 6.24. Transfer molding.

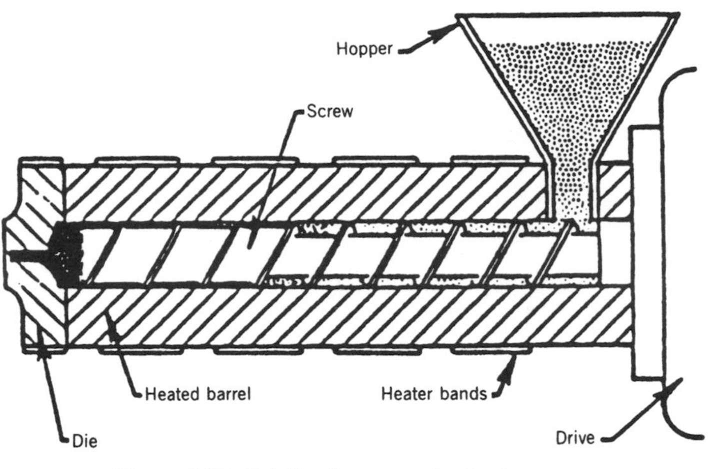

Figure 6.25. Details of screw and extruder zones.

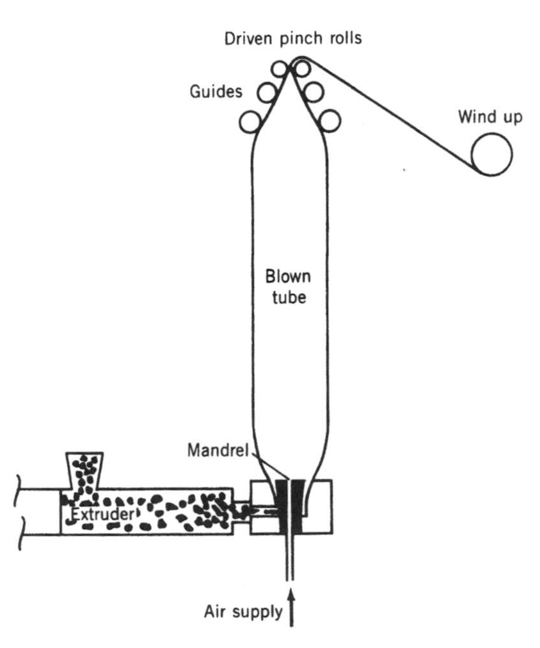

Figure 6.26. Diagram of film formation employing the extrusion process.

Figure 6.27. The entire extrusion line in Mobay's laboratories is controlled by state-of-the-art microprocessor technology, which was designed to optimize processing parameters and reduce start-up times. A specialist is shown here working at the central terminal.

Figure 6.28. An overview of the new Mobay multipurpose extrusion line.

rollers, which press the polymer into a continuous sheet of uniform thickness. The calendering process is used to produce sheeting for upholstery and for thermoforming.

I. Extrusion

The extrusion process is similar to squeezing toothpaste from its tube. As shown in Figure 6.25, pipe, rods, or profiles may be produced by the extrusion process. In this process, thermoplastic pellets are fed from a hopper to a rotating screw. The polymer is transported through heated, compacting, and softening zones and then forced through a die and cooled after it leaves the die. The extrusion process has been used to coat metal wire and to form coextruded sheet for packaging. Over 1 million tons of extruded pipe are produced annually in the United States. Figure 6.26 illustrates film formation employing the extrusion process, and Figures 6.27 and 6.28 show an automated extrusion production line.

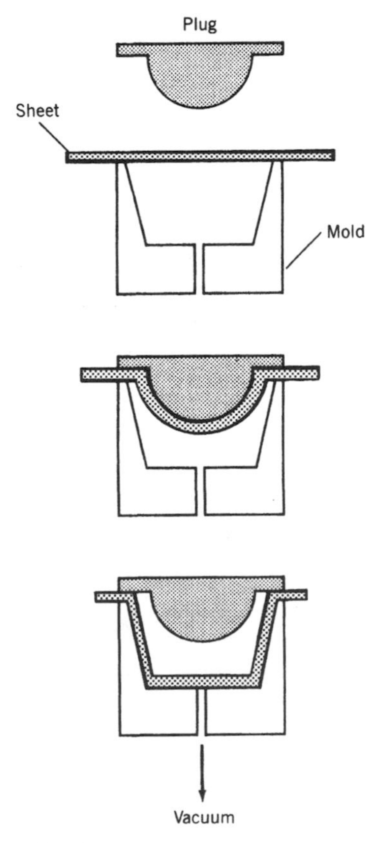

Figure 6.29. Steps in plug-assisted thermoforming.

J. Thermoforming

Thermoplastic sheet, produced by extrusion through a slit die, calendering, or hot pressing of several calendered sheets, is readily thermoformed by draping over a mold and using a plunger or vacuum to force the sheet into the shape of the mold. As illustrated in Figure 6.29, refrigerator liners or suitcases may be produced by vacuum sheet thermoforming.

K. Reinforced Plastics

Fiberglass-reinforced plastics (FRP) are fabricated by casting procedures using mixtures of the casting resins and glass or graphite fibers. In the simplest hand lay-up technique, a catalyzed resin, such as an unsaturated polyester resin, is placed on a male form. This gel coat formulation is followed by a sequential buildup of layers of catalyzed resin-impregnated glass mat. The composite is removed after it hardens. The curing step may be accelerated by heating. This technique may be modified by spraying a mixture of chopped fibers and the catalyzed prepolymer onto the form.

In a more sophisticated approach, a continuous resin-impregnated filament is wound around a rotating mandrel and cured as shown in Figure 6.30. In another modification, a bundle of resin-impregnated filaments is drawn through a heated die. Fishing rods and pipe are produced by this pultrusion technique.

L. Conclusion

A review of the many polymers and blends available and the many fabrication techniques that can be used to produce finished articles should demonstrate the

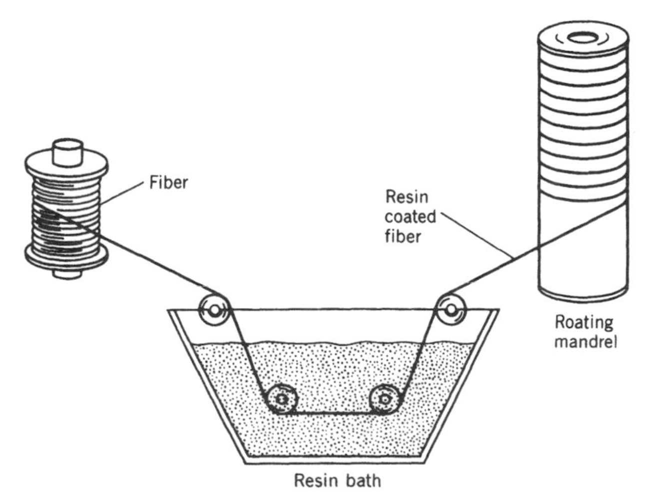

Figure 6.30. Illustration of the filament winding technique.

versatility of polymers. Because of this versatility, the polymer industry has grown at an unprecedented rate and will be the world's largest and most important industry into the beginning of the twenty-first century.

GLOSSARY

Accelerator: Catalyst for the vulcanization of rubber.

Banbury mixer: An intensive mixer.

Blow molding: The production of hollow articles, such as bottles, by air blowing a short section of pipe within a two-piece mold.

Blown film: Film produced by air blowing a warm pipe-like extrudate after it leaves the die.

Calendering: The formation of a sheet by passing a thermoplastic through a series of heated rolls.

Compounder: A technician who mixes polymers and additives.

Compression molding: The curing of a heated polymer under external pressure in a mold.

Elastomer: General term for rubbery materials.

Epoxy resin: The reaction product of bisphenol A and epichlorohydrin. The molecule has terminal epoxy groups

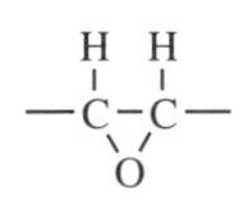

and multiple hydroxyl groups (OH).

Extrusion: The production of pipe and profiles by the continuous forcing of a heated, softened polymer through a die and cooling the extrudate.

Fiber: A threadlike structure with a length to diameter ratio of at least 100 to 1.

Filament winding: The winding and curing of a resin-impregnated filament around a rotating mandrel.

Gel coat: An unfilled and unreinforced polymer usually formed as the first (outer layer) of a fiber-reinforced composite.

GRP: Glass-reinforced plastic.

Gutta-percha: A nonelastic naturally occurring *trans*-polyisoprene.

***Hevea* rubber:** The most widely used natural rubber.

Injection molding: A rapid process for the production of molded articles by injecting a heat-softened thermoplastic into a closed, cooled split mold, ejecting the solid molded part, and repeating this process.

Latex: A stable suspension of a polymer in water.

Melamine-formaldehyde prepolymers: A low-molecular-weight product of the condensation of melamine and formaldehyde.

Oleoresinous paint: Coating based on polymerizable unsaturated oils.

Parison: A short section of thermoplastic pipe or tubing.

Plastisol: A dispersion of a finely divided polymer (usually PVC) in a liquid plasticizer.

Polyester: The product of the condensation of a diol ($R(OH_2)$) and a dicarboxylic acid ($R(COOH)_2$).

Polymer concrete: A solid formed by mixing an aggregate containing an initiator and a polymer that polymerizes *in situ*.

Pultrusion: The curing of a resin-impregnated bundle of filaments by drawing through a heated die.

Quaterary ammonium compound: R_4N^+, Cl^- or R_4N^+, OH^-.

Rayon: Regenerated cellulose fiber.

RIM: Reaction injection molding, in which polymerization takes place in the mold.

Rotational molding: The production of hollow articles by heating finely divided polymer particles in a rotating mold.

Simulated marble: A filled polyester produced by the room-temperature polymerization of a filled unsaturated polyester prepolymer.

Stabilizer: An additive used to retard degradation of polymers.

Thermoforming: The forming of trays and other three-dimensional articles by heating and pressing a sheet of thermoplastic over a mold.

Thermoplastic elastomer: An elastic polymer that does not require cross-linking for dimensional stability.

Thermoset: A cured polymer that cannot be softened by heat without decomposition.

Thinner: Solvent.

Transfer molding: An improved compression molding process in which a thermoset preform is preheated before being transformed to the hot mold.

Vehicle: The binder (resin) and solvent in a paint.

Vulcanization: Cross-linking, usually with sulfur.

REVIEW QUESTIONS

1. Are most commercial plastics thermoplastics or thermosets?
2. Is a cross-link a covalent bond or a hydrogen bond?
3. Which of the following natural polymers is not a hydrocarbon: gutta-percha, balata, casein?
4. Which is more highly crystalline: HDPE or LDPE?

5. Which is an alkane: polyethylene or paraffin?
6. What is the DP of HDPE with a molecular weight of 14,000?
7. Why is ABS used for making suitcases?
8. A PVC plastisol is a thermoplastic but is sometimes called a thermoset by nonpolymer technicians. Why?
9. How would you dissolve poly(acrylic acid)?
10. How do poly(methyl methacrylate) and poly(ethyl acrylate) differ in their empirical formulas?
11. How do poly(methyl methacrylate) and poly(ethyl acrylate) differ in softening point?
12. Why is poly(hydroxyethyl methacrylate) used for contact lenses?
13. What is the formula for the methyl radical?
14. Which has the larger specific volume: HDPE or LDPE?
15. Which would yield the larger area of film of similar thickness for a given weight: LPDE or HDPE?
16. Which has the larger bulky pendant group: polypropylene or TPX?
17. Which can be readily cross-linked: polyisobutylene or butyl rubber?
18. Which will have the higher melting point: atactic or isotactic PP?
19. Which would be more likely to crystallize: atactic PP or a block copolymer of ethylene and propylene?
20. Which would be more readily soluble in water: PVA or PVC?
21. In the early 1900s, most plastics were celluloid. Why is this no longer true?

BIBLIOGRAPHY

Avery, J. (2001). *Gas-Assist Injection Molding*, Hanser-Gardner, Cincinnati.

Baird, D., and Collias, D. (1998). *Polymer Processing Principles and Design*, Wiley, New York.

Baker, W., Scott, C, and Hu, G.-H. (2001). *Reactive Polymer Blending*, Hanser-Gardner, Cincinnati.

Beall, G. (1998). *Rotational Molding Design, Materials, Tooling, and Processing*, Hanser-Gardner, Cincinnati.

Beaumont, J., and Nagel, R. (2002). *Successful Injection Molding Through CAE*, Hanser-Gardner, Cincinnati.

Belcher, S. (1999). *Practical Extrusion Blow Molding*, Marcel Dekker, New York.

Belofsky, H. (1995). *Plastics: Product Design and Process Engineering*, Hanser-Gardner, Cincinnati.

Benedikt, G. (1999). *Metallocene Technology in Commercial Applications*, ChemTec, Toronto.

Bryce, D. (1999). *Plastic Injection Molding Manufacturing Startup and Management*, Society of Manufacturing Engineers, Dearborn, MI.

Carraher, C. ISO process and practices is given in a series of articles appearing in *Polymer News* 19(12), 373; 20(5), 147; 20(9), 278; 21(1), 21; 21(5), 167; 22(1), 16.

Calleja, F., and Roslaniec, Z. (2000). *Block Copolymers*, Marcel Dekker, New York.

Cheremisinoff, N. P. (1998). *Advanced Polymer Processing Operations*, Noyes, Westwood.

Chung, C. I. (2000). *Polymer Extrusion Theory and Practice*, Hanser-Gardner, Cincinnati.

Chung, C. (2001). *Extrusion of Polymers: Theory and Practice*, Hanser-Gardner, Cincinnati.

Craver, C., and Carraher, C. (2000). *Applied Polymer Science*, Elsevier, New York.

Dubois, P. (2000). *Advances in Ring Opening Polymerizations*, Wiley, New York.

Ebnesajjad, S. (2000). Fluoroplastics, ChemTec, Toronto.

Gruenwald, G. (1998). *Thermoforming, A Plastics Processing Guide*, Technomic, Lancaster.

Heim, H., Potente, H. (2001). *Specialized Molding Techniques*, ChemTec, Toronto.

Jaques, R. (2000). *Plastics and Technology*, Cambridge University Press, New York.

Kanai, T., and Campbell, G. (1999). *Film Processing*, Hanser-Gardner, Cincinnati.

Klempner, D, and Frisch, K. (2002). *Polymeric Foams*, 2nd ed., Hanser-Gardner, Cincinnati.

Lee, N. C. (2000). *Understanding Blow Molding*, Hanser-Gardner, Cincinnati.

Linder, E., Unger, P. (2002). *Injection Molds*, 3rd ed., Hanser-Gardner, Cincinnati.

Malloy, R. (2002). *Manufacturing with Recycled Thermoplastics*, Hanser-Gardner, Cincinnati.

Menges, G., Michaeli, W., and Mohren, P. (2002). *How To Make Injection Molds*, Hanser-Gardner, Cincinnati.

Moore, E. P. (1998). *The Rebirth of Polypropylene: Supported Catalysts*, Hanser-Gardner, Cincinnati.

Moore, E., ed. (1996). *Polypropylene Handbook*, Hanser-Gardner, Cincinnati.

O'Brian, K. T. (1999). *Applications of Computer Modeling for Extrusion and Other Continuous Polymer Processes*, Oxford University Press, New York.

O'Connor, L. (2000). *Advanced Catalysts*, Wiley, New York.

Olmsted, B. (2001). *Practical Injection Molding*, Marcel Dekker, New York.

Osswald, T. A. (1998). *Polymer Processing Fundamentals*, Hanser-Gardner, Cincinnati.

Osswald, T., Turng, L-S., and Gramann, P. (2001). *Injection Molding Handbook*, Hanser-Gardner, Cincinnati.

Paul, D. (2000). *Polymer Blends*, Vols. 1 and 2, Wiley, New York.

Peacock, A. (2000). *Handbook of Polyethylene*, Marcel Dekker, New York.

Pohanish, R. P. (2001). *Sittig's Handbook of Toxic and Hazardous Cehmicals and Carcinogens*, 4th ed., ChemTec, Toronto.

Ram, A. (1997). *Fundamentals of Polymer Engineering*, Plenum, New York.

Rauwendaal, C. (2000). *SPC Statistical Process Control in Injection Molding and Extrusion*, Hanser-Gardner, Cincinnati.

Rauwendaal, C. (2001). *Polymer Extrusion*, 4th ed., Hanser-Gardner, Cincinnati.

Rees, H. (2002). *Understanding Injection Mold Design*, Hanser-Gardner, Cincinnati.

Roovers, J. (1999). *Branched Polymers*, Vols. I and II, Springer-Verlag, New York.

Rosato, D., Rosato, D., and Rosato, M. (2000). *Injection Molding Handbook*, Kluwer Academic, New York.

Rudin, A. (1998). *Elements of Polymer Science and Engineering: An Introductory Text and Reference for Engineers and Chemists*, 2nd ed., Academic Press, Orlando, FL.

Sandler, S. R., and Karo, W. (1998). *Polymer Synthesis*, Academic Press, Orlando, FL.

Scheirs, J. (2000). *Metallocene-Based Polyolefins*, Wiley, New York.

Semlyen, J. (2000). *Cyclic Polymers*, Kluwer, New York.

Shonaike, G., and Simon, G. (1999). *Polymer Blends and Alloys*, Marcel Dekker, New York.

Stann, M. (2000). *Polymer Blends*, Wiley, New York.

Throne, J. L., and Crawford, R. J. (2001). *Rotational Molding Technology*, ChemTec, Toronto.

Todd, D. B. (1998). *Plastics Compounding Equipment and Processing*, Hanser-Gardner, Cincinnati.

Tres, P. A. (1998). *Designing Plastic Parts for Assembly*, Hanser-Gardner, Cincinnati.

Ultracki, L. (1998). *Commercial Polymer Blends*, Kluwer, Hingham, MA.

Vasile, C. (2000). *Handbook of Polyolefins*, Marcel Dekker, New York.

White, J., Coran, A., and Moet, A. (2001). *Polymer Mixing Technology and Engineering*, Hanser-Gardner, Cincinnati.

Wilks, E. (2001). *Industrial Polymers Handbook: Products, Process and Applications*, Wiley, New York.

ANSWERS TO REVIEW QUESTIONS

1. Thermoplastic.
2. Covalent bond.
3. Casein is a polyamide (protein).
4. HDPE.
5. Both have the empirical formula $H(CH_2)_nH$.
6. $14,000/28 = 500$.
7. ABS is a tough plastic.
8. It undergoes an irreversible physical change from a liquid to a solid when heated.
9. In an alkaline solution such as aqueous sodium hydroxide.
10. They are identical ($C_5H_8O_2$).
11. The softening point of poly(ethyl acrylate) is much lower than that of PMMA.
12. The hydrophilic hydroxy group absorbs water and keeps the polymer soft.
13. $\cdot CH_3$.

14. LDPE, the specific volume is the reciprocal of the density.

15. LDPE.

16. TPX (polymethylpentene).

17. Butyl rubber has a few double bonds that can be cross-linked; polyisobutylene has no double bonds.

18. Isotactic PP is a solid; atactic PP is a soft gummy substance.

19. The block copolymer providing the sequences were linear and ordered.

20. PVA because of the hydroxyl pendant groups present.

21. Many other less costly, more readily moldable, and less flammable thermoplastics are available commercially.

7

ENGINEERING PLASTICS

7.1 INTRODUCTION

The most common engineering thermosets are materials such as the phenol and amino plastics that are cross-linked three-dimensional networks. These materials

Giant Molecules: Essential Materials for Everyday Living and Problem Solving, Second Edition,
by Charles E. Carraher, Jr.
ISBN 0-471-27399-6

are covered in Chapter 8. Here we will be looking at only engineering thermoplastics. As noted in Chapter 6, the line between engineering thermoplastics and simply thermoplastics is not clear and varies with application and the person making the distinction.

The terms high-performance, engineering, and advanced thermoplastics are often used interchangeably. They generally contain no or only light cross-linking.

Engineering thermoplastics are bulk materials that can be easily and readily machined, milled, drilled, or otherwise have its shape modified while remaining in the solid state—much like metals. They retain their mechanical functionality even when subjected to mechanical stress/strain, vibration, friction, flexure, and so on over a general temperature range of 32–212°F. They deform when the weight that is added is too great, yielding and deforming rather than simply cracking or breaking in two. This property is called impact resistance—that is, the ability to withstand shock without undergoing brittle failure. These materials are rapidly replacing metals because they offer advantages such as lightness per strength, corrosion resistance, self-lubrication, economy and breath in fabrication, and in some cases transparency and ease in decoration. The need for engineering plastics is increasing. In 2001, more than 15 million tons of engineering thermoplastics were used in the United States.

7.2 NYLONS

Wallace Hume Carothers was brought to DuPont because his fellow researchers at Harvard and the University of Illinois called him the best synthetic chemist they knew. He started a program aimed at understanding the composition of natural polymers such as silk, cellulose, and rubber. Many of his efforts related to condensation polymers were based on his belief that if a monofunctional reactant reacted in a certain manner forming a small molecule, then similar reactions except employing reactants with two reactive groups would form polymers.

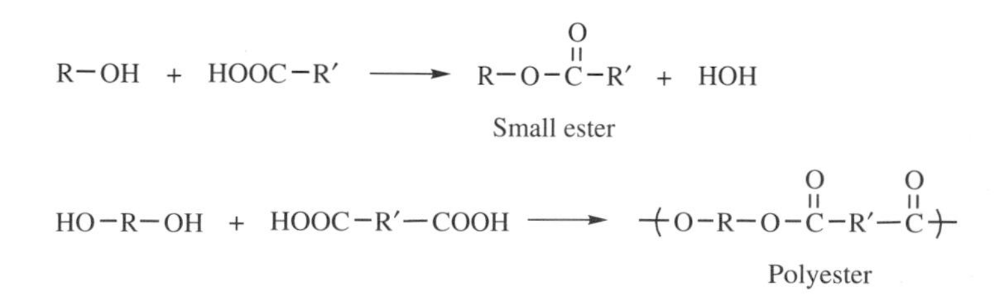

While the Carothers group had made both polyesters and polyamides, they initially emphasized work on the polyesters since they were more soluble and easier to work with. One of Carothers co-workers, Julian Hill, noticed that he could form fibers if he took a soft polyester material on a glass stirring rod and pulled some of it away from the clump. Because the polyesters had too low softening points for use as textiles, the group returned to work with the polyamides. They found

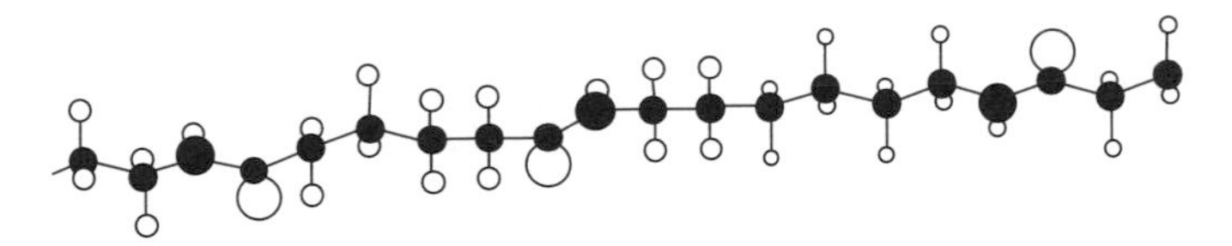

Figure 7.1. Ball-and-stick model of nylon 6,6.

that fibers could also be formed by the polyamides similar to those formed by the polyesters. These fibers allowed the formation of fibers that approached, and in some cases surpassed, the strength of natural fibers. This new miracle fiber was introduced at the 1939 New York World's Fair in an exhibit that announced the synthesis of this wonder fiber from "coal, air, and water"—an exaggeration but nevertheless eye-catching. These polyamides were given the name "nylons." When the polyamides, nylons, were first offered for sale in New York City on May 15, 1940, over 4 million pairs were sold in the first few hours. Nylon sales took a large drop when nylon was used to produce the parachute material so critical to WW II.

Although nylon 6,6 (Figure 7.1) and nylon 6 are used primarily as fibers, they are also used as engineering plastics. In fact, nylon 6,6 was the first engineering thermoplastic and until 1953 it represented the entire annual engineering thermoplastic sales. Nylon 6,6 is tough and rigid and does not need to be lubricated. It has a relatively high use temperature (to about 520°F or 270°C) and is used in the manufacture of items ranging from automotive gears to hair combs.

Most polymers progress from a glass solid to a softer solid and then to a viscous "taffy-like" stage allowing easy heat-associated fabrication. Nylon 6,6 has an unusually sharp transition from the solid to the soft stage requiring that fabrication be closely watched.

Nylon 4,6 was developed by DSM Engineering Plastics in 1990 and sold under the trade name Stanyl, giving a nylon that has a higher heat and chemical resistance for use in the automotive industry and in electrical applications. It has a T_m of 560°F (295°C) and can be made more crystalline than nylon 6,6.

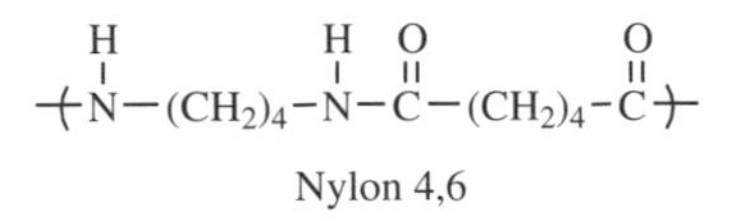

Nylon 4,6

A number of aromatic polyamides, aramids, have been produced that are strong, can operate under high temperatures, and have good flame-retardant properties. Nomex™ is used in flame-resistant clothing and in the form of thin pads to protect sintered silica-fiber mats from stress and vibrations during the flight of the space shuttles. Kevlar™ is structurally similar, and by weight it has a higher strength and modulus than steel and is used in the manufacture of so-called bullet resistant clothing. Because of its outstanding strength/weight ratio, it was used as the skin covering of the Gossamer Albatross, which was flown over the English Channel using only human power.

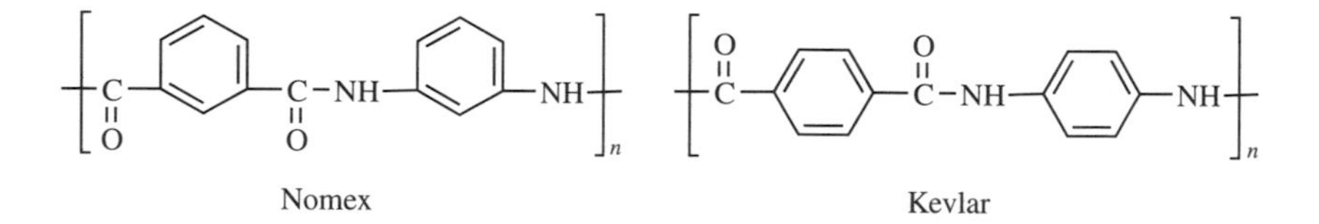

Nylons offered new challenges to the chemical industry. Because of the presence of polar groups, the attractive forces between chains was high in comparison to vinyl polymers. Nylons are generally semicrystalline, meaning they have a good amount of order. Thus, while they have a T_g, the main physical transition is the T_m so that they undergo a sharper transition from solid to melt in comparison to many of the vinyl polymers discussed in the preceding chapter. Thus, the processing temperature window is more narrow. If melt flow is required for processing, then the temperature must be high enough to allow for ready flow but low enough so as not to break primary bonds within the processed material. Even so, processing techniques have been developed that allow nylons to be readily processed using most of the standard techniques.

The presence of the polar groups result in materials with relatively high T_g and T_m values so that unlike many vinyl polymers that must be above their T_g to allow needed flexibility, nylons and many condensation polymers function best where strength, and not flexibility, is the desired behavior.

Because of the presence of these polar groups that also allow for internal hydrogen bonding, nylons and most condensation polymers are stronger, more rigid and brittle, and tougher in comparison to most vinyl polymers. Nylons are also "lubrication-free," meaning they do not need a lubricant for easy mobility so that they can be used as mechanical bearings and gears without need for periodic lubrication. Typical properties of nylon 6,6 are given in Table 7.1.

In general, more crystalline nylons are fibrous, whereas less crystalline nylon materials are more plastic in behavior. The amount of crystallinity is controlled through a variety of means including introduction of bulky groups and asymmetric units, rapid cooling of nonaligned melts, and introduction of plasticizing materials.

Table 7.1 Comparative data for nylon 6,6 and reinforced nylon 6,6 (Zytel, Vydine)

Property Reinforced	Unfilled	30% Glass
Tensile strength (psi)	8,500	26,000
Elongation (%)	150	4
Compressive strength (psi)	4,900	24,000
Flexural strength (psi)	15,000	41,000
Flexural modulus (psi)	290,000	1,300,000
Notched Izod impact (ft-lb/in. of notch)	1	2.4
Coefficient of expansion (cm/cm, °C)$(10)^{-6}$	50	17.5
Heat deflection temperature (°F)	135	485

7.3 POLYESTERS

Carothers and his research group at DuPont began to investigate the formation of polymers from the reaction of aliphatic diacids with diols, generally adipic acid and ethylene glycol (derived from reaction of ethylene oxide with water; major ingredient in most antifreeze), in search of materials that would give them fibers. They were only able to form syrupy mixtures. This is because unlike reactions with diamines (Section 7.2), the equilibrium reaction greatly disfavors ester formation. Furthermore, the ability to have almost equal amounts of functional groups is easily achieved with the amines through formation of salts with the amines, but diols do not form such salts. The critical need to have the reactants present in equal molar amounts for equilibrium determined reactions is clearly seen in Section 3.2. Carothers' group understood the principle of "driving" an equilibrium reaction so sought to remove water, thus forcing the reaction toward ester formation. For this they developed a so-called molecular still, which was simply heating the mixture and applying a vacuum coupled with a "cold-finger" that allowed evacuated water to condense and be removed from the reaction system. Since the fractional conversion (p) was only 0.95, the average chain length of these polyesters was less than 20.

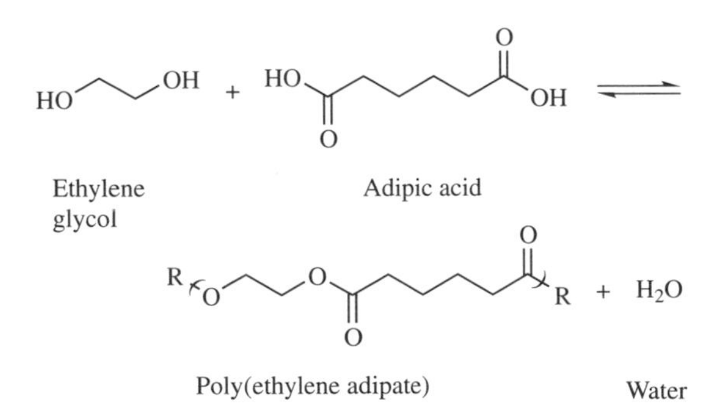

The DuPont research turned from the synthesis of polyesters to tackle, more successfully, the synthesis of the first synthetic fiber material, nylon, that approached, and in some cases exceeded, the physical properties of natural analogues (Chapters 13 and 14). The initial experience with polyesters was put to use in the nylon venture.

The initial polyester formation actually occurred much earlier and is attributed to Gay Lussac and Pelouze in 1833 and Berzelius in 1847. These polyesters are called glyptals and alkyds, and they are useful as coatings materials and not for fiber production. While these reactions had low fractional conversions, they formed high-molecular-weight materials because they had funtionalities (that is, number of reactive groups on a single reactant) greater than two, resulting in cross-linking.

The heat resistance of Carothers' polyesters was not sufficient to withstand the temperature of the hot ironing process. Expanding on the work of Carothers and

Hill on polyesters, Whinfield and Dickson, in England, overcame the problems of Carothers and co-workers by employing an ester interchange reaction between ethylene glycol and the methyl ester of terephthalic acid, forming the polyester poly(ethylene terephthalate), PET, with the first plant coming on line in 1953. This classic reaction producing Dacron, Kodel, and Terylene fibers and Dacron fibers is shown below.

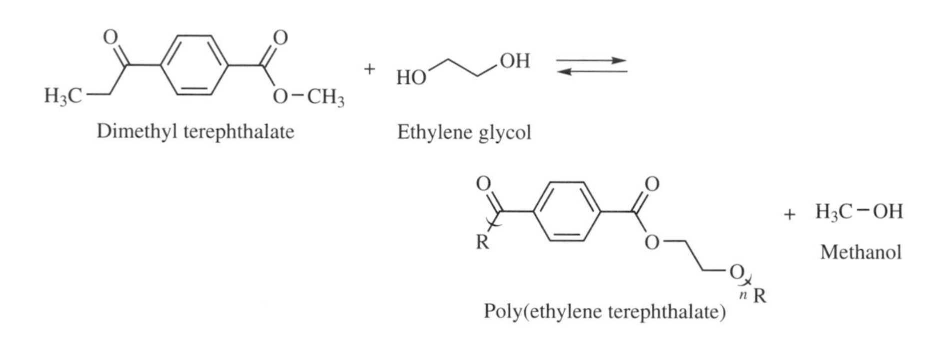

Methyl alcohol, methanol, is lower boiling than water (149°F compared with 212°F) and thus more easily removed, allowing the reaction to be forced toward polymer formation more easily. This illustrates how similar materials can be made from more than one chemical reaction. While the poly(aryl esters), now simply called polyesters, produced by Whinfield and Dickson met most of the specifications for a useful synthetic fiber but because of inferior molding machines and inadequate plastic technology, it was not possible to injection mold these materials until more recently.

Since the ease of processing and fabricating polyesters is related to the number of methylene groups (CH_2) in the repeating units, polymer chemists in several firms produced poly(butylene terephthalate) (PBT) for use as a moldable engineering plastic. PBT has four methylene groups, whereas PET has only two of these flexibilizing groups in each repeating unit.

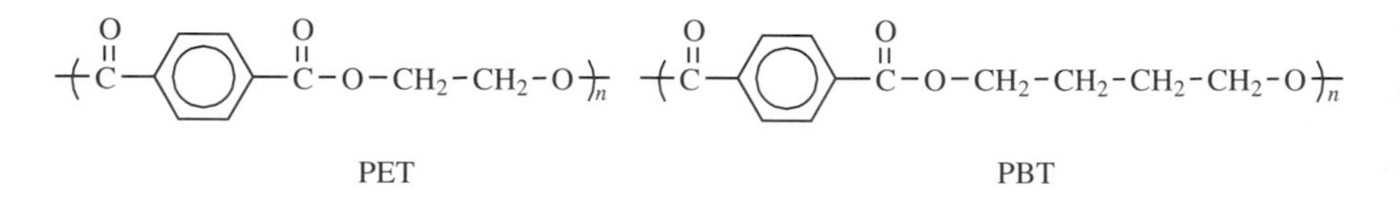

PET is now extensively used as bottling material for soft drinks because of its low carbon dioxide permeability (Figure 7.2). Carbon dioxide permeability decreases with increasing film thickness and crystallinity. To achieve the necessary crystallinity, partially crystalline PET is employed in the stretch blow molding process with the molding process carried out to promote further crystalline formation. It is also used for molded automobile parts. Over 500,000 tons of polyester engineering plastics are produced annually in the United States. The properties of PET and PBT are given in Tables 7.2 and 7.3.

Table 7.2 Comparative data for poly(ethylene terephthalate) and reinforced PET (Rynite)

Property Reinforced	Unfilled	30% Glass
Tensile strength (psi)	9,500	23,000
Elongation (%)	150	3
Compressive strength (psi)	13,000	23,000
Flexural strength (psi)	16,000	33,000
Flexural modulus (psi)	400,000	1,300,000
Notched Izod impact (ft-lb/in. of notch)	0.5	2
Coefficient of expansion (cm/cm, °C) $(10)^{-6}$	65	30
Heat deflection temperature (°F)	100	435

Figure 7.2. A 1.5-L bottle molded from Kodapak PET, a polyester manufactured by Eastman Chemical Products, Inc.

Table 7.3 Comparative data for poly(butylene terephthalate) and reinforced PBT (Valox, Celanex, Petra)

Property	Unfilled	30% Glass Reinforced
Tensile strength (psi)	8,000	18,000
Elongation (%)	150	3
Compressive strength (psi)	12,000	21,000
Flexural strength (psi)	14,000	27,000
Flexural modulus (psi)	350,000	1,100,000
Notched Izod impact (ft-lb/in. of notch)	1	1.5
Coefficient of expansion (cm/cm, °C) $(10)^{-6}$	75	25
Heat deflection temperature (°F)	150	425

PBT has a melting point around 435°F and is generally processed at about 480°F, whereas PET has a melting point around 520°F. This is a direct consequence of the presence of the addition of two methylene units in PBT, which allows easier fabrication of PBT by injection molding and extrusion procedures and through blow molding. PBT is employed for under-the-hood automotive parts, including fuse cables, pump housings, and electrical connectors, and for selected automotive exterior parts.

A more hydrophobic, stiffer polyester was introduced in 1958 by Eastman–Kodak as Kodel polyester. It contains a cyclohexanedimethanol moiety in place of the simple methylene moiety present in PET and PBT. Copolyesters based on cyclohexanedimethanol and ethylene glycol as the diols are blow-molded into bottles for shampoos and liquid detergents. Other, similar copolyesters are processed by extrusion into tough, clear fibers employed to package hardware and other heavy items.

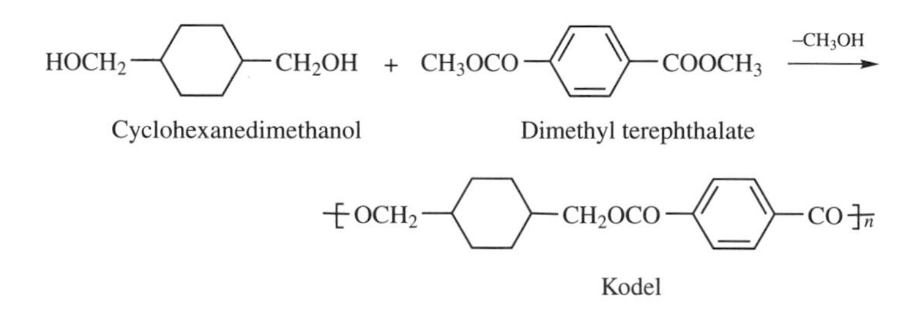

Several "wholly" aromatic polyesters are available. As expected, they are more difficult to process and are stiffer and less soluble, but they are employed because of their good high thermal performance. Ekonol is the homopolymer formed from *p*-hydroxybenzoic acid (below). Ekonol has a T_g in excess of 930°F (500°C). It is highly crystalline and offers good strength.

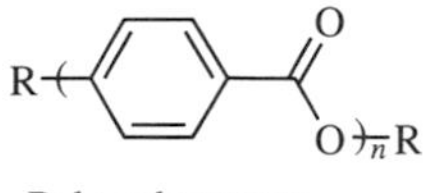

Poly-*p*-benzoate

It is not unexpected that such aromatic polyesters have properties similar to those of polycarbonates because of their structural similarities.

7.4 POLYCARBONATES

Einhorn produced a high-melting, clear polyester of carbonic acid in 1898 by the reaction of phosgene and hydroquinone or resorcinol. Commercial polycarbonates (PC) were produced in the 1950s by the General Electric Company and Bayer Company by the condensation of bisphenol A and phosgene ($COCl_2$). This tough, transparent polymer (Lexan, Merlon) is produced at an annual rate of 130,000 tons.

Polycarbonates are processed by all the standard plastic methods (Section 6.19). They are used in glazing (40%), appliances (15%), signs, returnable bottles, solar collectors, business machines, and electronics. They show good creep resistance, good thermal stability, and a wide range of use temperatures (about – 60°F to 270°F). Coatings are generally used on PC sheets to improve mar and chemical resistance.

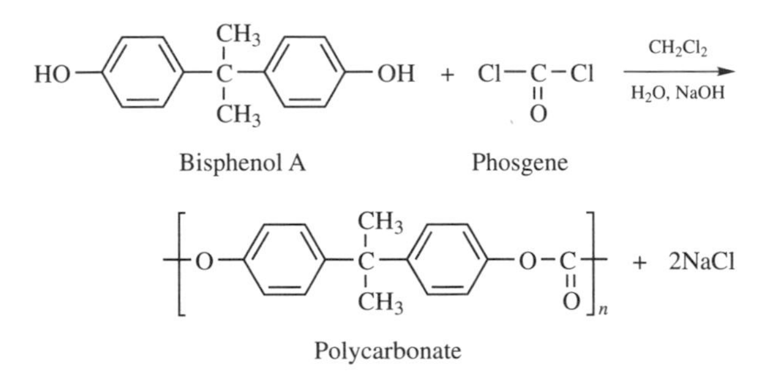

Blends of PC and ABS (Bayblend) or with poly(butylene terephthalate) (Xenoy) are tough, heat-resistant (HDUL 390°F) plastics. The properties of polycarbonates are shown in Table 7.4.

Nonrecordable compact discs (CDs) are made of rigid, transparent polycarbonates with a reflective metal coating on top of the polycarbonate. A laser is used to encode information through creation of physical features sometimes referred to as "pits and lands" of different reflectivity at the polycarbonate–metal interface.

Table 7.4 Comparative data for polycarbonate and reinforced polycarbonate (Lexan, Merlon)

Property 30% Glass	Unfilled	Reinforced
Tensile strength (psi)	9,500	19,000
Elongation (%)	110	4
Compressive strength (psi)	12,500	18,000
Flexural strength (psi)	13,500	23,000
Flexural modulus (psi)	340,000	11,000,000
Notched Izod impact (ft-lb/in. of notch)	14	2
Coefficient of expansion (cm/cm, °C) $(10)^{-6}$	68	22
Heat deflection temperature (°F)	270	295

Recordable CDs contain an organic dye between the polycarbonate and metal film. Here, a laser creates areas of differing reflectiveness in the dye layer through photochemical reactions.

A beam from a semiconductor diode laser "interrogates" the undersides of both types of CDs seeking out areas of reflected light, corresponding to the binary "one," and unreflected light, corresponding to the binary "zero." The ability to "read" information is dependent on the wavelength of the laser. Today, most of the CD players use a near-infrared laser because of the stability of such lasers. Efforts are underway to develop stable and inexpensive lasers of shorter wavelengths that will allow the holding of more information within the same space.

7.5 POLYACETALS/POLYETHERS

Aliphatic polyethers are also referred to as polyacetals. Polyoxymethylene (POM) precipitates spontaneously from uninhibited aqueous solutions of formaldehyde and was isolated by Butlerov in 1859. POM is also called poly(methylene oxide). Staudinger, in the 1920s and 1930s, experimented with the polymerization of formaldehyde but failed to produce chains of sufficient length to be useful. While pure formaldehyde readily polymerizes, it also spontaneously depolymerizes—that is, unzippers.

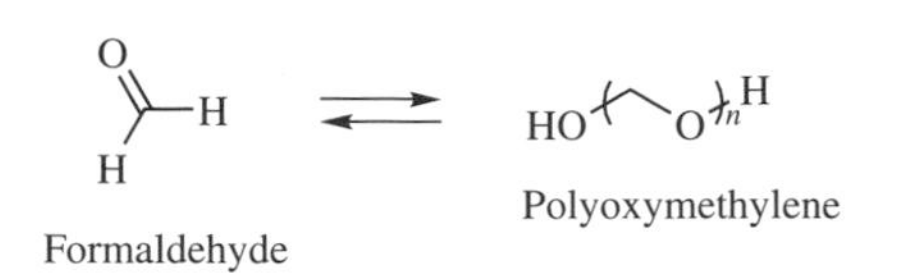

In 1947 DuPont began a program to make useful polymers from formaldehyde since formaldehyde is inexpensive and readily available. After 12 years they announced the commercialization of the polymer from formaldehyde, polyoxymethylene, under the trade name of Delrin. The "secret" was capping the end groups by acetylation of the hydroxyl end groups, thus preventing the ready unzipping of the polymer chain.

$$CH_3-\overset{O}{\overset{\|}{C}}-O-\overset{O}{\overset{\|}{C}}-CH_3 \quad + \quad H\overset{*}{O}CH_2\{CH_2-O\}_n CH_2-\overset{*}{O}H$$

Acetic anhydride

Polyoxymethylene (showing the end group noted by asterisks)

$$CH_3-\overset{O}{\overset{\|}{C}}-O-CH_2[CH_2-O]_n CH_2-O-\overset{O}{\overset{\|}{C}}-CH_3$$

Capped POM (Delrin)

Celanese came out a year latter with a similar product under the trademark of Celcon. Celanese circumvented Dupont's patent on the basis of employing a copolymer variation that allowed enhanced stabilization against thermal depolymerization. The copolymer has a T_m of 340°F (170°C).

$$HCHO + (CH_2CH_2O) \longrightarrow R\{CH_2-O\}\{CH_2CH_2-O\}H$$

POMs are employed in plumbing and irrigation because they resist scale accumulation and have good thread strength, torque retention, and creep resistance. Polyacetyls are used for molded door handles, tea kettles, pump impellers, shoe heels, and plumbing fixtures. Properties of these polymers are shown in Table 7.5.

Table 7.5 Properties of high-performance plastics

Property	ASTM Method	Acetal Copolymer	Acetal Homopolymer
Specific gravity (g/cm)	D792	1.410	1.425
Tensile strength at yield (psi)	D638	8,800	10,000
Elongation at break (%)	D638	60	25
Tensile modulus (psi, × 10)^5	D638	4.10	5.20
Flexural strength (psi)	D790	13,000	14,100
Flexural modulus (psi, × 10)	D790	3.75	4.10
Fatigue endurance limit (psi/no. of cycles)	D671	4200 per	5000 per
Compressive stress at 10% dilation (psi)	D695	16,000	18,000
Rockwell hardness (M)	D785	80	94
Notched Izod impact (ft-lb/in. of notch)	D256	1.3	1.4
Tensile impact (ft-lb/in.^{2})	D1822	70	94
Water absorption, 24 h immersion (%)	D570	0.22	0.26
Tabor abrasion, 1000-g load, Cs-17 wheel (mp/1000 cycles)	D1044	14	20

Table 7.6 Comparative data for poly(phenylene oxide) and reinforced PPO (Noryl)

Property	Unfilled	30% Glass Reinforced
Tensile strength (psi)	8,000	17,500
Elongation (%)	50	4
Compressive strength (psi)	14,000	18,000
Flexural strength (psi)	13,000	21,500
Flexural modulus (psi)	380,000	1,100,000
Notched Izod impact (ft-lb/in. of notch)	5	2
Coefficient of expansion (cm/cm, °C) $(10)^{-6}$	50	20
Heat deflection temperature (°F)	230	290

7.6 POLY(PHENYLENE OXIDE)

Poly(phenylene oxide) (PPO, Noryl), which is produced by the copper chloride-catalyzed oxidative coupling of a disubstituted phenol, was invented by A. Hay in 1956. The polymeric product is difficult to mold but the blend of PPO and polystyrene is readily injection-molded. This modified PPO is produced at an annual rate of 90,000 tons in the United States. This unusual engineering plastic, which is also called poly(phenylene ether), is used for window frames, beverage glasses, electrical switches, business machines, solar energy collectors, and wheel covers. The properties of PPO are shown in Table 7.6.

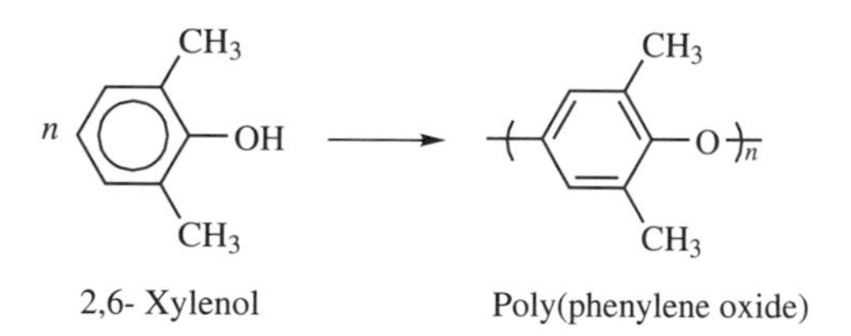

PPO-extruded sheet is being used for solar energy collectors, lifeguards on broadcasting towers, airline beverage cases, and window frames.

7.7 POLY(PHENYLENE SULFIDE)

Polyphenylene sulfide (PPS, Ryton) is produced by the condensation of *p*-dichlorobenzene and sodium sulfide at a rate of more than 35,000 tons annually. This high-melting polymer (550°F) is used for quartz halogen lamps, pistons, circuit boards, and appliances. The properties of PPS are shown in Table 7.7.

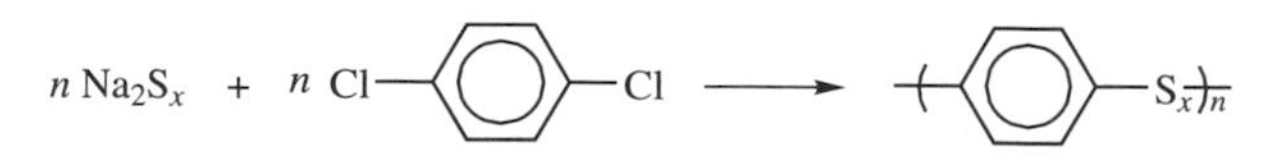

Table 7.7 Comparative data for poly(phenylene sulfide) and reinforced PPS (Ryton)

Property	Unfilled	30% Glass Reinforced
Tensile strength (psi)	9,500	19,500
Elongation (%)	1.5	1
Compressive strength (psi)	16,000	21,000
Flexural strength (psi)	14,000	29,000
Flexural modulus (psi)	550,000	1,200,000
Notched Izod impact (ft-lb/in. of notch)	0.5	1.4
Coefficient of expansion (cm/cm, °C) (10)	49	22
Heat deflection temperature (°F)	275	485

7.8 POLY(ARYL SULFONES)

Polysulfones exhibit excellent thermal oxidative resistance, as well as resistance to hydrolysis and other industrial solvents, and creep. The initial commercial polysulfones became commercially available in 1966 under the trade name Udel. It exhibits a reasonably high T_g of 375°F (Table 7.8).

Table 7.8 Commercially available polysulfones

Trade Name	Polymer Unit	T_g (°F)
Astrel (3m Corp.)	$-[C_6H_4-C_6H_4-SO_2]_n-$ $-[C_6H_4-O-C_6H_4]_n-SO_2-$	545
Poly(ether sulfone) 720 P (ICI)	$-[C_6H_4-C_6H_4-SO_2]_n-$ $-[C_6H_4-O-C_6H_4-SO_2]_n-$	480
Poly(ether sulfone) 200 P (ICI)	$-[C_6H_4-O-C_6H_4-SO_2]_n-$	445
Udel (Union Carbide)	$-[C_6H_4-C(CH_3)_2-C_6H_4-O-C_6H_4-SO_2-C_6H_4-O]_n-$	375

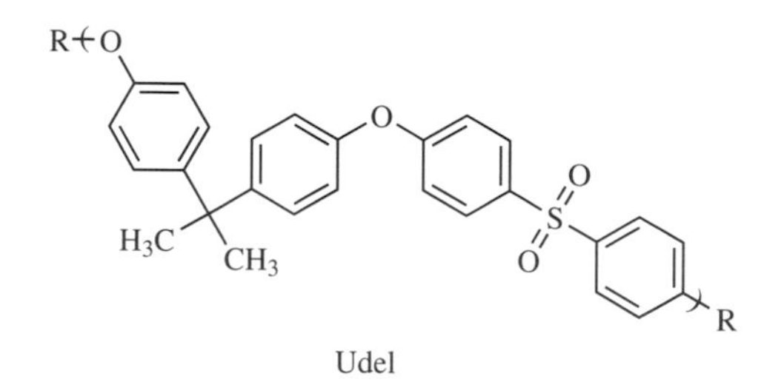

Udel

In 1976, Union Carbide made available a second-generation polysulfone under the trade name of Radel. This polysulfone exhibited greater chemical/solvent resistance, a greater T_g of 430°F, greater oxidative stability, and good toughness.

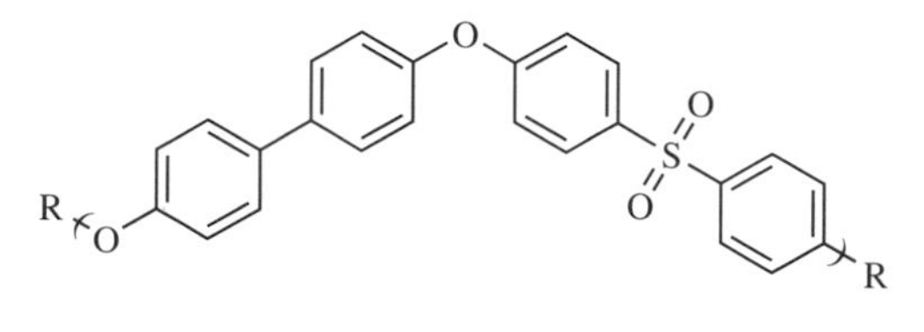

Radel

Complex-shaped objects can be made through injection molding without need for additional machining and other procedures. Films and foil are used for flexible printed circuitry. Polysulfones are also used for ignition components, hair dryers, cook ware, and structural foams. Because of their good hydrolytic stability, good mechanical properties, and high thermal endurance they are good candidate materials for hot water and food handling equipment, alkaline battery cases, surgical and laboratory equipment, life support parts, autoclavable trays, tissue culture bottles, and surgical hollow shapes, and film for hot transparencies. Their low flammability and smoke production, again because of their tendency for polycyclic formation on thermolysis and presence of moieties that are partially oxidized, makes them useful as materials for aircraft and the automotive industry.

Table 7.9 Comparative data for polysulfone and reinforced polysulfone (Vitrex, Udel)

Property	Unfilled	30% Glass Reinforced
Tensile strength (psi)	—	14,500
Elongation (%)	3	1.5
Compressive strength (psi)	40,000	19,000
Flexural strength (psi)	—	20,000
Flexural modulus (psi)	400,000	1,000,000
Notched Izod impact (ft-lb/in. of notch)	1	1
Coefficient of expansion (cm/cm, °C) (10)	55	25
Heat deflection temperature (°F)	345	350

Table 7.10 Ceiling (top) use temperatures for selected engineering thermoplastics

Polymer	Ceiling Use Temperature[a] °C	°F
Polybenzimidazole	400	752
Polyimides	260 (prolonged)	500
	480 (short)	895
Kevlar	500	932
Nomex	360	680
Udel	160 (prolonged)	320
	800 (short)	1472
PPS	150	302
PPO	130	266
PC	130	266
Nylon 6,6	100	212
PBT	90	194
ABS	80	176

[a]Unless noted, the temperatures are for extended use.

Typical properties are given in Table 7.9. Table 7.10 contains ceiling or upper use temperatures for selected engineering thermoplastics.

For complex-shaped objects since the polysulfones can be injection-molded into these shapes without the machining and other procedures that are required for metals. Films and foil of polysulfones are used for flexible printed circuitry.

7.9 POLYIMIDES

Intractable polyimides (PI) produced by the condensation of pyromellitic anhydride and various polyimides have been available for several years under the trade names of Kaptan and Kinel (Table 7.11). Moldable PI is now available. These products are high-melting and offer good stiffness, transparency, impact and mechanical strength, high flame resistance, low smoke generation, and broad chemical resistance. Some of these properties are expected. The high flame resistance is at least in part derived from the presence of already partially or largely oxidized atoms in the product. The low smoke generation is partially derived from the largely cyclic structure with other cyclic structures predictable from the product structure if exposed to sufficient heat. These cyclic structures often give products that are not evolved with good char formation when the material is exposed to ordinary flame conditions. The general good mechanical properties are a result of the

Table 7.11 Comparative data for polyimide and reinforced PI

Property	Unfilled	30% Graphite Filled
Tensile strength (psi)	13,000	7,500
Elongation (%)	9	3
Compressive strength (psi)	35,000	17,500
Flexural strength (psi)	24,000	14,000
Flexural modulus (psi)	500,000	700,000
Notched Izod impact (ft-lb/in. of notch)	1.5	0.7
Coefficient of expansion (cm/cm, °C) $(10)^{-6}$	50	38
Heat deflection temperature (°F)	600	680

presence of strong double bonds present within polycyclic structures composing the polymer backbone plus the presence of strongly polar bonding units that allow the formation of good interactions between chains. Furthermore, the structure is largely rigid with good dimensional stability along the polymer backbone. Any flexibility is gained largely from the presence of the ether linkages for the polyetherimides and the presence of methylene units for the polyimides. These products offer good stable melt viscosities even after recycling several times. They can be processed using a variety of techniques including formation of sections as thin as 5 mils.

A polyether imide (PEI, Ultem) and a polyamide imide (PAI, Torlon) are available commercially (Table 7.12). The latter has been used to fabricate a Ford prototype engine.

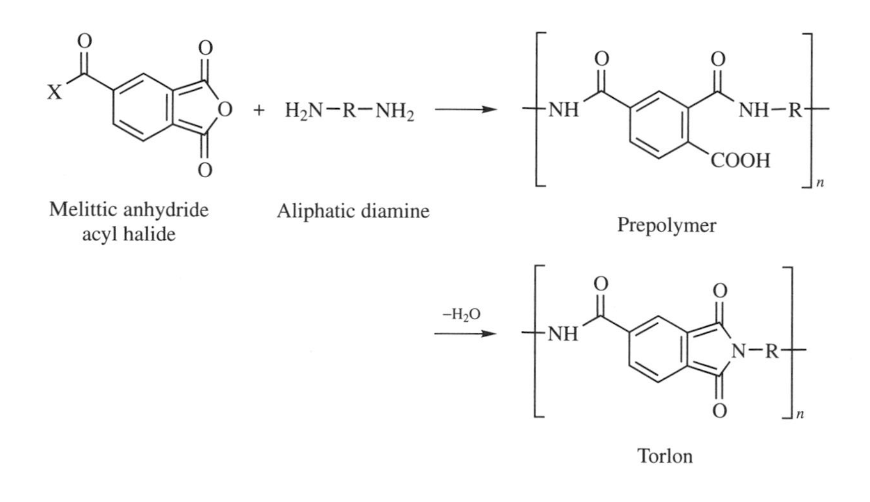

Table 7.12 Comparative data for poly(ether imide) and reinforced PEI

Property	Unfilled	30% Glass Filled
Tensile strength (psi)	15,200	24,500
Elongation (%)	7.5	0.2
Compressive strength (psi)	20,300	23,500
Flexural strength (psi)	21,000	33,000
Flexural modulus (psi)	480,000	1,200,000
Notched Izod impact (ft-lb/in. of notch)	1.0	2.0
Head deflection temperature (°F)	392	410

7.10 POLY(ETHER ETHER KETONE) AND POLYKETONES

Aromatic polyketones are semicrystalline materials that contain ketone groups generally flanked by aromatic units. They have good thermal stabilities, as well as offering good mechanical properties, flame resistance, impact resistance, and resistance to the environment.

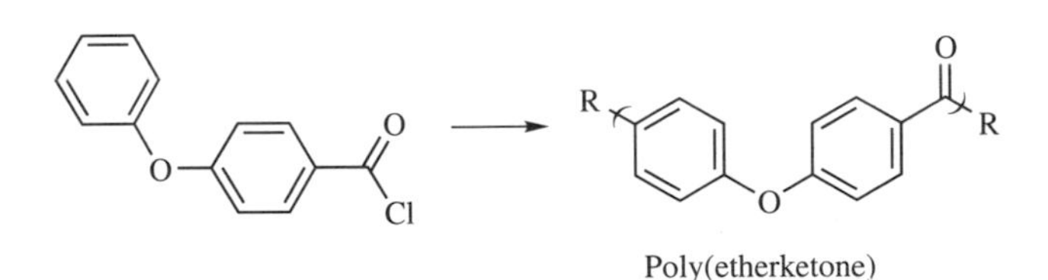

Poly(etherketone)

Poly(ether ether ketone), PEEK, was invented by J. Rose, who is also the inventor of poly(ether sulfones). Because of the presence of the polar carbonyl (C=O) stiffening groups and the aromatic stiffening groups, aromatic polyketones have an excellent resistance to elevated temperatures. The ether group(s) contribute flexibility and moldability, and the presence of the propyl (CH_3—C—CH_3) group also imparts flexibility and discourages crystallization. The structure of PEEK again illustrates how combinations of structures contribute to overall properties.

Poly(etherketone), PEK, was introduced by Raychem in the 1970s. A good solvent or other conditions are employed to keep the polymer in solution, allowing polymer growth to occur. Most polymerizations require that the reactants remain mobile, through solution or being melted, so that the individual units involved in the reaction can get together. Rapid precipitation of growing polymer chains often results in the formation of only small chains.

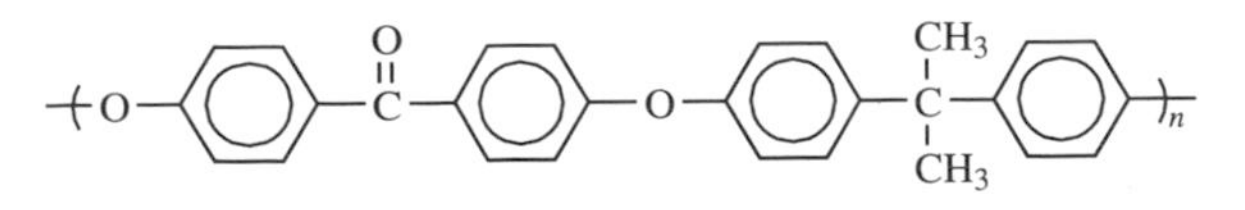

PEEK

Table 7.13 Comparative data for poly(ether ether ketone) and reinforced PEEK

Property	Unfilled	30% Graphite Filled
Tensile strength (psi)	10,044	23,500
Elongation (%)	100	3
Compressive strength (psi)	—	22,500
Flexural strength (psi)	—	39,500
Flexural modulus (psi)	—	1,250,000
Notched Izod impact (ft-lb/in. of notch)	1.6	2.7
Heat deflection temperature (°F)	320	594

Applications in the chemical industry include use as compressor plates, valve seats, thrust washers, bearing cages, and pump impellers. In the aerospace industry they are employed as aircraft fairings, fuel valves, and ducting. They are also used in the electrical industry as wire coating and semiconductor wafer carriers. Properties are given in Table 7.13.

Aliphatic polyketones are made from the reaction of olefin monomers and carbon monoxide using a variety of catalysts. Shell commercialized a terpolymer of carbon monoxide, ethylene, and a small amount of propylene in 1996 under the trade name of Carilon. They have a useful range of T_g (60°F, 15°C) and T_m (390°F, 200°C) that corresponds to the general useful range of use temperatures for most industrial applications. The presence of polar groups cause the materials to be tough with the starting materials readily available.

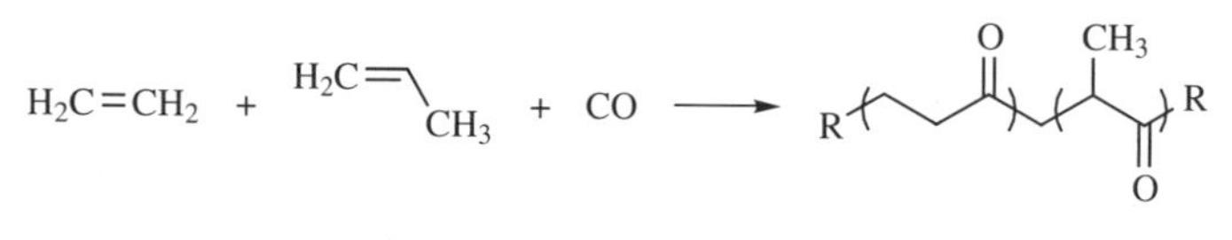

Aromatic polyketone

7.11 POLYSILOXANES

Polysiloxanes, also called silicones, offer a combination of electrical, chemical, and mechanical properties not found for any other class of polymers. They exhibit relatively high oxidative and thermal stability, low power loss, and unique flow and stress/strain properties, are inert to most ionic and inert reagents, exhibit low flow changes as temperature varies, are nonflammable, and have good shear stability, high compressability, low surface tension, and so on. They have an exceptionally wide use temperature from about −120°C to 200°C or an operating temperature range of about 300 degrees. This temperature range allows their use in extreme temperatures from Nome, Alaska to the Sharaha. The first footprints on the moon were made with polysiloxane elastomeric boots.

The first polysiloxanes were unstable, but this instability was overcome by capping the end groups. Almost all of the commercial polysiloxanes are based on polydimethylsiloxane, with trimethylsiloxy end groups.

$$(H_3C)_3Si-O-[Si(CH_3)_2-O]_n-Si(CH_3)_3$$

The reason for the low-temperature flexibility is because of a very low T_g, about $-120°C$, which is the result of the methyl groups attached to the silicon atoms being free to rotate, causing the oxygen and other surrounding atoms to "stay away" creating a flexible chain.

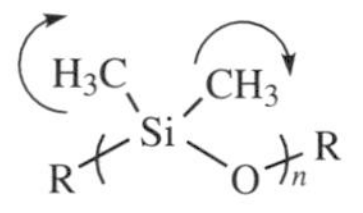

The viscosity or resistance to flow increases as the number of repeat units increase, but physical properties such as surface tension and density remain about the same after a DP of about 25. The liquid surface tension is lower than the critical surface tension of wetting, resulting in the polymer spreading over its own absorbed films. The forces of attraction between polysiloxane films is low, resulting in the formation of porous films that allow oxygen and nitrogen to readily pass though, but not water. Thus, semipermeable membranes, films, have been developed that allow drivers to "breath air under water" for short periods.

As noted above, viscosity increases with DP, allowing many of the uses to be grouped according to chain length. Low-viscosity fluids with DP values of 2–30 are used in antifoams and in the flow control of coatings applications. These applications are the direct consequence of the low attractions between polysiloxane chains, which, in turn, are responsible for their low surface tension. Thus, they encourage a coatings material to flow across the surface, thereby filling voids, corners, and crevices. Their good thermal conductivity and fluidicity at low temperatures allows their use as low-temperature heat exchangers and in low-temperature baths and thermostats.

Viscous fluids correspond to a DP range of about 50–400. These materials are employed as mold release agents for glass, plastic, and rubber parts. They are good lubricants for most metal to nonmetal contacts. They are used as dielectric fluids (liquids) in a variety of electrical applications including transformers and capacitors; as hydraulic fluids in vacuum and hydraulic pumps; in delicate timing and photographic devices; as antifoam agents; as components in protective hand creams; as toners in photocopiers; in oil formulations when mixed with thickeners; and in inertial guidance systems. High-performance greases are formed by mixing the polysiloxane fluids with polytetrafluoroethylene or molybdenum disulfide.

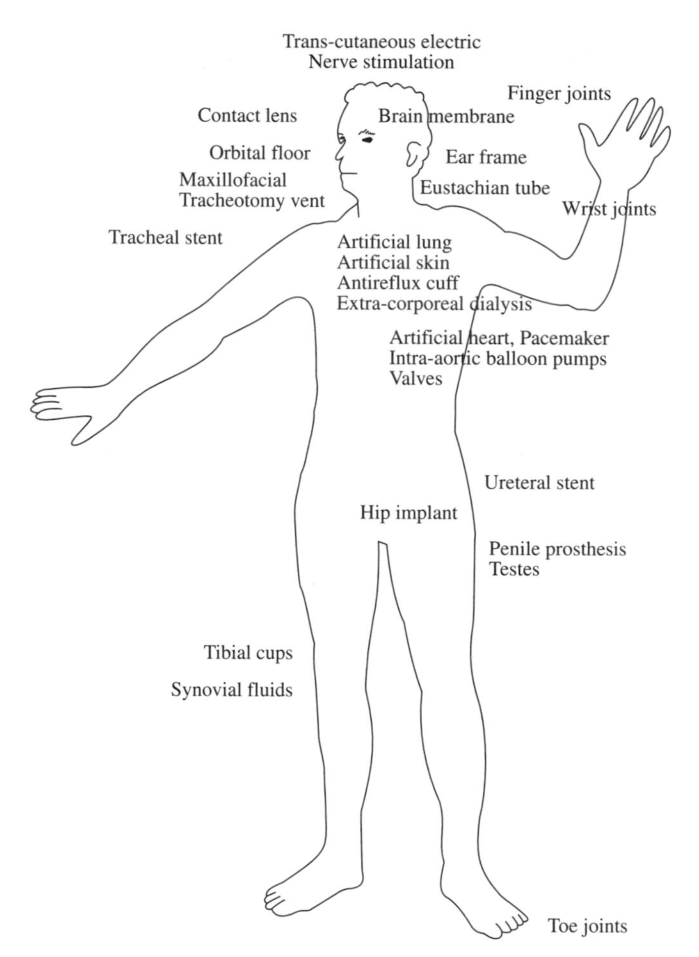

Figure 7.3. Polysiloxanes used as human body parts.

Brake fluids are formulated from polydimethylsiloxane fluids with DP values of about 50. High-viscosity fluids with DP values of about 700–6000 are used as damping fluids for weighting meters at truck stops. They act as liquid springs in shock absorbers. The longer-chained fluids are used as impact modifiers for thermoplastic resins and as stationary phases in gas chromatography.

As with the alkanes, even longer chains form the basis for solid polysiloxanes that according to design can be classified as thermoplastics, engineering thermoplastics, elastomers, and (when cross-linked) thermosets. Solid polysiloxanes are used in a variety of applications including use as sealants, thermostripping, caulking, dampening, O-rings, and window gaskets. Weatherstripping on cooling units, trucks, and automobiles is often made of polysiloxanes.

Room-temperature-vulcanizing (RTV) silicon rubbers make use of the room-temperature reaction of certain groups that can be placed on polydimethylsiloxanes that react with water. When exposed to water, such as that normally present in the atmosphere, cross-links are formed, thereby creating an elastomeric product.

The first contact lenses were based on poly(methyl methacrylate). While they could be polished and machined, they did not permit gas exchange and were rigid. By early 1970s these were replaced by soft contact lenses containing cross-linked poly(2-hydroxyethyl methacrylate), HEMA. These so-called disposable lenses do permit gas exchange. More recently, Salamone and co-workers developed contact lenses based on the presence of siloxane units. Polysiloxanes have good gas permeability. These polymers are referred to as Tris materials and are generally copolymers containing units as shown below.

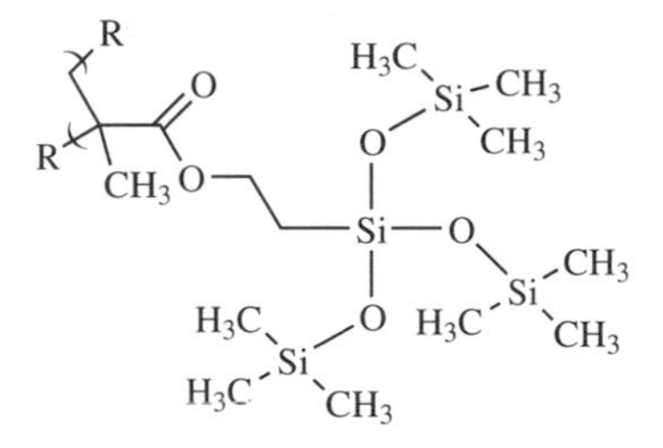

Polysiloxanes are widely employed as biomaterials. Figure 7.3 shows some of the applications where polysiloxanes play an integral role. Artificial skin can be fabricated from a bilayer fabricated from a cross-linked mixture of bovine hide, collagen, and chondroitin sulfate derived from shark cartilage with a thin top layer of polysiloxane. The polysiloxane acts as a moisture and oxygen-permeable support and to protect the lower layer from the "outer world." A number of drug delivery systems use polysiloxanes because of the flexibility and porous nature of the material.

7.12 OTHER ENGINEERING THERMOPLASTICS

Polybenzimidazoles, which were developed for aerospace applications, possess good stiffness and strength but are generally difficult to process. They have one of the highest use temperatures (about 750°F) of any organic polymer.

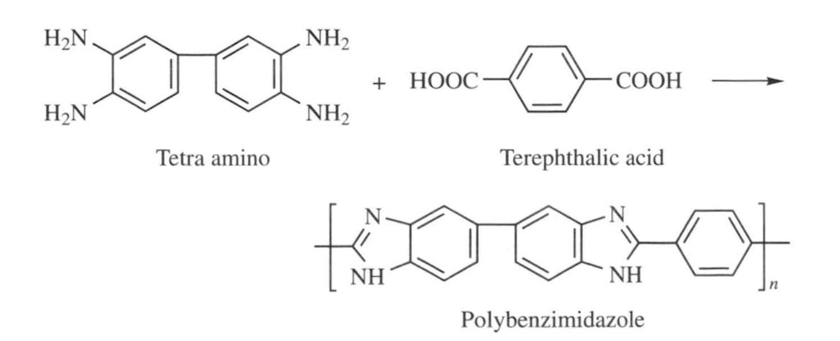

Polyphosphazenes are in the early stages of development and have many potential and actual uses. They exhibit a broad service range temperature (–85°F to 250°F), have outstanding resistance to fuels, oils, and chemicals, and possess good mechanical properties.

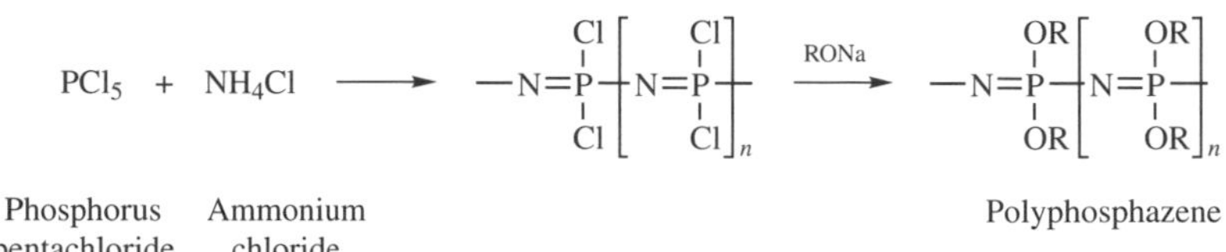

Polytetrafluoroethylene and other highly halogenated (i.e., containing Group 7A substituents like chlorine, fluorine, and bromine) polymers are used as coatings and as thermoplastics. Polytetrafluoroethylene (PTFE, Teflon) is insoluble in all organic solvents and is processed only by ram extrusion, cutting, machining, and sinter molding techniques.

$$-\!\!\left[CF_2CF_2\right]\!\!-_n$$

Polytetrafluoroethylene (PTFE)

PTFE has outstanding resistance to chemical attack. In fact, pipes that are employed to convey molten, liquid sodium metal are made from PTFE. Although PTFE is normally stable to metallic sodium, a very violent exothermic reaction can occur when sodium strips off the fluoride ions on PTFE to form sodium fluoride, which may create a fire and/or explosion. PTFE has a high impact strength, but it cold flows (creeps) and has low wear resistance and tensile strength. Because of its excellent "lubricity" (low friction) and outstanding hydrophobic (water-hating) nature, it is widely employed in easy-clean and nonstick cookware, such as frying pans, muffin pans, and cake pans.

PTFE is also used in high-temperature cable insulation and molded electrical applications. Reinforced PTFE is used as seals and bushings in compressor hydraulic applications, pipe lines, and automotive applications, as a specialty tape to ensure closure of pipe fittings, as a seal for gasket applications, and in the laboratory as a covering for magnetic stirrers, on stopcocks in liquid delivery devices, and on laboratory ware such as beakers, flasks, and condensers.

GLOSSARY

Ablative: A process in which the surface of the plastic is degraded and removed.

ABS: A tough copolymer containing repeating units of acrylonitrile, butadiene, and styrene.

Arylate: Trade name for the reaction product of bisphenol A and terephthalic acid.

Astrel: Trade name for a poly(aryl sulfone).

Bayblend: Trade name for a blend of PC and flexibilizing polymers.

Bisphenol A:

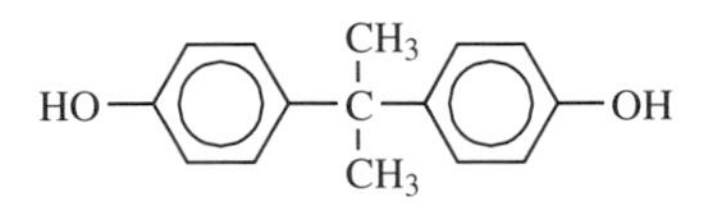

Cadon: Trade name for a heat-resistant terpolymer of styrene, acrylonitrile, and maleic anhydride.

Capping: Reaction with end groups in a polymer.

Carbonyl group: C=O.

Celcon: Trade name for an acetal copolymer.

Delrin: Trade name for POM.

Engineering plastic: A plastic with a high modulus and high melting point that can be used in place of metals in some applications.

Heat deflection temperature (HDUL): The temperature at which a heated beam deflects a specific distance.

Heterocyclic: A cyclic compound consisting of carbon and other atoms such as nitrogen.

Impact resistance: Toughness, the ability to withstand mechanical shock.

Inorganic polymer: A polymer, such as siloxanes and phosphazenes, that does not have carbon atoms in its backbone.

Kaptan: Trade name for PI.

Kevlar: Trade name for an aromatic nylon.

Kinel: Trade name for PI.

Ladder polymer: A polymer in which the backbone is a double chain.

Noryl: Trade name for a blend of polystyrene and PPO.

Nylon: A polyamide produced by the condensation of a diamine and a dicarboxylic acid or by the polymerization of a lactam.

Nylon 6: The product of the polymerization of caprolactam.

Nylon 6,6: The reaction product of hexamethylenediamine and adipic acid.

Phosgene: $COCl_2$.

Polyacetal (POM): A polymer of formaldehyde.

Polybenzimidazole (PBI): A heat-resistant heterocyclic polymer.

Polybutylene terephthalate (PBT): The reaction product of butylene glycol and terephthalic acid.

Polycarbonate (PC): The reaction product of phosgene and bisphenol A.

Poly(ether ether ketone) (PEEK): An aromatic thermoplastic having ether and carbonyl groups.

Poly(ethylene terephthalate) (PET): The reaction product of ethylene glycol and terephthalic acid.

Polyimide (PI): A polymer produced by the condensation of pyromellitic anhydride and diamines.

Polyoxymethylene (POM): Polyacetal.

Poly(phenylene oxide) (PPO): A polymer produced by the copper chloride-catalyzed oxidative coupling of 2,6-xylenol.

Poly(phenylene sulfide) (PPS): A polymer produced by the condensation of sodium sulfide (NaS) and *p*-dichlorobenzene.

Polyurethane (PUR): The reaction product of a diisocyanate and a dihydric alcohol.

Quartz: Silicon dioxide.

Reaction injection molding (RIM): A process in which the reactants are introduced and polymerized in the mold.

Ryton: Trade name for PPS.

Silicone: Siloxane.

Siloxane:

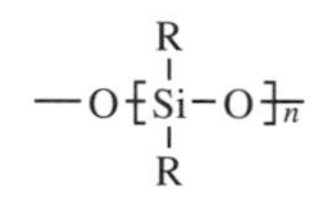

Tolylene diisocyanate (TDI):

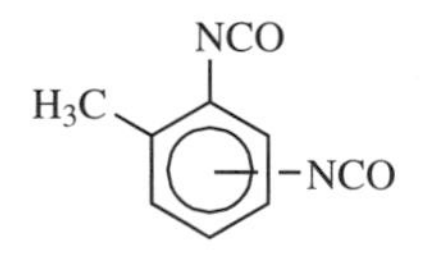

Torlon: Trade name for a poly(amide imide).

Udel: Trade name for a poly(aryl sulfone).

Ultem: Trade name for a poly(ether imide).

Victrex: Trade name for a poly(aryl sulfone).

Xenoy: Trade name for a blend of PC and PBT.

REVIEW QUESTIONS

1. Which of the following are engineering resins: polystyrene, polyimide, nylon, polycarbonate?
2. Which has the higher melting point: nylon 6,6 or Kevlar?
3. What is the big advantage of RIM?
4. Which has the higher melting point: PET or PBT?
5. Why must polyacetal (POM) be capped?

6. What is the advantage of a clear polycarbonate sheet over a sheet of poly(methyl methacrylate)?
7. What is the advantage of polymer blends?
8. Which of the following is a stiffening group in a polymer chain: SO_2, CO, CH_2, O?
9. What polymer is used in prototype automobile combustion engines?
10. What is a ladder polymer?
11. What is the principal difference in the structure of a quartz or silica sand and a siloxane polymer (silicone)?
12. Why is polyphosphazene preferred over natural rubber in the Alaskan oil fields?

BIBLIOGRAPHY

Amato, I. (1997). *Stuff: The Materials the World Is Made Of*, Harper Collins, New York.

Ball, P. (1994). *Designing the Molecular World*, Princeton University Press, Princeton, NJ.

Bottenbruch, L., ed. (1996). *Engineering Thermoplastics: Polycarbonates, Polyacetals, Polyesters, Cellulose Esters*, Hanser-Gardner, Cincinnati.

Carraher, C. (2003). *Polymer Chemistry*, Marcel Dekker, New York.

Carraher, C., and Moore, J. A., (1983). *Modification of Polymers*, Plenum, New York.

Carraher, C., and Preston, J. (1982). *Interfacial Synthesis*, Vol. 3, *Recent Advances*, Marcel Dekker, New York.

Carraher, C., Swift, G., and Bowman, C. (1997). *Polymer Modification*, Plenum, New York.

Collier, B. (2000). *Understanding Textiles*, Prentice-Hall, Englewood Cliffs, NJ.

Craver, C., and Carraher. C. (2000). *Applied Polymer Science*, Elsevier, New York.

Fakirov, S. (1999). *Transreactions in Condensation Polymers*, Wiley, New York.

Fourne, F. (1998). *Synthetic Fibers*, Hanser-Gardner, Cincinnati.

Kadolph, S., Langford, A. (2001). *Textiles*, Prentice-Hall, Englewood Cliffs, NJ.

Mann, D. (1999). *Automotive Plastics and Composites*, Elsevier, New York.

Mathias, L., and Carraher, C. (1984). *Crown Ethers and Phase Transe Catalysis in Polymer Science*, Plenum, New York.

Millich, F., and Carraher, C., (1977). *Interfacial Synthesis*, Vols. I and II, Marcel Dekker, New York.

Mittal, K. L. (2001). *Polyimides and Other High Temperature Polymers*, Leiden, Netherlands.

Mittal, K. L. (2001). *Metallized Plastics, VSP*, Leiden, Netherlands.

Pritchard, G. (1995). *Anti-Corrosion Polymers: PEEK, PEKK and Other Polyaryls*, Rapra Technology, Charlotte, NC.

Sandler, S. R., and Karo, W. (1998). *Polymer Synthesis* (three volumes), Orlando, FL.

Thompson, T. (2000). *Design and Applications of Hydrophillic Polyurethanes*, Technomics, Lancaster, PA.

ANSWERS TO REVIEW QUESTIONS

1. Polyimide, nylon, and polycarbonate.
2. Kevlar, it is an aromatic nylon.
3. Large molded parts can be made in relatively inexpensive molds in one step.
4. PET. (PBT has more methylene flexibilizing groups in its repeating unit.)
5. The uncapped polymer decomposes to formaldehyde when heated.
6. Polycarbonate is tougher.
7. They are more readily molded and produced in available processing equipment.
8. SO_2 and CO; CH_2 and O are flexibilizing groups.
9. Poly(amide imide).
10. A polymer with a double polymer chain.
11. They have similar backbones, but the siloxane polymer has organic pendant groups on the silicon atoms.
12. The temperature in the Alaskan oil fields is often below the T_g of natural rubber but above the T_g of polyphosphazenes. The latter are flexible at very low temperatures at which natural rubber is brittle.

8

THERMOSETS

8.1 INTRODUCTION

Thermoplastics and thermosets form the two major groups of plastics. They share many common processing sequences (see Section 6.19), and many are plastic in having properties offering flexible dimensional stability; that is, they can be bent to some extent, yet are also rigid. "Plastic" cups, rulers, bottle caps, and

Giant Molecules: Essential Materials for Everyday Living and Problem Solving, Second Edition,
by Charles E. Carraher, Jr.
ISBN 0-471-27399-6

Table 8.1 U.S. production of thermosetting resins, 2002

Thermosetting Resin	Production (Millions of Pounds)
Epoxies	660
Melamines	290
Phenolics	3940
Polyesters	2400
Ureas	2580

Source: American Plastics Council.

so on, are representative examples of plastics. Thermoplastics can be further divided according to general-purpose plastics and engineering plastics. The topics of general-purpose plastics and engineering plastics are covered in Chapters 6 and 7. Thermosets are discussed in this chapter.

The commercialization of phenol–formaldehyde plastics preceded the large-scale production of most of the commercial synthetic thermoplastics. Vulcanized rubber, which was introduced by Charles Goodyear in 1838, was a low-density cross-linked elastomer. Hard rubber, which was invented by Nelson Goodyear, was a high-density cross-linked plastic. The cross-linking agent in both of these thermosets was sulfur.

Glyptal coatings, developed by W. Smith at the beginning of the twentieth century, were produced by the condensation of glycerol ($HOCH_2CH(OH)CH_2OH$) and phthalic anhydride ($C_6H_4C_2O_3$). Since the secondary hydroxyl group in glycerol was less reactive than the terminal primary hydroxyl groups, the prepolymer was a linear prepolymer with one unreacted hydroxyl group in each repeating unit. This thermoplastic prepolymer was converted to a thermoset by heating after it was applied to the surface of metals.

Oleoresinous paints were also applied as thermoplastic prepolymers. These prepolymers were cross-linked by an addition polymerization reaction with oxygen in the presence of driers. The term thermoset applies to all cross-linked polymers, regardless of whether the cross-links were formed by heating, irradiation, or chemical reaction. Table 8.1 contains the U.S. production of thermosets for the year 2002.

8.2 PHENOLIC RESINS

Products that at the time were called "goos, gunks, and messes" were produced by some of the world's most renowned chemists in the late 1890s. Unfortunately, these chemists were unfamiliar with the importance of functionality. Later chemists, such as Leo Baekeland, recognized that the combination of reactants with difunctional groups such as diols ($R(OH)_2$) and dicarboxylic acids ($R(COOH)_2$) produced linear polymers. They also knew that a trifunctional reactant, such as phenol (C_6H_5OH), when reacted with a difunctional reactant, such as formaldehyde, ($H_2C{=}O$), produced an infusible cross-linked polymer. The three reactive sites in phenol are in the 2, 4, and 6 positions for reaction with formaldehyde.

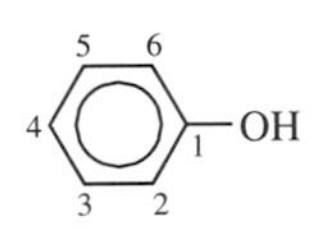

The problem of uncontrolled cross-linking was solved by Baekeland, who used an insufficient amount of the difunctional reactant (formaldehyde) with the trifunctional phenol, in the presence of an acid, to produce a linear prepolymer. This prepolymer contained reactive centers that could react with additional formaldehyde from hexamethylenetetramine in a controlled secondary reaction.

Thus, Baekeland was able to place a mixture of the prepolymer and hexamethylenetetramine in a mold and obtain a thermoset article by heating the mixture. The concept recognized by Baekeland may be demonstrated by the following reactions:

The thermoplastic prepolymer obtained from the reaction of phenol with an insufficient amount of formaldehyde is called a novolak resin. Most phenolic molding powders contain novolak PF, hexamethylenetetramine, pigment, and a filler, such as wood flour. Wood flour is a finely divided, fibrous wood filler obtained by the attrition grinding of debarked wood.

A linear product called a resole resin is also obtained by reacting equivalent amounts of the reactants in the presence of an alkali. This product is a viscous liquid while it is cooled. This liquid prepolymer is converted to a solid when heated or when an acid is added (Figure 8.1).

Linear PF resins are also obtained by the condensation of formaldehyde with a para-substituted phenol, such as *p*-phenylphenol, which has a functionality of 2.

The properties of phenolic resins are shown in Table 8.2.

About 2 million tons of phenolic polymers are produced annually in the United States, using the same formulations developed by Baekeland in the early 1900s. Molded wood flour-filled phenolic resins are used for electrical insulators. Almost 700,000 tons of phenolic resins are also used annually in the United States as adhesives for plywood. Phenofoam, which is produced by adding a gaseous propellant to PF, has the lowest flammability of all commercial plastic foams.

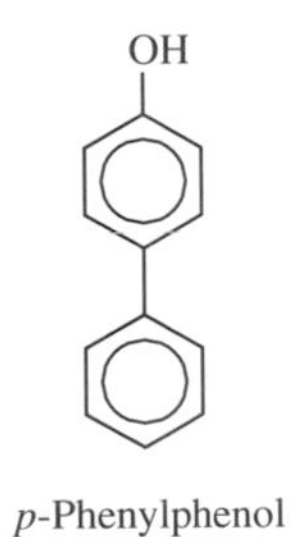

p-Phenylphenol

PF resins are used as molding resins (about 35%), laminating resins (about 10%), bonding resins, coatings and adhesives (about 35%), and ion-exchange

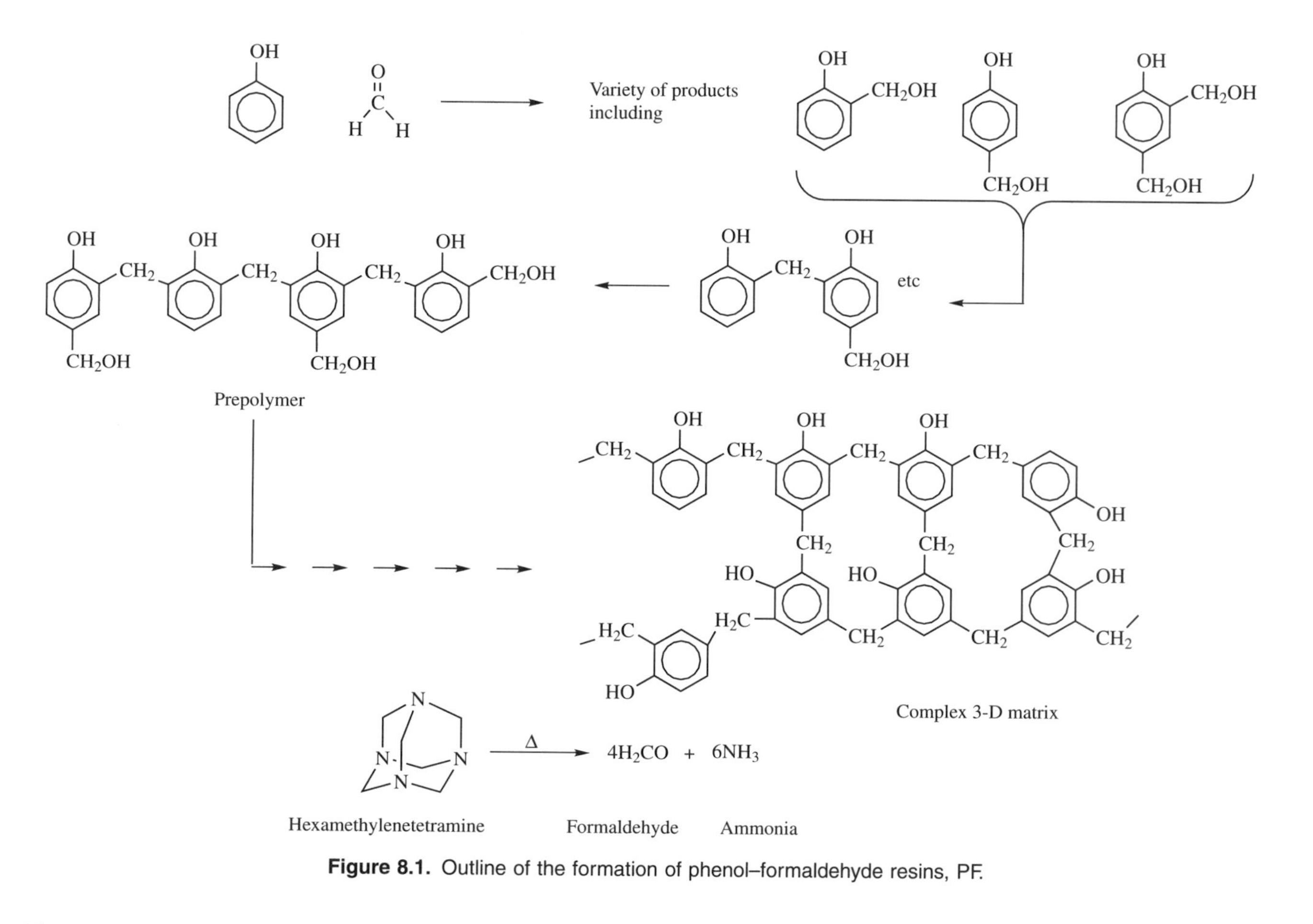

Figure 8.1. Outline of the formation of phenol–formaldehyde resins, PF.

Table 8.2 Properties of typical phenolic and amino plastics

Property	PF, Wood Flour Filled	UF, Cellulose Filled	MF, Cellulose Filled
Processing temperature (°F)	350	300	350
Molding pressure (10^3 psi)[a]	15	15	15
Mold shrinkage (10^{-3} in./in.)	6	10	10
Heat deflection temperature under flexural load of 264 psi (°F)	350	275	300
Maximum resistance to continuous heat (°F)	300	260	250
Coefficient of linear expansion (10^{-6} in./in., °F)	20	15	15
Compressive strength (10^3 psi)	25	35	28
Impact strength Izod (ft-lb/in. of notch)[b]	0.4	0.3	0.3
Tensile strength (10^3 psi)	7	10	8
Flexural strength (10^3 psi)	10	12	9
% elongation	0.5	1	3
Tensile modulus (10^3 psi)	1200	1200	1200
Flexural modulus (10^3 psi)	1000	1400	1100
Rockwell hardness	M105	E90	E95
Specific gravity	1.4	1.5	1.6
% water absorption	0.7	0.5	0.5
Dielectric constant	6	6	5
Dielectric strength (V/mil)	325	350	275
Resistance to chemicals at 75°F[c]			
Nonoxidizing acids (20% H_2SO_4)	S	Q	Q
Oxidizing acids (10% HNO_3)	S	U	U
Aqueous salt solutions (NaCl)	S	S	S
Polar solvents (C_2H_5OH)	S	S	S
Nonpolar solvents (C_6H_6)	S	Q	Q
Water	S	S	S
Aqueous alkaline solutions (NaOH)	U	Q	Q

[a] psi/0.145 = kPa (kilopascals).
[b] ft-lb/in. of notch/0.0187 = cm · N/cm of notch.
[c] S, satisfactory; Q, questionable; U, unsatisfactory.

resins. The major adhesive use is in the manufacture of plywood. Phenolic resins are widely used in the coatings industry in varnishes. They are employed as bonding resins in the production of abrasive (grinding) wheels, sandpaper, and brake linings. Decorative laminates formed from impregnating wood and paper for countertops, printed circuits, and wall coverings account for about 10% of its uses. The impregnated material is dried in an oven, then hot pressed and molded into the desired shape. Fillers are typically used in molding applications to improve impact properties and to reduce cost. Common fillers are fiberglass, nylon and other fibers, cloth, and cellulosic materials, such as cotton and wood flour. Because of its nonconductive nature, PF is used in molding TV and radio cabinets, appliance parts, and automotive parts.

The ability of PF resins to withstand high temperatures for a brief time has led to their use in the construction of missile nose cones. Metals vaporize, but PF resins decompose, leaving a carbon (similar to graphite) protective coating.

8.3 UREA RESINS

Resins are produced by the reaction of tetrafunctional urea and formaldehyde and were described by Holzer in 1884 and John and Pollak in 1918. In spite of Baekeland's early work with thermosets, these polymers were not commercialized until the mid-1920s.

$$\left(H_2N-\overset{\overset{\displaystyle O}{\|}}{C}-NH_2\right)$$

Urea–formaldehyde resin (UF) is produced by the condensation of urea with an insufficient amount of formaldehyde under alkaline conditions. The liquid resin is mixed with fillers, such as wood flour, α-cellulose, and an acid. The mixture is dried and densified on a rubber mill, cooled, and granulated. Almost 1.2 million tons of UF are produced annually in the United States.

$$H_2NCONH_2 \xrightarrow{CH_2O} HOCH_2NHCONH_2 \xrightarrow{CH_2O} HOCH_2NHCONHCH_2OH$$

$$\longrightarrow \text{Urea–formaldehyde resin (UF)}$$

Urea resins are much lighter in color than the dark phenolic resins. They are used as adhesives for particleboard and laminated paper and foams. α-Cellulose-filled urea resins are used as molding resins.

Unless stabilizers are present, UF adhesives and foams may release some formaldehyde, which is considered to be toxic. The properties of UF and melamine formaldehyde (MF) are summarized in Table 8.2. UF and MF are called amino plastics.

Urea resins offer less heat and moisture resistance and are softer than the melamine resins, but they are also generally less expensive (about 50¢/lb compared to about 75¢/lb for PF resins). Almost all urea-molded products are cellulose-filled. Because of their good solvent and grease resistance, surface hardness, mar resistance, and easy colorability, they are widely employed for bonding wood in furniture and plywood. They are also added to cotton and rayon textiles to impart crease resistance, shrinkage control, water repellency, fire retardance, and stiffness. Urea-based enamels are used for coating kitchen appliances, including dishwashers and refrigerators. These enamels are called baked enamels and contain urea resins as well as alkyd resins. About 60% of the urea–formaldehyde resins are now employed in the production of particleboard.

8.4 MELAMINE RESINS

Melamine, which is shown by the following structure, has six reactive sites (two on each amine group). It may be condensed with formaldehyde to produce light-colored, heat-resistant plastics (Figure 8.2).

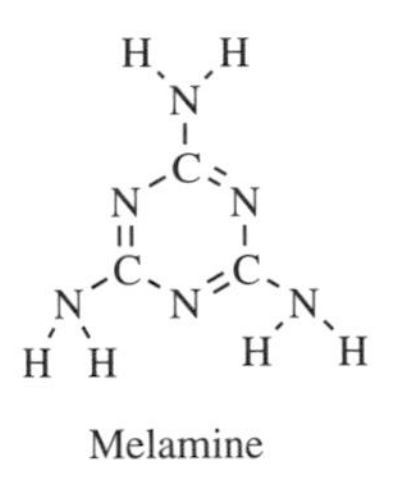

Melamine

Melamine was characterized by Liebig in 1843, but in spite of the information available on phenolic and urea resins, melamine resins (MF) were not produced commercially until the mid-1900s. MF resins are used for the production of decorative laminates. α-Cellulose-filled MF resins are used for molding dinnerware (Melmac). The properties of a cellulose-filled MF molded plastic are shown in Table 8.2. The annual production of MF in the United States is about 100,000 tons.

A distinct advantage of urea and melamine resins over phenolic PF resins is the fact that they are clear and colorless, thus aiding in ease of coloration. Melamine resins have poorer impact strength and moisture and heat resistance compared to PFs, but they are harder. Melamine resins are predominantly cellulose-filled, although fiberglass and cotton fabrics are also employed. The production of dinnerware (such as plates, cups, and serving bowls) using cellulose-filled UF resins is the single largest use for these resins. Melamine resins are also used for the production

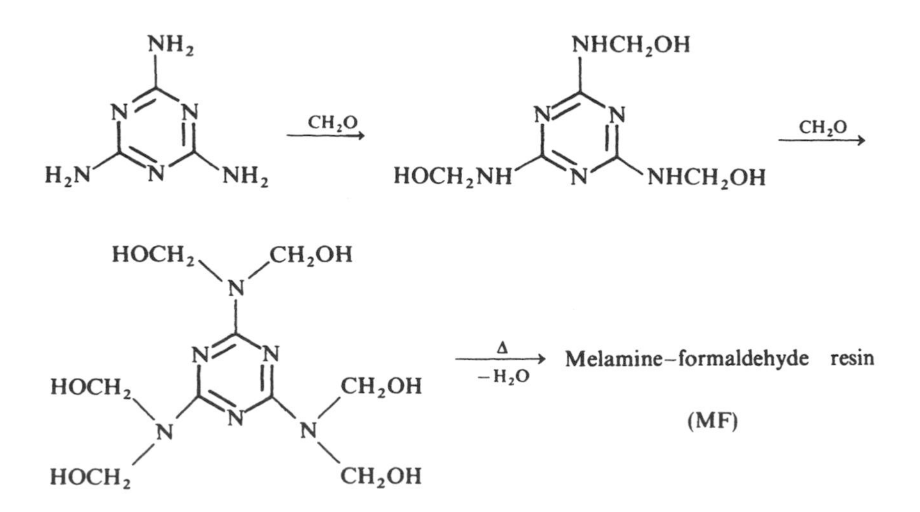

Figure 8.2. Condensation of melamine with formaldehyde to form thermoset plastics.

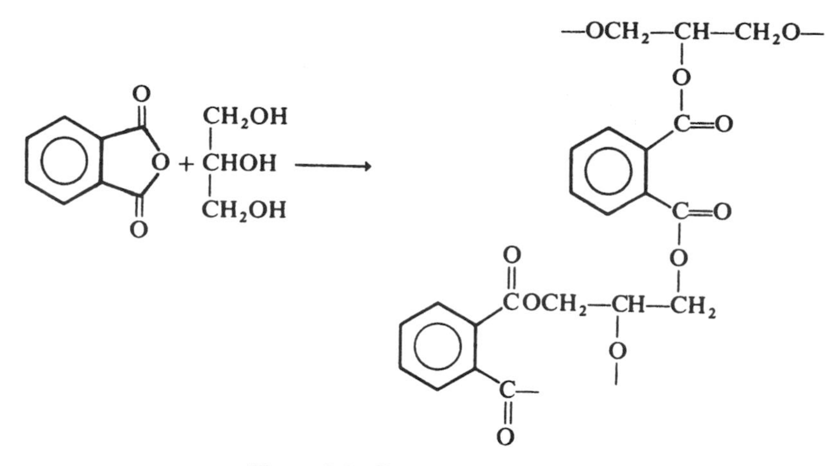

Figure 8.3. Glyptal resin formation.

of laminates, such as tops for counters, cabinets, and tables. Typically, a core of phenolic-impregnated paper is overlaid with melamine-impregnated sheets to produce Formica. Melamine formulations are also employed as automotive finishes.

8.5 ALKYDS-POLYESTER RESINS

The first polyester plastic was a cross-linked resin produced by Berzelius in 1847 by the condensation of difunctional tartaric acid and trifunctional glycerol. Commercial resinous coatings, called Glyptal resins, were produced by W. Smith in 1902 by the condensation of difunctional phthalic acid and trifunctional glycerol. The linear polyester prepolymer obtained at moderate temperatures by the reaction with the two primary hydroxyl groups was converted to a cross-linked plastic by a reaction with the residual (secondary) hydroxyl group after the reaction temperature was increased (Figure 8.3).

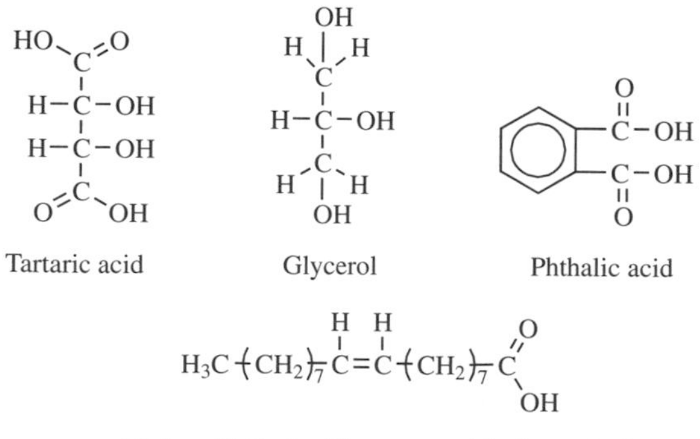

In the 1920s, R. Kienle coined the term alkyd for these polyester resins, which he obtained by the condensation of alcohols and acids. Many alkyd resins are produced by the condensation of difunctional reactants, alcohol and carboxylic acid, in the presence of unsaturated acids, such as oleic acid. The double bonds in these alkyds serve as sites for cross-linking in the presence of oxygen. Commercial alkyd molding powders were produced in 1948.

The most widely used polyester plastics are fiberglass-reinforced polyester plastics (FRP), which were introduced by Ellis in 1940. These FRP materials are based on unsaturated alkyd-type polyester prepolymer (polyethylene maleate), which is dissolved in styrene monomer. The prepolymer solution is cross-linked by the addition of peroxide-type initiators, such as benzoyl peroxide.

Table 8.3 Properties of typical reinforced polyesters

Property	Alkyd Mineral-Filled	BMC Polyester	SMC Polyester
Processing temperature (°F)	300	300	300
Molding pressure (10^3 psi)[a]	15	1	1
Mold shrinkage (10^{-3} in./in.)	2	4	2
Heat deflection temperature under flexural load of 264 psi (°F)	425	375	425
Maximum resistance to continuous heat (°F)	400	350	400
Coefficient of linear expansion (10^{-6} in./in., °F)	15	10	15
Compressive strength (10^3 psi)	25	20	20
Impact strength Izod (ft-lb/in. of notch)[b]	0.5	10	12
Tensile strength (10^3 psi)	6	10	12
Flexural strength (10^3 psi)	2	2.0	2.5
% elongation	1	4	3
Tensile modulus (10^3 psi)	1500	2000	2000
Flexural modulus (10^3 psi)	2000	2000	2000
Rockwell hardness	E98	60 (Barcol)	60 (Barcol)
Specific gravity	2	1.9	2
% water absorption	0.5	0.5	0.5
Dielectric constant	5	4	4
Dielectric strength (V/mil)	400	400	400
Resistance to chemicals at 75°F[c]			
Nonoxidizing acids (20% H_2SO_4)	S	Q	Q
Oxidizing acids (10% HNO_3)	U	U	U
Aqueous salt solutions (NaCl)	S	S	S
Polar solvents (C_2H_5OH)	S	Q	Q
Nonpolar solvents (C_6H_6)	U	U	U
Water	S	S	S
Aqueous alkaline solutions (NaOH)	Q	U	U

[a]psi/0.145 = kPa (kilopascals).
[b]ft-lb/in. of notch/0.0187 = cm · N/cm of notch.
[c]S, satisfactory; Q, questionable; U, unsatisfactory.

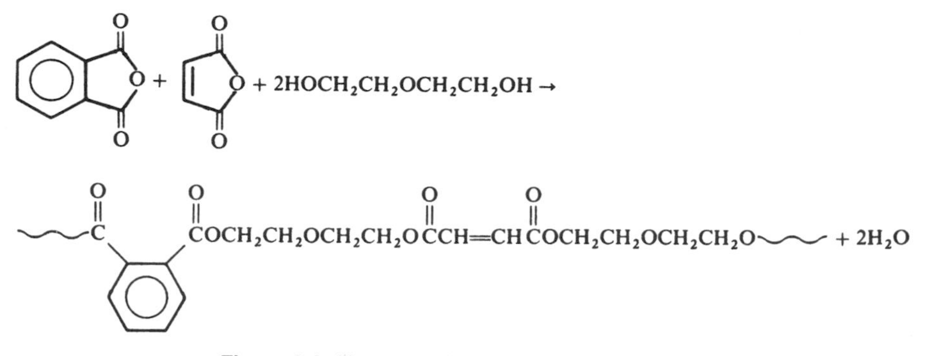

Figure 8.4. Unsaturated polyester resin formation.

Mixtures of chopped fiberglass and polyester prepolymers and fiberglass mat impregnated with polyester prepolymer are called bulk molding compounds (BMC) and sheet molding compounds (SMC), respectively.

The so-called vinyl ester resins (Derakane) are also used for the production of FRP. These vinyl esters are formed by the reaction of bisphenol A, acrylic acid, and ethylene oxide. Other polyesters used in FRP are produced from esters of bisphenol A and fumaric acid, isophthalic acid (*m*-phthalic acid), ethylene glycol, and maleic anhydride. The properties of a typical fiberglass-reinforced polyester laminate are shown in Table 8.3. Over 1 million tons of FRP are used annually in the United States in boats, panels, and automotive components.

Alkyd resins are primarily used in organic coating applications. They are used in some lacquers along with natural resins such as shellac and in varnish-type coatings as drying oils or resins. Prepolymers (partially polymerized but still thermoplastics) are used as molding resins in the production of fiberglass-reinforced laminates.

About 80% of unsaturated polyesters (excluding alkyd resins) are used to produce reinforced products, including electrical, marine, and transportation applications (Figure 8.4). Speedboat and motorboat hulls are generally produced by the SMC process, in which a mixture of unsaturated polyester resin, fibers (often fiberglass), and fillers is held between sheets of polyethylene film until it thickens to a leathery sheet. These sheets are molded under pressure to give fiber-reinforced plastic hulls. Shower stalls and industrial tubs are also made by the SMC process.

8.6 EPOXY RESINS

The word epoxy is derived from the Greek prefix *epi*, meaning between, and the English suffix of oxygen. Epoxy resins, which are obtained by the condensation of bisphenol A and epichlorhydrin, were described by Linderman in the last part of the nineteenth century and patented by P. Schlack in 1934. These linear prepolymers contain hydroxyl (OH) and epoxy

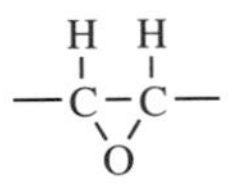

groups, which may be cross-linked by reaction with cyclic anhydrides, such as maleic anhydride, at elevated temperatures or with polyamines ($R(NH_2)_n$) at moderate temperatures.

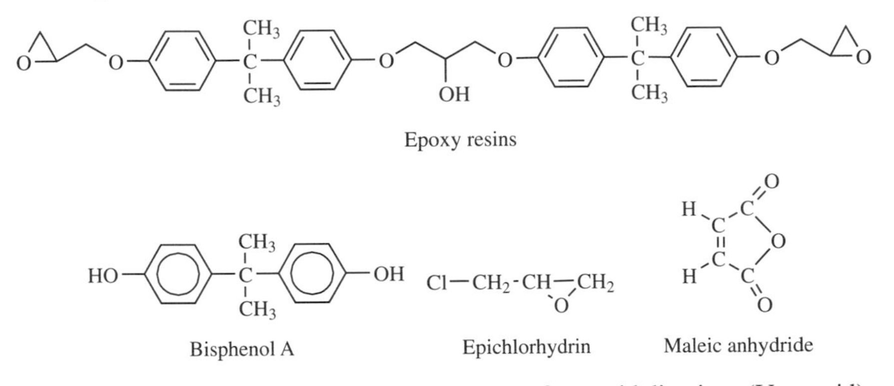

Epoxy resins (EP) may be flexibilized by using fatty acid diamines (Versamid) as the cross-linking or curing agent. The properties of typical epoxy resins are shown in Table 8.4.

Over 225,000 tons of epoxy resins are used annually in the United States as adhesives, coatings, and encapsulating compositions. Fiberglass-reinforced epoxy resins are used as aircraft components. About 110,000 tons of EP are used annually as coatings, and 45,000 tons are used for GRP (FRP).

The wide range of applications of EP is a result of the versatility of the system. By varying ratios of materials present in the prepolymers, large differences in curing rate and hardness are possible. There are two major types of surface coatings. The first is room (ambient)-temperature-cured. These are cross-linked using polyamides and polyamines. The second type of coating is heat-cured. These are cured using formaldehyde resin, anhydrides, and polycarboxylic acids. Phenol–formaldehyde resin–epoxy products include drum and tank linings, wire coatings, impregnation varnishes, and food and beverage can coatings. Epoxy coatings can be used as powder coatings, thus eliminating the use of solvents.

Printed circuit boards are made from fiberglass laminates and epoxy resins. Epoxy resins are used as binders for floor surfaces that are subject to heavy traffic, in patching concrete, and in casting, encapsulation, and potting electrical equipment. The common adhesive sold in hardware stores is generally a two-component liquid or paste that cures to give an epoxy resin.

8.7 SILICONES

Polysiloxanes (—Si—O—Si—) were investigated by F. Kipping prior to World War I. Since he believed that these compounds were ketones, he called them silicones.

Table 8.4 Properties of typical epoxy and silicone resins

Property	Glass Fiber-Reinforced EP	Mineral-Filled Silicone
Processing temperature (°F)	300	300
Molding pressure (10^3 psi)[a]	2	3
Mold shrinkage (10^{-3} in./in.)	1	1
Heat deflection temperature under flexural load of 264 psi (°F)	350	500
Maximum resistance to continuous heat (°F)	325	500
Coefficient of linear expansion (10^{-6} in./in., °F)	20	15
Compressive strength (10^3 psi)	25	12
Impact strength Izod (ft-lb/in. of notch)[b]	2.5	4
Tensile strength (10^3 psi)	12	6
Flexural strength (10^3 psi)	20	12
% elongation	4	5
Tensile modulus (10^3 psi)	3000	—
Flexural modulus (10^3 psi)	3000	1500
Rockwell hardness	M105	M85
Specific gravity	1.8	1.9
% water absorption	0.1	0.2
Dielectric constant	4	3
Dielectric strength (V/mil)	300	300
Resistance to chemicals at 75°F[c]		
Nonoxidizing acids (20% H_2SO_4)	S	Q
Oxidizing acids (10% HNO_3)	U	U
Aqueous salt solutions (NaCl)	S	S
Polar solvents (C_2H_5OH)	S	S
Nonpolar solvents (C_6H_6)	S	Q
Water	S	S
Aqueous alkaline solutions (NaOH)	S	S

[a]psi/0.145 = kPa (kilopascals).
[b]ft-lb/in. of notch/0.0187 = cm·N/cm of notch.
[c]S, satisfactory; Q, questionable; U, unsatisfactory.

Unfortunately, he did not think these inorganic polymers were useful and did not attempt to make them on a large scale. Nevertheless, these silicones were commercialized by G.E. and Dow–Corning Corporation during the early 1940s.

The first silicones were produced by the hydrolysis of chloromethylsilanes. The dichlorodimethyl silane ($Cl_2Si(CH_3)_2$) produces linear polymers that cross-link when some trichloromethyl silane ($Cl_3Si(CH_3)$) is present. The original silicones were produced by the reaction of magnesium with methyl chloride (CH_3Cl) in the Grignard reaction. They are now produced by heating methyl chloride with silicon in the presence of a copper catalyst. Either chloroalkylsilanes or methoxyalkysilanes will polymerize in the presence of water to produce silicones.

Silicones are available as low-molecular-weight fluids, molding resins, and elastomers. As shown in Table 8.4, these polymers have excellent resistance to solvents and retain their properties at elevated temperatures. Also see Section 7.11 for more about silicones.

8.8 POLYURETHANES

Organic isocyanates (RNCO) were synthesized by Wurtz in 1819. These active materials, particularly phenyl isocyanate (C_6H_5NCO), were used in qualitative organic chemistry to characterize alcohols by a reaction that produced urethane ($C_6H_5NHCOOR$). In the mid-1930s, Otto Bayer used difunctional reactants to produce polyurethanes (PUR). Fibers, plastics, coatings, adhesives, and foams were

Table 8.5 Properties of typical polyurethanes

Property	RIM (PUR)	PUR, 50% Mineral Filled
Processing temperature (°F)	25	25
Molding pressure (10^3 psi)[a]	—	—
Mold shrinkage (10^{-3} in./in.)	2.0	2.0
Heat deflection temperature under flexural load of 264 psi (°F)		
Maximum resistance to continuous heat (°F)		
Coefficient of linear expansion (10^{-6} in./in., °F)	60	40
Compressive strength (10^3 psi)	20	—
Impact strength Izod (ft-lb/in. of notch)[b]	25	5
Tensile strength (10^3 psi)	10	5
Flexural strength (10^3 psi)	20	5
% elongation	50	10
Tensile modulus (10^3 psi)	25	5
Flexural modulus (10^3 psi)	100	30
Rockwell hardness	D90 (Shore)	R40
Specific gravity	1.05	1.7
% water absorption	0.2	0.4
Dielectric constant	6	6
Dielectric strength (V/mil)	400	600
Resistance to chemicals at 75°F[c]		
Nonoxidizing acids (20% H_2SO_4)	Q	Q
Oxidizing acids (10% HNO_3)	U	U
Aqueous salt solutions (NaCl)	S	S
Polar solvents (C_2H_5OH)	U	U
Nonpolar solvents (C_6H_6)	Q	Q
Water	S	S
Aqueous alkaline solutions (NaOH)	Q	Q

[a] psi/0.145 = kPa (kilopascals).
[b] ft-lb/in. of notch/0.0187 = cm·N/cm of notch.
[c] S, satisfactory; Q, questionable; U, unsatisfactory.

produced from this versatile reaction. The adhesive properties were accidentally discovered when a PUR molding stuck tenaciously to the metal mold. The foam, which was called "imitation Swiss cheese" by Bayer's critics, was discovered when an organic acid was present in the reaction. Traces of water in the reactants will also produce carbon dioxide (CO_2), which causes foaming to take place.

The most widely used diisocyanate is tolylene diisocyanate (TDI, $H_3C{-}C_6H_3(NCO)_2$). Hydroxyl-terminated low-molecular-weight polyesters and polyethers are used as the diols $HO({-}RO{-})_nH$. The extent of cross-linking is controlled by the amount of triol, such as glycerol, present in the reactants.

Because of its versatility, PUR is used in a wide variety of applications ranging from foundation garments to bowling pin coatings to upholstery to automobile tires. Over 750,000 tons of flexible PUR foam, over 450,000 tons of rigid PUR foam, and almost 50,000 tons of PUR elastomers are used commercially each year in the United States. One of the fastest growing applications of PUR is in reaction injection molding (RIM) in which the reactants (diisocyanate and diol) are mixed and the polymer is formed rapidly under very little pressure in a mold. The properties of PURs are summarized in Table 8.5.

8.9 PLASTIC COMPOSITES

Although all mixtures of polymers and additives are composites, the term composite is used primarily for reinforced plastics. Asbestos-filled phenolics and α-cellulose-filled ureas and melamines are also plastic composites, but the emphasis is on those composites containing fibers with greater aspect ratios (l/d).

It was fortuitous that commercial fiberglass and commercial unsaturated polyester resins were introduced almost simultaneously in the late 1930s by Slater and Thomas and by Foster and Ellis, respectively. Fiberglass has limited use by itself, and unsaturated polyesters are too brittle for commercial use as plastics, but the combination (FRP) has unusually good properties and is now produced in the United States at an annual rate of over 1 million tons.

Composites containing fiber glass are erroneously referred to as fiberglass. Of course, the composite would be useless without both the resinous continuous phase and the discontinuous reinforming, fiberglass fiber-containing, phase. While improved unsaturated polyesters dominate much of the composite market, other materials are employed in the construction of composites. The topic of composites is more fully covered in Chapter 12.

The original FRP composites were made by impregnating fiberglass mat with a catalyzed (initiated) prepolymer. This hand lay-up technique has been supplemented by a spray-up technique in which chopped fibers and prepolymers are applied by a special spray gun. More uniform FRP composites are produced by bulk molding and sheet molding. The production of FRP articles may be automated by filament winding and pultrusion.

Asbestos and cellulose fibers have been used, to a limited extent, as reinforcements for thermosetting resins. Because of its toxicity, asbestos is being displaced by other fibers, and the use of cellulose fibers is limited because of their high water absorption and poor resistance to elevated temperatures.

Aromatic polyamides (aramids) and boron fibers are also used as reinforcements for thermosets. Boron filaments are produced by the chemical vapor decomposition (CVD) process in which boron trichloride is heated with hydrogen to produce hydrogen chloride and boron. The latter is deposited uniformly on a tungsten or graphite filament.

The second most widely used reinforcing filament is graphite. This strong fiber may be produced by the pyrolysis of polyacrylonitrile filaments (PAN) or by the thermal treatment of pitch. Graphite-reinforced resinous composites have outstanding strength given their light weight. Because of the high cost of graphite fibers, they are sometimes mixed with glass fibers. Composites based on these hybrid fibers have properties that are superior to FRP.

The early research emphasis was on reinforced thermosets, and some plastic technologists believed that the use of reinforcements in thermoplastics was not advantageous. However, molding compounds of chopped fiberglass and most thermoplastics are now available commercially. The properties of high-performance plastics have been upgraded considerably by the addition of glass or graphite fibers. Over 100 million tons of reinforced thermoplastics (RTP) are now used annually in the United States.

GLOSSARY

α-Cellulose: High-molecular-weight cellulose, insoluble in 17.5% NaOH.

Alkyd: A generic name for unsaturated polyester resins.

Amino plastic: Urea and melamine plastics.

Baekeland, Leo: Inventor of phenolic resins.

Bakelite: Trade name for PF.

Bayer, Otto: Inventor of polyurethanes.

Benzoyl peroxide: $C_6H_5COOOOCC_6H_5$.

Cross-link: Intermolecular primary valence bond.

Diamine: A compound with two amino (NH_2) groups.

Drier: A paint catalyst, usually the heavy metal salt of an organic acid.

Epoxy group:

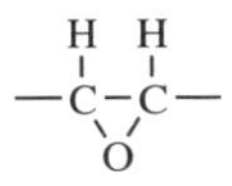

Epoxy resin: A resin obtained by the reaction of bisphenol A and epichlorohydrin. The terminal groups in epoxy resins are epoxy groups.

Ethylene oxide:

$$H_2C{-}CH_2 \text{ (bridged by } O\text{)}$$

Fiberglass: Fibers obtained by the melt spinning of glass.

Formaldehyde: H_2CO.

FRP: Fiberglass-reinforced plastics.

Fumaric acid: The trans isomer of maleic acid,

$$HOOC{-}HC{=}CH{-}COOH$$

Glycerol: $(H_2COH)CHOHH_2COH$.

Glyptal: A resin obtained by the reaction of glycerol and phthalic anhydride.

Goodyear, Charles: Inventor of vulcanized rubber.

Goodyear, Nelson: Inventor of hard rubber.

Grignard reagent: RMgX, where R = an alkyl or aryl group.

Hexamethylenetetramine: The reaction product of ammonia (NH_3) and formaldehyde.

Isocyanate group: NCO.

Isophthalic acid: *m*-Phthalic acid.

Laminate: A composite consisting of resin bonded to reinforcing sheets.

Linear polymer: A polymer with a continuous chain (thermoplastic).

Melamine resin (MF): The reaction product of melamine and formaldehyde.

Meta group: A group in position 3 or 5 on a substituted benzene.

Methacrylic acid: $H_2C{=}C(CH_3)COOH$.

Novolak: A thermoplastic prepolymer produced by the reaction of an insufficient amount of formaldehyde and phenol under acidic conditions.

Oleic acid: A monounsaturated acid, $C_{17}H_{33}COOH$.

Oleoresinous paint: A paint based on curable unsaturated oils.

Ortho group: A group in position 2 or 6 in a substituted benzene.

***P*-Phenyl phenol:**

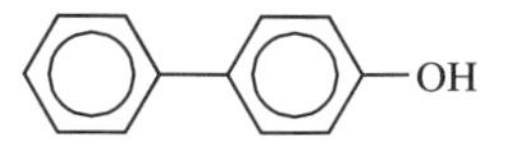

Phenol: C_6H_5OH.

Phenol–formaldehyde resin (PF): The reaction product of phenol and formaldehyde produced under controlled conditions.

Phenolic resin (PF): The reaction product of phenol and formaldehyde.

Plywood: A composite consisting of thin sheets of wood bonded together by an adhesive, such as a phenolic resin.

Polyester: A product obtained by the reaction of a dihydric alcohol and a dicarboxylic acid.

Polyurethane (PUR): The reaction product of a diisocyanate ($R(NCO)_2$) and a diol ($R(OH)_2$).

Prepolymer: A low-molecular-weight polymer that can be converted to a useful higher-molecular-weight polymer by heat or by the addition of a catalyst.

Primary hydroxyl group: A hydroxy group bonded to a carbon atom that is joined to two hydrogen atoms,

$$-\overset{\displaystyle H}{\underset{\displaystyle H}{\overset{|}{\underset{|}{C}}}}-OH$$

Silane: SiH_4, the simplest silicon hydride.

Silicone: A polysiloxane,

$$\left(\overset{\displaystyle R}{\underset{\displaystyle R}{\overset{|}{\underset{|}{Si}}}}-O\right)_n$$

Styrene: $C_6H_5CH{=}CH_2$.

Tartaric acid: A dihydroxy, dicarboxylic four-carbon compound.

TDI: Tolylene diisocyanate,

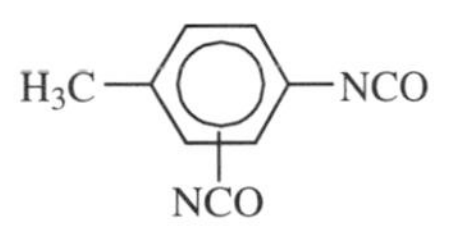

Thermoplastic: A linear or branched fusible polymer.

Thermoset: A cross-linked polymer.

Urea:

$$H_2N-\underset{\underset{\displaystyle O}{\|}}{C}-NH_2$$

Urea resin (UF): The reaction product of urea and formaldehyde.

Vinyl ester resin: A product obtained by the reaction of bisphenol A, ethylene oxide, and methacrylic acid.

Wood flour: Finely divided wood fibers obtained by attrition grinding.

REVIEW QUESTIONS

1. Is the reaction product of *p*-phenylphenol and formaldehyde a thermoset?
2. Which is more reactive in ester formation: a primary or a secondary alcohol?

3. What is the functionality of glycerol?
4. What is the functionality of ethanol?
5. What is the functionality of *m*-phenylphenol with formaldehyde?
6. What is the function of hexamethylenetetramine in a novolak molding compound?
7. Which will produce a thermoset when heated: a novolak or a resole resin?
8. What is the difference between wood flour and sawdust?
9. What is the functionality of urea?
10. How many carbon–carbon double bonds are in a molecule of oleic acid?
11. What is the functionality of melamine?
12. What is the repeating group in a polyurethane?
13. What is the formula for bisphenol A?
14. What is the structural difference between fumaric acid and maleic acid?
15. Silicon dioxide,

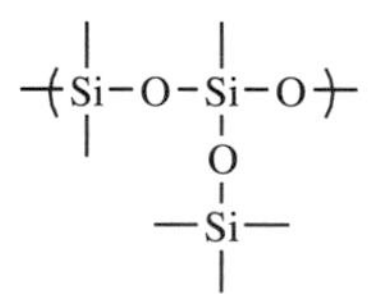

sand is abrasive but silicones are lubricants. Why?

BIBLIOGRAPHY

Aharoni, S. (1992). *Synthesis, Characterization, and Theory of Polymer Networks and Gels*, Plenum, New York.

Bayer, A. (1878). Phenol–Formaldehyde Condensates, *Ber. Bunsenges, Phy. Chem.* 5:280, 1094.

Carraher, C. (2003). *Polymer Chemistry*, Marcel Dekker, New York.

Hurley, S. (2000). *Uses of Epoxy, Polyester and Similar Reactive Polymers in Construction Materials Technology*, Construction Industry Research and Information Association, London.

Pascault, J., Sautereau, H., Verdu, J., and Williams, R. (2002). *Thermosetting Polymers*, Marcel Dekker, New York.

Schwartz, M. (1996). *Emerging Engineering Materials*, Technomic, Lancaster, PA.

Ward, T., Coates, P., and Dumoulin, M. (2000). *Solid Phase Processing of Polymers*, Hanser-Gardner, Cincinnati.

ANSWERS TO REVIEW QUESTIONS

1. No, *p*-phenylphenol is bifunctional.
2. Primary alcohol.
3. 3.
4. 1.
5. 3, the two ortho and para groups are reactive.
6. It supplies formaldehyde when heated.
7. A resole resin.
8. Wood flour has a fibrous structure.
9. 4.
10. 1.
11. 6.
12. RNHCOOR.
13. $HOC_6H_4C(CH_3)_2C_6H_4OH$.
14. Maleic acid is a cis isomer and fumaric acid is a trans isomer.
15. The groups in silicones in contact with another surface are oily alkyl groups and not abrasive siloxane groups (—Si—O—Si—O) which are highly cross-linked and rigid.

9

FIBERS

9.1 INTRODUCTION

Before the advent of man-made fibers, clothing was made from natural fibers, that is, plant fibers (cotton, hemp, and jute) and animal fibers (fur and hair; wool and silk). The modification of natural fibers began in the early 1800s and continues today.

Giant Molecules: Essential Materials for Everyday Living and Problem Solving, Second Edition, by Charles E. Carraher, Jr.
ISBN 0-471-27399-6

Fibers are threadlike strands with a length (l) to thickness (d) ratio of at least 100:1. Fibers are usually spun into yarns, which are made into textiles, which are then fabricated into finished products such as rugs, clothing, and tire cords. In this chapter, the basic concepts and processing techniques of fibers will be described. Individual fiber groups will also be discussed.

Fiber properties such as high tensile strength and high modulus are characteristic of polymers having good molecular symmetry, which allows the chains to be closely associated with one another in order to enhance dipolar and hydrogen bonding interactions between the chains.

Even though polystyrene chains may exhibit high symmetry, polymers such as polystyrene are not fibers because the forces associated with inter- and intrachain attractions have low energy. Likewise, a nylon that is produced by the condensation of 2-*tert*-butylterephthalic acid and 1-phenyl-1-*n*-butylhexamethylenediamine is not a good fiber because of the bulky butyl and phenyl groups, which prevent close chain association. Branched nylon 6,6 is also a poor fiber since the branching also prevents close chain association. These structures are shown in Figure 9.1.

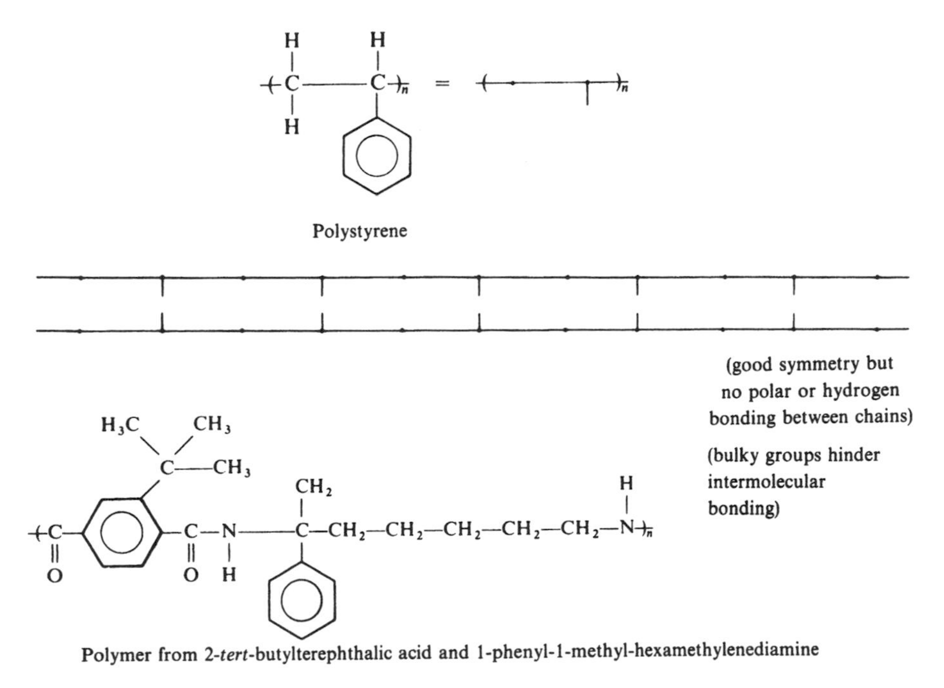

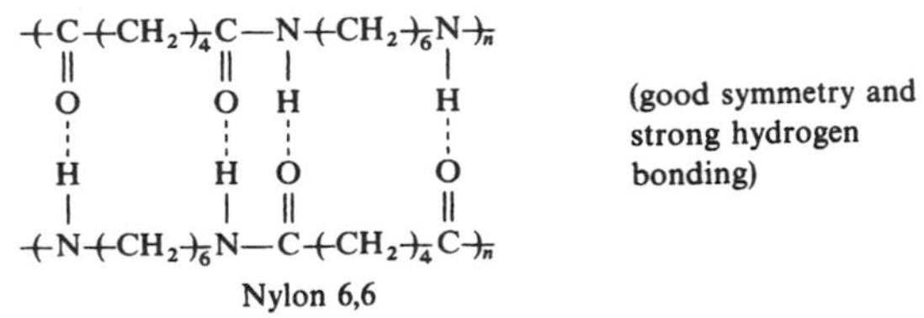

Figure 9.1. Fibrous and nonfibrous structures.

In contrast, wool, silk, and cellulose contain both the essential strong secondary bonding forces and unit symmetry, which are necessary for good fibers. Polymers such as nylon 6,6 also exhibit the required molecular symmetry and good bonding. In fact, nylon was the first man-made fiber that performed as well as its natural proteinaceous counterparts, wool and silk.

Fibers generally exhibit the following characteristics: They are thermoplastics (able to be molded or shaped through application of heat), abrasion-resistant (withstand surface wear), resilient (spring back when deformed), strong, and relatively nonabsorbent. In addition, synthetic fibers are usually resistant to mildew, rot, and moths. Some polymers, such as nylon 6,6, may be used as both fibers and plastics.

It is important to note that fiber producers make the fibers but usually do not make the finished products. Thus, the fiber producers, like the farmers who grow our food, depend on others for the processing of their products.

Staining is an interesting problem. Most items that we worry about, such as tomato paste, coffee, and foods, are hydrophilic. They must be hydrophilic to be metabolized by our bodies, which are "walking water containers"; that is, proteins, nucleic acids, carbohydrates, and water are themselves hydrophilic.

The second major group of staining materials is human excretions such as perspiration, which are also hydrophilic. Thus, fabrics such as nylons and polyesters that hydrogen-bond and interact favorably with these wastelike materials are typically easily stained. Even so, rugs and clothing made from nylons and polyesters feel softer to the touch and "breathe" when worn and thus offer advantages that more than offset the "staining problem." Staining is often resisted by application of a surface coating that is somewhat hydrophobic, but its major role is in preventing the staining material from penetrating the inner fibers of the fabrics. Scotchgard® is one of the many available surface treatments.

Hydrophobic (water-repellant) fabrics, such as polyolefin textiles, do not accept stains readily, and thus the staining material is readily washed away. However, such fabrics must undergo additional treatments before they are soft to the touch and even then may be inferior in this regard to nylons and polyesters.

Most polymeric materials are controlled by the Federal Trade Commission (FTC) with respect to the relationship between the name and content. This includes

Table 9.1 U.S. production of fibers for 2000

Fiber	Production (Millions of Pounds)
Cellulosic: Acetate and Rayon	350
Fiberous glass	2000
Noncellulosic	
Acrylics	340
Polyamides, nylons	2610
Olefins	3180
Polyesters	3870

Source: Fiber Economics Bureau.

fibers. While the FTC controls industry in the United States, the international standards are generally determined by the International Organization for Standardization (ISO).

The U.S. production of fibers is given in Table 9.1.

Many of the materials covered in this chapter can also act as plastics. These aspects are covered in Chapters 6 and 7.

9.2 PRODUCTION TECHNIQUES

Fibers can be produced as continuous filaments, as staple yarns, as the staple (short fibers) itself, or as filament yarn, depending on the processing and intended end use. Most natural fibers are of the staple variety and are made into staple yarn. These fiber types are illustrated in Figure 9.2.

Most man-made fibers (including regenerated cellulose and cellulosic derivatives) are formed by forcing a solution of the polymer through tiny holes (called spinnerets). Depending on the shape of the spinneret, the fiber can be circular, star-shaped, oval, square, and so on, and it can be hollow or solid. These somewhat subtle changes in the fiber shape influence the overall fiber property. As the amount of surface area (for a given weight of fiber) increases, the fiber becomes more flexible, pliable, and absorbate. The shape of the fiber can also influence the "feel" of the fiber. Rugs made from so-called olefin fiber like polypropylene with thick square "fibers" used for in-house putting greens "feel" like shredded thin pieces of plastic which in fact is what they are. By comparison, polypropylene fibers with high surface areas can be made to look and act more like nylon in rug applications.

"Feel" is important to the use of fabrics. As noted before, nonpolar polymers such as polypropylene and polyethylene composing the "olefins" category do not generally "feel" nice to our touch because of the polar nature of our skin

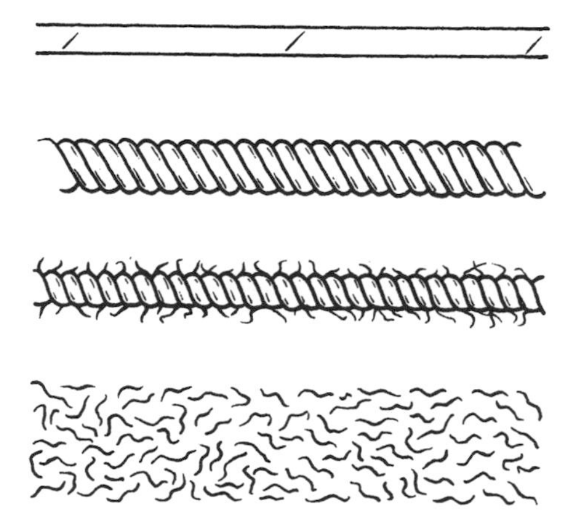

Figure 9.2. Illustrations of fiber types. *Top to bottom*: Continuous monofilament, filament yarn, staple yarn, and staple.

composed of proteins (Chapter 14), but polymers derived from more polar polymers such as nylons, polyesters, and plant-derived polymers such as cellulose, acetate, and rayon (Chapter 13) feel better because of their polar nature. Even so, olefin fibers can be made to feel better through fiber design and surface treatment. In general, "naturally feel-good" fabrics from more polar polymers such as nylons, cellulose, and polyacrylamide are used for dress-wear material such as suits, dresses, skirts, socks, and undergarments, whereas nonpolar polymer products are generally used for non-wear applications.

Some fibers, such as rayon, are produced by wet spinning, in which the filaments from the spinnerets are passed through chemical baths that insolubilize the soluble polymer. The solvent is evaporated by passing warm air by the spinneret as the polymer solution exits. Other fibers, such as cellulose acetate, are produced by evaporation of the solvated filament as it passes through the spinneret. In the melt spinning technique, the feedstock, such as pellets of nylon 6,6, are melted, forced through a spinneret, and cooled to form continuous nylon fibers. These processes, which should not be confused with the process of "spinning" of fibers to form yarn, are illustrated in Figures 9.3 and 9.4.

Filaments of polymers from spinnerets are then stretched to reduce their diameter, orient the polymer chains, and permit the fibers to arrange themselves along the pull axis. A wide variety of fiber sizes and strengths can be achieved by adjusting the pull rate and spinneret size. Polypropylene fibers are also produced by a unique fibrillation process in which strips of the polymer film are twisted and stretched.

The denier of a filament or yarn is the weight of 9000 m expressed in grams. The lower the denier, the lighter and finer the yarn. For example, 15-denier filaments are often used in women's hosiery, whereas 840-denier filaments are used in tires.

Today the dimensions of a filament or yarn are expressed in terms of a unit called the "tex," which is a measure of the fineness or linear density. One tex is 1 gram per 1000 meters. The tex has replaced denier as a measure of the density of the fiber. One denier is 1 gram per 9000 meters, so 1 denier = 0.1111 tex.

Monofilaments, which are obtained by continuous "pulls" from the spinneret, are employed for a variety of uses, such as fishline. The monofilaments can also

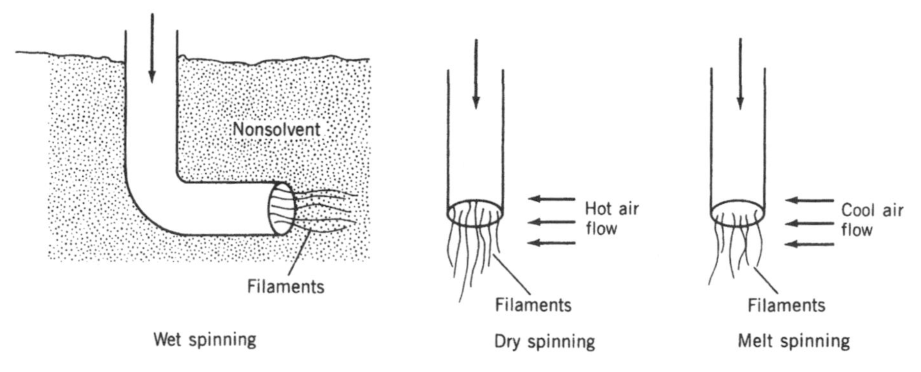

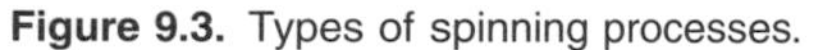
Figure 9.3. Types of spinning processes.

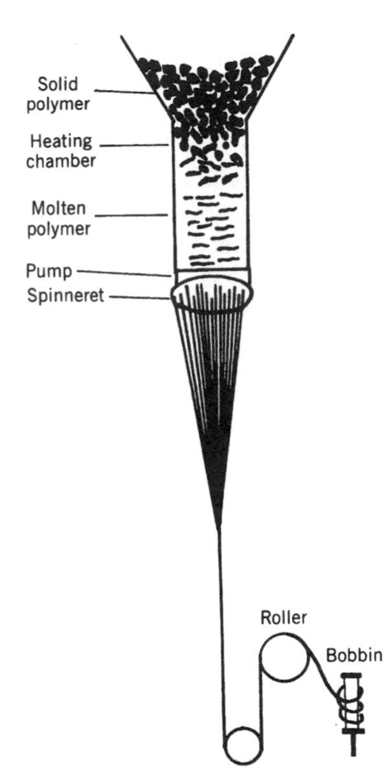

Figure 9.4. Diagram of the melt spinning process for the production of thread and filaments.

be woven into items such as sheer curtains or knitted to form products such as hosiery. Continuous strands of two or more filaments can be twisted or braided together to form filament yarns.

Large groups of continuous untwisted filaments are called "tow" and are often cut or broken to produce short segments, that is, staple fibers. Natural fibers such as wool and cotton are also staple fibers. The staple fiber can be twisted or spun or used without textile spinning as filling in mattresses, comforters, pillows, and sleeping bags.

Yarns spun from staple are more irregular than filament yarns, since the short ends of the fibers projecting from the yarn surface produce a fuzzy effect. Spun yarns are also more bulky than filament yarns of the same weight. Therefore, they are often used for porous, warm fabrics and for the production of nonsmooth fabric surfaces.

Textured filament yarns are made by twisting (throwing) the yarn in a designated manner. New filament yarns are being produced by untwisting, false twisting, deknitting, knitting, and crimping. Different bulk and stretch properties provide new fabrics for the fashion designer.

Both natural and synthetic fibers are be dyed for cosmetic effects. Cellulose fibers are often scoured with alkaline solutions, treated with an agent to prevent

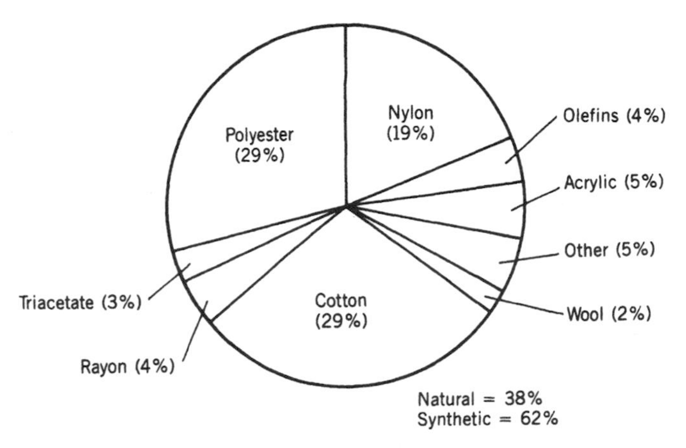

Figure 9.5. Relative production of commercial fibers.

mildew, soaked in copper aqueous or mercury salt solutions, and treated with water repellents such as quaternary ammonium compounds such as $R_4N^+Cl^-$. Wool is scoured and is often made crease- and wrinkle-resistant by immersing in aqueous solutions of melamine–formaldehyde prepolymers.

Cotton and other fibers or textiles are dyed by immersing in a solution or dispersion of the selected dye. Direct dyes, such as azo dyes, are used on cotton and rayon; reactive dyes, such as those with chlorotriazinyl groups, and triarylmethane dyes are used on polyamides; and bis azo disperse dyes and basic dyes are used on polyester and acrylic fibers. Fibers can be colored at various steps during their production. Fibers that are colored before spinning are called spun-dyed, dope-dyed, and solution-dyed. Fibers can also be dyed after being made into finished textiles.

The characteristics of two or more staple fiber types can be achieved by blending the fibers together prior to spinning into yarn. Various types of monofilament or filament yarns may also be combined and twisted together to form a "combination" filament yarn. The relative production of various commercial fibers is shown in Figure 9.5.

Fiberous glass is the most important inorganic fiber. It is produced by melt spinning in both a continuous filament and staple form. The molten glass is fed directly from the furnace, or melted from rods or marbles, to the spinneret. As the fiberous glass emerges, it is attenuated, quenched, lubricated, and wound, forming a yarn or continuous filament. The temperature for spinning is on the order of 1200–1500°C (2000–2700°F). This temperature is important since it controls the output and, in conjunction with the removal speed, helps control the properties of the resultant fiber, including thickness and density.

9.3 NYLONS

As noted in 7.2 Wallace Hume Carothers was brought to DuPont because his fellow researchers at Harvard and the University of Illinois called him the best synthetic chemist they knew. He started a program aimed at understanding the composition of natural polymers such as silk, cellose and rubber. Many of his efforts related to condensation polymers was based on his belief that if a monofunctional reactant reacted in a certain manner forming a small molecule, then similar reactions except employing reactants with two reactive groups would form polymers.

$$\underset{\text{Alcohol}}{R{-}OH} + \underset{\text{Acid}}{HO\overset{O}{\overset{\|}{C}}{-}R'} \rightleftharpoons \underset{\text{Small ester}}{R{-}O{-}\overset{O}{\overset{\|}{C}}{-}R'} + \underset{\text{Water}}{H_2O}$$

This reasoning that giant molecules would be formed from reaction with materials with two reactive groups, one at each end, was correct and allowed the first synthesis of polyesters—that is, giant molecules containing within their backbone chain ester groups as below.

$$\underset{\text{Dialcohol or diol}}{HO{-}R{-}OH} + \underset{\text{Diacid}}{HO\overset{O}{\overset{\|}{C}}{-}R'{-}\overset{O}{\overset{\|}{C}}OH} \rightleftharpoons \underset{\text{Polyester}}{\left[O{-}R{-}O{-}\overset{O}{\overset{\|}{C}}{-}R'{-}\overset{O}{\overset{\|}{C}}\right]} + \underset{\text{Water}}{H_2O}$$

W. Carothers and J. Hill produced low-melting polyester fibers in the DuPont laboratories in 1932 but shelved this project in favor of the development of the more heat-resistant polyamide fibers. The first polyamide fiber, poly-ω-aminononanoic acid (or nylon-9) $[NH(CH_2)_8CO]_m$, had a higher melting point (380°F) than the original polyester fibers but was softened at the temperature used for ironing. These polyamides were given a special name "nylons."

These investigators then synthesized polyhexamethylenadipamide, which had a melting point of 505°F. This pioneer synthetic fiber was produced commercially by DuPont at Seaford, Delaware, in 1939. Carothers recognized that high-molecular-weight polymers could not be synthesized unless the reactants were extremely pure. Accordingly, he made a salt from the diamine and diacid, removed the impurities by crystallization, and then heated the purified salt to produce a polymer, which he called a "super polyamide." This polymer is now called nylon 6,6 because both reactants have six carbon atoms. The equation showing that the degree of polymerization (DP) is a function of the purity of the reactants ($DP = (1 - P)^{-1}$) is called the Carothers equation. P is defined as the extent of reaction, which must be at least 0.995 for polymer formation.

$$H_2N(CH_2)_6NH_2 + HOOC(CH_2)_4COOH \longrightarrow H_2N(CH_2)_6NH_3^+\ {}^-OOC(CH_2)_4COOH$$

$$nH_2N(CH_2)_6NH_3^+\ {}^-OOC(CH_2)_4COOH \overset{\Delta}{\rightleftharpoons} \left[NH{-}(CH_2)_6{-}NH{-}\overset{O}{\overset{\|}{C}}{-}(CH_2)_4{-}\overset{O}{\overset{\|}{C}}\right]_n + nH_2O$$

Nylon 6,6
(polyhexamethyleneadipamide)

Carothers also found from his earlier work with polyesters that for the reaction to produce long chain, water must be removed, resulting in a shift of the equilibrium to favor larger chains. This was accomplished using what was called a molecular still, actually a hot plate assembly fitted with vacuum to help in the removal of water. This is an example of applying old knowledge to new problems.

Before nylon could be produced for public consumption, scientists needed to find large, inexpensive sources of the reactants—hexamethylenediamine and adipic acid. The DuPont Company scientists devised a scheme for producing these two reactants from coal, air, and water. The process was laborious, and later scientists developed procedures to make these two reactants from agricultural byproducts such as rice hulls and corn cobs. Several chemical equations are involved in these synthetic procedures:

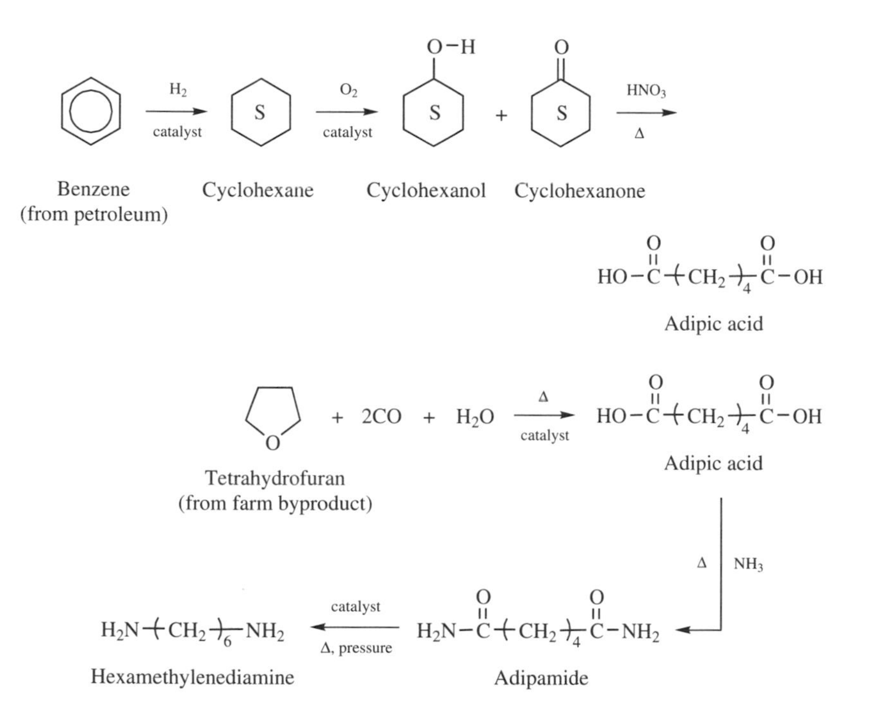

Nylon 6,6 allowed the formation of fibers that approached, and in some cases surpassed, the strength of natural fibers. This new miracle fiber was introduced at the 1939 New York World's Fair in an exhibit that announced the synthesis of this wonder fiber from "coal, air, and water"—an exaggeration but nevertheless eye-catching. When the polyamides (nylons) were first offered for sale in New York City, on May 15, 1940 over 4 million pairs were sold in the first few hours. Nylon sales took a large drop when it was noted that nylon was needed to produce the parachute material so critical to WW II.

Nylon is one of our most important polymers and it can be formed into sheets, rods, fibers, bristles, tubes, and coatings and used as a molding powder. Nylon 6,6 still accounts for the majority of nylons synthesized, with fiber applications including uses in dresses, lace, parachutes, tire cord, upholstery, underwear, carpets, ties, suits, socks, fishing line, bristles for brushes, and thread (including surgical thread).

In the early 1950s George deMestral was walking in the Swiss countryside. When he got home he noticed that his jacket had a lot of cockleburs on them. He examined the cockleburs and noticed that they had a lot of tiny "hooks." His cotton jacket had loops that "held" the cockleburs. He began putting into practice his observations, making combinations of materials with rigid hooks and flexible loops or eyes. The initial hook and eye for commercial use was made in France. Today, Velcro™, the name given to the hook-and-eye combination, is often based on nylon as both the hook and eye material. Remember that nylon can be made to behave as both a fiber and as a plastic. Polyester is blended with the nylon to make it stronger. Polyesters have also been employed to make hook-and-eye material. The hook-and-eye material is used to fasten shoes, close space suits and in many other applications.

Aromatic polyesters had been successfully synthesized from reaction of ethylene glycol and various aromatic diacids, but commercialization awaited a ready inexpensive source aromatic diacides (Section 7.3). An inexpensive process was discovered for the separation of the various xylene isomers by crystallization. The availability of inexpensive xylene isomers allowed the formation of terephthalic acid through the air oxidation of the *p*-xylene isomer. In 1953, DuPont produced polyester fibers from melt spinning, but it was not until the 1970s that DuPont-produced polyester fibers became commercially available.

The hard–soft block copolymer approach (Section 9.8) employed to produce segmental polyurethanes has also been used with polyesters with the hard block formed from 1,4-butadienediol and terephthalic acid, while the soft block is provided from oligomeric (approximate molecular weight of 2000 daltons) poly(tetramethylene glycol) and is sold under the trade name of Hytrel.

Along with nylons, polyester fibers approach and exceed common natural fibers such as cotton and wool in heat stability, wash-and-wear properties, and wrinkle resistance. Blended textiles from polyester and cotton and wool also can be made to be permanent press and wrinkle-resistant. The fibers are typically formed from melt or solvent spinning. Chemical and physical modification are often employed to produce differing fiber appearances from the same basic fiber material. Self-crimping textiles are made by combining materials with differing shrinkage properties. Different-shaped dyes produce materials with varying contours and properties including hollow fibers.

Other nylons have also been developed. Nylon-6 (polycaprolactam, Perlon), which was patented by P. Schlack in 1937, is produced on a large scale in Europe, but only on a moderate scale in the United States.

$$\left[NH(CH_2)_5-\underset{\underset{O}{\|}}{C} \right]_n$$

Nylon 6

Nylon 4,6 (Stanyl), which is more hydrophilic than nylon 6,6, and nylon-12 are also available commercially.

Aromatic nylons, prepared from the condensation of terephthalic acid and aliphatic diamines, have high melting points and are tough and strong. These nylons, which are called aramids, are second-generation fibers and are utilized in the construction of radial tires, bulletproof clothing, and fiber-reinforced composites. The principal aramid fibers are poly-*p*-benzamide (Kevlar) and polyphenyleneisophthalamide (Nomex). Over 10,000 tons of Kevlar are produced annually.

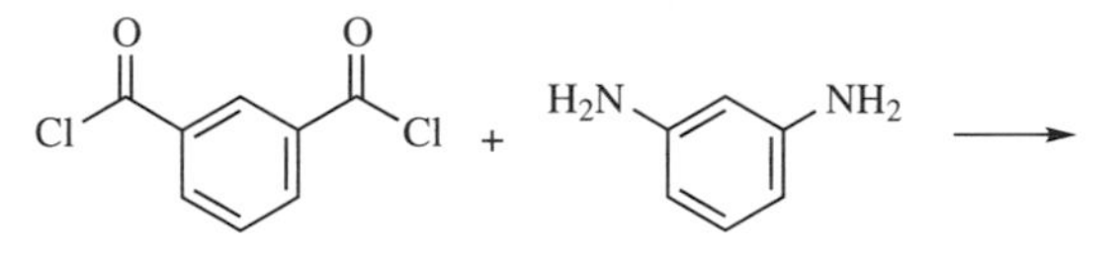

Phthaloyl dichloride

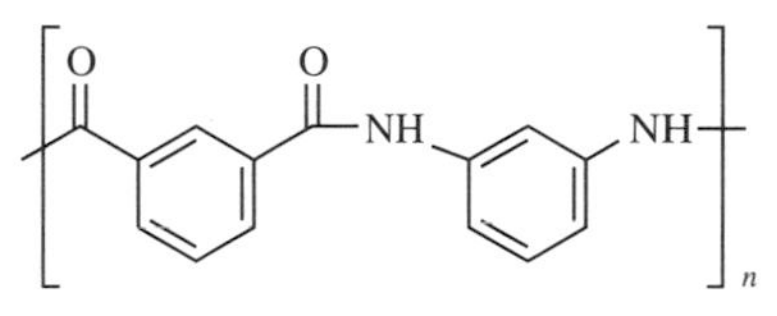

Nomex

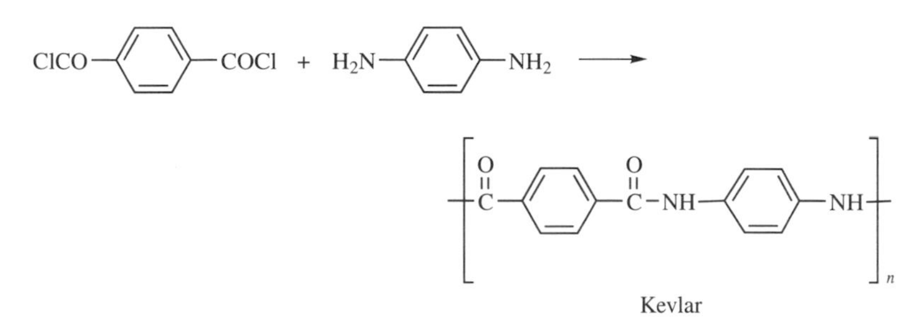

Kevlar

A number of other nylon fibers have also been made. Quiana fibers were developed as a synthetic silk material and are now employed in the production of blouses, dresses, and shirts. Quiana fibers are spun from the nylon made from condensing dodecanedioic acid and di-4-aminocyclohexylmethane. Interestingly, it has a lower melting point (205°C) than nylon 6,6 but a higher glass transition temperature (135°C). The high T_g enables fabrics made from Quiana to resist wrinkles and creasing during laundering.

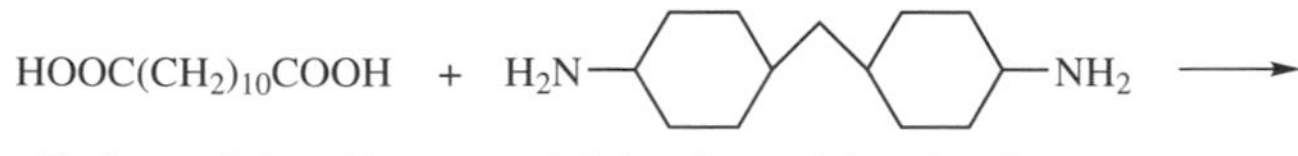

Dodecanedioic acid Di-4-aminocyclohexylmethane

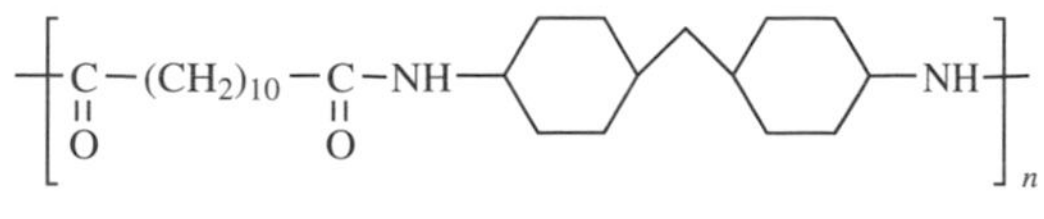

Quiana

9.4 POLYESTERS

J. R. Whinfield and J. T. Dickson substituted phthalic acid for Carother's adipic acid and produced a relatively high melting (507°F) polyester fiber in 1940. This synthetic fiber was produced in England in 1940 and in the United States by DuPont in 1945. Polyester fibers are now produced by a number of companies and are sold under a variety of trade names, such as Dacron, Kodel, Terylene, and Trevira.

The most popular polyester is obtained by the condensation of ethylene glycol (the major ingredient in antifreeze) and terephthalic acid. These two reactants can be made from a number of petrochemical feedstocks. As shown by the following equation, terephthalic acid is produced by the catalytic oxidation of *p*-xylene.

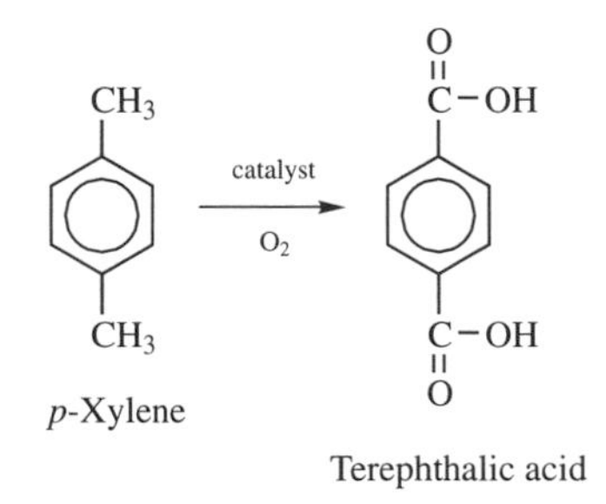

Ethylene glycol is produced by the hydrolysis of ethylene oxide, which is obtained by the catalytic oxidation of ethylene produced from natural gas or petroleum crude. Both xylene and ethylene can also be made from coal and oil shale, and ethylene has been made by the destructive distillation of alcohol, corn, wheat, and other natural, renewable products. Thus, the source of the starting material is variable.

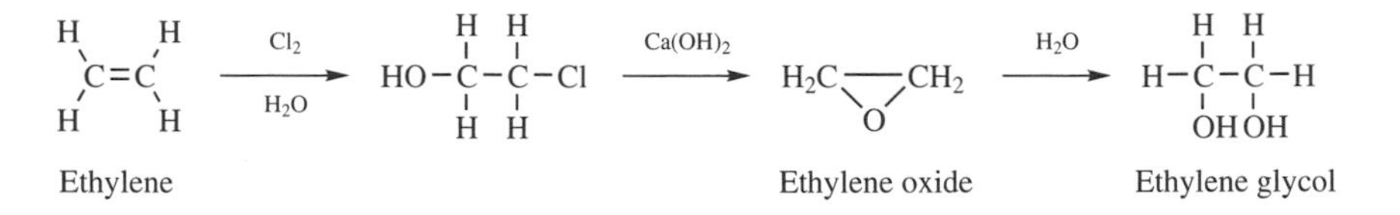

Polyesters are manufactured as films, plastics, and fibers. Polyester fibers are mainly used in making fabrics such as carpets, clothing, upholstery, and underwear. The fibers are also employed in the construction of tire cord. Polyester fabrics are easy to care for and resist mildew, rot, and fading. Most of the outerwear garments are permanent-press textiles.

The formula for the repeating unit in most polyester fibers is

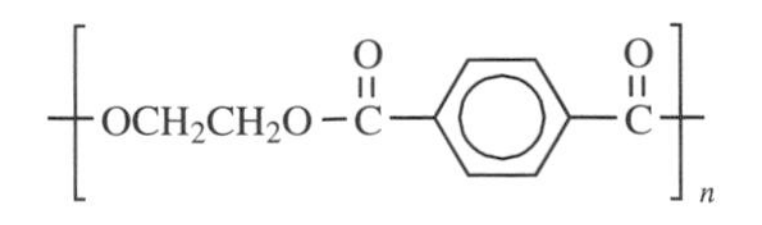

Poly(ethylene terephthalate)(PET)

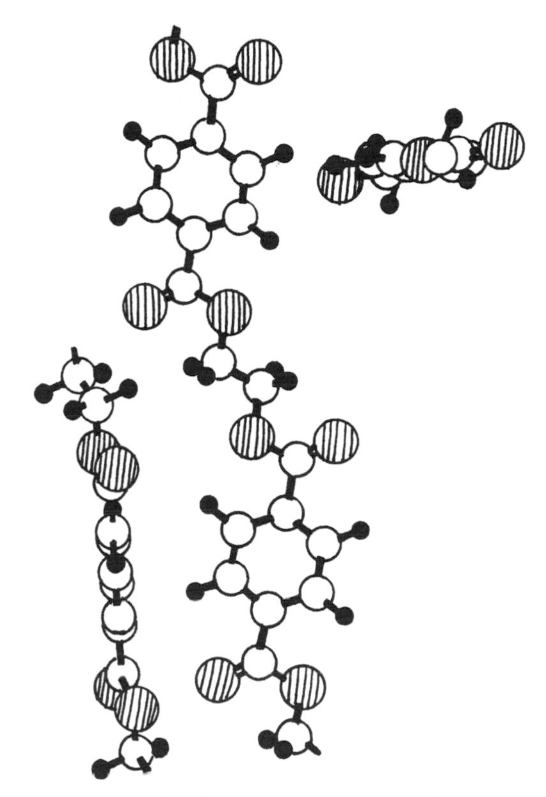

Figure 9.6. Ball-and-stick models of poly(ethylene terephthalate) showing various views.

Figure 9.6 shows ball-and-stick models of poly(ethylene terephthalate).

9.5 ACRYLIC FIBERS

In the 1940s, chemists at Monsanto and DuPont dissolved polyacrylonitrile in dimethylacetamide (DMAC) and produced unique fibers by passing these solutions through spinnerets and evaporating the solvent. Modacrylic fibers, such as copolymers of vinylidene chloride, are related but more easily dyed than are copolymers of acrylonitrile and other monomers. These were produced commercially in the 1950s.

Fibers from both polyacrylonitrile and copolymers derived from acrylonitrile are classified jointly as acrylic fibers. The Textile Fiber Products Identification Act divides these fibers into two categories: Acrylic fibers are those containing at least 85% by weight of acrylonitrile, and modacrylic fibers contain less than 85% but at least 35% acrylonitrile.

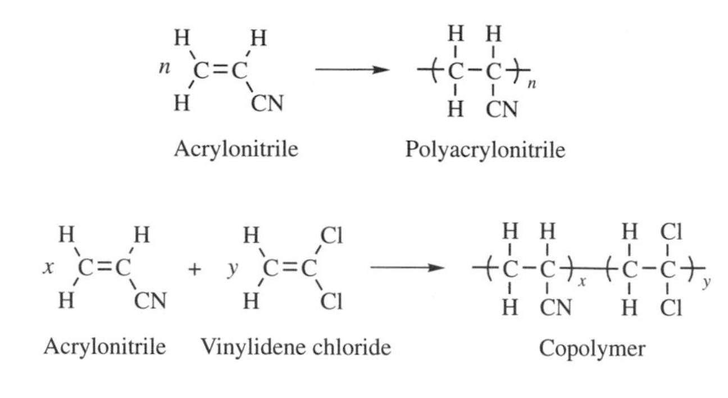

These fibers are long-lasting, dry rapidly, and resist fading, wrinkling, and mildew. They are used in making carpets, sportswear, sweaters, and blankets as well as many other products. As in the case of nylon and polyesters, acrylics are also being employed as coatings and plastics.

The monomer acrylonitrile is synthesized at an annual rate in the United States of 700,000 tons by the ammoxidation of propylene, the latter being a product obtained from petroleum.

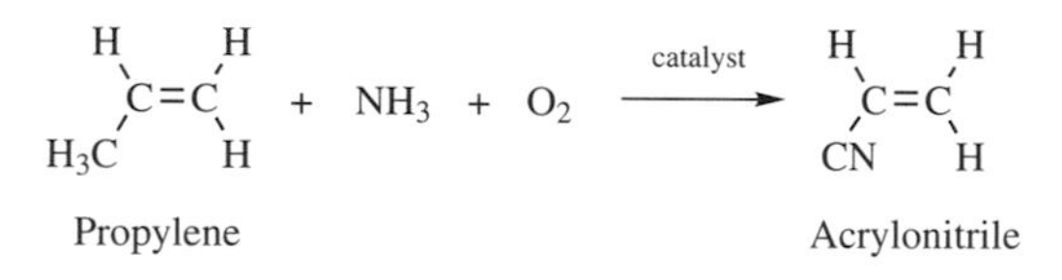

9.6 GLASS FIBERS

Glass fibers may be thinner than human hair and may look and feel like silk. These flexible glass fibers are stronger than steel on a weight basis and will not burn, stretch, or rot. The ancient Egyptians used coarse glass fibers for decorative purposes.

Edward Libbey, an American glass manufacturer, exhibited a dress made from fiberglass and silk at the Columbian Exposition in Chicago in 1893. Fiberglass was made in Germany as a substitute for asbestos during World War I and, because of the toxicity of asbestos, is being used again for this purpose. The Owens Illinois Glass Company and the Corning Glass Works developed practical methods for making fiberglass commercially in the 1930s.

Fiberglass, which is an inorganic polymeric fiber, is made from the same raw materials as those used to make ordinary glass. As shown in Figure 9.7, glass marbles are melted in special furnaces and molten glass passes through spinnerets (bushings) at the furnace bottom. The hot filaments are wound on a spinning drum, which stretches and orients the glass filaments. The principal use of fiberglass is as a reinforcement for polyester and epoxy resin.

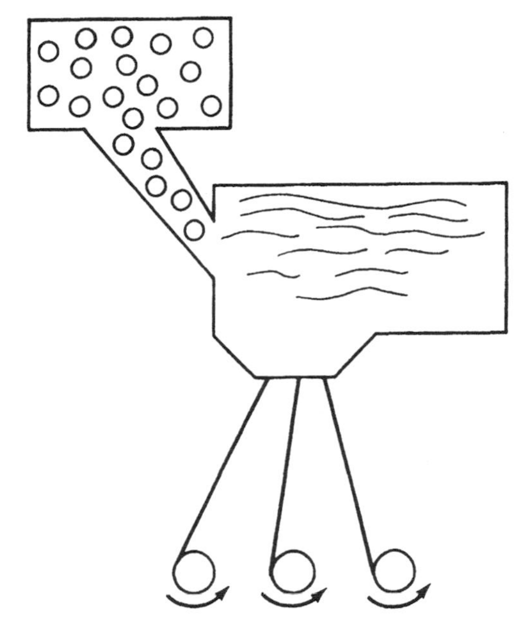

Figure 9.7. Illustration of one procedure for producing fiberglass.

Fiberglass wool is produced by exposing the hot filaments to high-pressure steam jets to produce fibers that are gathered together to form a white wool-like mass, which is used for insulation.

Fiberglass can also be woven into a fabric to make tablecloths and curtains. The end products can be dyed, do not wrinkle or soil easily, and need no ironing after washing. Textile material is also used for electrical insulation. In bulk form, it is used for heat and sound insulation and for air filters. The insulation properties are a result of the high bulk property of fiberglass whereby still air also acts as a good buffer to thermal changes.

Fiberglass is also employed in the manufacture of reinforced plastics that are strong yet lightweight. Car bodies, ship hulls, building panels, aircraft parts, and fishing rods are popular examples. The fibers can be woven, matted together, or used as individual strands depending on the nature and price of the final product. About 400,000 tons of fiberglass are produced annually in the United States.

9.7 POLYOLEFINS

A number of olefinic polymers are being used as fibers. The most popular is polypropylene fiber, which is used in the production of outdoor–indoor carpets, cordage, and upholstery. Polypropylene fibers are low-melting (below 355°F) and are degraded by sunlight. Yet, because of their resistance to soiling and ease in cleaning, stabilized polypropylene fibers are widely accepted as carpeting for heavily traveled, dirt-attracting areas, such as store and home entrances, patios, and swimming pool areas. The formulas for polyolefin and polypropylene are

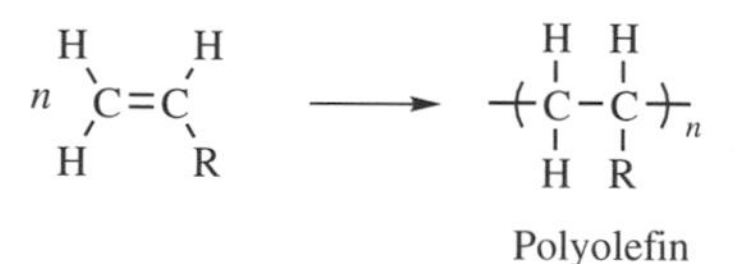

Polyolefin

where R=H, CH_3, or other hydrocarbon group, and

$$\left[CH_2\underset{\displaystyle CH_3}{\underset{|}{CH}} \right]_n$$

Polypropylene (PP)

The strength of polypropylene filaments (Spectra) has been increased dramatically by chain extension, that is, stretching solvent-swollen filaments.

9.8 POLYURETHANES

Joseph Shivers, a DuPont chemist, invented spandex in 1959 after about a 10-year search. It was first named Fiber K, but DuPont chose the more appealing, smooth-like trade name of Lycra. Spandex (Lycra) is an elastomeric fiber popularized by "cross-your-heart" bra commercials. It was introduced by DuPont in 1958. Lycra is a segmented copolymer, with each segment contributing its own properties to the whole material. The soft segment is composed of flexible macroglycols, whereas the rigid segment is formed from 4,4-diisocyanatodiphenylmethane (MDI) and hydrazine (Figure 9.8).

Macroglycols can be made from polyesters, polyethers, and polycaprolactones. The key is that the end groups are both alcohols (hydroxyls). The macroglycols are generally short chains with DP values around 40. The hydroxyl end groups are reacted with an excess of the MDI diisocyanate to form urethane linkages with isocyanate end groups. These in turn are further reacted with hydrazine to form urea-like and urethane linkages, giving segmented elastomeric fibers, that is, Lycra. This process illustrates the ability of scientists to design molecules with specific, desired properties. Lycra is used extensively in the manufacture of foundation garments, swimsuits, and running and exercise suits.

9.9 OTHER FIBERS

There are a number of other fibers being produced for specialty applications. For example, polyimides have good thermal stability and are employed where resistance to high temperature is required.

Carbon fibers (also called graphite fibers) are used for applications where great strength, rigidity, and light weight are required. Graphite fibers are used to reinforce polymers in the construction of lightweight, highly durable bicycles, car bodies, golf

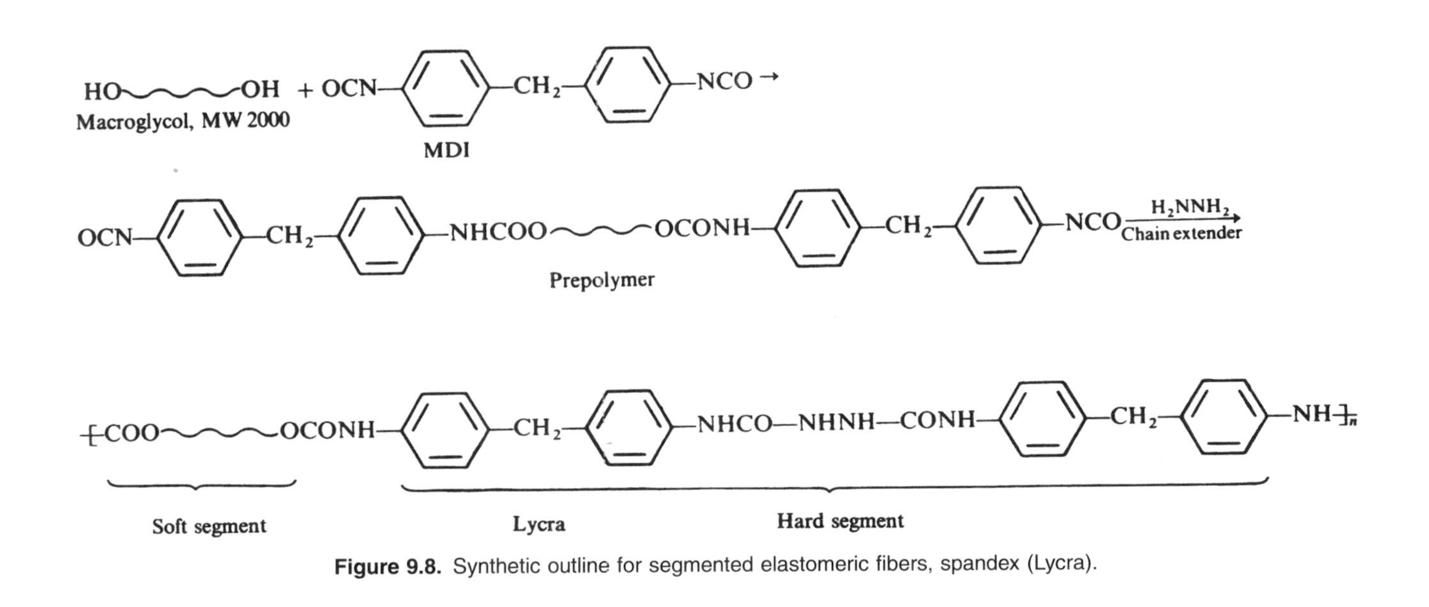

Figure 9.8. Synthetic outline for segmented elastomeric fibers, spandex (Lycra).

club shafts, fishing rods, firemen's suits, and aircraft for space exploration. The Voyager, which flew around the world without refueling in 1986, was constructed from graphite and aramid-reinforced plastics.

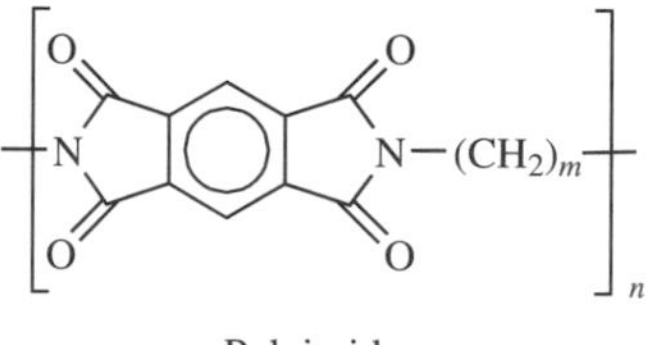

Polyimide

As shown by the following equation, carbon fibers are produced by the pyrolysis of polyacrylonitrile. The polyacrylonitrile cyclizes at about 570°F and remains stable up to about 1290°F. This product, originally reported by Hautz in 1950, was dubbed black orlon after DuPont's trade name for its light-colored polyacrylonitrile fiber. Graphite fibers are also produced from pitch and other materials.

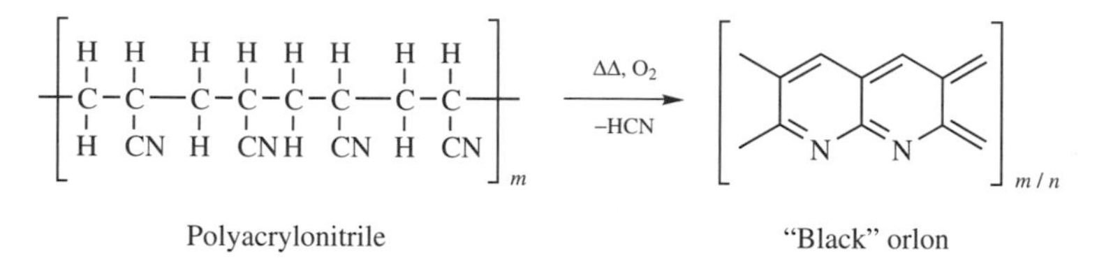

Polyacrylonitrile "Black" orlon

Table 9.2 Physical properties of typical fibers

Polymer	Tenacity (g/denier)	Tensile Strength (kg/cm^2)	Elongation (%)
Cellulose			
Cotton	2.1–6.3	3000–9000	3–10
Rayon	1.5–2.4	2000–3000	15–30
High-tenacity rayon	3.0–5.0	5000–6000	9–20
Cellulose diacetate	1.1–1.4	1000–1500	25–45
Cellulose triacetate	1.2–1.4	1000–1500	25–40
Protein			
Silk	2.8–3.0	3000–6000	13–31
Wool	1.0–1.7	1000–2000	20–50
Vicara	1.1–1.2	1000–1000	30–35
Nylon 6,6	4.5–6.0	4000–6000	26
Polyester	4.4–5.0	5000–6000	19–23
Polyacrylonitrile	2.3–2.6	2000–3000	20–28
Saran	1.1–2.9	1500–4000	20–35
Polyurethane (Spandex)	0.7	630	575
Polypropylene	7.0	5600	25
Asbestos	1.3	2100	25
Glass	7.7	2100	3.0

Boron filaments are produced by a unique process in which boron produced by the reaction of boron trichloride and hydrogen is deposited on a tungsten or graphite filament.

The physical properties of typical fibers are compared in Table 9.2.

GLOSSARY

Acrylic fiber: Fiber containing more than 85% repeating units of acrylonitrile ($\left(\!\!+CH_2CH(CN)+\!\!\right)$).

Adipic acid: $HOOC\!\left(CH_2\right)_4COOH$.

Aminononanoic acid: $H_2N\!\left(CH_2\right)_8COOH$.

Ammoxidation: Oxidation in the presence of ammonia (NH_3).

Black orlon: Pyrolyzed Orlon.

Borazole: Boron nitride ($B\equiv N$).

Bulky group: Large substituent on a polymer chain.

Carothers, W.: The inventor of nylon 6,6.

Composite: A mixture of polymer and an addition such as fiberglass.

Copolymer: A macromolecule consisting of more than one repeating unit in the chain.

Dacron: Trade name for polyester fibers.

Denier: The weight in grams of 9000 m of a fiber; the finer the fiber, the lower the denier.

Dry spinning: The production of filaments by evaporation of the solvent from the solution, which was exuded from the spinneret.

Ethylene glycol: $HO\!\left(CH_2\right)_2OH$.

Fiber: Strong strands of the polymer with a length to diameter ratio of at least 100 to 1.

Fiberglass: Glass in the form of fine fibers.

Filament: A very long fiber.

Hexamethylenediamine: $H_2N\!\left(CH_2\right)_6NH_2$.

Hydrophobic: Water-hating or water-repellent.

Kodel: Trade name for polyester fibers.

Lycra: Trade name for spandex fibers.

Melt spinning: The production of filaments by cooling a molten exudate after it leaves the spinneret.

Modacrylic fiber: Acrylic fiber containing less than 85% repeating units of acrylonitrile ($\left(\!\!+CH_2CH(CN)+\!\!\right)$).

Nylon: A generic term for synthetic polyamides.

Nylon-6: Polycaprolactam.

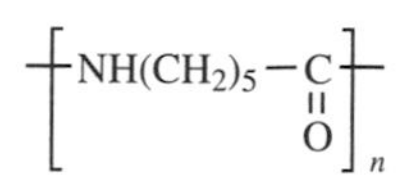

Nylon 6,6: The reaction product of hexamethylenediamine ($H_2N(CH_2)_6NH_2$) and adipic acid ($HOOC(CH_2)_4COOH$).

$$\left[NH(CH_2)_6NH-\underset{\overset{\|}{O}}{C}-(CH_2)_4-\underset{\overset{\|}{O}}{C} \right]_n$$

Olefin: A hydrocarbon that is a member of the alkene homologous series ($H(CH_2)_nCH{=}CH_2$).

Orlon: An acrylic fiber.

Polyamide: A polymer produced by the condensation of a diamine (H_2NRNH_2) and a dicarboxylic acid (HOOCRCOOH).

Polyester: A copolymer produced by the condensation of a dicarboxylic acid (HOOCRCOOH) and a diol (HOROH).

Polyimide: A heat-resistant heterocyclic polymer.

Polypropylene: $[-CH(CH_3)CH_2]_n$.

Polystyrene: $[CH_2-CH(C_6H_5)]_n$.

Propylene: $HC(CH_3){=}CH_2$.

Pyrolysis: Thermal degradation.

Spandex: An elastic fiber consisting of sequences of repeating units of polyesters and polyurethanes.

Spinnerets: Small, uniform holes used for the extrusion of filaments.

Spun-dyed fiber: Fiber that is dyed before the spinning process.

Staple fiber: A short fiber.

Tenacity: Tensile strength of a fiber expressed as g/denier. If a 100-denier yarn fails under a 300-g load, its tenacity is 300/100 or 3 g/denier.

Tensile strength: The maximum stress that a material can withstand without failure when stretched.

Textured yarn: Twisted filament yarn.

Thermoplastic: A linear or branched polymer that can be softened by heat and cooled to reform the solid.

Throwing: Twisting of filaments.

Tow: Several twisted filaments gathered together.

Urethane linkage:

$$(-\overset{\overset{H}{|}}{N}-\overset{\overset{O}{\|}}{C}O-)$$

Yarn: Spun fibers used in weaving of fabrics.

Wet spinning: The production of filaments by passing the polymer solution, exuded from the spinneret, into a bath to insolubilize the polymer.

REVIEW QUESTIONS

1. What is the minimum length to diameter ratio for a substance to be classified as a fiber?
2. What atoms are involved in hydrogen bonding in nylons and proteins?
3. Which is stronger: hydrogen bonding or dipole–dipole interactions?
4. Besides hydrogen bonding, what else is characteristic of fiber molecules?
5. In the nylon nomenclature, which number is the number of carbon atoms in the diamine?
6. Which is more hydrophobic: cellulose or nylon 6,6?
7. How would you convert a long filament into a staple fiber?
8. How could you convert cellulose staple fiber to a long filament?
9. Is rayon produced by wet or dry spinning?
10. Are acrylic fibers produced by dry or melt spinning?
11. Name a fiber that is produced by melt spinning.
12. How are polypropylene fibers produced?
13. Which has the lower denier value: a nylon fishline or nylon fiber used to manufacture hosiery?
14. How do the polyester fibers produced originally by Carothers differ from today's polyester fibers, such as Dacron?
15. What type of spinning is used to produce glass fibers?
16. What is the precursor for some of the high-grade graphite fibers?

BIBLIOGRAPHY

Carothers, W., and Arvin, J. (1929). Polyesters, *J. Am. Chem. Soc.* 51:2560.

Fourne, F. (1999). *Synthetic Fibers*, Hanser-Gardner, Cincinnati.

Mark, H., and Whitby, G., eds. (1940). *The Collected Papers of Wallace Hume Carothers*, Interscience, New York.

Morgan, P., and Kwolek, S. (1959). The Nylon Rope Trick, *J. Chem. Ed.* 36:182.

Salem, David (2001). *Structure Formation in Polymeric Fibers*, Hanser-Gardner, Cincinnati.

ANSWERS TO REVIEW QUESTIONS

1. 100 to 1.
2. Hydrogen, nitrogen, and oxygen.
3. Hydrogen bonding.
4. Structural symmetry.
5. The first integer.
6. Nylon 6,6.
7. By cutting it into small sections.
8. By converting it to rayon (regenerated cellulose).
9. Wet spinning.
10. Dry spinning.
11. Nylon.
12. By fibrillation and melt spinning.
13. Hosiery fiber.
14. Carothers produced an aliphatic polyester with a low softening point. Today's polyester fibers are aromatic polyesters with high softening points.
15. Melt spinning.
16. Acrylic fiber.

10

RUBBER (ELASTOMERS)

10.1 EARLY HISTORY

Athletes of the Mayan civilization used a ball made from rubber for their national game, called tlachti, over 1000 years ago. This game resembled modern basketball

Giant Molecules: Essential Materials for Everyday Living and Problem Solving, Second Edition,
by Charles E. Carraher, Jr.
ISBN 0-471-27399-6

in that the ball was thrown through a circular stone hole. However, this game differed from the modern sport since only one goal was scored in each game and the members of the losing team could be executed.

The Indians of Mexico called the rubber tree *ule*, and the rubbery product was called *ulei*. Some of the South American Indians called the tree *heve*, but others called it *caaochu* or "weeping wood." The name caoutchouc is still in use in France, but *Hevea braziliensis* is the more widely used term for natural rubber (NR). A comparable elastomeric product is also present in the domesticated rubber plant (*Ficus elastica*), the guayule shrub (*Parthenium argentatum*), goldenrod (*Solidago*), and the dandelion (*Koksaghyz*). *Ficus elastica* was used unsuccessfully as a source of rubber in Malaysia, and *Castilloa elastica* and *Castilloa ulai* were used as the original sources of rubber in Brazil, but *H. braziliensis* from Indonesian plantations is now the principal source of natural rubber.

The American Indians made waterproof boots and containers by dipping in the rubber latex. Latex, which is the liquid exuded by the rubber tree, is an aqueous emulsion of rubber. The name rubber was given to the sticky elastomeric material by Joseph Priestley, who used it to erase pencil marks from paper.

The use of natural rubber was limited by its characteristic stickiness. However, MacIntosh made a cloth sandwich from a solution of rubber in naphtha in 1823, and this type of construction is still used for waterproofing garments. However, there was little use for rubber until Charles Goodyear vulcanized (cross-linked) the crude product by heating it with sulfur. This discovery and other accidental discoveries are called serendipity after a name coined by Walpole. This author described three princes of Serendip or Sri Lanka who were seeking potential princesses but accidentally made many apparently more valuable discoveries.

The only major source of rubber in the nineteenth century was the wild rubber tree from Brazil, Central America, the west coast of Africa, and Madagascar. However, this supply was insufficient to meet the demands brought on by the introduction of the automobile, each of which required four pneumatic tires.

Since Brazil prohibited the export of rubber seeds or seedlings, H. A. Wickham smuggled 70,000 rubber seeds hidden in banana leaves and brought them to England in 1876. The 1900 seedlings that germinated and survived were used to start the rubber plantations in Malaya late in the nineteenth century. The first year's production of four tons of plantation rubber was small compared to the production of 50,000 tons of wild rubber obtained in 1900. However, the source of wild rubber continued to decrease with further exploitation, but over 1 million tons of plantation rubber were produced annually just prior to World War II.

A small amount of wild rubber is still obtained from Brazil, but over 90% of today's natural rubber supply is obtained from plantations of about 14 million acres in Indonesia, Malaysia, Thailand, Sri Lanka, India, Vietnam, Cambodia, and Sarawak. The latex from the rubber plant contains 36–40% of rubber.

Charles Goodyear was born in 1800 in New Haven, Connecticut. He became driven to work with rubber to try to make it more temperature-stable. This passion affected his health and took what little money he had. On more than one occasion he lived in debtor's prison. One of his jobs was to supply the U.S. government with waterproof mailbags, but the mailbags he prepared were sticky and malformed. Another failure. After many unsuccessful attempts, one of which was to mix the rubber with sulfur, he accidently allowed a mixture of sulfur and rubber to touch a hot stove. The rubber did not melt but only charred a little. As are many of the so-called discoveries by chance or accident, his mind was ready for the result and by 1844 had been given a patent for a process he called "vulcanization" after the Roman god of fire, Vulcan.

Vulcanization is the cross-linking reaction between the rubber chains and the sulfur.

Goodyear had trouble defending his patent, piling up huge debts before he died in 1860. Daniel Webster defended him in one of his patent infringement cases. By 1858 the value of rubber goods was about $5 million. The major rubber-producing plants clustered about Akron, Ohio, with the Goodyear Company founded in 1870.

Chemists learned about the structure of rubber through degrading it through heating and analyzing the evolved products. One of the evolved products was isoprene, a five-carbon hydrocarbon containing a double bond. Isoprene is a basic building block in Nature, serving as the "repeat" unit in rubber and also as the building block of steroids such as cholesterol.

$$H_2C{=}\underset{\displaystyle CH_3}{\underset{|}{C}}{-}CH_2{-}CH_3$$

Isoprene

With knowledge that natural rubber had isoprene units, chemists worked to duplicate the synthesis of rubber except using synthetic monomers. These attempts failed until two factors were realized. First, after much effort it was discovered that the methyl groups were present in a "cis" arrangement. Second, it was not until the discovery of stereoregular catalysts that the chemists had the ability to form natural rubberlike material from butadiene.

The synthesis of a purely synthetic rubber, structurally similar to natural rubber, involved a number of scientists building upon one another's work—along with a little creativity. Nieuwland, a Catholic priest, President of Notre Dame University, and a chemist, did extensive work on acetylene. He found that acetylene could be made to add to itself, forming dimers and trimers.

Calcott, a DuPont chemist, attempted to make polymers from acetylene, reasoning that if acetylene formed dimers and trimers, conditions could be found to produce polymers. He failed but went to Carothers, who had one of his chemists, Arnold Collins, work on the project. Collins ran the reaction described by Nieuwland, purifying the reaction mixture. He found a small amount of material that was not vinylacetylene or divinylacetylene. He set the liquid aside. When he came back, the liquid had solidified giving a material that seemed rubbery and even bounced. They analyzed the rubbery material and found that it was not a hydrocarbon, but it had chlorine in it. The chlorine had come from HCl that was used in Nieuwland's procedure to make the dimers and trimers, adding to the vinylacetylene forming chloroprene.

Table 10.1 U.S. Production of synthetic rubber, 2000

Rubber	Production (Millions of Pounds)
Ethylene–propylene	700
Nitrile	180
Polybutylene	1210
Styrene–butadiene	1750
Other	1100

Source: International Institute of Synthetic Rubber Producers.

$$\underset{\text{Acetylene}}{HC{\equiv}CH} \longrightarrow \underset{\text{Vinylacetylene}}{H_2C{=}CH{-}C{=}CH} + \underset{\text{Divinylacetylene}}{H_2C{=}CH{-}C{=}C{-}CH{=}CH_2}$$

This new rubber was given the name Neoprene. Neoprene had outstanding resistance to gasoline, ozone, and oil, in contrast to natural rubber. Today, Neoprene is used in a variety of applications such as electrical cable jacketing, window gaskets, shoe soles, industrial hose, and heavy-duty drive belts.

$$\underset{\text{Chloroprene}}{H_2C{=}\underset{Cl}{\underset{|}{C}}{-}CH{=}CH_2} \longrightarrow \underset{\text{Polychloroprene (Neoprene)}}{-\left(-CH_2{-}\underset{Cl}{\underset{|}{C}}{=}CH{-}CH_2-\right)-}$$

Table 10.1 contains the annual production of synthetic rubber in the United States for 2000.

10.2 GENERAL PROPERTIES OF ELASTOMERS

The individual polymer chains of elastomers are held together by weak intermolecular bonding forces—that is, London dispersion forces—which allow rapid chain slippage when a moderate pulling force is employed. Cross-links, which are introduced during vulcanization, permit rapid elongation of the principal sections, to a point where the chains are stretched to their elastic limit. Any additional elongation probably causes a combination of breakage of primary bonds and breakup of crystalline regions (Figure 10.1). As elongation occurs, so does formation of ordered structures leading to crystallization, which results in a stronger material. The cross-links, which are the boundaries for the principal sections, permit the rubber to "remember" its original shape, that is, the original positions of the chains.

10.3 STRUCTURE OF NATURAL RUBBER (NR)

In 1826, Faraday used carbon–hydrogen analysis to show that rubber was a hydrocarbon with the empirical formula of C_5H_8. Subsequently, it was shown

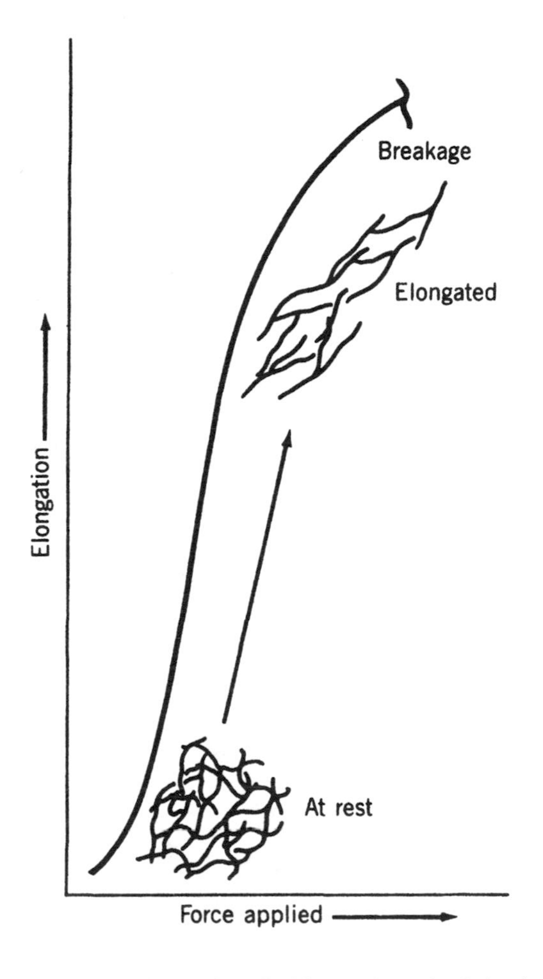

Figure 10.1. Illustration of typical applied-force–elongation behavior of rubber.

that the pyrolysis of natural rubber produced isoprene, which had the skeletal formula of C=C(C)–C=C. In the early part of the twentieth century, Harries added ozone (O_3) to natural rubber and showed that this elastomer consisted of repetitive units of 2-methyl-2-butene, $\left(\text{C–C(C)=C–C}\right)_n$, where n was equal to several hundred.

These high-molecular-weight molecules are giant molecules. If we could see molecules of rubber (without the cross-links), they would look like strands of cooked spaghetti, and the polymer chains would be entangled much like the spaghetti strands. However, since these chains are in constant motion at room temperature, a can of worms serves as a more realistic model.

It is of interest to point out that the butene units in natural rubber have a cis arrangement; that is, the carbon–carbon chain extensions are on the same side of each ethylene unit (C=C). Thus, the skeletal chain of *Hevea* rubber would look like

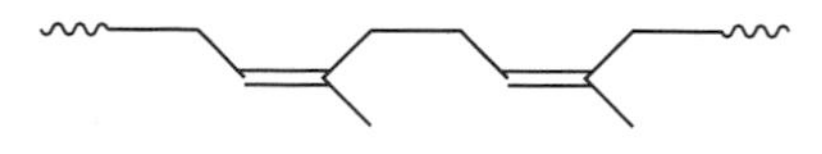

In contrast, another naturally occurring polymer of isoprene called balata or gutta-percha has a trans arrangement, as shown by

Ball-and-stick models of *Hevea* rubber and gutta-percha are shown in Figures 10.2 and 10.3. The trans arrangements permit these chains to fit closely together so that gutta-percha is a hard plastic, in contrast to the flexible cis polyisoprene. Staudinger

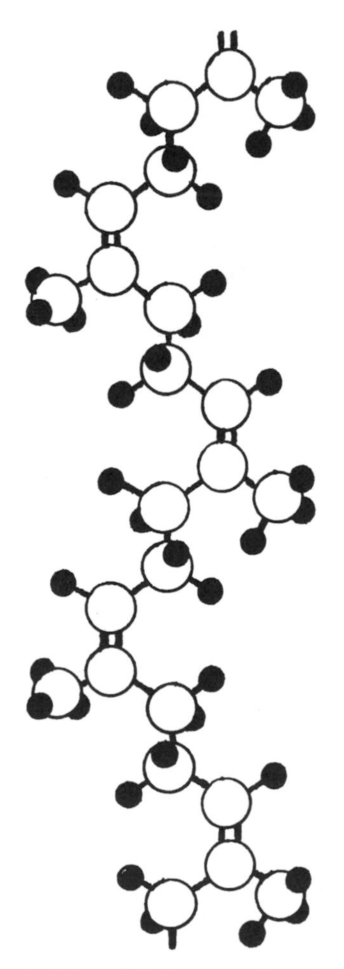

Figure 10.2. Ball-and-stick model of *Hevea* rubber (*cis*-1,4-polyisoprene).

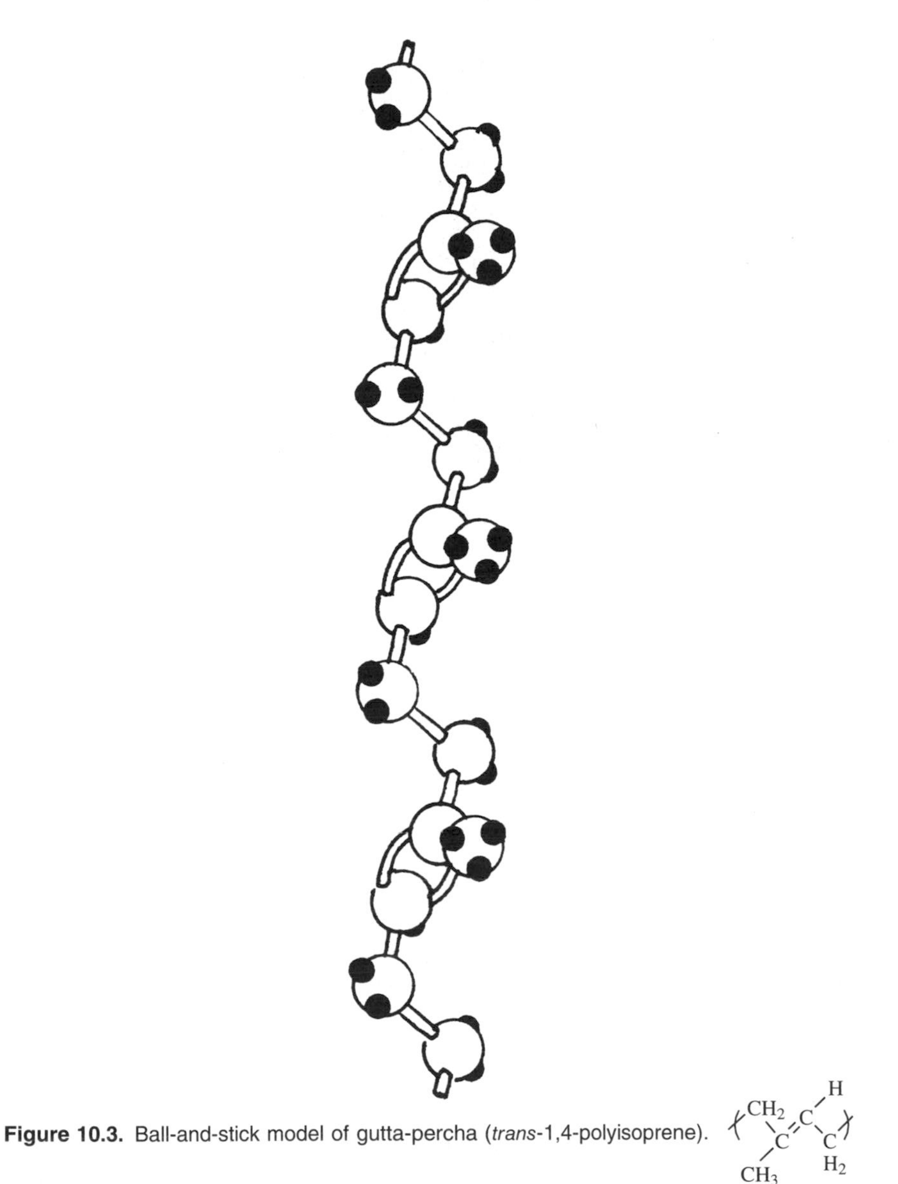

Figure 10.3. Ball-and-stick model of gutta-percha (*trans*-1,4-polyisoprene).

reccived the Nobel prize for his interpretation of the correct structure of rubber and other macromolecules.

10.4 HARVESTING NATURAL RUBBER

The *Hevea* tree grows best in hot, moist climates in acidic, well-drained soils. The cultivated rubber tree grows to 60–70 feet tall. The rubber latex flows through a

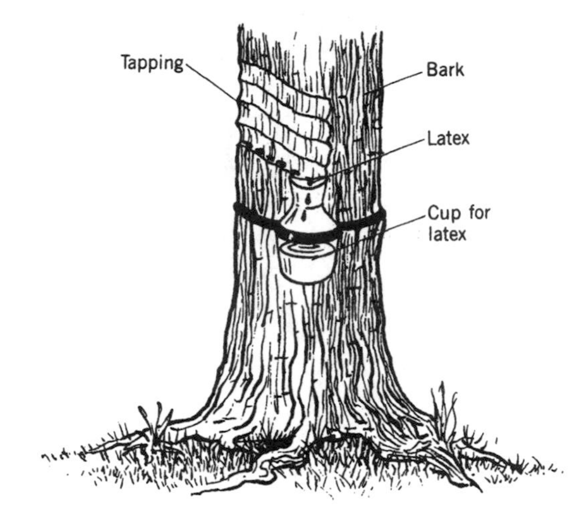

Figure 10.4. Illustration of how latex is collected from rubber trees.

series of tubes in the tree's cambium layer; that is, the outer wood layer directly beneath the bark. The latex oozes out when this layer is pierced.

Botanists continue to work on improving the tree and, through grafting and breeding, have grown trees that produce 1000% more rubber than the wild *Hevea* trees. Further work is being done on obtaining rubber from other plants, such as guayule, which will grow in the American southwestern desert area and can be harvested mechanically. Thus, even though natural rubber has been utilized by humans for several thousand years and cultivated for over a hundred years, research continues on the improvements of tree yields and the development of alternative sources.

Rubber tappers cut a narrow diagonal groove in the bark about 4 feet from the ground with a long, curved knife called a gouge (Figure 10.4). A U-shaped metal spout with a small cup to catch the latex is attached at the bottom of the cut. Trees are tapped for 25–30 years, commencing when the young tree is 5–7 years of age.

Crude rubber is obtained by coagulation of the latex by the addition of formic acid (HCOOH) or acetic acid (H_3CCOOH). Some of the latex is not coagulated but is concentrated in machines called separators, which are similar to the cream separators employed in dairies. This rubber latex is used to make articles such as surgeon's gloves, condoms, tubing, elastic thread, and foam-backed carpeting.

10.5 STYRENE–BUTADIENE RUBBER (SBR)

Hofman synthesized isoprene in Germany in 1909, and the English chemists Matthews and Strange and the German chemist Harries converted methylisoprene to a rubbery product in 1910 by use of sodium metal. Over 2500 tons of this type of synthetic rubber were produced in Germany during World War I. Kaiser Wilhelm equipped his Mercedes-Benz with synthetic rubber tires in 1912 and was impressed with their utility. However, since the methyl rubber was not reinforced by carbon black, these tires were not satisfactory when used on heavier equipment by the German army in World War I.

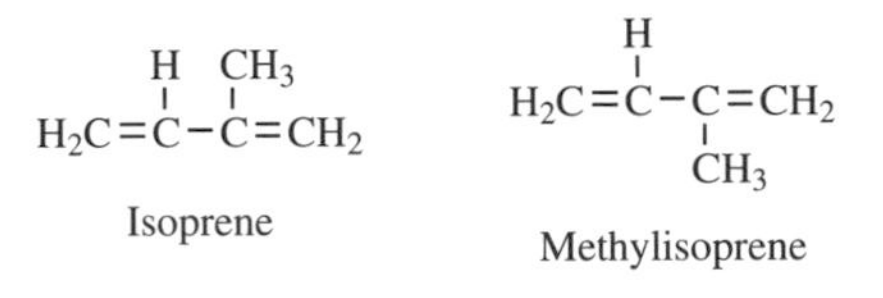

In the late 1920s, Tschunker and Bock patented a method for producing a copolymer of 1,3-butadiene and styrene in an aqueous emulsion. The synthetic rubber molecule, which was called Buna-S, contained repeating units from both butadiene and styrene in a ratio of about 3:1. Most of the SBR now produced contains about 20% of the 1,2 configuration, 20% of the *cis*-1,4, and 60% of the *trans*-1,4 configuration. The precise structure of SBR is varied, but it is reproducible. The irregular structure prevents the chains from close contact with one another and promotes rapid slippage of chain segments past one another.

The name Buna-S was derived from the first letters of butadiene (Bu) and styrene (S) and the chemical symbol for sodium (Na). Metallic sodium was employed to initiate the first polymerization of dimethylbutadiene. The German chemists obtained U.S. patents in which this novel polymerization process was described in detail. This synthetic rubber (Buna-S) was initially synthesized on an industrial scale by Germany's I. G. Farbenindustrie in 1933.

Prior to the bombing of Pearl Harbor in 1941, the Germans had an annual production capacity of 175,000 tons of Buna-S, and the Russians had an annual production capacity of 90,000 tons of sodium-catalyzed polybutadiene rubber. In contrast, annual American production of synthetic rubber, prior to 1942, was less than 10,000 tons, and most of this was specialty oil-resistant rubber that was not suitable for the manufacture of pneumatic tires.

Nevertheless, the production of Buna-S was duplicated in the United States during World War II, and the product was called GRS (Government rubber styrene). Over 50 GRS plants were constructed and operated in North America during the early 1940s, and the annual production of GRS reached 700,000 tons before the end of World War II.

After the war ended, the U.S. synthetic rubber production facilities were acquired by private industry, and the name for this synthetic rubber was changed by the American Society for Testing Materials (ASTM) from GRS to SBR. Although this rubber was not as good as natural rubber, it was readily available and produced from inexpensive petroleum feedstocks as shown in Figure 10.5.

10.6 POLYMERS FROM 1,4-DIENES

There are three commercially important monomeric 1,4-dienes. These are polybutadiene, polyisoprene, and polychloroprene. For all of these products, cross-linking

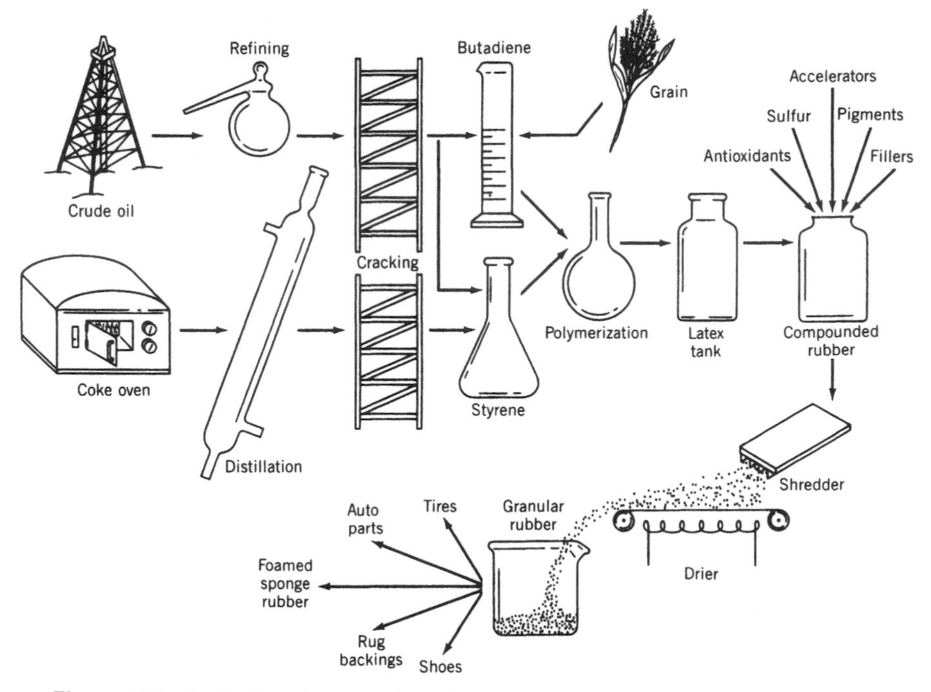

Figure 10.5. Production of styrene–butadiene rubber (SBR) from common feedstocks.

and grafting sites are available through the remaining double bond. Thus, all three are often used in the production of elastomeric materials.

1,4-Butadiene can form three repeat units, the 1,2; *cis*-1,4; and *trans*-1,4. Commercial polybutadiene is mainly composed of the 1,4-*cis* isomer and is known as butadiene rubber (BR). As noted before, polybutadiene is made from the use of stereoregulating catalysts. The composition of the resulting polybutadiene is quite dependent on the nature of the catalyst such that almost total *trans*-1,4 units, or *cis*-1,4 units, or 1,2 units can be formed as well as almost any combination of these units. The most important single application of polybutadiene polymers is its use in automotive tires where over ten million tons are used yearly in the U.S. manufacture of automobile tires. BR is usually blended with natural rubber (NR) or styrene–butadiene rubber (SBR) to improve tire tread performance, particularly wear resistance.

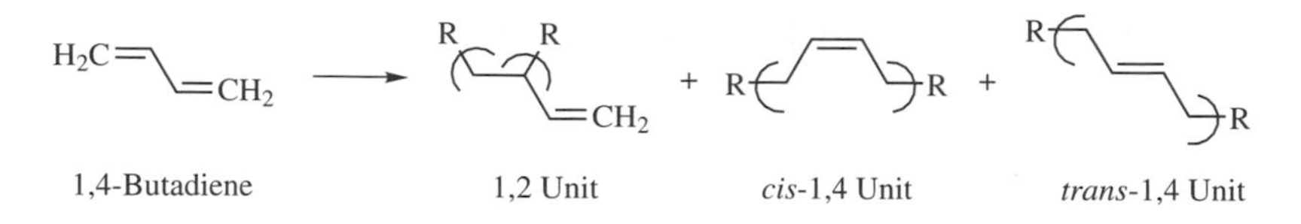

A second use is in the manufacture of ABS copolymers where the stereogeometry is also important. A polybutadiene composition of about 60% *trans*-1,4; 20% *cis*-1,4;

and 20% 1,2 configuration is generally employed in the production of ABS. The good low-temperature impact strength is achieved in part because of the low T_g values for the compositions. For instance, the T_g for *trans*-1,4-polybutadiene is about −24°F, while the T_g for *cis*-1,4-polybutadiene is about −160°F. Most of the ABS rubber is made employing an emulsion process where the butadiene is initially polymerized forming submicron particles. The styrene–acrylonitrile copolymer is then grafted onto the outside of the BR rubber particles. ABS rubbers are generally tougher than HIPS rubbers but are more difficult to process. ABS rubbers are used in a number of appliances including luggage, power tool housings, vacuum cleaner housings, toys, household piping, and automotive components such as interior trim.

Another major use of butadiene polymer is in the manufacture of high-impact polystyrene (HIPS). Most HIPS has about 4–12% polybutadiene in it so that HIPS is mainly a polystyrene-intense material. Here the polybutadiene polymer is dissolved in a liquid along with styrene monomer.

The polymerization process is unusual in that both a matrix composition of polystyrene and polybutadieneis is formed as well as a graft between the growing polystyrene onto the polybutadiene being formed. The grafting provides the needed compatibility between the matrix phase and the rubber phase. The grafting is also important in determining the structure and size of rubber particles that are formed.

Because of the presence of a methyl group replacing one of the hydrogens, polyisoprene is composed of four structures as shown below. As in the case of polybutadiene, it is the *cis*-1,4 structure that is emphasized commercially. The *cis*-1,4-polyisoprene is similar to the *cis*-1,4-polybutadiene material except it is lighter in color, more uniform, and less expensive to process. Composition-wise, polyisoprene is analogous to natural rubber. The complete *cis*-1,4 product has a T_g of about −100°F. Interestingly, isomer mixtures generally have higher T_g values. Thus an equal molar product containing *cis*-1,4, *trans*-1,4, and 3,4 units has a T_g of about −40°F.

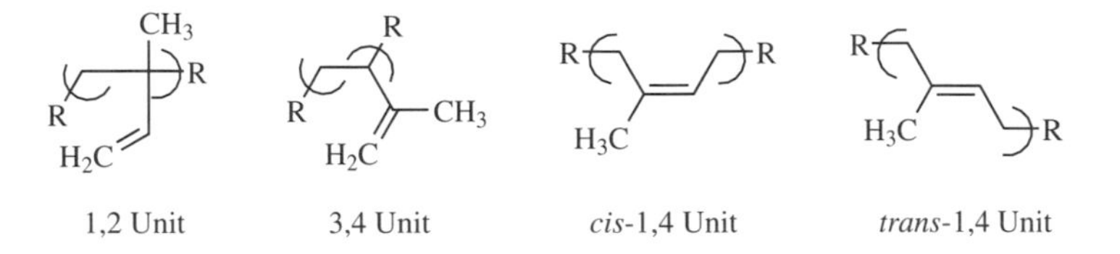

The double bond is often reacted, giving a cross-linked product. Polyisoprene rubbers are used in the construction of passenger, truck, and bus tires and inner liners as well as sealants and caulking compounds, sporting goods, gaskets, hoses, rubber sheeting, gloves, belts, and footwear. The polyisoprene elastomer is designated as IR by the ASTM and has been called by the incongruous trade name Natsyn, which stands for natural synthetic rubber.

Polychloroprene was the first commercially successful synthetic elastomer introduced in 1932 under the trade names of DuPrene and Neoprene by DuPont. It was discovered by Carothers and co-workers. Because of its early discovery, good synthetic routes were worked out prior to the advent of good steroregulating catalytic systems. Thus, polychloroprene is largely manufactured by emulsion polymerization using both batch and continuous systems. Free radical products contain mainly 1,4-*trans* units.

Compounding of polychloroprene is similar to that of natural rubber. Vulcanizing is achieved using a variety of agents including accelerators. Because of its durability, polychloroprene rubber is often used where deteriorating effects are present. It offers good resistance to oils, ozone, heat, oxygen, and flame (the latter because of the presence of the chlorine atom). In the automotive industry, it is used to manufacture hoses, V-belts, and weatherstripping. Rubber goods include gaskets, diaphragms, hoses, seals, conveyer belts, and gaskets. It is also used in construction for highway joint seals, bridge mounts and expansion joints, and soil-pipe gaskets. Finally, it is also used for wet-laminating and contact-bond adhesives, in coatings and dipped goods, as modifiers in elasticized bitumens and cements, and in fiber binders.

10.7 POLYISOBUTYLENE

Polyisobutylene (PIB) was initially synthesized in the 1920s. It is one of the few examples of the use of cationic catalysis to produce commercial scale polymers. Low-molecular-weight (about 90 units) PIB can be produced at room temperature, but large chains (about 20,000 units) are made at low temperatures where transfer reactions are suppressed.

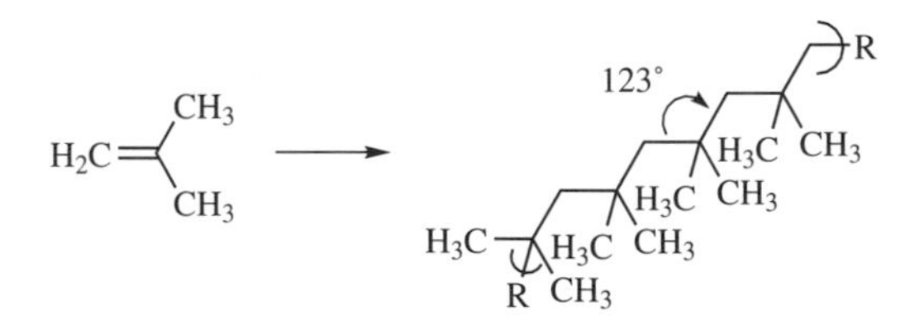

PIB itself is sticky. Because PIB is fully saturated, it is cured as a thermoset elastomer through inclusion of about 1–2% isoprene that supplies the needed double bonds used in the curing process. This copolymerization with a small amount of isoprene gives a material that is not sticky.

PIB and various copolymers are called butyl rubber and given the designation IIR by the ASTM. Butyl rubbers have lower permeability and higher damping than other elastomers, making them ideal materials for tire innerliners and engine mounts. PIB is also used in sealing applications and medical closures and sealants.

10.8 HEAT-SOFTENED ELASTOMERS

While we considered physical cross-linking occurring because of chain entanglement, it can be imposed on a system because of localized crystallization. Thus, above the T_g the amorphous segments are able to move about but crystalline sites are still tied together, limiting the mobility of the sites and surrounding segments. Materials known as thermoplastic elastomers make use of these crystalline cross-links. A number of copolymers employ this type of cross-linking. The crystalline sites are called the "hard" segments while the amorphous sites are called the "soft" segments. In styrene–butadiene–styrene block copolymers (SBS) under room conditions, the styrene blocks form crystalline (hard segments) regions that lock in the butadiene amorphous blocks (soft segments) that are above their T_g and are free to act in an elastic manner. The styrene blocks, with a T_g (polystyrene itself has a T_g of about 212°F) above room temperature, act as cross-links.

Spandex (Lycra™) (Section 9.8) is another example of the hard/soft strategy. Here the "macroglycol" portion is the soft segment and the rigid urethane segments act as the hard segment.

The first thermoplastic elastomers of commercial value were the plastisols formed from introducing plasticizers into poly(vinyl chloride). The plasticizers act to keep the PVC segments from forming large-scale crystalline regions, causing the plasticized PVC to act like it is above its T_g (PVC has a T_g of about 180°F). PVC pipe and other PVC plastics are generally plastisols.

Ionomers are thermosets that can be processed as thermoplastics. Inomers are cross-linked through introduction of metal ions that bind the acid portions of the poly(ethylene-co-methacrylic acid). Here, the ionomer can be processed because the ethylene portion of the copolymer gives mobility to allow localized movement sufficient to allow movement and reforming through application of heat and pressure. Ionomers are flame-retardant and tough and are used as the covers of many of the golf balls, shoe soles, and weather stripping.

These elastomeric materials can be recycled and reformed through application of sufficient heat to melt the crystalline regions.

10.9 OTHER SYNTHETIC ELASTOMERS

Although the bulk of synthetic rubber is of the SBR variety, several other elastomers have been synthesized for special-purpose applications. These rubber products generally are more costly than natural and SBR rubber, but their special properties justify their higher costs.

In addition to synthesizing Buna-S, Tschunker and Bock also patented a process for the aqueous emulsion copolymerization of butadiene and acrylonitrile. They called this oil-resistant elastomer Buna-N. This copolymer is now produced under the ASTM name NBR.

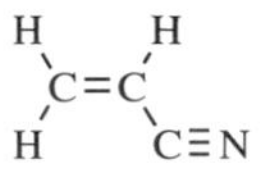

```
H     H
 \   /
  C=C
 /   \
H     C≡N
```

Acrylonitrile

The first American synthetic elastomer was synthesized by Patrick in 1927. Since the product was an organic polysulfide, it was called Thiokol. The prefix thio is the Greek word for sulfur. The ol suffix was used because the original objective was to produce a permanent antifreeze, that is, ethylene glycol.

$$2n\,\mathrm{Cl{-}\!\left(CH_2\right)_2\!{-}Cl} \; + \; 2n\,\mathrm{NaS_xNa} \longrightarrow \mathrm{\left[\left(CH_2\right)_2 S_x {-} \left(CH_2\right)_2 S_x\right]_n}$$

Ethylene dichloride — Sodium polysulfide — Thiokol

Heat-resistant elastomers, called silicone rubbers and designated as SI by ASTM, are produced by a repetitive condensation of dimethyldichlorosilane in the presence of water. The backbone of this polymer consists of siloxane units (–Si–O–Si–O–), which are characterized by a high bond strength.

$$n\,\mathrm{Cl{-}Si(CH_3)_2{-}Cl} \xrightarrow{H_2O} \mathrm{\left(Si(CH_3)_2{-}O\right)_n}$$

The methyl groups on the chain act like oil or paraffin and provide nonstick and water-repellent properties in these elastomers.

Neither HDPE nor polypropylene (PP) is elastomeric. However, the copolymer of ethylene and propylene (EP) is an amorphous copolymer with elastomeric characteristics. A commercial vulcanizable elastomer (EPDM) is produced when a diene is added to ethylene and propylene before polymerization by a Ziegler–Natta catalyst.

Polyphosphazenes are useful as elastomers over an unusually broad temperature range. Polydichlorophosphazene, $\mathrm{\left(N{=}P(Cl_2)\right)_m}$, is unstable in humid atmospheres, but stable elastomers are produced when the chlorine substituents are replaced by phenoxy groups ($-OC_6H_5$) or other organic materials.

Elastomeric polyurethanes are produced by the reaction of a flexible polyester or polyether diol (HO–R–OH) with a diisocyanate (OCNRNCO).

Other important classes of elastomers are also available. Polyurethanes represent a broad range of elastomeric materials. Most polyurethanes are either hydroxyl or isocyanate terminated. Three groups of urethane elastomers are commercially produced. Millabile elastomers are produced from the curing of the isocyanate group using trifunctional glycols. These elastomers are made from high polymers made by the chain extension of the polyurethane through reaction of the terminal isocyanate groups with a polyether or polyester. Low molecular weight isocyanate terminated polyurethanes are cured through a combination of chain extension by reaction with a hydroxyl-terminated polyether or polyester and trifunctional glycols giving cast elastomers. Thermoplastic elastomers are block copolymers formed from the reaction of isocyanate-terminated polyurethanes with hydroxyl-terminated polyethers or polyesters. These are generally processed as thermoplastic materials as are the thermoplastic elastomers. Many of these materials have little or no chemical cross-linking. The elastomeric behavior is due to the presence of physical hard domains that act as cross-links. Thus, SBR consists of soft butadiene blocks sandwiched between polystyrene hard blocks. These hard blocks also act as a well-dispersed fine-particle reinforcing material increasing the tensile strength and modulus. The effectiveness of these hard blocks greatly decreases above the T_g (about 100°C) of polystyrene.

As described before, silicons form another group of important elastomers. Again, processing typically does not involve either carbon black or sulfur.

Unforseen complications can arise. The Ford Motor Company was using EPDM for their radiator hose, but it kept deteriorating prematurely. The pre-EPDM contains unreacted double bonds that are cross-linked using sulfur, thereby producing EPDM rubber. The scientists replaced EPDM with other rubbers, but the radiator hoses continued to prematurely deteriorate. Eventually, a Ford scientist noticed that the voltage difference between the radiator and car block was about 0.5 V, which was sufficient to oxidize the sulfur cross-links and thereby reduce the material to a pliable pre-cross-linked form. (Remember that most rubbers and other general-use giant molecules are nonconductive.) Elimination of the voltage difference through grounding allowed the radiator hoses to operate as originally planned.

10.10 PROCESSING OF ELASTOMERS

With only a few exceptions, the general steps involved in processing natural and synthetic rubber are the same. The exact steps vary according to the polymer utilized and its intended application.

Manufacture using bulk natural and synthetic material can be divided into four steps:

- Mastication
- Incorporation or compounding
- Shaping
- Vulcanization

The shaping and vulcanization steps are combined in a number of processes such as transfer or injection molding or may be separated as in the extrusion and subsequent vulcanization sequence. An outline of these steps is given in Figure 10.6.

Mastication is intended to bring the material to the necessary consistency to accept the compounding ingredients. Mastication results in a lowering of chain length. Two basic types of internal mixers are in use. The Banbury has rotors rotating at different speeds, creating a "kneading" action such as that employed in handling bread dough. A shearing action between the rotors and the walls of the mixer is also achieved. The Shaw Intermix employs rotors that turn at the same speed and closely intermesh, thereby causing an intracompound friction for mixing, thus closely resembling a mill's mixing action.

Compounding includes the incorporation of various additives. The chief ingredients of compounded rubber are (a) sulfur, (b) accelerators, (c) pigments, (d) antioxidants, (e) reclaimed (recycled) rubber, and (f) fillers, such as carbon black. Each

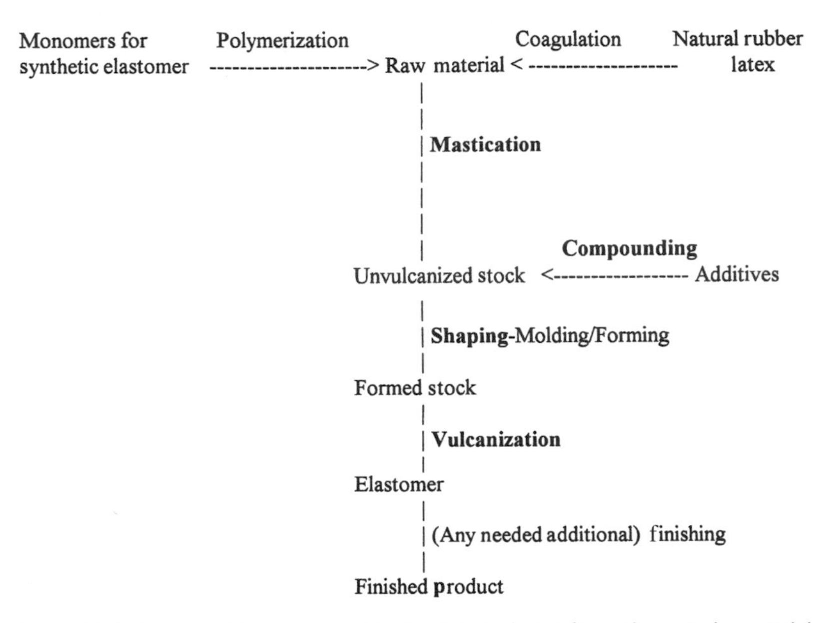

Figure 10.6. Outline of steps involved in the processing to form elastomeric materials.

of these additives performs a special function. Sulfur is added early in the processing and is largely responsible for the formation of cross-links in the vulcanization process. Accelerators (catalysts) are added to enhance the rate of vulcanization. Reinforcing pigments, such as carbon black, make rubber stronger and more resistant to wear. Antioxidants protect the rubber against chemical changes and the harmful effects of air, moisture, heat, and sunlight. Reclaimed rubber is recycled rubber that has been pretreated to make it compatible with the rubber mix.

The rubber mixture (compound) is shaped by (a) calendering, (b) extrusion, (c) molding, or (d) dipping. Calendering means rolling the rubber material into sheets. In extrusion, tube machines push the soft rubber material through different-sized holes, similar to pushing toothpaste out of the tube. Extruded products include hoses, inner tubes, rubber stripping for use on refrigerator doors, and automobile windshields. Extruded products are vulcanized after they have been formed.

The material is now heated to cure, set, or **vulcanize** (all terms are appropriate) the material into the (typically) finished shape. During vulcanization, the heat causes the sulfur to combine with the rubber, thereby forming cross-links. Between 1% and 5% of sulfur (by weight) is added in typical black rubber mixes, giving a vulcanized material with an average of about 500 carbon atoms between cross-links. Generally, the more sulfur added, the greater the number of cross-links giving a harder rubber. Ebonite is formed from a compound containing one-third sulfur. Today, many other vulcanizing agents are used, but sulfur is still the major vulcanizing agent because of its low cost, availability, and familiarity.

Familiarity includes both knowledge and instrumentation. It is important because replacing a material may include retraining of technicians, determining optimum conditions, developing new quality control systems, and new instrumentation. Thus, there are many considerations when adopting new procedures and chemicals.

Most products are molded and vulcanized in the same step. Molded products include rubber tires, mattresses, and hard-rubber articles such as rubber hammers, gaskets, fittings, shoe soles, and heels.

Sometimes additional finishing may be desirable including painting, machining, grinding, and cutting.

Dipping is used to make products such as rubber gloves and balloons from liquid latex. Forms, typically made of glass, metal, or ceramic, are dipped into vats of latex. Repeated dipping increases the thickness of the product.

Recently, the use of thin rubber gloves has greatly increased. Most of the rubber gloves are made from natural rubber, where the gloves are formed from simply dipping a hand mold into the natural latex. The latex also contains small amounts of cross-linking agents such as organic disulfides. After the latex is dried, it is heated and lightly cross-linked to the extent of about one cross-link per 80 isoprene units (for comparison, the cross-link density for a typical rubber band is about one-half of this).

Natural latex contains about 92% rubber and 3–4% protein, with the rest being other ingredients such as lipids. Some people are allergic to the proteins in the latex gloves. This is generally overcome by either using gloves made of synthetic materials such as polyurethane and nitrile gloves or subjecting the gloves to an additional washing that removes surface protein.

Latexes are also used to make thread for the garment industry and to make adhesives for shoes, carpets, and tape.

Thermoplastic elastomers that may be used in place of cross-linked elastomers are not vulcanized but are simply molded, like thermoplastics in a heated mold. Rubbery products, such as solid tires and bumpers, are produced by reaction injection molding (RIM) in which the reactants are injected into the mold, where the polymerization reaction takes place.

Lattices of elastomers are processed like water-borne coatings. The curing or cross-linking agents and accelerators, stabilizers, and pigments are added as aqueous dispersions prior to curing.

Sponge rubber can be made from either dry rubber or latex. For dry rubber, chemicals such as sodium bicarbonate ($NaHCO_3$) are added that will form gas when heated during the vulcanization process and produce bubbles. Foam rubber, which is used for upholstery and foam strips for surgical use, is obtained by "whipping" air into the latex.

10.11 TIRES

The pneumatic (air-filled) tire was invented in 1845 by a Scottish engineer, Robert Thomson, but it was not strong enough for regular use. In 1888, John Dunlop, a Scottish veterinarian, developed usable rubber tubes for his son's tricycle.

The early automobile tires were single air-filled rubber tubes. Although they made movement easier and the ride smoother, they tended to develop many leaks. A two-piece tire consisting of a flexible, thin "inner tube" and a tough "outer tube" was developed in the early 1900s. The modern tubeless tire was developed in 1948. The basic parts of a tire are shown in Figure 10.7.

Modern tires are built on a slowly rotating roller called a drum. Initially, an inner liner of soft rubber is wrapped about the drum. A rubberized cord fabric is then laid, typically perpendicular to the initial inner liner. This fabric is called ply. A four-ply tire has four layers of cord fabric; a two-ply tire has two layers of cord. Next, layers of fibers [normally nylon, aramid (aromatic nylon), rayon, or fiberglass] or steel are laid. These layers are called belts. Most tires contain two belts.

An inner and outer ridge, called a bead, is then constructed. Each bead contains steel wire strands wound together into a hoop and covered with hard rubber. The ends of each ply are wrapped about the bead, forming the connective points of the tire to the wheel rim. Rubber sidewalls and outer tread material are then added. The individual parts—sidewalls, beads, plies, belts, and inner liner—are connected in a process called stitching.

The tire parts are now "permanently" cemented together in a process called vulcanization. The tire is removed from the drum and placed in a mold (curing press) that has the appropriate tread pattern. The mold acts like a large waffle

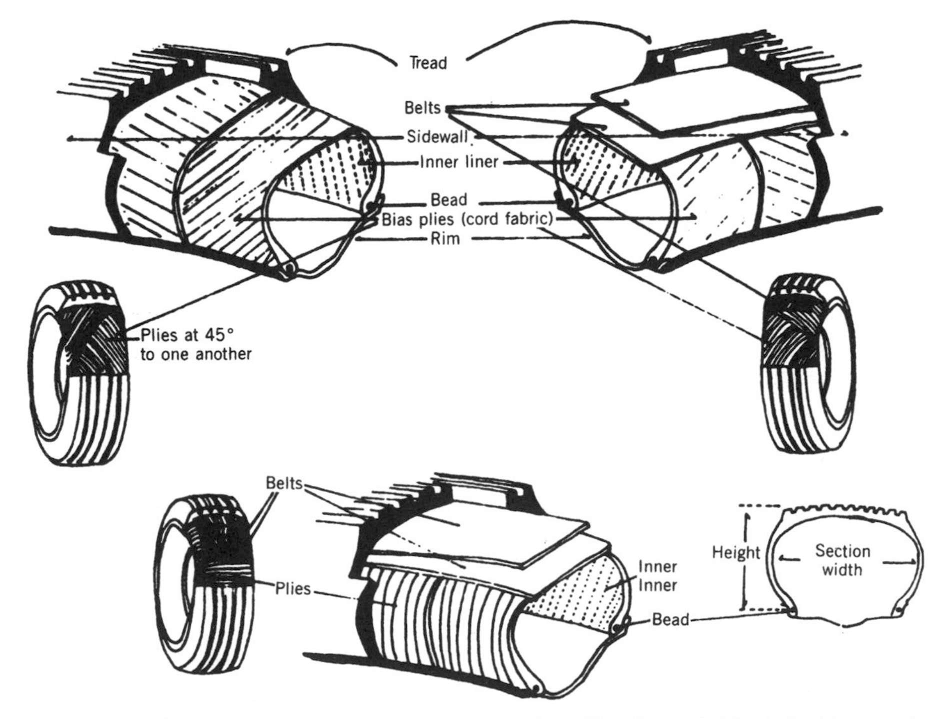

Figure 10.7. Cut-away views of modern passenger tires: Bias (*top left*), bias-belted (*top right*), and radial (*bottom*).

iron: The "raw" tire is inserted in the mold, the mold is closed, heat is applied, and an inner air bag is filled with steam. The filled air bag pushes the tire against the mold sidewall that causes the tread pattern.

Bias-belted tires are constructed similarly to bias tires except belts are placed between the plies and tread. The belts act to coordinate the movement of the plies and treads, and thus resist tread squirm and puncture. Fiberglass belts are used in bias-belted tires.

All radial tires are belted. The cord fabric is added with the cord running radially—that is, from bead to bead. The combination of belting and addition of radial plies results in less flex and squirm and longer wear than either the bias or bias-belted tires. The belts are mainly steel, aramid, fiberglass, or rayon. These types of passenger tires are illustrated in Figure 10.7.

The physical properties of typical elastomers are shown in Table 10.2. While there are a variety of materials employed to make the modern tire, many of them contain a chlorobutyl or bromobutyl rubber inner core; the body of the tire is a blend of NR and SBR; and the tread is a different blend of NR and SBR. The bead and belts are steel wiring coated with brass (a copper–zinc alloy) and the cord is made of PET polyester. Once tire blowouts were common, but the chlorobutyl (or bromobutyl) rubber inner liner has a low diffusion that absorbs the pressure drop from about three times the outside pressure (like 30 to 45 pounds of tire pressure) to the pressure found outside the tire (about 15 pounds/square inch) for the remainder of the tire composition. The brass coating for the steel bead and belting actually forms a covalent bond between the metal bead/belts and the rubber, R, through formation of R-S-Zinc-Brass (steel) bonding.

Table 10.2 Physical properties of typical elastomers

	Pure Gum Vulcanizates			Carbon-Black-Reinforced Vulcanizates	
	T_g (°F)	Tensile Strength (kg/cm^{-2})	Elongation (%)	Tensile Strength (kg/cm^{-2})	Elongation (%)
Natural rubber (NR)	−100	210	700	315	600
Styrene–butadiene rubber (NBR)	−60	28	800	265	550
Acrylonitrile–butadiene rubber (NBR)		42	600	210	550
Polyacrylates (ACM)				175	400
Thiokol (T)		21	300	85	400
Neoprene (CR)		245	800	245	700
Butyl rubber (IIR)	−81	210	1000	210	400
Polyisoprene (IR), polybutadiene (PB)	−140	210	700	315	600
Polyurethane elastomers (AU)		350	600	420	500

10.12 THE BOUNCE

We know why balls bounce, don't we? Is the bounce of a ball related to energy or probability factors? The answer to the second question, of course, is "Yes." The bounce depends on both factors. This dependency can be illustrated through the following demonstration, which uses a metal ball bearing and a "superball." When dropped onto a hard surface, the balls hit the surface and rebound, but why?

The rebound of the ball bearing is largely due to the deformation of metal bonds upon striking the hard surface. This collision pushes the metal atoms into a higher energy situation. The metal atoms then move back to the original, lower energy sites, resulting in a push against the surface and the "bounce." For the ball bearing there is little unoccupied or free volume so the applied force is used to disrupt the primary (metal) bonded iron atoms with little change in the overall order of the iron atoms. Energy is the principal driving force here.

Polymers are most tightly packed when they are arranged in an ordered fashion such as a folded clothesline or thread on a spool. The "superball" is a semitough rubber that is solidified so that the polymer chains are arranged in a highly disorganized, random manner. When the "superball" hits the surface, the decreased space is largely accommodated by a reorganization of the polymer chains into a more ordered, less probable configuration. When the polymer chains return to their original, highly disorganized state, the "push" to occupy the original, predeformation volume is translated into a push against the surface, resulting in the "bounce." In this situation, probability is the major driving force in the bounce.

As noted above, the rubber ball goes from its indented organized, ordered arrangement of giant molecules back to the original less ordered arrangement. The move toward disordered situations is a natural or more probable occurrence. For instance, consider your bedroom "where every thing begins in its place." After a while, chairs are moved, cloths are on the floor, the bed is unmade, and so on. This is what is referred to as a natural move toward disorder. Because this tendency is so widespread, it has been incorporated as part of a scientific law that has been summarized by some as "things go from bad to worse," or another statement of this same idea is that in the absence of any intervening factors, disorder increases. The intervening factor is energy. Thus, we can overcome the room becoming more disorganized by using energy to clean up and reorder the room.

In summary, the driving force for the metal ball bounce is energy-related while the driving force for the rubber ball bouncing is a drive toward disorder or a more probably orientation. Thus, an apparently similar phenomenon, the "bounce," mainly results from two different factors in the metal and rubber balls.

GLOSSARY

Accelerator: Vulcanization catalyst.

Aluminum chloride: $AlCl_3$.

Aramid: An aromatic nylon.

Boric acid: H_3BO_3.

Buna-N: Copolymer of butadiene and acrylonitrile.

Buna-S: Original name for SBR.

Butadiene: C=C–C=C.

Butene: C–C=C–C.

Butyl rubber: Copolymer of isobutylene and isoprene (IIR).

Caoutchouc: Natural rubber.

Chloroprene: 2-Chlorobutadiene

$$\text{(—C=}\overset{\text{Cl}}{\overset{|}{\text{C}}}\text{–C=C—)}$$

Cis arrangement: Configuration in which substituents or chain extensions are on the same side of the ethylene double bond.

1,2 Configuration:

$$\left[\text{C(–CH=C)–C}\right]_n$$

CR: Neoprene.

Cross-link: Primary bonds joining polymer chains.

Elastomer: A general term for natural and synthetic rubbers.

Empirical formula: Simplest formula.

EPDM: Vulcanizable elastomeric copolymer of ethylene and propylene.

Ethylene–propylene copolymers (EP):

$$\left[\text{C–}\underset{\underset{\text{C}}{|}}{\text{C}}\text{–C–C}\right]_n$$

Formic acid: HCOOH.

Goodyear, Charles: Discoveries of vulcanization (cross-linking of rubber).

GRS: Government rubber styrene; name used for SBR during World War II.

***Hevea braziliensis*:** Natural rubber.

IIR: Butyl rubber.

Intermolecular force: Attraction between atoms on different polymer chains.

IR: Polyisoprene.

Isobutylene:

$$\text{(C)}_2\text{C=C}$$

Isoprene:

$$\text{(C=}\overset{\text{C}}{\overset{|}{\text{C}}}\text{–C=C)}$$

Latex: An aqueous emulsion.

Memory: The process whereby stretched elastomers return to their original dimensions when tension is released.

Methylisoprene:

```
  C C
  | |
C=C-C=C
```

***n*:** Number of repeating units in a polymer (DP).

Natsyn: A trade name for polyisoprene.

NBR: Copolymer of butadiene and acrylonitrile.

Neoprene: Polychlorobutadiene.

NR: Natural rubber.

Ozone: O_3.

Polyphosphazene: Inorganic elastomer with the repeating unit $-N{=}P(R_2)-$.

Primary bond: Covalent bond of carbon–carbon atoms.

Pyrolysis: Process of thermal degradation.

Random copolymer: A macromolecule containing randomly arranged repeating units of two different monomers.

SBR: Styrene–butadiene elastomers.

Silane: $Si(CH_3)_4$.

Silicon: Si.

Silicone: Incorrect name for polysiloxanes.

```
  R     R
  |     |
-[Si-O-Si-O]n-
  |     |
  R     R
```

Siloxane: Compounds containing one or more Si—O unit such as

```
  |     |
—Si-O-(Si-O)n
  |     |
```

Skeletal formula: Structural formula showing carbon–carbon bonds and omitting the hydrogen atoms such as used in this glossary.

Sodium: Na.

SR: Synthetic rubber.

Trans arrangement: Configuration in which substituents or chain extensions are on opposite sides of the ethylene double bond.

***Ule, ulei*:** Names used by Aztecs for the rubber tree and rubber, respectively.

Vinylacetylene: $C{=}C{-}C{\equiv}C$.

Ziegler–Natta catalyst: Catalyst system that produces linear polyethylene and stereoregular vinyl polymers, usually $TiCl_3$ and $Al(CH_3)_3$.

REVIEW QUESTIONS

1. What are the advantages of using the guayule shrub as a source of natural rubber?
2. What does rubber latex have in common with milk?
3. What substance did Charles Goodyear use as a cross-linking agent in his vulcanization process?
4. What is stronger: the sulfur bonds that cross-link the polymer chains or the intermolecular forces between polymer chains in elastomers?
5. A stretched rubber band returns to its original dimensions when the tension is released because of what characteristic quality of elastomers?
6. What is the difference between Buna-S and SBR?
7. Why is butyl rubber more resistant to ozone degradation than natural rubber?
8. What is the difference between DP and n?
9. Which of the following configurations is cis?

(a) H(R)C=C(R)(H) — upper substituents H and R, lower substituents R and H

(b) R(H)C=C(R)(H) — upper substituents R and R, lower substituents H and H

10. What is the difference in the structure of isoprene and butadiene?
11. Which repeating unit provides more elasticity in SBR?
12. What compounds are present in the Ziegler–Natta catalyst?
13. Why is a silicone elastomer more heat resistant than *Hevea braziliensis*?
14. Why is ethylene–propylene copolymer (EPDM) used in place of *Hevea* rubber in white sidewalls of tires?
15. What is the advantage of using an accelerator in the vulcanization of *Hevea* rubber?
16. Define an accelerator.

BIBLIOGRAPHY

Bhowmick, A., and Stephens, H. (2000). *Handbook of Elastomers*, Marcel Dekker, New York.

Brown, R. P. (2001). *Practical Guide to the Assessment of the Useful Life of Rubbers*, ChemTec, Toronto.

Brown, R. P., Butler, T., and Hawley, S. W. (2001). *Ageing of Rubber-Accelerated Weathering and Ozone Test Results*, ChemTec, Toronto.

Ciullo, P., and Hewitt, N. (1999). *Rubber Formulary*, ChemTec, Toronto.

Coling, J-P. (1997). *Silicon-On-Insulator Technology*, 2nd ed., Kluwer, Hingham, MA.

Craver, C., and Carraher, C. (2000). *Applied Polymer Science*, Elsevier, New York.

Holden, G., Legge, N., Quirk, R., and Schroeder, H. (1996). *Thermoplastic Elastomers*, 2nd ed., Hanser-Gardner, Cincinnati.

Johnson, P. (2001). *Rubber Processing: An Introduction*, Hanser-Gardner, Cincinnati.

Scheirs, J. (2000). *Practical Polymer Analysis: Techniques and Strategies for the Compositional and Failure Analysis of Polymers, Elastomers and Composites*, Wiley, New York.

Wright, R. C. (2001). *Failure of Plastics and Rubber Products*, ChemTec, Toronto.

ANSWERS TO REVIEW QUESTIONS

1. Guayule can be grown in the arid areas of northern Mexico and southwestern United States. It can be harvested mechanically and no overseas shipment is involved.
2. They are both emulsions of polymers, that is, *Hevea* rubber and casein.
3. Sulfur (S).
4. The sulfur bonds. These are the primary covalent bonds, which are at least 25 times stronger than the weak intermolecular forces (London or dispersion forces, i.e., secondary bonds).
5. Memory.
6. They are the same—that is, copolymers of butadiene and styrene.
7. Butyl rubber has fewer double bonds. Ozone attacks the ethylene double bonds.
8. They are identical. Each is equal to the number of repeating units in a macromolecule.
9. (b) Both substituents are on the same side of the plane of the ethylenic double bond.
10. Isoprene is 2-methylbutadiene.
11. Butadiene.
12. $TiCl_3$ and $Al(C_2H_5)_3$.
13. The siloxane bonds are stronger than carbon–carbon bonds.
14. EPDM has fewer carbon–carbon double bonds and hence is more resistant to ozone, which causes cracking of *Hevea* rubber.
15. Decreases curing time.
16. Accelerators are catalysts that speed up the vulcanization process.

11

PAINTS, COATINGS, SEALANTS, AND ADHESIVES

Giant Molecules: Essential Materials for Everyday Living and Problem Solving, Second Edition,
by Charles E. Carraher, Jr.
ISBN 0-471-27399-6

11.1 HISTORY OF PAINTS

Early humans used crude paintings as a means of communication. The mineral-based paintings of a bison in the Altimara cave in Spain and of a Chinese horse at Laucaux, in France, are at least 15,000 years old. Aboriginal mineral-based paintings, such as the Obiri Rock sketches at Arnhem Land in northern Australia, are at least 5000 years old.

Lacquer, which includes a polymer as its principal component, originated in China at the time of the Chou Dynasty, sometime before the Christian era. The lacquers in China and Japan were based on the sap of specific trees, whereas the lacquers in India and Burma were based on shellac, which is a resinous material exuded by insects. The early painters also used exudates from trees, such as copals, and these resins are still used today. A lacquer is a solution that forms a film by evaporation of the solvent. The Egyptians employed pitch and balsam resins as sealants for ships.

Because of high costs and lack of knowledge, improvements in the lacquer and sealant art were slow. However, pigments, such as white lead ($2PbCO_3 \cdot Pb(OH)_2$), zinc oxide (ZnO), litharge (PbO), red lead (Pb_3O_4), and carbon black (C), as well as naturally occurring polyunsaturated vegetable oils, were produced prior to the Industrial Revolution.

11.2 PAINT

Oleoresinous coatings or paint have been produced from flax seed (linseed oil, *Lininum usitatissium*) and finely divided pigments since the fourteenth century using recipes supplied by the monk Theophilus. Other vegetable oils used in the paint industry are soybean, safflower, tung, oiticica (*Licania rigida*), and menhaden oils. All of these oils contain monounsaturated oleic acid ($C_{17}H_{33}COOH$) and diunsaturated linoleic acid ($C_{17}H_{31}COOH$). Linseed and soybean oil also contain triunsaturated linolenic acid ($C_{17}H_{29}COOH$). The polymerizable unsaturated oil, which is called a binder, polymerizes by cross-linking in the presence of oxygen, and this reaction is catalyzed (accelerated) in the presence of soluble organic acid salts of heavy metals, such as lead or cobalt.

The polymerization (hardening or drying) of these unsaturated oils is a chain reaction, which is similar to the initiation, propagation, and termination that occurs in the addition polymerization of vinyl monomers (Section 3.2). In the initiation reaction, the unsaturated oil adds oxygen to produce a hydroperoxide, which decomposes to produce a free radical. These free radicals are responsible for the propagation, which also involves cross-linking. The binder is the film former. The liquid, which includes the binder, is called the vehicle or medium. The unpigmented paint is called a varnish. The solvent that dissolves the binder is sometimes called a "thinner." Although these oil-based paints are still in use today, they have been displaced, to some extent, by lacquers based on man-made resins and waterborne coatings.

Table 11.1 U.S. production of paints and coatings, 2000

Painting/Coating	Production (Millions of Gallons)
Architectural	668
Product	449
Special	189

Source: Department of Commerce.

The first widely used commercial lacquer was based on pigmented cellulose nitrate. This man-made resin, as well as natural resins such as kauri, was dissolved in ester solvents, such as butyl acetate, and used as an automotive finish (Duco) and textile finish (Pyroxylin).

Although the resins or resin-forming compounds used in paints may be cured or cross-linked after application, they must be linear and flexible when applied to the surface or substrate. The glass transition temperature (T_g) is used to define the temperature above which a polymer is flexible because of the segmental motion of the polymer chains. All coatings must be applied at temperatures above their characteristic glass transition temperatures.

The product mix for coatings is changing from organic solvent-borne to waterborne or high-solids coatings, and the market for industrial and residential coatings continues to grow. Over 1 billion (10^9) gallons of coatings with a value of almost \$10 billion (10^{10}) are produced annually in the United States by over 1200 paint manufacturers. However, the 10 leading manufacturers produce about 40% of all paint sold in the United States.

The principal steps in the production of coatings are mixing, grinding, and thinning. In most cases, the pigment is mixed with the vehicle to form a heavy paste, which is then ground in a ball mill or by a high-speed impeller. Appropriate solvents (thinners) are then added to these dispersions. The mixing and grinding step for waterborne coatings is similar to that used for solution coatings, but the liquid dispersion for waterborne coatings is water. The dispersion of the pigments and other additives are then mixed with an aqueous dispersion of the resin. The components of the five major types of coatings (paint) are shown in Figure 11.1.

Table 11.1 contains a listing of the production of paints for the year 2000.

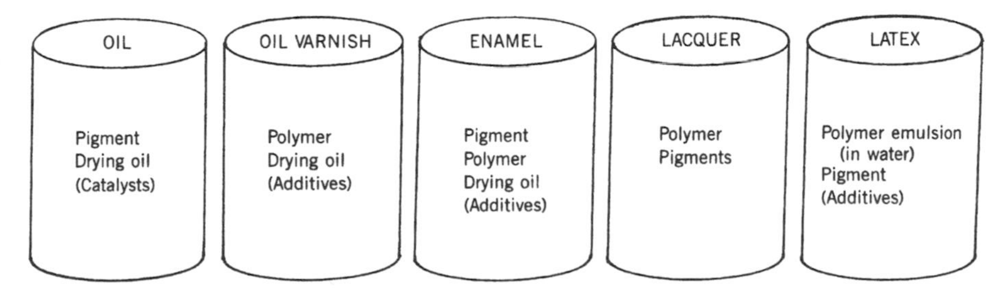

Figure 11.1. Components of the five major types of coatings (paint).

Since the primary cost of most commercial application of coatings is labor, the market will allow price increases for products that give added positive properties.

Major driving forces in coatings continues to be a move toward water-based and solids (with little or no solvent) coatings. Another is to eliminate the "odor" of the coating. Most waterborne coatings actually have about 8–10% non-water solvent. The odor we get as the coating is drying is mainly due to the non-water solvent evaporating. Work continues to develop the right balance of properties and materials that allow the latex particles to flow together and coalesce into suitable films without the need of non-water liquids.

Another area of active research is the development of paints that dry under extreme or unusual conditions including under water and on cool substrates. The latter allows the painting season for exterior coating to be extended, particularly in the northern states.

Work continues on making more durable exterior paints. Remember that there is a difference in requirements of exterior and interior paints. For instance, interior paints are generally required to be faster-drying and more durable against scraps and punctures since it is the inside of the house that generally experiences such traumatic events. By comparison, exterior paints need to remain flexible and adhered under a wide variety of humidity and temperature. A more durable exterior coating should allow it a longer lifetime because it can better withstand stress caused by the pounding of the rain, sticks, and human afflicted dings and dents.

11.3 PAINT RESINS

The first synthetic paint resin was introduced by Leo Baekeland early in the twentieth century. This soluble resin was produced by heating phenol (C_6H_5OH) and formaldehyde ($H_2C{=}O$) in the presence of rosin. The latter is one of the constituents of the exudate from pine trees.

The most widely used paint resin, which is called an alkyd, was introduced in 1925 by R. Kienle. It was obtained by the reaction of an alcohol, such as glycerol, and an acid, such as phthalic acid. Unsaturated acids, such as linoleic acid, are also incorporated in the alkyd reactants so that this oil-modified resinous product is unsaturated. Hence, oil-modified alkyds will cure or dry in air much like the oil paints. Another ester type of paint called glyptal was introduced by W. Smith in 1901. Glyptal is produced by the condensation of glycerol and phthalic anhydride.

Many other synthetic polymers, such as chlorinated rubber, polyvinyl chloride, polystyrene, melamine–formaldehyde, silicone, and epoxy resins, are also used as paint resins. In the past, these coatings have been applied as solutions of the resins in volatile solvents. Since the solvents usually are evaporated into the atmosphere, they contribute to atmospheric pollution. Accordingly, alternate methods of application, such as powdered resins and aqueous emulsions of resins, are preferred today in place of the solvent-based application of coatings. The solvent content of coating solutions is also being reduced. These higher-solids coatings, waterborne

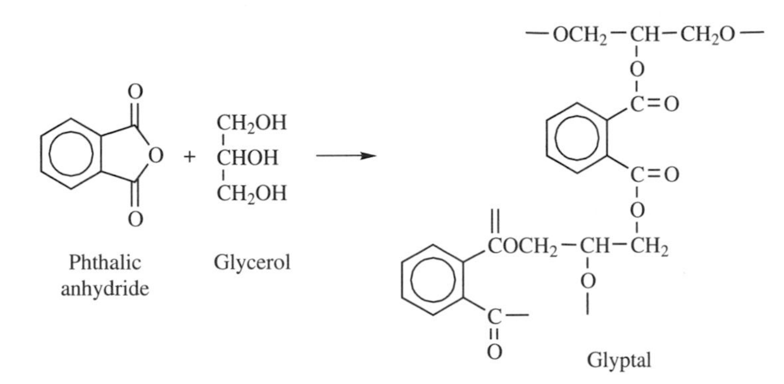

coatings, powder coatings, and two compound coating systems now account for over 60% of the total paint market.

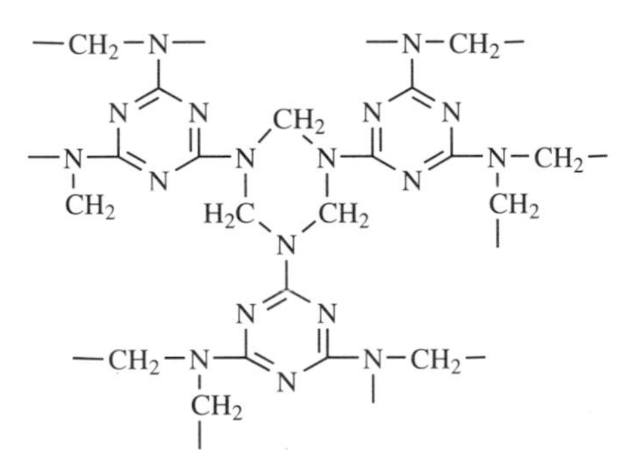

Melamine–formaldehyde resins (MF)

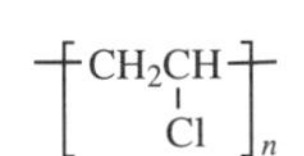

Poly(vinyl chloride) (PVC)

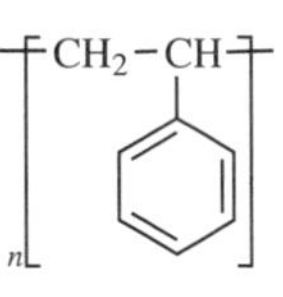

Polystyrene (PS)

11.4 WATER-BASED PAINTS

Primitive humans used aqueous suspensions of colored clays to decorate their cave dwellings. Tempera paint was also a water-based coating in which eggs were used as the binder. Whitewash, also called whiting, consisted of a dispersion of calcium hydroxide ($Ca(OH)_2$) in water. Since this was an inferior and temporary coating, the term whitewash is now used to describe a cover-up of vices or crimes.

Starch is water-soluble and has been used for centuries as a coating. Over 50,000 tons of starch and chemically modified starch are now used annually in the United States as coatings, primarily for textile sizes. The first commercial water-based paint, which consisted of an ammonia (NH_3)-stabilized solution of casein and dispersed pigments, was introduced in the 1930s.

Emulsions of poly(vinyl acetate) $(CH_2CHOOCCH_3)_n$ containing aqueous dispersions of pigments were used as paint substitutes in Germany during World War II. These so-called water paints were introduced in the United States in 1948 and are now in wide use. Pigmented emulsions of polymethyl methacrylate, $\lbrack CH_2C(CH_3)COOCH_3 \rbrack_n$, and of copolymers of styrene ($H_2C{=}CH{-}C_6H_5$) are also used as water-based paints. Unlike the previously discussed SBR synthetic rubber, the principal constituent in this copolymer is styrene instead of butadiene. Coalescent agents that contain hydrophilic (water-loving) and lyophilic (resin-loving) groups are usually added to aqueous resin emulsions to assure the formation of continuous films.

Waterborne coatings based on resins with water-soluble groups are also available. For example, an alkyd resin with a large number of water-soluble hydroxyl (OH) groups may be produced from the reaction of phthalic anhydride and an excess of pentaerythritol ($(HOCH_2)_4C$) or by the condensation of ethylene glycol with an excess of phthalic anhydride or the anhydride of trimellitic acid ($HOOC(C_6H_3)C_2O_3$). Aqueous suspensions of these resins with residual water-soluble groups may be used as waterborne coatings.

11.5 PIGMENTS

Pigments used by the paint industry include iron blue (Prussian blue), which is produced by the precipitation of a soluble ferrocyanide salt, such as yellow prussiate ($K_4Fe(CN)_6 \cdot 3H_2O$), with iron sulfate ($FeSO_4$). The ferroferricyanide precipitate is oxidized by air to form a ferriferricyanide.

Chrome yellow ($PbCrO_4$), ultramarine blue (lapis lazuli), sodium aluminum silicate, white lithopone [barium sulfate, ($BaSO_4$, zinc sulfide (ZnS)], chromic oxide (green cinnabar, Cr_2O_3), white titanium dioxide (TiO_2), and many organic pigments are also used as colorants by the coatings industry. Orr's zinc white was introduced in 1874 by J. Orr. This mixture of barium sulfate and zinc sulfide is now known as lithopone.

It is of interest to note that titanium ore, called ilmenite, was proposed as a black pigment by J. Ryland in 1865. Jebson and Farup extracted white titanium dioxide from this ore in 1880, and this pigment was made commercially in 1927. Titanium dioxide is the most widely used white pigment today. The annual volume of inorganic and organic pigments consumed by the American paint industry is valued at \$600 million and \$125 million, respectively.

11.6 APPLICATION TECHNIQUES FOR COATINGS

The classic tempera and oil-based paints were applied by brushing, and this technique continues to be used by artists and, to a lesser degree, by house painters.

This labor-intensive process has been replaced in many instances by less labor-intensive methods, such as dipping, flow coating, curtain coating, roll coating, spraying, powder coating, and electrodeposition.

Dipping is a simple immersion coating process that can be automated. In flow coating, the part to be coated is sprayed or showered with an excess of the coating and then allowed to drain before drying or curing. In curtain coating, the part to be coated is passed through a "curtain" of the coating material. This process is repeated if both sides of the part are to be protected by the coating.

Roll coating is used by "do-it-yourself" applicators and in industrial coating processes. In this process, the coating is transferred from a roller to a flat surface. Air spraying of coatings is fast but inefficient, with much of the coating being wasted by overspray. Nevertheless, this technique is widely used in industry. This system is similar to that used when one applies a coating from an aerosol paint can. Less overspray is encountered when one uses hydrostatic spraying techniques.

A minimum amount of overspray is encountered when the spray is electrically charged and the surface to be coated has the opposite charge. Electrostatic spraying is being used industrially to coat metal parts and nonmetallic objects, such as golf balls. The hiding power is a measure of the ability of the coating to achieve a specified degree of "hiding" or obliteration. Industrially, it is often tested by comparing the reflectance of the coated surface overpainting a black surface (that is, the tested paint applied over a black surface) with white panels. The ability to cover or hide is related to the scattering of incident light hitting the surface and returning to the observer or light meter. As the film surface increases, the ability of light to penetrate the surface coating and be scattered from the (black for tests) undercoating lessens. For a simple white latex paint, no absorption occurs and we can consider the scattering occurring at the interfaces of the transparent polymer matrix and the dispersed pigment particles. While the refractive indices for most polymers do not widely vary (generally about 1.5), the scattering can vary widely. For good scattering, the refractive index of the polymer should differ from that of the pigment. For instance, while calcium carbonate, with a refractive index of about 1.6, is often used as a pigment in paints, it has a much lower hiding power than titanium dioxide, with a refractive index of about 2.8.

11.7 END USES FOR COATINGS

Table 11.1 contains the major use areas for coatings. Here we will explore some of the more unusual uses. The use of tin plate on food cans has been replaced, to some extent, by the use of clear lacquers. A newer application of coatings is for the protection of onshore and offshore installations against marine atmosphere, the protection of steel and concrete against radiation in nuclear reactor environments, and the protection of objects in outer space. Space exploration would have been impossible without thermal control surface coatings. The silicone rubber thermal control coatings on the polyester film used in the Skylab mission reflected 75% of the solar energy.

Intumescent paints are also used to protect wood from burning. These coatings contain borax ($Na_2B_4O_7$), boric acid (H_3BO_3), sodium silicate, aluminum sulfate ($Al_2(SO_4)_3 \cdot 18H_2O$), or sodium carbonate ($Na_2CO_3 \cdot 10H_2O$), which release water of hydration, sodium bicarbonate ($NaHCO_3$), which releases carbon dioxide (CO_2), and diammonium phosphate ($(NH_4)_2HPO_4$) and melamine–formaldehyde resin, which contribute to intumescence or foam formation.

11.8 SOLVENT SELECTION

Before the advent of waterborne and powder coatings, paint technicians spent considerable time trying to discover appropriate solvents for resins used in coatings. It is of interest to note that although cellulose nitrate is insoluble in ethanol (C_2H_5OH) and ethyl ether ($(C_2H_5)_2O$), Menard produced collodion by dissolving this polymer in an equimolar mixture of these two solvents in the 1950s. Other solvent systems have also been developed empirically, but more scientific guidelines are now available.

Paint chemists obtained useful solvency data by determining the temperature (aniline point) at which equal volumes of aniline ($C_6H_5NH_2$) and an unknown solvent became turbid when cooled. Paint chemists also obtained kauri–butanol values by determining the volume of unknown solvent that would produce turbidity when added to a solution of kauri copal resin in *n*-butanol ($H(CH_2)_4OH$).

These and other empirical tests have been largely replaced by solubility parameters developed by J. Hildebrand in the 1920s. These values can be calculated from the square root of the cohesive energy density [$(CED)^{1/2}$]. CED is a measure of the intermolecular forces present in 1 mol of liquid. The solubility parameter values can be used to select solvents or mixtures of solvents for polymers.

11.9 SEALANTS

Sealants, such as classic putty, are paintlike products that are formulated for the filling of cracks and voids. The classic putty that was used as a sealant for window glass was based on linseed oil. It hardened in the presence of a drier by reacting with oxygen from the atmosphere.

Modern sealants are based on butyl rubber, acrylic polymers, polyurethane, polyolefin sulfides (Thiokol), neoprene, silicones, and chlorosulfonated polyethylene. These sealants are used for sealing fabricated building units and fuel tanks. Thiokol is used as a highway and bridge sealant and was the sealant used in the construction of the destroyed World Trade Building in New York City.

$$\left[CH_2-CH_2-SSSS \right]_n$$

Polyethylene tetrasulfide (Thiokol)

Hot melt butyl rubber sealants are used in automotive windshields and as automotive sealants. Both solvent-borne and waterborne sealants are available. Thiokol sealant is used as the binder in solid rocket propellants. This low-molecular-weight liquid (LP2) is obtained by reduction of polyethylene sulfide (Thiokol) to produce a prepolymer with thio terminal groups (SH). The prepolymer (LP2) is oxidized *in situ* to produce a high-molecular-weight polymer. This process is similar to that used in the cold waving of hair.

$$\left[-CH_2\overset{CH_3}{\underset{CH_3}{\overset{|}{\underset{|}{C}}}}CH_2CH{=}\underset{CH_3}{\underset{|}{C}}CH_2- \right]_n$$

Butyl rubber

11.10 HISTORY OF ADHESIVES

In contrast to coatings, which must adhere to one surface only, adhesives are used to join two surfaces together. Resinous adhesives were used by the Egyptians at least 6000 years ago for bonding ceramic vessels. Other adhesives, such as casein (from milk), starch and sugar (from plants), and glues (from animals and fish), were first used about 3500 years ago.

Combinations of egg white and lime ($Ca(OH)_2$) as well as sodium silicate (Na_2SiO_3) were used in the first century, and a glue works or factory was built in Holland in 1690. Animal glue is produced by dissolving the calcium phosphate ($Ca_3(PO_4)_2$) and calcium carbonate ($CaCO_3$) in bones by heating with hydrochloric acid (HCl). The residue, ossein, plus collagen from animal skins, is treated with lime, extracted with hot water, and concentrated by evaporation to form an adhesive.

Adhesives are also made from fish skins, dextrins (degraded starch), and gum arabic. An aqueous solution of the latter is called mucilage. Animal glue continues to be used for gummed tapes, labels, and match heads. Casein is still used as an adhesive in wallboard, and starch and dextrin are still used for making corrugated board, but the use of these natural adhesives is decreasing, accounting for less than 10% of the entire adhesives market. Over 5 million tons of adhesives at a cost of over $3 billion are used annually in the United States.

In 1903 Edouard Benedictus, a French chemist, dropped a glass flask on the floor. It broke, but the broken pieces retained the shape of the original flask instead of breaking into many pieces and scattering over the flood. He found that the inside of the flask had a thin film that was doing the "holding together". The film was the result of the evaporation of cellulose nitrate prepared from cotton and nitric acid. Shortly after the laboratory accident, he read about a girl that had been badly cut from flying glass resulting from an automobile accident. Latter he read about other persons being cut from flying glass in automobile accidents. He remembered the dropped flask that did not break. He experimented with placing some of the cellulose nitrate between sheets of glass, using pressure to help adhere the glass with the cellulose nitrate. This was the first safety glass and it was called "Triplex" since it consisted of outer layers of glass and an inner layer of polymer. By 1909 Benedictus had patented the material and it began to be used in automobiles.

The first safety glass turned yellow after several years exposure to light. The bonding layer was replaced in 1933 by cellulose acetate, made from the reaction of cotton with acetic acid. By 1939 this was replaced by poly(vinyl butyral), PVB. PVB is still used today in our automotive safety glass and represents one of the longest used materials that remains an important technical material today.

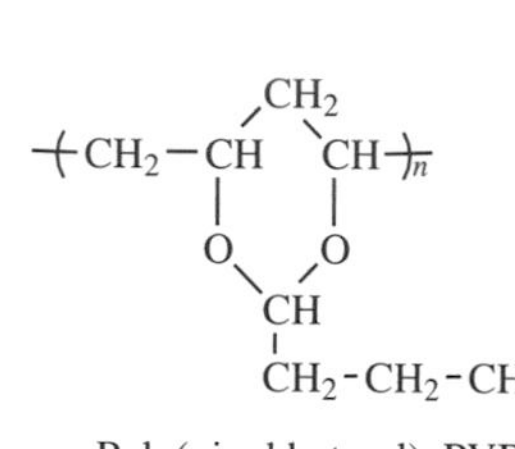

Poly(vinyl butyral), PVB

11.11 ADHESION

An adhesive is an agent that binds together two or more surfaces. The surfaces adhered may be as smooth as steel or as rough as masonry blocks; these surfaces may require one or many coatings. Secondary bonds between the adhesive and adhered surfaces are required for good adhesion. Primary bonds may be formed by the addition of cross-linking agents. Hydrogen and polar bonds may also bond two surfaces together. Polar groups are present in many adhesives, such as cyano- and acrylic-based glues.

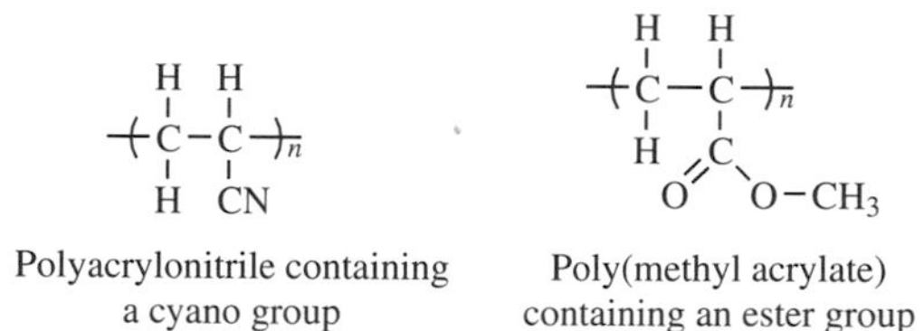

Polyacrylonitrile containing a cyano group

Poly(methyl acrylate) containing an ester group

11.12 TYPES OF ADHESIVES

In solvent-based adhesives, the polymer is dissolved in an appropriate solvent. Solidification occurs after the evaporation of the solvent. A good bond is formed if the solvent attacks or actually dissolves some of the plastic (adherend).

Adhesives may be solvent-based, latex-based, pressure-sensitive, or reactive adhesives. Solvent-based adhesives, such as model airplane glue, depend on the evaporation of the solvent for the formation of a bond (solvent weld) between the polymer (adherend) and the surface to be adhered. This type of adhesive is used to join poly(vinyl chloride) pipe in a process called solvent-welded pipe (SWP).

Latex-based adhesives should be used at temperatures above the glass transition temperature of the adhesive resin. This type of adhesive is widely used for bonding pile to the backing of carpets.

Holt melt adhesives, which may be used in electric "glue guns," are applied as molten polymers. Plywood is produced by the impregnation of thin sheets of wood by a reactive resin adhesive that cures after it has been applied. Phenolic, urea, melamine, and epoxy resins are used as the reactive adhesives.

Pressure-sensitive adhesives must also be applied as a highly viscous solution at a temperature above the glass transition temperature of the polymer. The application of pressure causes the adhesive to flow to the surface to be adhered—for example, adhesive tape.

The important factors involved with pressure sensitive adhesion is a balance between allowing molecular interaction between the adhesive and the adherent (often referred to as "wetting") and dynamic modulus of the adhesive mixture. This also involves a balance between "pull-off rate" and "wetting rate." Mechanical adhesion with interlocking and diffusion factors are less important than for permanent adhesion.

Pressure-sensitive adhesives such as those present in "pull-off" tabs such as Post-it® notes contain components similar to those present in more permanent "Scotch-tape" (TM) except that particles of emulsified glassy polymer are added to reduce the contact area between the adhesive and the substrate.

Some polymers such as polyethylene might appear to be a decent adhesive material, but, even in its melt, it is not exceptionally tacky. This is believed to be because of the high degree of chain entanglement. Since the dynamic modulus increases with increasing chain entanglement, PE is not "tacky" (does not easily contact and wet a substrate) and it is not useful as a pressure-sensitive adhesive.

11.13 RESINOUS ADHESIVES

Unsaturated polyester resins and polyurethanes are used for automobile body repair and for bonding polyester cord to the rubber in tires. Both polyester and epoxy resins are used to bond fibrous glass and aramid fibers in reinforced plastic composites.

A solution of natural rubber in naphtha was used by MacIntosh to produce a waterproof cloth laminate in the nineteenth century, and comparable systems continue to be used today. Blends of neoprene and phenolic resins are used as contact adhesives, in which the adhesive is applied to both surfaces, which are then pressed together.

"White glue," which is used as a general-purpose adhesive, consists of a polyvinyl acetate emulsion. Copolymers of ethylene and vinyl acetate are used as hot melt adhesives. Anaerobic adhesives, which cure when air is excluded, consist of mixtures of dimethacrylates and hydroperoxide. "Super-glue" or "Krazy-glue"

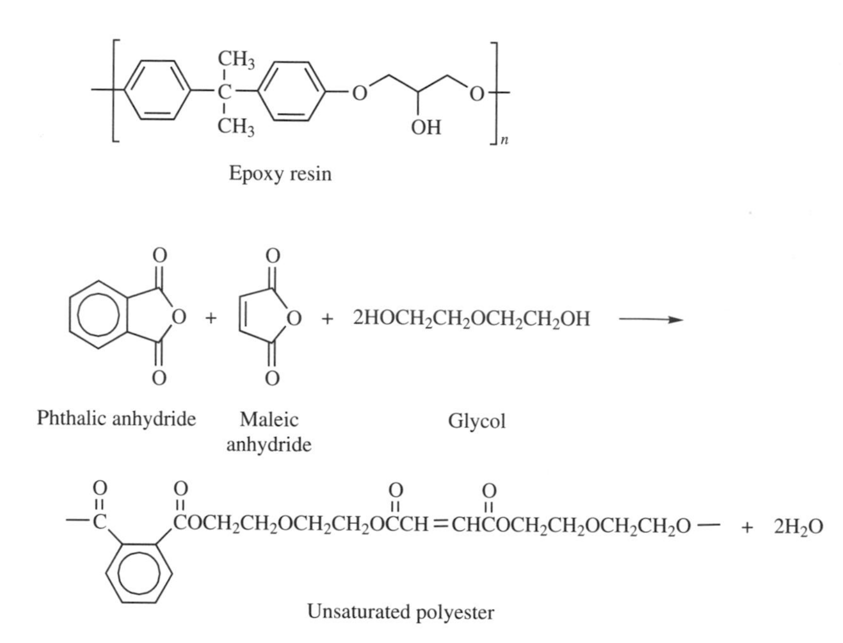

contains butyl-α-cyanoacrylate, which polymerizes spontaneously in the presence of moist air or on dry glass surfaces. This superior adhesive is used in surgery and for mechanical assemblies.

$$CH_2{=}C(CN){-}COO{-}C_4H_9$$

Butyl-α-cyanoacrylate

$$CH_2{=}C(CN)COOCH_3$$

Methyl-α-cyanoacrylate

GLOSSARY

Acrylic polymer: Poly(methyl methacrylate) or poly(ethyl acrylate).

Adhesion: The degree of attachment between a film and another surface.

Adhesive: An agent that binds two surfaces together.

Alkyd: Resin produced by the condensation of glycerol and phthalic acid.

Anaerobic: Free of oxygen.

Aniline point: Temperature at which a 50–50 mixture of aniline and unknown solvent becomes turbid when cooled.

Aqueous: Watery.

Aramid fiber: Aromatic nylon fibers.

Baekeland, Leo: Inventor of commercial phenolic resins.

Balsam: An aromatic exudate from trees or shrubs, such as Canadian balsam.

Binder: Film-forming constituent of a coating system.

Borax: $Na_2B_4O_7$.

Boric acid: H_3BO_3.

Butyl rubber: A copolymer of isobutylene ($H_2C{=}C(CH_3)_2$) and isoprene ($H_2C{=}C(CH_3)CH{=}CH_2$).

Casein: Milk protein.

Chlorinated rubber: Product of the reaction of chlorine and natural rubber.

Chrome yellow: $PbCrO_4$.

Chromium oxide: Cr_2O_3, a green pigment.

Coalescent agent: Substance added to emulsions to ensure the formation of continuous films.

Coating, curtain: The deposition of a curtain of paint on a flat surface followed by draining.

Cohesive energy density (CED): Energy of intermolecular forces between molecules.

Collagen: Gelatinlike protein.

Collodion: A solution of cellulose nitrate in a mixture of ethanol and ethyl ether.

Copal: A natural resin obtained from tropical trees.

Copolymer: A polymer with more than one type of repeating units in its chain.

Dextrin: Degraded starch.

Diammonium phosphate: $(NH_4)_2HPO_4$.

Drier: A soluble salt of a heavy metal and an organic acid (a catalyst for the polymerization of unsaturated oils).

Drying: Polymerization of unsaturated oil in the presence of oxygen.

Electrodeposition: Deposition of a coating from a waterborne system on an object of opposite charge.

Emulsion: A permanent aqueous suspension of a polymer.

Enamel: A term used for ceramic coatings and also for polymer solution coatings.

Ester gum: Glycerol ester of abietic acid (rosin).

Ethylene glycol: $HO(CH_2)_2OH$.

Formaldehyde: H_2CO.

Glass transition temperature: The temperature at which a glassy polymer becomes flexible when the temperature is increased.

Glue: Adhesive usually derived from animals or fish.

Glycerol: $HOCH_2CH(OH)CH_2OH$.

Glyptal: Resin produced by the condensation of glycerol and phthalic anhydride.

Gum arabic: Salts of arabic acid obtained from mimosa plants (acacia).

Hildebrand, J.: Developer of the solubility parameter concept.

Hydroperoxide: ROOH.

Ilmenite: Titanium ore.

Intumescent paint: One that forms a protective tarry sponge when burned.

Iron blue: Prussian blue; ferriferricyanide.

Kauri: A copal resin.

Kauri–butanol value: Volume of unknown solvent that causes turbidity when added to a solution of kauri copal resin in 1-butanol.

Lacquer: A solution of a film-forming resin (binder).

Linolenic acid: A diunsaturated acid ($C_{17}H_{31}COOH$).

Linseed oil: Oil from flax seed.

Lithopone: $BaSO_4$ and ZnS.

MacIntosh: A laminate of natural rubber and cloth.

Menhaden oil: An unsaturated (drying) oil obtained from menhaden (Moss bunker) fish.

Mucilage: An aqueous solution of gum arabic.

Neoprene: Polychloroprene,

$$\left[\begin{array}{c} H \\ | \\ C \\ | \\ H \end{array} - \begin{array}{c} Cl \\ | \\ C \end{array} = \begin{array}{c} H \\ | \\ C \end{array} - \begin{array}{c} H \\ | \\ C \\ | \\ H \end{array} \right]_n$$

Oiticica oil: An unsaturated (drying) oil obtained from the Brazilian oiticica tree (*Licania rigida*).

Oleic acid: A monounsaturated acid ($C_{17}H_{33}COOH$).

Oleoresinous: A material based on unsaturated vegetable oils and a drier.

Paint: A liquid system consisting of a solid (pigment) and a liquid (vehicle).

Pentaerythritol: $(HOCH_2)_4C$.

Phenol: C_6H_5OH.

Phthalic acid: $C_6H_4(COOH)_2$.

Pigment: A colorant.

Pitch: A bituminous substance based on asphalt, wood tar, or coal tar.

Polyurethane: The reaction product of a diol (HOROH) and a diisocyanate (OCNRNCO).

Pressure-sensitive adhesive: A viscous solution that flows under pressure to produce an adhered system, for example, adhesive tape.

Putty: A sealant or caulking material based on a mixture of filler, drier, and linseed oil.

Reinforced plastic composite: Polyester-bonded fiberglass composite.

Rosin: Pine resin.

Safflower oil: An unsaturated (drying) oil obtained from safflower seed (*Carthamus*).

Sealant: A crack filler.

Shellac: A resinous material secreted by insects that feed on the lac tree.

Silicone: An inorganic polymer with the repeating unit $-Si(R_2)-O-$.

Sodium bicarbonate: $NaHCO_3$.

Sodium carbonate: Na_2CO_3.

Sodium silicate: Na_2SiO_3.

Solubility parameter: A measure of solvency; these values generally increase with the polarity of the solvent.

Substrate: A surface to be coated.

Super-glue: An adhesive based on butyl-α-cyanoacrylate.

SWP: Solvent welded pipe.

Teflon: Polytetrafluoroethylene.

Tempera: A paint based on egg binder.

Thinner: Paint solvent.

Thiokol: The first American synthetic rubber produced by the condensation of ethylene dichloride and sodium polysulfide.

Titanium dioxide: TiO_2, a white pigment.

Trimellitic acid: $C_6H_3(COOH)_3$.

Tung oil: China wood oil (*Aleutites cordata*), an unsaturated (drying) oil.

Ultramarine blue: Lapis lazuli (sodium aluminum silicate).

Unsaturated oil: Oil with carbon–carbon double bonds.

Varnish: An unpigmented oil-based paint.

White glue: Polyvinyl acetate.

Whitewash: Aqueous dispersion of calcium hydroxide ($Ca(OH)_2$).

REVIEW QUESTIONS

1. Which will produce a tack-free film first: a layer of lacquer or a layer of classic paint?
2. Why must a classic paint be applied in thin layers?
3. Which is more highly unsaturated: linseed oil or mineral oil?
4. What color is lithopone?
5. What polymer is present in collodion?
6. Describe the relationship between chain length and viscosity.
7. What is the function of a paint drier?
8. Is a polymer more ductile or more brittle when it is cooled below the glass transition temperature?
9. Why is whitewash not permanent?

10. Why must a coalescent agent be present in emulsion coatings?

11. What are the names of two polyester coatings based on phthalic acid or anhydride and glycerol or ethylene glycol?

12. Why does putty harden?

13. What is mucilage?

14. What is ester gum?

15. Why is butyl rubber called a copolymer?

16. Why must butyl-α-cyanoacrylate be kept in a moisture-free container?

17. Will an anaerobic adhesive polymerize in the presence of moist air?

18. What polar groups are present in "Super-glue"?

BIBLIOGRAPHY

Bauer, D. R., and Martin, J. (1999). *Service Life Prediction of Organic Coatings—A Systemic Approach*, Oxford University Press, New York.

Benedek, I. (2000). *Pressure-Sensitive Formulation*, VSP, Leiden, Netherlands.

Bieleman, J. (2000). *Additives for Coatings*, Wiley, New York.

Bierwagen, G. (1998). *Organic Coatings for Corrosion Control*, Oxford University Press, New York.

Blunt, J. (1998). *Engineering Coatings Design and Application*, William Andrew Publishers, Norwich, CT.

Bunshah, R. (1999). *Handbook of Hard Coatings*, Noyes Publishers, Park Ridge, NJ.

Craver, C., and Carraher, C. (2000). *Applied Polymer Science*, Elsevier, New York.

Davison, G., and Skuse, D. (1999). *Advances in Additives for Water-Based Coatings*, Royal Society of Chemistry, London.

Konstandt, F. (2000). *Organic Coating Properties and Evaluation*, Chemical Publishers, New York.

Lambourne, R., and Stivens, T. (1999). *Paint and Surface Coatings Theory and Practice*, Woodhead Publishers, Cambridge, England.

Mittal, K. L. (2000). *Polymer Surface Modification: Relevance to Adhesion*, VSP, Leiden, Netherlands.

Mittal, K. L. (2001). *Adhesion Aspects of Thin Films*, VSP, Leiden, Netherlands.

Munger, C. G. (1999). *Corrosion Prevention by Protective Coatings*, NACE International, Houston.

Pocius, A. (2001). *Adhesion and Adhesives Technology*, Hanser-Gardner, Cincinnati.

Pulker, H. K. (1999). *Coatings on Glass*, Elsevier, New York.

Rao, C. (1998). *Handbook of Metallurgical Coatings*, CRC, Boca Raton, FL.

Ryntz, R. (2001). *Plastics and Coatings*, Hauser-Gardner, Cincinnati.

Sudarshan, T., and Dahorte, N. (1999). *High-Temperature Coatings*, Marcel Dekker, New York.
Wicks, Z., Jones, F., and Papas, S. P. (1999). *Organic Coatings Science and Technology*, Wiley, New York.

ANSWERS TO REVIEW QUESTIONS

1. The lacquer. It hardens by evaporation of the solvent.
2. Because it hardens by the reaction of oxygen from the air. The oxygen cannot diffuse readily through thick layers of paint.
3. Linseed oil contains linolenic acid.
4. White.
5. Cellulose nitrate.
6. Viscosity increases as chain length increases.
7. It catalyzes the polymerization (hardening or drying) of the unsaturated oil in the paint.
8. More brittle. It is glasslike below T_g.
9. There is no resinous binder present.
10. The polymer will not form a continuous film below its T_g in the absence of a coalescent agent.
11. Alkyd and glyptal.
12. It polymerizes (dries) when exposed to air.
13. A solution of gum arabic in water.
14. A glyceryl ester of rosin.
15. The repeating units of both isobutylene and isoprene are present in the backbone.
16. It polymerizes in the presence of moisture.
17. No.
18. Cyano and ester groups.

12

COMPOSITES

12.1 INTRODUCTION

Composites are all about us. In some ways they are the materials of the twenty-first century. For instance, they are used where high-temperature stability is needed as in

Giant Molecules: Essential Materials for Everyday Living and Problem Solving, Second Edition,
by Charles E. Carraher, Jr.
ISBN 0-471-27399-6

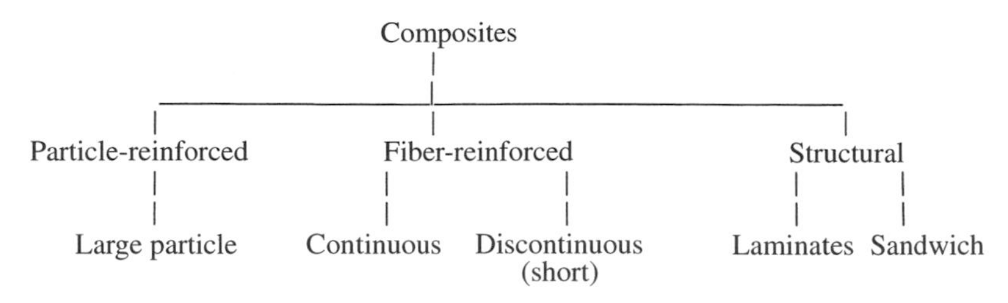

Figure 12.1. Classification of polymer-intense composites.

the space shuttle. They are also used where weight is important such as in the airplanes. Succeeding families of Boeing aircraft have increased the use of composites and other giant molecules. The Stealth fighter's exterior is essentially all graphite composite. The snow ski, with the exception of steel edges, is a showcase of polymers and composites that work together giving a material that has a balance of flexibility, strength, weight, and cost.

Welcome to the wonderful world of composites.

12.2 GENERAL

Composites are generally composed of two phases, one called the continuous or matrix phase that surrounds the discontinuous or dispersed phase. There are a variety of polymer-intense composites that can be classified as shown in Figure 12.1.

12.3 THEORY

Composites are multicomponent (made up of more than one material) materials that contain strong fibers embedded in a continuous phase (Figure 12.2). The fibers are called reinforcement fibers, and the continuous phase is called the matrix. The continuous phase is generally made up of an organic polymer called a resin. While many naturally occurring materials such as wood are reinforced composites consisting of a resinous continuous phase and a discontinuous fibrous reinforcing phase, we will emphasize what are called space age and advanced materials composites.

Generally, the matrix material itself is not particularly strong relative to the composite. The overall strength of a single fiber is also not great. It is the combination that is strong. The resin acts as a transfer agent, transferring and distributing applied stress to the fibers.

Fibers generally have a ratio of length to diameter greater than 100. Most fibers are thin, about a tenth of the thickness of a human hair. Fibers should be stiff and strong.

There is a relationship between the ideal length of a fiber and the amount of adhesion between the matrix and the fiber. Let us assume that only the tip (one

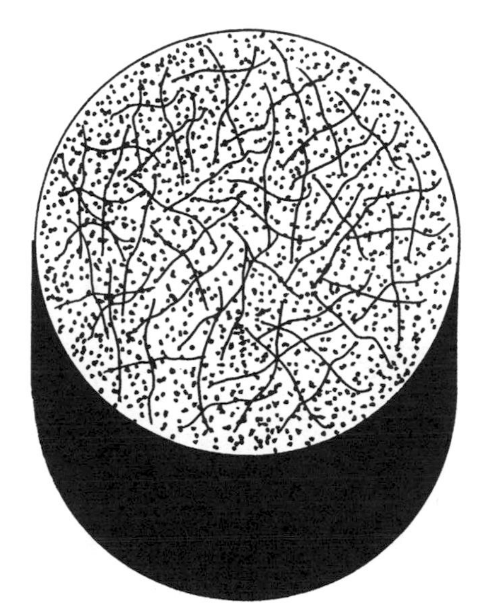

Figure 12.2. Illustration of fiber embedded in a matrix.

end) of a fiber is placed in a resin (Figure 12.3, top). The fiber is pulled. The adhesion is broken and the fiber is pulled unbroken from the matrix. Next, let us repeat the experiment except placing the fiber deeper into the matrix. Again, the fiber is pulled. The adhesion is again not sufficient to retain the fiber. This experiment is repeated until the fiber is broken before it can be pulled from the matrix (Figure 12.3, bottom). Somewhere between the depth where the fiber can be pulled without breaking it and the depth where the fiber is broken is a fiber length; the fiber depth into the matrix is a fiber length, called the critical fiber length, where there is a balance between the strength of the fiber and the adhesion between the fiber and matrix. Remember that the strength of a fiber increases as the fiber thickness is increased, so a balance is sought between fiber thickness, length, and matrix adhesion.

Fiber breakage is generally a catastrophic process where failure is sudden, This is common for giant molecules where the weak link(s) determines the overall strength. Thus, in composite parts of an aircraft, one does not look at cracks before becoming alarmed, but rather technicians look for voids indicating separation of the fiber from the matrix.

12.4 FIBER-REINFORCED COMPOSITES

A. Fibers

Fibers where the fiber length is greater than the critical fiber length are called continuous fibers, while those that are less than this critical length are called discontinuous or short fibers. Little transference of stress and thus little reinforcement is

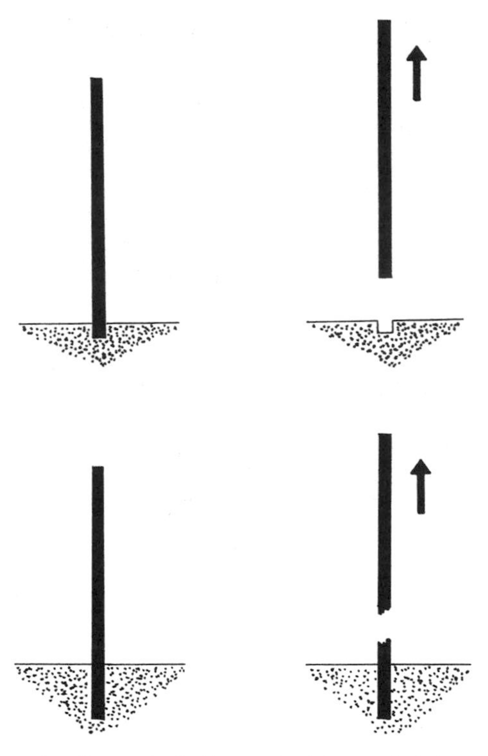

Figure 12.3. Illustration of pulling a single fiber embedded partially in a matrix. The left side is prior to pulling, and the right side is after pulling.

achieved for short fibers. Thus, fibers whose lengths exceed the critical fiber length are generally used.

Fibers can be divided according to their diameters. Whiskers are very thin single crystals that have large length-to-diameter ratios. They have a high degree of crystalline perfection and are essentially flaw-free. They are some of the strongest material known. Whisker materials include graphite, silicon carbide, aluminum oxide, and silicon nitride. Fine wires of tungsten, steel, and molybdenum are also used, but here, even though they are fine relatively to other metal wires, they have large diameters. The most frequently used fibers are "fibers" that are either crystalline or amorphous or semicrystalline with small diameters.

Most fibers used in today's composites are of three general varieties: glass (16.5), carbon (graphite; 9.9), and aromatic nylons (aramides; 9.3). Asbestos, which once was widely used, is little used, holding less than a 1% fiber-composite market share.

Fiberglass or fibrous glass, is manufactured from largely silicon dioxide glass that is cooled below its melting point so that is it largely amorphous. The glass fibers are pulled from the melted glass. This pulling orients the structure, giving a material that is stronger in the direction of the pull. As with other three-dimensional materials, the limits of strength are due to the presence of voids. In the case of fiberglass, these voids mostly occupy the fiber surface.

While carbon can be made to form many allotropic forms from linear short-chained carbon black to complex carbon nanotubes, it also forms high-strength fibers and whiskers known as carbon or graphite fibers. Carbon whiskers are sheets of hexagonal carbon atoms layered like a laminate, one on top of another in an ordered array. Leslie Phillips, one of the inventors of carbon fibers, describes them as bundles of oriented crystalline carbon held in a matrix of amorphous carbon.

As in the case of fiberglass, the carbon fibers have surface voids. These voids are often filled by surface treatment with a low-chain-length epoxy resin.

Two types of aromatic nylon fibers are generally used. One is more flexible and is employed in situations where flexibility is important. The second type is stiffer and is used when greater strength is needed.

Generally, good adhesion between the fiber and the matrix is desirable. In some cases though, poor adhesion is favored such as in body armor where the separation of the fiber from the matrix is a useful mode of absorbing an impact.

B. Matrixes (Resins)

Both thermosetting and thermoplastic matrixes are used. Thermosetting resins are generally used as prepolymers where the resin is still mobile. After the fiber is laid or mixed in with the resin, the resin is further reacted, forming a complex cross-linked material. Unsaturated polyesters, epoxys, phenol-formaldehyde, amino-formaldehyde, and polyimides give thermoset composites.

Often-used thermoplastic matrix materials are the engineering thermoplastics covered in Chapter 7, namely, nylons (Section 7.2), poly(phenylene sulfide) (Section 7.7), polycarbonates (Section 7.4), poly(ether ether ketone) (Section 7.10), poly(ether ketone) (Section 7.10), and poly(ether sulfones) (Section 7.8).

12.5 PARTICLE-REINFORCED COMPOSITES—LARGE-PARTICLE COMPOSITES

If sufficient adhesion does not occur between a fiber and matrix, the fiber acts merely as a filler. Because many filler materials are inexpensive, they may be added to simply increase the bulk of a material without significant increase in overall material strength.

Some materials to which fillers have been added can be considered as low-grade composites. These include a number of the so-called cements such as concrete (Section 16.2). As long as the added particles are relatively small, of roughly the same size, and evenly distributed throughout the mixture, there can be a reinforcing effect. The major materials in Portland cement concrete are the Portland cement, a fine aggregate (sand), course aggregate (gravel and small rocks), and water. The aggregate particles act as inexpensive fillers. The water is also inexpensive. The relatively expensive material is the Portland cement. Good strength is gained by having a mixture of these such that there is a dense packing of the aggregates and good interfacial contact, both achieved by having a mixture of aggregate sizes—thus the use of large gravel and small sand. The sand helps fill the voids between the various larger gravel particles. Mixing and contact is achieved with the correct amount of water. Enough water must be present to allow a wetting of the surfaces to occur along with providing some of the reactants for the setting up of the cement. Too much water creates large voids and weakens the concrete.

12.6 APPLICATIONS

Many of the applications of composite materials involve their light weight, resistance to weathering and chemicals, and ability to be easily fabricated and machined. While they are relatively inexpensive, the cost is increased when speciality fabrication is necessary and when special properties are necessary.

One of the oldest and largest uses for composites is the construction of water-going vessels ranging from rowboats, sailboats, racing boats, and motor craft to large seagoing ships. Boats can be made at local home operations to large shipyards. Most boats are composed of fiberglass and fiberglass/carbon fiber composites. Aromatic nylon is also widely used.

Because of the large amount of fuel required to propel spacecraft into outer space, weight reduction is an essential consideration. For the space shuttles, some of the solid propellant tanks are composite generally fiberglass and fiberglass/carbon fiber. The cargo bay doors are sandwich composites composed of carbon fibers/epoxy/honeycombmaterials. The manipulation arm used for loading the payload bay is made of a number of composites including carbon/epoxy composite laminates, aromatic nylon laminates, and sandwich materials. Composites are used for the construction and mounting of mirrors, telescopes, solar panels, and antennae reflectors. They are also used where excessive high heat stability is needed. (Figure 12.4).

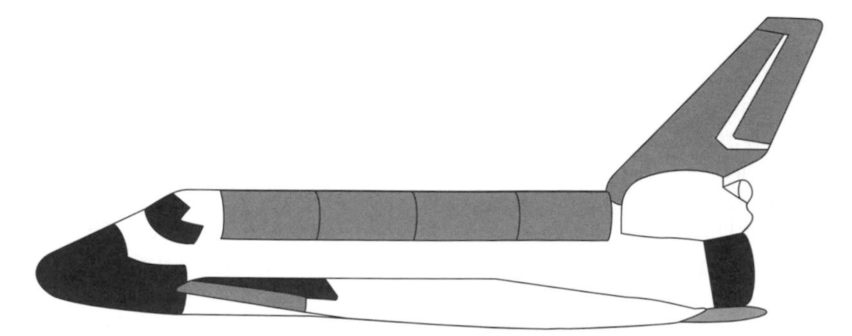

Figure 12.4. Locations (*shaded areas*) of various advanced materials, including composites, employed for heat protection in the Space Shuttle Orbiter.

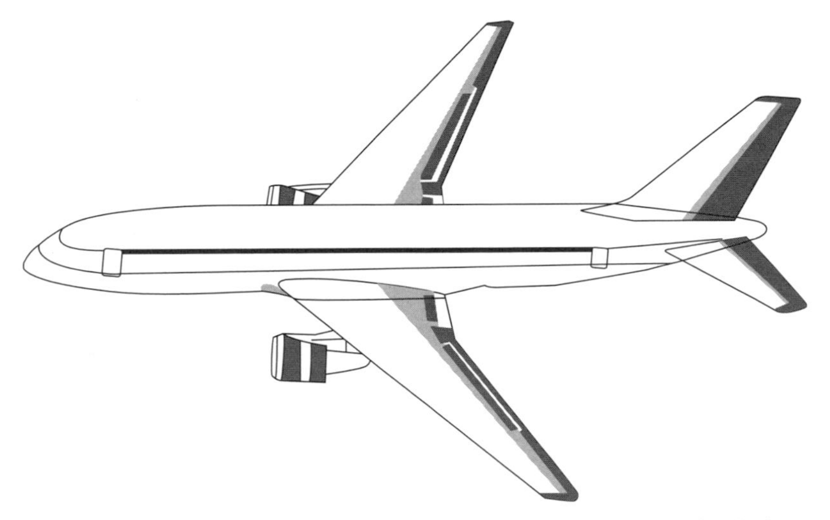

Figure 12.5. Use of graphite (*solid*) and graphite/Kevlar (TM) (*dotted*) composites in the exterior of the Boeing 767 passenger jet.

The Gulf and Iraqi Wars spotlighted the use of composite materials in new-age aircraft. The bodies of both the Stealth fighter and bomber are mainly carbon fiber composites. The Gossamer Albatross, the first plane to crossed the English Channel with only human poser, consisted largely of composite material including a carbon fiber/epoxy and aromatic nylon composite body and propellers containing a carbon fiber composite core.

The use of fiberglass in the Boeing aircraft has increased from 20 square yards for the 707, to 200 square yards for the 727, to 300 square yards for the 737, and over 1000 square yards for the 747, and over 1200 square yards for the 767 (Figure 12.5).

The use of composite material is also increasing for small aircraft. The McDonnell Douglas F-18 has about 50% of its outer body made up of composite material.

Composites are widely used in sports, ranging from gold clubs, baseball bats, bicycle frames, basketball backboards, and fishing rods.

The modern ski is a good example of the use of composites to make a product with unique properties (Figure 12.6). The top and sides are composed of ABS polymer that has a low T_g allowing it to remain flexible even at low temperatures. It is employed for cosmetic and containment purposes. Polyurethane forms the core and a damping layer that acts as a filler and to improve chatter resistance. The base is a carbon-impregnated matrix composite that is hard, strong, and with good abrasion resistance. There are numerous layers of fiberous glass that are a mixture of bidirectional layers to provide torsional stiffness, unidirectional layers that provide longitudinal stiffness with bidirectional layers of fiberglass acting as outer layers to the polyurethane layers composing a torsional box. The only major noncomposite

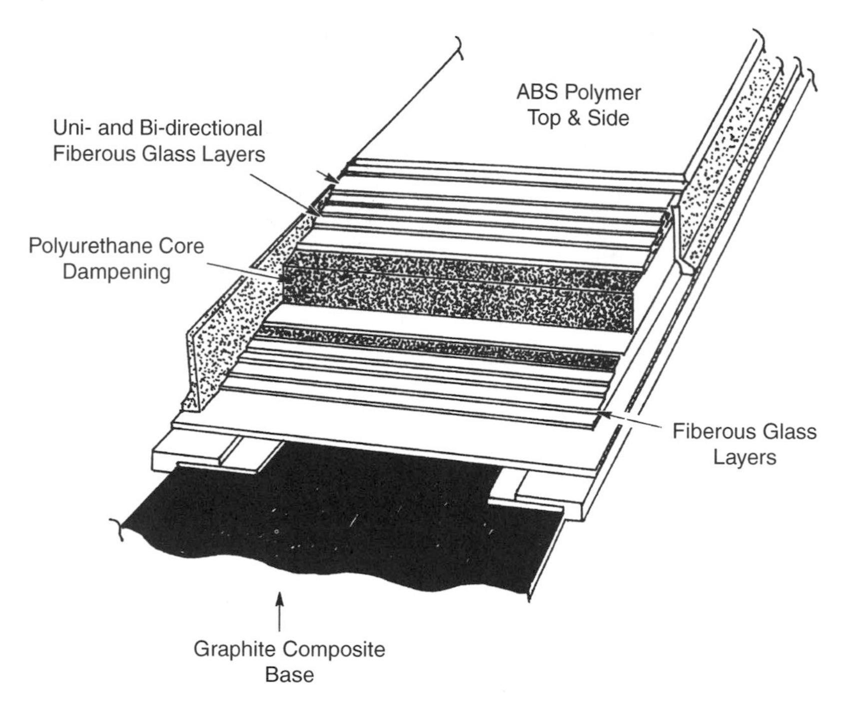

Figure 12.6. Cut-away illustration of a modern ski.

material is the hardened steel edge that assists in turning by cutting into the ice. They all work together to give a light, flexible, shock absorbing, tough ski.

Composites are used in a number of automotive applications and in industry.

They are also important as biomaterials. Bones are relatively light compared to metals. Composite structures approach the density of bones, and they are inert and strong so they possess the necessary criteria to be employed as body-part substitutes. Power-assisted arms have been made by placing hot-form strips of closed-cell polyethylene foam over the cast of an arm. Grooves are cut into these strips prior to application and carbon fiber/resin added to the grooves. The resulting product is strong and light, and the cushioned PE strips soften the attachment site of the arm to the living body. Carbon fiber/epoxy plates are replacing titanium plates used in bone surgery.

12.7 PROCESSING—FIBER-REINFORCED COMPOSITES

There exist a wide variety of particular operations, but briefly they can be described in terms of filament winding, preimpregnation of the fiber with the partially cured resin, and pultrusion. Pultrusion is used to produce rods, tubes, beams, and so on, with continuous fibers that have a constant cross-sectional shape. The fiber (as a continuous fiber bundle, weave, or tow) is impregnated with a thermosetting resin and pulled through a die that shapes and establishes the fiber-to-resin ratio. This stock is then pulled though a curing die that can machine or cut, producing the final shape such as filled and hollow tubes and sheets.

The term used for continuous fiber reinforcement preimpregnation with a polymer resin that is only partially cured is "prepreg." Prepreg material is generally delivered to the customer in the form of a tape. The customer then molds and forms the tape material into the desired shape, finally curing the material without having to add any additional resin. Preparation of the prepreg can be carried out using a calendering process. Briefly, fiber from many spools are sandwiched and pressed between sheets of heated resin, with the resin heated to allow impregnation but not so high as to be very fluid.

Thus, the fiber is impregnated in the partially cured resin. Depending upon the assembly, the fiber is usually unidirectonal, but can be made so that the fibers are bidirectional or some other combination. The process of fitting the pregpreg into, generally onto, the mold is called "lay-up." Generally, a number of layers of prepreg are used. The lay-up may be done by hand, called hand lay-up, or done automatically, or some combination of automatic and hand lay-up. As expected, hand lay-up is more costly but is needed where one-of-a-kind products are produced and by the occasional customer.

In filament winding the fiber is wound to form a desired pattern, usually but not necessarily hollow and cylindrical. The fiber is passed through the resin and then spun onto a mandrel. After the desired number of layers of fiber is added, it is cured. Prepregs can be filament-wound. With the advent of new machinery, complex shapes and designs of the filament can be readily wound.

12.8 PROCESSING—STRUCTURAL COMPOSITES

Structural composites can be combinations of homogeneous and composite materials. Laminar composites are composed of two-dimensional sheets that generally have a preferred high-strength direction. The layers are stacked so that the preferred high-strength directions are different, generally at right angles to one another. The composition is held together by a resin. This resin can be applied as simply as an adhesive to the various surfaces of the individual sheets, or the sheet can be soaked in the resin prior to laying the sheets together. In either case, the bonding is usually of a physical type. Plywood is an example of a laminar composite. Laminar fiberous glass sheets are included as part of the modern ski construction. These fiberous glass sheets are fiber-reinforced composites used together as laminar composites.

Laminar materials are produced by a variety of techniques. Coextrusion blow molding produces a number of common food containers that consist of multilayers such as layers consisting of polypropylene/adhesive/poly(vinyl alcohol)/adhesive/ adhesive/ polypropylene.

Sandwich composites are combinations where a central core(s) is surrounded generally by stronger outer layers. Sandwich composites are present in the modern ski and as high temperature-stable materials used in the space program. Some cores are very light, acting something like a filler with respect to high strength, with the strength provided by the outer panels. Simple corrugated cardboard is an example of a honeycomb core sandwich structure except that the outer paper-intense layers are not particularly strong. Even in the case of similar polyethylene and polypropylene corrugated structures, the outer layers are not appreciatively stronger than the inner layer. In these cases the combination acts to give a lightweight somewhat strong combination, but they are not truly composites but simply exploit a common construction.

12.9 PROCESSING—LAMINATES

Laminating is a simple binding together of different layers of materials. The binding materials are often thermosetting plastics and resins. The materials to be bound together can be paper, cloth, wood, or fibers like fiberglass. Typically, sheets, impregnated by a binding material, are stacked between metal plates and heated and a high pressure is applied. This produces a bonded product that can be milled, cut, machined, and so on.

Reinforced plastics differ from high-pressure laminates in that little or no pressure is used. Here impregnated reinforced materials are added to a mold and the mold is heated.

12.10 NANOCOMPOSITES

Nature has employed nanomaterials since the beginning of time. Much of the inorganic part of our soil is a nanomaterial with the ability to filter out particles often on a molecular or nano level. The driving force toward nanomaterials is that they can offer new properties or enhanced properties unobtainable with so-called traditional bulk materials. The nanoworld is often defined for materials where some dimension is on the order of 1–100 nm. In a real way, single-polymer chains are nanomaterials since the distances between the chains in less than 100 nm.

The ultimate strength and properties of many materials is dependent on the intimate contact between the various members. This is true for composites. Our bones are examples of nanocomposites. The reinforcement material is platelike crystals of hydroxyapatite, $Ca_{10}(PO_4)_6(OH)_2$, with a continuous phase of collagen fibers. The shell of a mollusk is microlaminated, containing as the reinforcement aragonite (a crystalline form of calcium carbonate), and the matrix is a rubbery material.

Nanofibers allow more contact between the fibers and matrix because of increased surface area per volume of fiber. A number of inorganic/organic nanocomposites have been made. These include nanofibers and whiskers made from tungsten carbide, silicon nitride, and so on. This includes the use of special clays (layered silicates) mixed with nylons. These nylon–clay microcomposites are used to make some Toyota air intake covers.

Carbon fibers and carbon nanotubes are being used to create stronger composites. Compared with carbon fibers, carbon nanotubes, because of their flexibility, offer stronger composites.

Formica^{TM} coverings, hand-layered fiberglass boat hulls, and plywood are examples of laminate materials.

GLOSSARY

Aramid: Nylon produced from aromatic reactants.

Asbestos: Fiberous magnesium silicate.

Aspect ratio: Ratio of length to diameter.

Composite: Material that contains strong fibers embedded in a continuous phase called a matrix or resin.

Continuous phase: Resin or matrix in a composite.

Discontinuous phase: Discrete filler (fibers for composites) additive.

Extender: Inexpensive filler.

Fiber glass: Filaments made from molten glass.

Filament: A continuous thread.

Filament winding: Process where filaments are dipped in a prepolymer of polymer and wound on a mandrel and cured.

Graphite fibers: Fibers made from pyrolysis of a carbon-based giant molecule such as polyacrylonitrile; also often called carbon fibers.

Lamellar: Sheetlike.

Laminate: Composite consisting of layers adhered by a resin.

Lamination: Laying sheet on top of one another.

Pultrusion: Process in which filaments are dipped in a prepolymer or polymer, passed through a die, and cured.

Reinforced plastic: Composite whose additional strength is dependent on a fibrous additive.

Whiskers: Single crystals used as reinforcement fibers.

REVIEW QUESTIONS

1. Would you define raisin bread to be a composite with the raisins being the fibers and the dough being the resin?
2. Would palm and peanut hull fibers be useful as fibers in a composite?
3. What is the continuous phase in wood?
4. Name three laminated products.

5. Which is the continuous phase in cookware that is coated with composite made from Teflon and poly(phenylene sulfide)?
6. How might you increase the strength of a PVC pipe?
7. Why is fiber glass or aramid fibers not used in polyethylene store bags?
8. Name four general areas where composites are widely employed.

BIOGRAPHY

Carlsson, L. A., and Poipes, R. B. (1996). *Experimental Characterization of Advanced Composite Materials*, Technomic, Lancaster, PA.

Composites Institute (1998). *Introduction to Composites*, 4th ed., Technomic, Lancaster, PA.

Dave, R., and Loos, A. (2000). *Processing of Composites*, Hanser-Gardner, Cincinnati.

De, S., and White, J. (1996). *Short Fiber Composites*, Technomic, Lancaster, PA.

Eklund, P., and Rao, A. (2000). *Fullerene Polymers and Fullerene Polymer Composites*, Springer-Verlag, New York.

Gupta, R. (2000). *Polymer and Composite Rheology*, Marcel Dekker, New York.

Holloway, H. (2000). *Advanced Polymer Composites in Engineering*, Elsevier, New York.

Jones, R., and Jones, M. (1998). *Guide to Short Fiber Reinforced Plastics*, Hanser-Gardner, Cincinnati.

Karian, H. (1999). *Handbook of Polypropylene and Polypropylene Composites*, Marcel Dekker, New York.

Peters, S. T. (1998). *Handbook of Composites*, 2nd ed., Kluwer, Hingham, MA.

Rosato, D. (1997). *Designing Reinforced Composites*, Hanser-Gardner, Cincinnati.

ANSWERS TO REVIEW QUESTIONS

1. No, if for no other reason that the aspect ratio of the raisin is well less than 100. Furthermore, the adhesion between the raisin and the dough is not particularly high.
2. It is possible that the correct resin could be found that would give a reasonable composite. Some of the fibers in palms and peanut hulls have decent aspect ratios.
3. Lignin.
4. Formica™ table tops, plywood, hand-layered speed boat hulls, and so on.
5. Teflon.
6. Could use fiber windings or include fibers in with the PVC making a composite material. Could also make the pipe thicker.

7. Give-away bags are inexpensive; and use of these fibers, while making them stronger, would also increase the cost because of the cost of the fiber and increased production steps. Furthermore, the bags are generally, but not always, strong enough now. Stronger bags can more easily be achieved by simply making the bags a little thicker.

8. Boating, aircraft, biomedical, athletics, and so on.

13

NATURE'S GIANT MOLECULES: THE PLANT KINGDOM

13.1 INTRODUCTION

Natural polymers (giant molecules) are well known and have been essential for human life for thousands of years. Starch, cellulose, lignin, and rubber are polymers

Giant Molecules: Essential Materials for Everyday Living and Problem Solving, Second Edition, by Charles E. Carraher, Jr.
ISBN 0-471-27399-6

of the plant kingdom and are essential for food and shelter. These polymers are produced at an annual rate of millions of tons. Natural rubber will be discussed in Chapter 12.

Polymers of the animal kingdom, namely, proteins, nucleic acids, chitin, and glycogen, which are also essential products, are produced at annual rates of over a million tons. Some of these naturally occurring giant molecules are classified as biopolymers, but their behavior follows the same laws as those followed by synthetic polymers. These naturally occurring giant molecules of the animal kingdom are discussed in Chapter 14. While the topics of nucleic acids and proteins are dealt with in the next chapter, they are also essential plant material. For instance, nucleic acids are the gene material for both plants and animals.

13.2 SIMPLE CARBOHYDRATES (SMALL MOLECULES)

In 1812, Gay-Lussac showed that carbohydrates, such as starch and cellulose, contained 45% carbon, 49% oxygen, and 6% hydrogen by weight. After dividing these percentages by the appropriate mass numbers, one obtains the empirical or simplest formula of CH_2O. Hence, Gay-Lussac called starch and cellulose "watered carbon" or carbohydrates. Both starch and cellulose are polymers of D-glucose, but they differ in the manner in which these building blocks or repeating units are joined together in these giant molecules.

As shown by the chemical and skeletal formulas for glyceraldehyde (2,3-dihydroxypropanal), this simple compound and all higher-molecular-weight aldocarbohydrates contain hydroxyl (OH) and carbonyl (C=O) groups. The OH is represented by — in the skeletal formulas.

```
    H
    |
1 H-C=O        C=O
    |          |
2 H-C-OH       C—
    |          |
3  H2COH       C—
```

Glyceraldehyde   Skeletal formula

If one were to make a model of glyceraldehyde using toothpicks and gumdrops, one would discover that there are two possible arrangements for the hydroxyl group and hydrogen atom around carbon atom No. 2. The two arrangements are not as obvious as those in other isomers unless one looks at three-dimensional models, which differ like right- and left-hand gloves; that is, they are mirror images of each other.

Since they rotate the plane of polarized light in equal but opposite directions, these so-called stereoisomers are optical isomers. Ordinary light waves vibrate in all directions, but polarized light, such as that passing through a Polaroid lens, vibrates in a single plane that is perpendicular to the ray of light.

The original optical isomers were labeled dextro after the Latin word *dexter*, meaning right, and *levo* after the Latin word *laevus*, meaning left. Accordingly, the trivial names dextrose and levulose have been used for the principal D-hexoses, namely, D-glucose and D-fructose.

Nature synthesizes dextro-carbohydrates almost exclusively. When we discuss proteins in Chapter 14, we will note that nature also synthesizes L-amino acids almost exclusively. The skeletal formulas that are mirror images for the simplest hydroxy aldehyde (glyceraldehyde) are

Since carbon No. 2 in glyceraldehyde has four different groups, namely, H, HC=O, OH, and H_2COH, which can be arranged in two different ways, it is called an asymmetric or chiral carbon atom. The aldoses containing three carbon atoms are called trioses.

There are two possible arrangements for each chiral atom, and there are four (2^2), eight (2^3), and sixteen (2^4) possible arrangements (isomers) for tetroses, pentoses, and hexoses, respectively. The general formula for describing the number of possible optical isomers is 2^n, where n is equal to the number of chiral atoms in a molecule.

We will direct our attention to pentoses and hexoses and specifically to the D isomers, that is, those with the hydroxyl group on the right-hand side of the chiral carbon atom farthest away from the carbonyl group. The OH on carbon No. 5 in D-glucose is on the right-hand side, as shown by the skeletal molecular structure

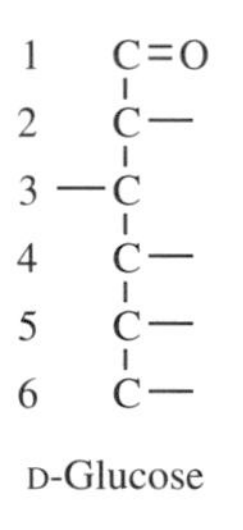

For aqueous (water) solutions of small sugars the linear form is in equilibrium with the cyclic or ring form as shown in Figure 13.1, but for larger structures such as cellulose and starch the compounds exist in the cyclic form.

The hydroxyl group on carbon No. 1 in the cyclic form of D-glucose, which is called the anomeric carbon atom, may be arranged in two different ways. The two forms are called α- and β-D-glucopyranose or simply the α and β forms of D-glucose. These glycopyranoses and all other low-molecular-weight aldoses and ketoses are called monosaccharides.

Condensation of two monosaccharides produces a disaccharide. The everyday compound we call sugar is a disaccharide with the common name of sucrose.

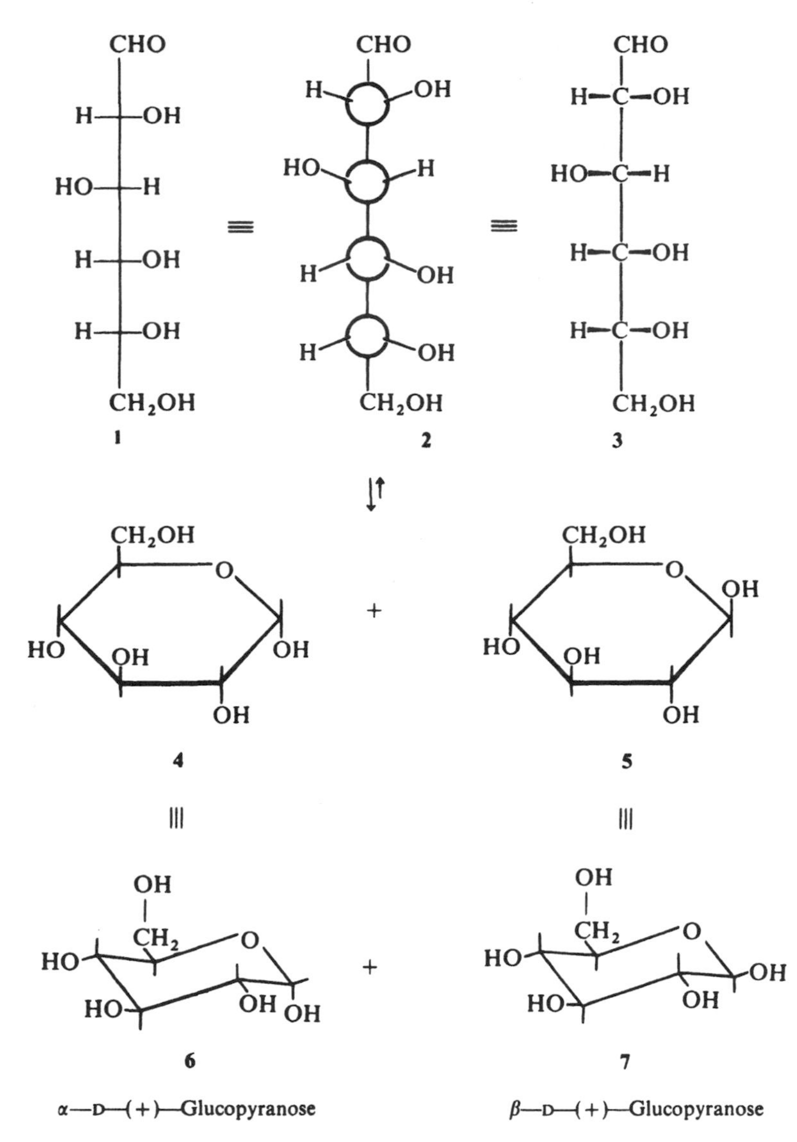

Figure 13.1. Open (*top*) and cyclic (*middle* and *bottom*) formulas for glucose. The top left form is called a Fischer projection. Structures **4** and **5** are called Haworth structures, and structures **6** and **7** are called "chair" structures.

Sucrose contains one unit of α-D-glucose and one of β-D-fructose. The C—O—C ether linkage connecting the two monosugar units is called a glycosidic linkage or bond. The geometry of the C—O—C linkage of α-D-glucose with β-D-fructose is called an alpha linkage whereas the C—O—C linkage formed through condensation of with the β-D-glucose is called a beta linkage.

13.3 CELLULOSE

D-Glucose is the building block or repeating unit in the principal polysaccharides, starch and cellulose. These giant molecules serve as the reserve carbohydrates in plants. The reserve carbohydrates in humans and many other animals is a polysaccharide called glycogen.

Glycogen is synthesized rapidly in a process called glycogenesis and hydrolyzed with equal rapidity to D-glucose in a process called glycogenolysis. These polymerization (building-up) and depolymerization (breaking-down) processes are regulated by the hormone insulin, which is excreted by the pancreas.

Two molecules of α-D-glucopyranose may join together and produce a molecule of a disaccharide called α-maltose. The mechanism for this dimerization is, of course, more complicated, but this general statement will be adequate for our discussion. As shown in the following structural formula, the two α-D-glucopyranose monomer units in α-maltose are joined through an oxygen or acetal linkage between carbons 1 and 4:

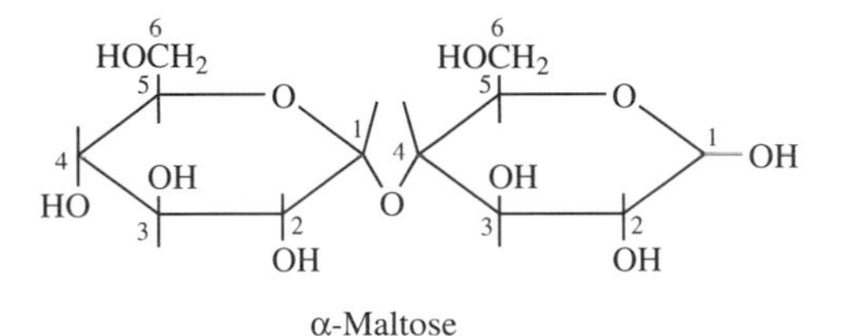

α-Maltose

When this building-up process is continued in plants, a giant molecule or polymer called amylose is produced. This polymeric chain, which may contain a thousand or more maltose units, is relatively flexible and tends to form a spiral or helix, as shown in Figure 13.2.

An iodine molecule fits well in this helix and gives a characteristic blue color test for starch. The other form of starch, called amylopectin, is a highly branched structure in which branches or chain extensions are formed on the No. 6 carbon atoms of the repeating D-glucose units.

When two molecules of β-D-glycopyranose are joined together through carbon atoms 1 and 4, the product is called cellobiose. As shown in Figure 13.3, cellulose is a polymer made up of many cellobiose units.

Cellulose was originally "discovered" by Payen in 1838. For thousands of years, impure cellulose formed the basis of much of our fuel and construction systems in the form of wood, lumber (cut wood), and dried plant material; served as the vehicle for the retention and conveying of knowledge and information in the form of paper; and provided clothing in the form of cotton, ramie, and flax. Much of the earliest research was aimed at developing stronger materials with greater resistance to the natural elements (including cleaning) and to improve dyeability so that the color of choice by common people for their clothing material could be other than a drab white. In fact, the dyeing of textile materials, mainly

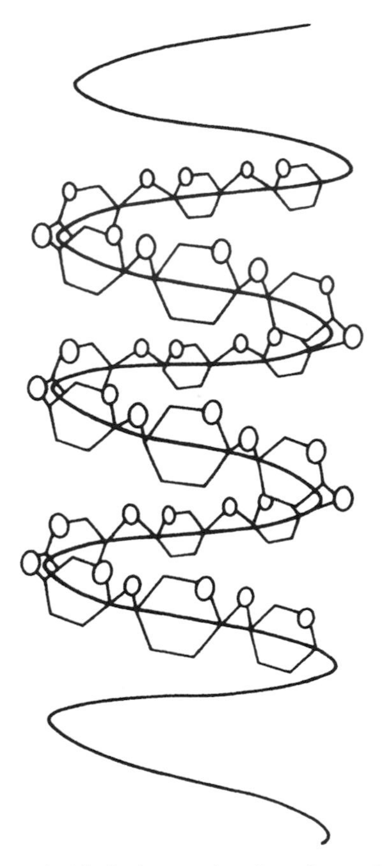

Figure 13.2. The loose, somewhat helical geometry of amylose, which is one of the two major components of starch.

cotton, was a major driving force in the expansion of the chemical industry in the latter part of the nineteenth century.

The top structure bottom, next page, is most commonly employed as a description of the repeat unit of cellulose, but the lower structure more nearly represents the actual three-dimensional structure with each D-glucosyl unit rotated 180 degrees. We will employ a combination of these two structural representations. Numbering is shown above and the type of linkage is written as $1 \rightarrow 4$ since the units are connected through oxygens contained on carbon 1 and 4.

Cellulose is connected through beta (β) linkages described as $\beta\ 1 \rightarrow 4$ linkages. The other, similar $1 \rightarrow 4$ linkage found in starch is called an alpha (α) linkage. The geometric consequence of this difference is great. The linear arrangement of cellulose with the β linkage (Figure 13.3) forming a puckered sheet structure gives an arrangement where the OH groups reside somewhat uniformly on the outside of the chain allowing close contact and ready hydrogen bond (secondary bonding between hydrogen and oxygen) formation between chains. This arrangement results in a

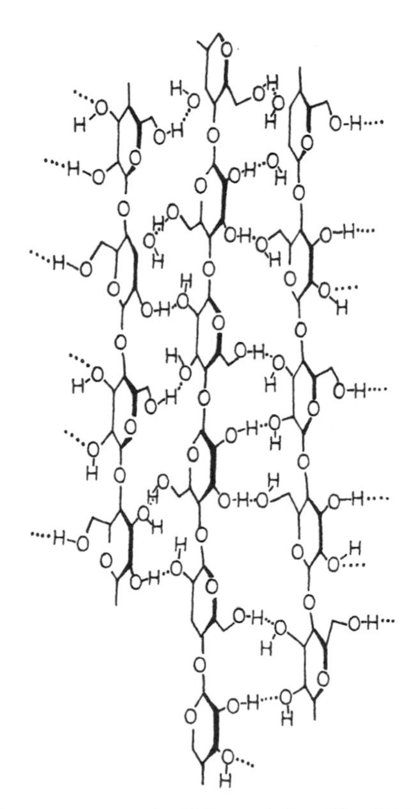

Figure 13.3. Puckered sheet structure of cellulose chains. The dots denote hydrogen bonding.

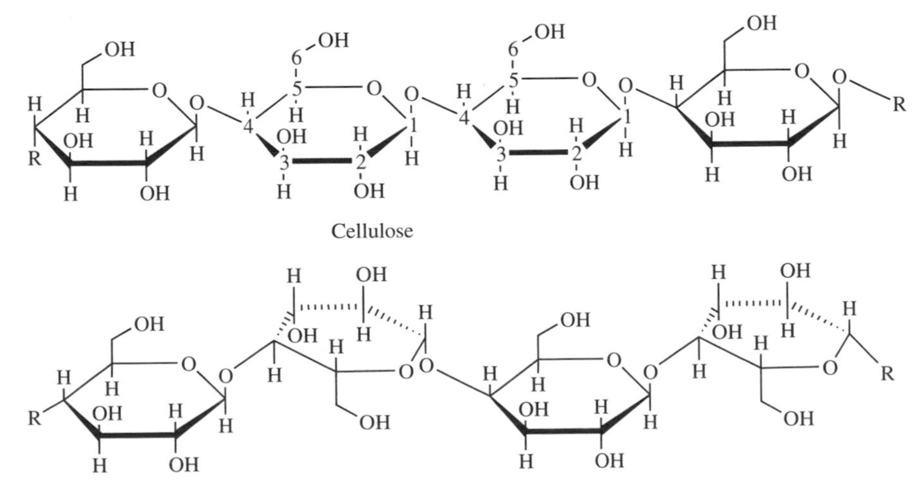

Cellulose

Cellulose three-dimensional structure

tough, insoluble, rigid, and fibrous material that is well-suited as cell wall material for plants. By comparison, as noted before, the α linkage of starch (namely amylose) results in a helical structure (Figure 13.2) where the hydrogen bonding is both interior and exterior to the chain allowing better wettability.

The intermolecular hydrogen bonds in cellulose are so strong that it is insoluble in water, whereas the linear chains of alpha-linked glucose units in amylose are water-soluble. Furthermore, because of the lack of intermolecular bonding, amylose is flexible in contrast to cellulose, which is rigid.

The difference between the alpha and beta linkages in these carbohydrate polymers shows up not only in varying physical properties, but also in their digestibility. Humans possess a gut enzyme that will break down alpha linkages specifically. Thus, we can eat potatoes and gain food value from them. However, humans cannot digest cellulose, but termites have enzymes that can digest this β-polysaccharide. Thus, they are able to feast on our timberland and wooden portions of our houses.

The rigidity of wood is also due to hydrogen bonding between the cellulose molecules and lignin. Wood exposed to ammonia (NH_3), a base, will begin to degrade as a result of the rupture of the hydrogen bonds. The partially degraded cellulose can then be shaped. When the ammonia is washed away, new hydrogen bonding occurs, locking the molecules into a new form, similar to that which occurs when we "set" our hair. Longer exposure to bases disrupts the glycosidic bonding and causes a permanent loss of strength.

Acids act similarly but faster to disrupt the glycosidic (acetal) bonds. Thus, polysaccharides degrade to their original monosaccharide units (D-glucose) when heated with acids. Enzymes catalyze this degradation and provide both plants and humans with a source of D-glucose.

The precise structure varies with plant source, location in the plant, plant age, season, seasonal conditions, treatment, and so on, and in turn the precise physical properties also vary within limits. Thus, in general, bulk properties of polysaccharides are generally measured with average values and tendencies given. These variations are sufficient for most applications but possibly not for specific biological applications where the polysaccharide is employed as a drug, within a drug delivery system, or as a biomaterial within the body.

Cellulose comprises more than one-third of all vegetable matter, and thus is the world's most abundant organic compound. Approximately 50 billion tons of this renewable resource is produced annually by land plants and probably more than double this is produced in the oceans yearly. Cellulose is often present as thread-like strands or bundles called fibrils. It is not found in pure form but rather is associated with other materials such as lignin and the so-called hemicelluloses. Cotton is the purest form of cellulose. Dried wood contains about 40–55% cellulose, 15–35% lignin, and 25–40% hemicellulose. Plant pulp is the major source of commercial cellulose with the extraction of cellulose from plants called pulping. The major source for nontextile fibers is wood pulp. The major source of textile cellulose is cotton.

13.4 COTTON

Cotton is grown in semitropical regions throughout the world. This essentially pure cellulose has been harvested for many thousands of years in China, Egypt, and Mexico. There are several species of *Gossypium* plants, but *Gossypium hirsutum* is the most common cotton plant.

Large-scale production of cotton was hampered until the invention of the cotton gin by Eli Whitney in 1793. This machine, which consists of a rotating spiked cylinder, displaced the labor-intensive hand process for separating the cotton from the cotton seeds. The production of cotton was also increased by the invention of the mechanical cotton picker in the twentieth century. Cotton was "king" in the cotton-producing states for almost a century, but has now been partially displaced by synthetic fibers, such as polyesters and nylon. Over 85 million bales (16.7 million tons) of cotton are produced annually worldwide.

Cotton, which is a cellulosic fiber, readily absorbs water and is stronger when wet than dry. Cotton is a good source of cellulose, and it is the source for most of the cellulose employed in the synthesis of rayon, cellulose acetate, and cellulose nitrate.

13.5 PAPER

It is believed that paper was invented by Ts'ai Lun in China in the second century A.D. Paper was first produced in the United States by William Rittenhouse in Germantown, Pennsylvania, in 1690. The original Chinese paper was a mixture of bark and hemp, but prior to the eighteenth century, much of the paper was made from rags. Paper was named after the papyrus plant, which is no longer used for the production of paper.

Modern paper is made from wood pulp (cellulose), which is obtained by the removal of lignin from debarked wood chips by use of chemicals, such as sodium hydroxide, sodium sulfite, or sodium sulfate. Newsprint and paperboard, which is thicker than paper, may contain some residual lignin. Lignin is the structural support and adhesive matter of the plant world. Wood contains cellulose bonded by at least 25% lignin.

The book you are reading, the newspaper, materials used to write notes on, and even some clothing are made of paper. If you rip a piece of ordinary paper (not your book, please!), you will note that paper consists of small fibers. Most of these cellulosic fibers are randomly oriented, but a small percentage of fibers are aligned in the direction in which the paper was produced from a watery slurry to a water-free sheet on a series of heated rolls. The papermaking process is sketched in Figure 13.4.

Wood and other woody products contain mostly cellulose and lignin. In the simplest papermaking process, the wood is chopped (actually torn) into smaller fibrous particles as it is pressed against a rapidly moving pulpstone. A stream of water

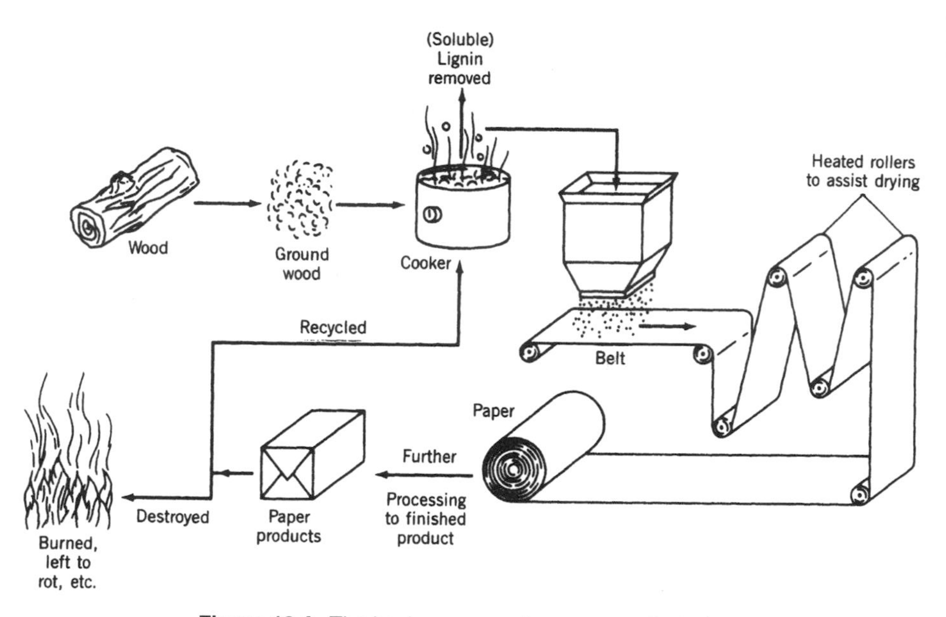

Figure 13.4. The basic process of paper manufacturing.

washes the fibers away and dissolves much of the water-soluble lignin. The fibrous material is concentrated into a paste called pulp. The pulp is layered into thin sheets and rollers are employed both to squeeze out water and to assist in achieving paper of a uniform thickness. Paper produced by this mechanism is not very white or strong because the remaining lignin is somewhat acidic and causes hydrolytic breakup of the cellulose chains. Most of the paper utilized for newsprint is of this type or regenerated, reused paper.

The sulfate process, also called the kraft process (kraft comes from the Swedish word for strong since good strength paper is produced), is more commonly employed. The kraft process is also favored over the sulfite process because of environmental considerations, since the sulfite process employs more chemicals that must be disposed of—particularly mercaptans (RSH), which are quite odorous, similar to the compounds emitted by a frightened skunk. Present research involves reclaiming and recycling these pulping chemicals, and so far more than a 10-fold decrease in the amount of chemical used per volume of paper produced has been attained.

If pure cellulose pulp were used, the fiber mat formed would be largely water soluble with only surface polar and hydrogen bonding acting to hold the fibers together. White pigments such as clay and titanium dioxide (TiO_2) are added to help "cement" the fibers together and to fill the voids, thus producing a firm, white writing surface. Resins, surface-coating agents, and other special surface treatments (such as coating with polypropylene and polyethylene) are employed for paper products intended for special purposes such as milk cartons, ice cream cartons, roofing paper, extra strength uses, light building materials, and drinking cups. The cellulose supplies the fundamental structure and most of the bulk (about 90% of the weight) and strength for the paper product whereas the additives provide special properties needed for special applications.

As costs rise, the interest in recycling paper also increases. Community recycling centers and paper drives organized by local Scout, church, or school groups help in this recycling process. Today about 20–25% of our paper products are being recycled. This percentage could be doubled if more emphasis was placed on the collection of waste paper. Unfortunately, recycled paper costs slightly more than virgin pulp. On the other hand, recycling can reduce the destruction of natural resources (by saving forests) and minimize the load on waste-disposal systems.

Most paper is coated to provide added strength and smoothness. The coating is basically an inexpensive paint that contains a pigment and a small amount of polymeric binder. Unlike most painted surfaces, most paper products are manufactured with a short lifetime in mind with moderate performance requirements. Typical pigments are inexpensive low-refractive index materials such as plate-like clay and ground natural calcium carbonate. Titanium dioxide is used only when high opacity is required. The binder may be a starch or latex or a combination of these. The latexes are usually copolymers of styrene, butadiene, acrylic, and vinyl acetate. Other additives and coloring agents may also be added for special performance papers. Resins in the form of surface coating agents and other special surface treatments (such as coating with polypropylene and polyethylene) are used for paper products intended for special uses such as milk cartons, ice cream cartons, light building materials, and drinking cups. The cellulose supplies the majority of the weight (typically about 90%) and strength, with the special additives and coatings providing special properties needed for the intended use.

13.6 STARCH

As noted before, starch can be divided into two general structures, branched amylopectin and largely linear amylose:

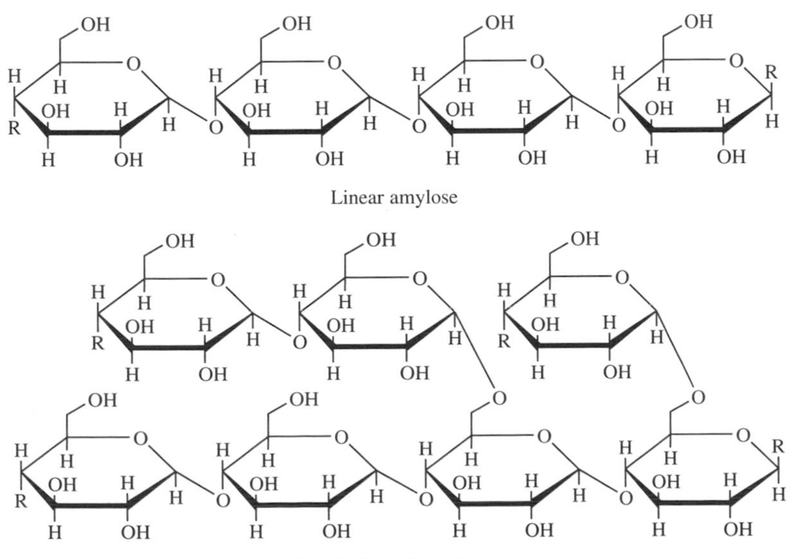

Linear amylose

Branched amylopectin

Amylose typically consists of over 1000 D-glucopyranoside units. Amylopectin is a larger molecule containing about 6000 to 1,000,000 hexose ("hexa" is used to signify "six," so a hexose is a six-membered-ring sugar structure) rings essentially connected with branching occurring at intervals of 20–30 glucose units. Branches also occur on these branches, giving amylopectin a fan or treelike structure similar to that of glycogen. Thus, amylopectin is a highly structurally complex material. Unlike nucleic acids and proteins where specificity and being identical are trademarks, most complex polysaccharides can boast of having the "mold broken" once a particular chain was made, so that the chances of finding two exact molecules is very small.

Starch granules are insoluble in cold water but swell in hot water. As the water temperature continues to increase to near 212°F, a starch dispersion is obtained. Oxygen must be avoided during heating or the chains break up. Both amylose and amylopectin are then water-soluble at high temperatures. Amylose chains tend to assume a helical arrangement (Figure 13.2), as already noted, giving it a compact structure. Each turn contains six glucose units.

Starch is the principal food-reserve polysaccharide in plants and serves as the main source of carbohydrate in our diet and the diet of many animals. It is found in seeds, fruits, tubers, roots, and stems of plants. Commercial sources include grains such as corn, tapioca, sorghum, rice, millet, barley, and wheat, as well as potatoes. Starch is a major source of corn syrup and corn sugar. It is also used in adhesive formulations for paper and as a textile-sizing agent. Cyclodextrins are formed when starch is treated with a particular enzyme, the amylase of *Bacillus macerans*.

Natural starch is a polydisperse macromolecule consisting of different ratios of amylose and amylopectin, depending on its source. For example, the so-called waxy starches contain over 98% amylopectin. The polydisperse nature appears to be characteristic of natural structural and storage materials (such as glycogen and starch), but not of those macromolecules whose shape, size, and electronic nature are critical in carrying out specific biological functions (such as DNA and enzymes).

13.7 OTHER CARBOHYDRATE POLYMERS

Glycogen is a food-reserve polysaccharide of animals that also occurs in some yeasts and fungi. It is found in the liver and muscles of animals, and its structure is similar to that of amylopectin, but it has a larger number of chain branches on carbon No. 6.

Dextran is a poly-α-D-glucose linked through the Nos. 1 and 6 carbon atoms with occasional branching at the No. 3 carbon atom. Dextran is an amorphous solid that forms random coils in aqueous (water) solutions. Dextrans have been used as blood plasma volume extenders, in pharmaceuticals, to increase the viscosity of foodstuffs (such as ice cream), and as emulsion stabilizers.

Many plants and some species of seaweed contain polyuronides in which the methylol group (CH_2OH) in the repeating unit of the polymeric carbohydrate is replaced by a carboxylic acid group (COOH). These gums, such as alginic acid, agar, pectic acid, and carrageenan, are water-soluble. Some polyuronides, such as galactomannans, carrageenans, agar, gum arabic, and alginates, are used as food additives in ice cream, pie fillings, gelled desserts, and salad dressings. Many are also used in the pharmaceutical industry as encapsulating materials and as emulsion stabilizers. Agar is used as a culture medium for bacteria. Pectic acid is widely used in the making of jams and jellies.

The shells of crustacea and some insects and fungi consist of celluloselike polymers in which the hydroxyl groups on carbon No. 2 of the repeating glucose unit are replaced by an acetylamino group ($NH(CO)CH_3$). This polymer, called chitin, has also been found in tact in fossils that are over 500 million years old.

13.8 LIGNIN

Lignin is the second most widely produced organic material, after the saccharides. It is found in essentially all living plants. It is produced at an annual rate of about 2×10^{10} tons with the biosphere containing a total of about 3×10^{11} tons.

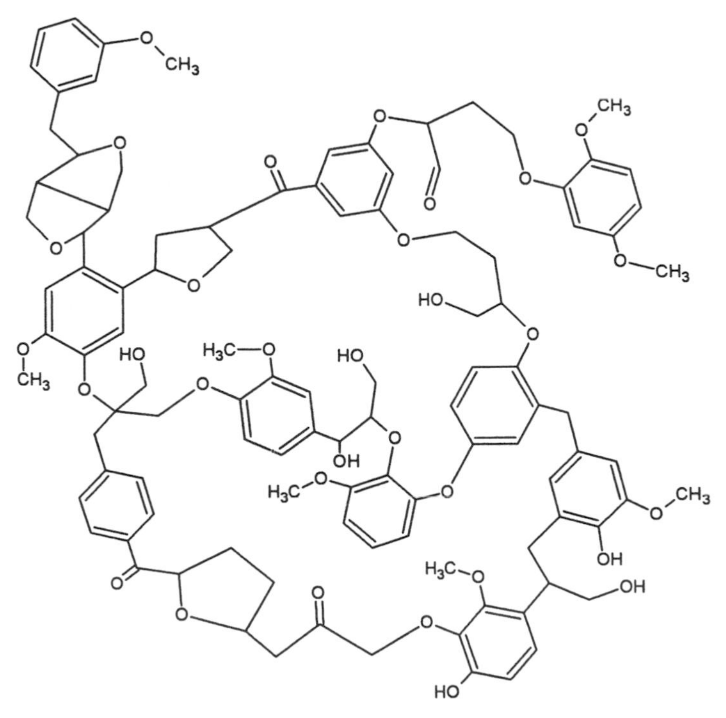

Figure 13.5. Representative structure of lignin.

It contains a variety of structural units including those pictured in Figure 13.5. Lignin functions as the structural support and cement material of the plant world. It constitutes about 25% of wood and is thus one of the most abundant natural polymers. Since lignin is relatively inert and is insoluble in the plant, it is not easily isolated in a pure and undegraded state. However, it can be extracted from wood by dissolving it in a dilute solution of acetic acid (H_3COOH) and ethyl acetate ($H_3CCOOC_2H_5$) in the "ester process."

Because of the synthetic sequence that occurs in plants, lignin appears to have a two-dimensional sheet structure. Polysaccharides, such as cellulose and hemicellulose, are deposited as plant cell walls are formed. Lignin is then synthesized to fill the spaces between the polysaccharide fibers. This process constrains the "three-dimensional" growth of lignin. The structure of lignin is complex and variable, but it contains ethers, aromatic and aliphatic alcohol functions, ketones, and aromatic rings.

Lignin sulfonate is used as a dispersant and wetting agent and is an important additive in the preparation of oil well drilling muds. Lignin sulfonates are used in road binders, industrial cleaners, and adhesives. Lignin obtained by alkali extraction is used as an additive for cement, filler for rubber, and as a dye dispersant. Nevertheless, considerable lignin is burned as fuel at paper mills. It is currently available at 20–30 cents per pound, making it a potentially attractive feedstock for other uses.

13.9 BITUMENS

The petroleum industry, including the commercial bitumen industry, was born in the United States in August 27, 1859 when Colonel Drake drilled about 70 feet near Titusville, Pennsylvania to "bring in" the first producing well. By 1908, Henry Ford began to mass-produce his Model "T" Ford, creating an additional need for this petroleum in the form of gasoline. The distillation residue became more plentiful and a need for large-scale usage of bitumens was increased.

Even so, the bitumens are a very old material. They were used in the waterproofing of the cradle that baby Moses was floated in. It was used by the ancient Egyptians in their mummification process. Bitumens were used in sand stabilization and for lighting the naval base by the Second Muslim Caliph, Omar ben Khattab, at Basra on Shattul-Arab on the West Coast of what is now Saudi Arabia around 640 A.D.

Bitumens occur naturally or are formed as the residue in the distillation of coal tar, petroleum, and so on. Industrially, the two most important bitumens are asphalt and coal tar. Gilsonite is the third important derivative and occurs naturally. Asphalt is a brown to black tar-like variety of bitumen that again occurs naturally or is the residue of distillation. Coal tar is the black, thick liquid obtained as the residue from the distillation of bituminous coal.

Bitumens are examples of materials that have only an approximate structure. Bitumens are carbon-intense small polymers with molecular weights from about 200 to 1000 daltons for coal tar with a calculated average number of carbons in a chain of about 15 to 70. Asphalt has a molecular weight averaging about 400 to 5000 daltons with a calculated average number of carbons in a chain of about 30 to about 400. Thus, they are generally oligomeric to short polymers. Asphalt has a C/H ratio of about 0.7 whereas coal tar has a C/H ratio of about 1.5, approaching that of a typical hydrocarbon where the C/H ratio is about 2.

As with most nonpolar hydrocarbon-intense polymers, bitumens exhibit good resistance to attack by inorganic salts and weak acids. They are dark, generally brown to black, with their color difficult to mask with pigments. They are thermoplastic materials with a narrow service temperature range unless modified with fiberous fillers and/or synthetic resins. They are abundant materials that are relatively inexpensive, thus their use in many bulk applications.

Bitumens are consumed at an annual rate in excess of 75 billion pounds in the United States. Bitumens are generally used in bulk such as pavements (about 75%), as well as in coatings for roofs (15%), driveways, adhesive applications, construction, metal protection, and so on, where the bitumen acts as a weather barrier. Bituminous coatings are generally applied either hot or cold. Hot-applied coatings are generally either filled or nonfilled. Cold-applied coatings are generally either non-water-containing or water-containing. In the hot-applied coatings, the solid is obtained through a combination of cooling and liquid evaporation which in the cold-applied coatings the solid material is arrived at through liquid evaporation. Onc often used coating employs aluminum pigments compounded along with solvents. These coatings are heat-reflective and decrease the energy needs of buildings using them. The aluminum–metallic appearance is generally more desirable than black, and the reflective nature of the aluminum reflects light that may damage the bitumen coating, thereby allowing the coating a longer useful life.

Today, many of the bitumen coatings contain epoxy resins, various rubbers, and urethane polymers.

13.10 OTHER NATURAL PRODUCTS FROM PLANTS

Many natural resins are fossil resins exuded from trees thousands of years ago. Exudates from living trees are called recent resins, and those obtained from dead trees are called semifossil resins. Humic acid is a fossil resin found with peat, brown coal, or lignite deposits throughout the world. It is a carboxylated phenolic-like polymer that is used as a soil conditioner, as a component of oil drilling muds, and as a scavenger for heavy metals.

Amber is a fossil resin found in the Baltic Sea regions, and sandarac and copals are found in Morocco and Oceania, respectively. Other fossil resins, called Manila, Congo, and Kauri, are named after their geographic source.

Frankincense and myrrh, which are mentioned in the *New Testament*, contain polyuronides. Bitumens, which were used by Noah for waterproofing the ark in the biblical story, occur as asphalt at Trinidad Lake (West Indies) and as gilsonite in Utah.

Natural rubber also comes from plants. This typic is covered in Chapter 10.

13.11 PHOTOSYNTHESIS

The sun is the source that winds the clock of life. Green plants absorb solar energy and converts it to carbohydrates in a process called photosynthesis. Photosynthesis is the process in which carbon dioxide (CO_2) combines with water (H_2O) to form glucose ($C_6H_{12}O_6$).

Photosynthesis begins with the absorption of light by chlorophyll in plants. The absorbed energy excites electrons in this green pigment to higher energy states. When the electrons return to their original "ground" state, the released energy is used to decompose water, produce a strong reducing agent, and energize a select phosphate ester from nucleic acid.

Reduction refers to processes whereby electrons are added in a chemical reaction, and oxidation refers to the process of giving up electrons:

$$2Fe^{3+} + 2e^- \longrightarrow 2Fe^{2+} \quad \text{(reduction)}$$

$$H_2 \longrightarrow 2H^+ + 2e^- \quad \text{(oxidation)}$$

The overall reaction, which is the sum of the two preceding reactions, is

$$H_2 + 2Fe^{3+} \longrightarrow 2H^+ + 2Fe^{2+} \quad \text{(net reaction)}$$

Detailed discussion of this topic is found in most introductory chemistry books.

The synthesis of glucose is described by the equation

$$6CO_2 + 6H_2O \xrightarrow{\text{light}} C_6H_{12}O_6$$

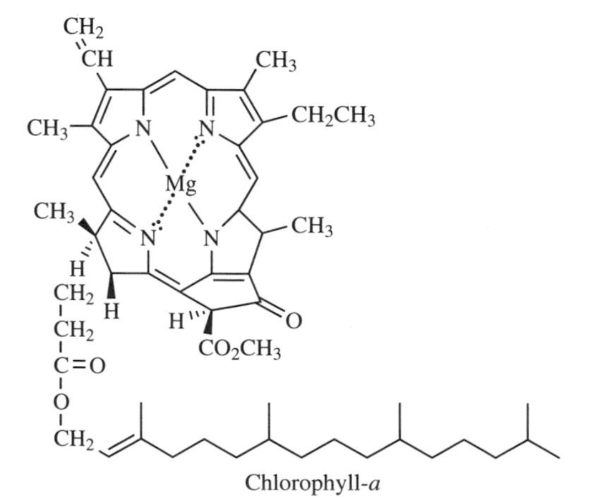

Chlorophyll-*a*

The most important and common plant pigment is chlorophyll, which is a polymer with a structure similar to that of heme proteins, except that iron is replaced by magnesium. A number of compounds are grouped under the name of chlorophyll. The preceding structure is the "active" part of chlorophyll-*a*, which was confirmed by H. Fischer, R. Willstatter, and James B. Conant.

The importance of photosynthesis in the biosphere cannot be overstated; without photosynthesis, there would be no biosphere. The photosynthetic process plays a major role in maintaining the desirable amounts of CO_2 and O_2 in the atmosphere. We breathe in O_2 and emit CO_2; conversely, plants absorb CO_2 and H_2O and form carbohydrates and emit O_2. This natural cycle emphasizes the harmony or balance that is often apparent in Nature and demonstrates the symbiotic relationship between the plant and animal kingdom.

GLOSSARY

Acetal: The oxygen linkage between monosaccharides in polysaccharides.

Aldo: Prefix for aldehyde compounds (—CHO).

Aldose: A compound containing hydroxyls and an aldehyde group, like glyceraldehyde.

Amino acid: Compound with the general formula

$$H_2N\overset{\displaystyle H}{\underset{\displaystyle R}{C}}COOH$$

Amylopectin: A highly branched starch molecule.

Amylose: A linear starch molecule.

Anomeric: A carbon atom on which the hydroxyl groups may be arranged in two different ways—that is, alpha and beta.

Asymmetric: A molecule that is not symmetrical; that is, the atoms or groups around the carbon atom may be arranged in two different ways.

Bitumen: Asphalt and gilsonite resins.

Carbonyl:

$$\rangle C{=}O$$

Chiral: An asymmetric carbon atom; derived from the Greek word *cheir*, meaning hard.

Chlorophyll: A green pigment in plants.

Dextro: Derived from the Latin word *dexter*, meaning right.

Dextrose: A trivial name for D-glucose.

Empirical formula: Simplest formula.

Fossil resin: Aged exudates from tropical trees.

Fructose: $C_6H_{12}O_6$, a ketohexose.

Glucose: $C_6H_{12}O_6$, an aldose with six carbon atoms and five hydroxyl groups.

Glyceraldehyde:

```
        H  H
        |  |
  H2C - C - C = O
     |  |
     OH OH
```

Glycogen: The reserve carbohydrate in animals.

Glycogenesis: The polymerization of glucose to glycogen.

Glycogenolysis: The hydrolysis of glycogen to form glucose (depolymerization).

Hexose: An aldose or ketose with six carbon atoms.

Humic acid: A carboxylated phenolic-like semifossil resin found in peat and lignite deposits.

Hydroxyl: —OH.

Insulin: A hormone that regulates glycogenesis and glycogenolysis.

Isomers: Molecules with the same empirical formulas.

Isomers, optical: Molecules with similar formulas but with arrangements in space that rotate the plane of optical light in equal and opposite directions.

Isomers, steric: Molecules with substituents arranged differently in space.

Levo: Derived from the Latin word *laevos*, meaning left.

Levulose: A trivial name for D-fructose.

Light, polarized: Light that is vibrating in a single plane.

Lignin: The noncellulosic material in wood.

Lignin sulfonate: The product of the reaction of lignin and sulfuric acid.

Maltose: The disaccharide made up of two molecules of D-glucose.

Monosaccharide: The simplest saccharide, that is, fructose and glucose.

Natural polymer: Giant molecules such as starch, cellulose, and proteins that occur in nature.

Oxidation: The loss of electrons by a molecule or ion.

Photosynthesis: The production of glucose by the catalytic combination of carbon dioxide and water.

Polydisperse: A mixture of polymers with different molecular weights.

Pyranose: A cyclic molecule consisting of five carbon atoms and one oxygen atom.

Recent resin: Exudate from live trees.

Reduction: The addition of electrons to a molecule or ion.

Semifossil resin: Exudate from dead trees.

Skeletal formula: A formula in which the hydrogen atoms are omitted.

Solar energy: Energy from the sun.

REVIEW QUESTIONS

1. What is a natural polymer?
2. What is the empirical formula for glucose ($C_6H_{12}O_6$)?
3. What optically active acid occurs in milk?
4. In the dextro–levo convention, what would your right hand be called?
5. How do dextrose (D-glucose) and levulose (D-fructose) differ chemically?
6. The simplest amino acid is glycine

$$(\mathrm{N_2H}\overset{\displaystyle \mathrm{H}}{\underset{\displaystyle \mathrm{H}}{\mathrm{C}}}\mathrm{COOH})$$

   Is glycine optically active?
7. What is the shape of the pyranose molecule?
8. How many anomeric carbon atoms are present on D-glucose?
9. How many monosaccharide repeating units are there in maltose?
10. Which is more linear: amylose or amylopectin?
11. Is cellulose mono- or polydisperse?
12. What polymer is present in paper?
13. Which will have less residual solvent: a fossil resin or a recent resin?
14. What is the generic name for asphalt and gilsonite?
15. Define oxidation.
16. Define reduction.
17. Which can you digest: starch or cellulose?

BIBLIOGRAPHY

Atkins, E. (1986). *Polysaccharides*, VCH, New York.

Carraher, C., and Sperling, L. (1983). *Polymer Applications of Renewable-Resource Materials*, Plenum, New York.

Carraher, C., and Tsuda, M. (1980). *Modification of Polymers*, ACS Symposium Series, ACS, Washington, D.C.

Gebelein, C. (1992). *Biotechnology and Polymers*, Plenum, New York.

Gebelein, C., and Carraher, C. (1994). *Biotechnology and Bioactive Polymers*, Plenum, New York.

Gebelein, C., and Carraher, C. (1995). *Industrial Biotechnological Polymers*, Technomic, Lancaster, PA.

Gebelein, C., and Carraher, C. (1985). *Bioactive Polymeric Systems*, Plenum, New York.

Gilbert, R. (1994). *Cellulosic Polymers*, Hanser-Gardner, Cincinnati.

Hecht, S. M. (1998). *Bioorganic Chemistry: Carbohydrates*, Oxford University Press, Cary, NC.

Kennedy, J., Mitchell, J., and Sandford, P. (1995). *Carbohydrate Polymers*, Elsevier, New York.

Paulsen, B. (2000). *Bioactive Carbohydrate Polymers*, Kluwer, New York.

Scholz, C., and Gross, R. (2000). *Polymers from Renewable Resources: Biopolyesters and Biocatalysis*, ACS, Washington, D.C.

Seeberger, P. (2001). *Solid Support Oligosaccharide Synthesis and Combinatorial Carbohydrate Libraries*, Wiley, New York.

Steinbuckel, A. (2001). *Lignin, Humic, and Coal*, Wiley, New York.

Steinbuchel, A. (2001). *Polyisoprenoides*, Wiley, New York.

Vigo, T. (2001). *Bioactive Fibers and Polymers*, ACS, Washington, D.C.

Woodings, C. (2001). *Regenerated Cellulose Fibers*, Woodhead Publishers, Cambridge, England.

ANSWERS TO REVIEW QUESTIONS

1. A giant molecule found in nature.
2. CH_2O.
3. Lactic acid

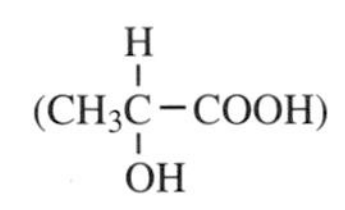

4. Dextro-hand.
5. Dextrose is an aldohexose and levulose is a ketohexose.
6. No. It does not contain a chiral carbon atom (only three different groups on carbon No. 1).
7. It has a ring or cyclic structure.
8. One.
9. Two (it is a disaccharide).
10. Amylose.
11. Polydisperse.

12. Cellulose. (Also some lignin.)

13. Fossil resin.

14. Bitumen.

15. Loss of electrons.

16. Gain of electrons.

17. Starch.

14

NATURE'S GIANT MOLECULES: THE ANIMAL KINGDOM

14.1 INTRODUCTION

One of the strongest and most rapidly growing areas of polymer science is that of natural polymers. Our bodies are largely composed of polymers: deoxyribonucleic

Giant Molecules: Essential Materials for Everyday Living and Problem Solving, Second Edition, by Charles E. Carraher, Jr.
ISBN 0-471-27399-6

acid (DNA), ribonucleic acid (RNA), proteins, and carbohydrate polymers. These polymers are related to aging, awareness, mobility, strength, and so on, all characteristics that contribute to our being "alive and well." Many medical, health, and biological projects and advances are concerned with materials that are polymeric. There is an ever-increasing emphasis on molecular biology—that is, science applied at the molecular level to biological systems. An understanding of natural polymers is advantageous to intelligent citizens and to those who desire to understand and contribute positively to advances in biology, but these will be discussed only briefly in this chapter.

Starch, cellulose, and lignin are the building blocks of the plant world, whereas proteins and nucleic acids serve a similar role in the animal kingdom. Natural rubber, resins, and gums are also polymeric and play an important role in our everyday activities. The shape and size of these natural polymers are critical to their ability to carry out their highly specialized functions.

We are witnessing a reemergence of the use of natural polymers in many new and old industrial applications, since natural polymers are renewable resources that nature continues to provide. There is no difference in the science and technology of natural and synthetic polymers, and manufacturing techniques suitable for application to synthetic polymers are normally equally applicable to natural polymers.

14.2 AMINO ACIDS

More than 200 amino acids exist, but only about 20 of them are necessary for the existence of animal life. Some of these amino acids can be synthesized in adequate quantities in the human body (Table 14.1). Each amino acid contains an amino functional group (NH_2) and a carboxylic acid functional group (COOH). These

Table 14.1 The twenty amino acids commonly found in proteins[a]

Glycine	Gly	$H-CH(NH_2)-COOH$
Alanine	Ala	$CH_3-CH(NH_2)-COOH$
*Valine	Val	$CH_3-CH(CH_3)-CH(NH_2)-COOH$
*Leucine	Leu	$CH_3CH(CH_3)-CH_2-CH(NH_2)-COOH$
*Isoleucine	Ile	$CH_3-CH_2-CH(CH_3)-CH(NH_2)-COOH$

Table 14.1 (*Continued*)

*Phenylalanine	Phe	(benzene ring)$-CH_2-CH-COOH$ NH_2 (on CH)
Tyrosine	Tyr	HO(benzene ring)$-CH_2-CH-COOH$ NH_2 (on CH)
Serine	Ser	$HO-CH_2-CH-COOH$ NH_2 (on CH)
*Threonine	Thr	$CH_3-CH-CH-COOH$ OH NH_2
Cysteine	Cys	$HS-CH_2-CH-COOH$ NH_2 (on CH)
*Methionine	Met	$CH_3-S-CH_2CH_2-CH-COOH$ NH_2 (on CH)
*Tryptophan	Trp	(indole ring, N–H)$-CH_2-CH-COOH$ NH_2 (on CH)
Proline	Pro	H_2C-CH_2 $H_2C-N(H)-CH-COOH$ (ring)
Asparagine	Asn	$NH_2-C-CH_2-CH-COOH$ $=O$ (on C), NH_2 (on CH)
Aspartic acid	Asp	$HO-C-CH_2-CH-COOH$ $=O$ (on C), NH_2 (on CH)
Glutamine	Gln	$NH_2-C-CH_2CH_2-CH-COOH$ $=O$ (on C), NH_2 (on CH)
Glutamic acid	Glu	$HO-C-CH_2CH_2-CH-COOH$ $=O$ (on C), NH_2 (on CH)
*Lysine	Lys	$NH_2-CH_2CH_2CH_2CH_2-CH-COOH$ NH_2 (on CH)
*Arginine	Arg	$NH_2-C-NH-CH_2CH_2CH_2-CH-COOH$ $=NH$ (on C), NH_2 (on CH)
*Histidine	His	(imidazole ring, N, N–H)$-CH_2-CH-COOH$ NH_2 (on CH)

[a]Amino acids not synthesized naturally by humans are called essential amino acids. These are denoted by an asterisk.

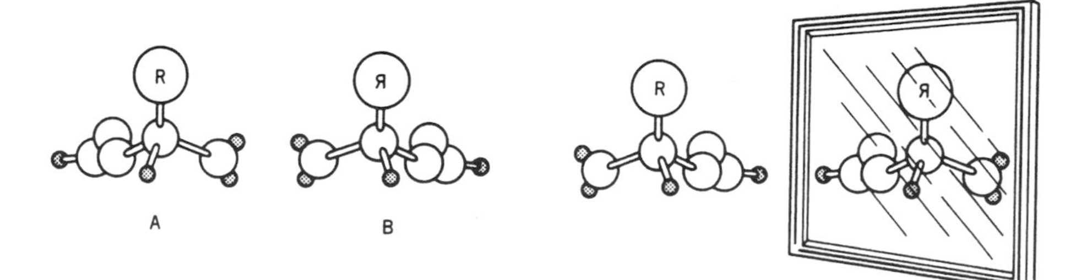

Figure 14.1. α-Amino acids. Representations illustrate optically active forms.

two functional groups are attached to the same carbon atom, which is called the alpha carbon atom.

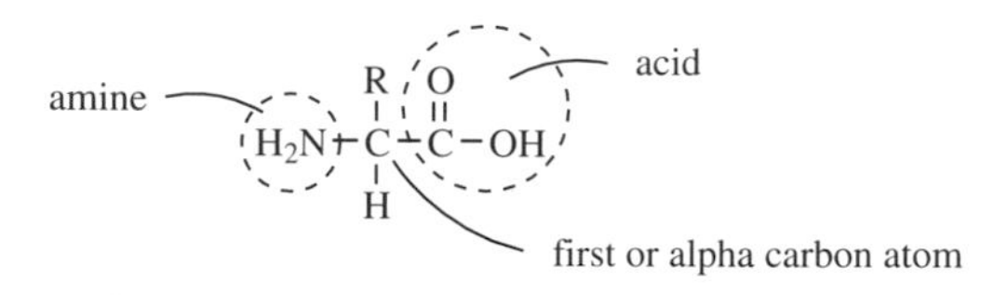

With the exception of glycine, in which the R group is a hydrogen atom, two optical isomers exist for each α-amino acid. As shown in Figure 14.1, these two forms are mirror images of each other. Similar phenomena are our right and left hands, which are approximately mirror images of each other (Figure 14.2). Structures A and B of Figure 14.1 are called optical isomers since they rotate the plane of polarized light in opposite directions. All the essential amino acids are of levo optical isomeric form, that is, the L form.

Since amino acids contain both acidic (RCOOH) and basic (RNH_2) groups in the same molecule, they are called zwitterions, after the Greek word *zwitter*, which means hybrid. Amino acids can have different electrical charges, depending on the pH of the solution.

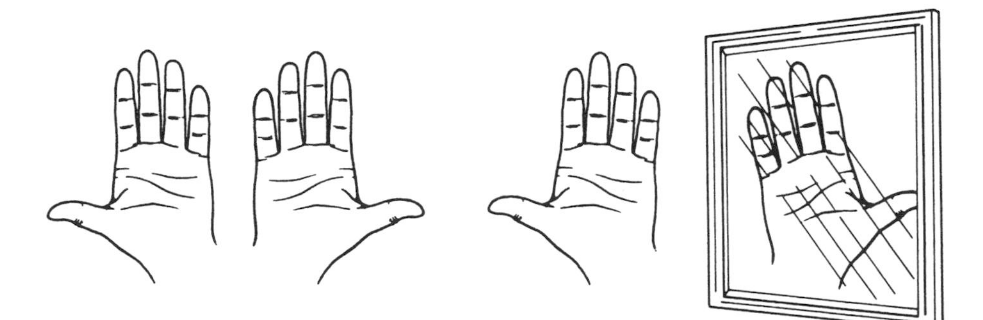

Figure 14.2. Left and right hands. Representations illustrate optically active forms.

The pH of a solution is a measure of acidity or basicity of that solution. A pH of 7 is neutral, that is, neither acidic nor basic. The solution is acidic at pH values less than 7, and the lower the pH value, the more acidic the solution. Basic solutions have pH values higher than 7, and the higher the value, the more basic the solution.

The pH at which the negative and positive charges are balanced (i.e., there is no movement toward either pole in an electric field) is characteristic for each amino acid and is called the isoelectric point. Shifts in the amount and form of the various amino acids are controlled in the body by variations in pH, thus allowing the body to effectively regulate the supply of specific amino acids. This assures essential supplies of specific amino acids for producing proteins for hair, skin, and so on. The changes in structure of phenylalanine (Phe) with changes in pH are shown in Figure 14.3.

While humans synthesize about a dozen of the 20 amino acids needed for good health, the other eight are obtained from outside our bodies, generally from eating foods that supply these essential amino acids. Different foods are good sources of different amino acids. Cereals are generally deficient in lysine. Thus, diets that emphasize cereals will also have other foods that can supply lysine. In the orient the combination of soybean and rice supply the essential amino acids while in Central Americas bean and corn are used.

Almost all of the sulfur needed for healthy bodies is found in amino acids as cysteine and methionine. Sulfur serves several important roles including as a cross-linking agent similar to that served by sulfur in the cross-linking, vulcanization of rubber. This cross-linking allows the various chains, that are connected

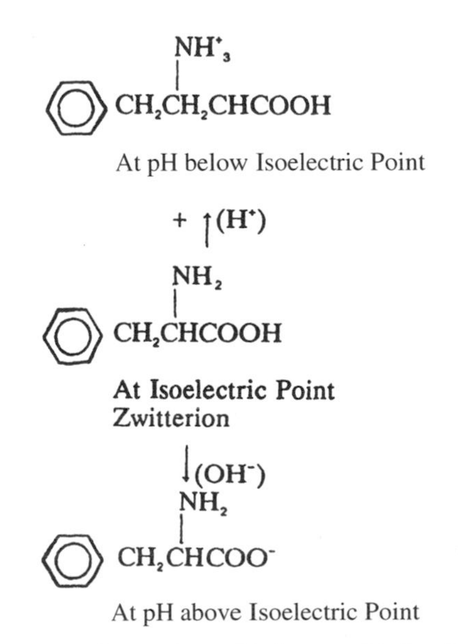

Figure 14.3. Structure of an amino acid above and below its isoelectric point.

by these cross-links, to "remember" where they are relative to one another. This cross-linking allows natural macromolecules to retain critical shapes to perform necessary roles.

14.3 PROTEINS

The name protein is derived from the Greek word *proteios*, meaning of first importance. G. V. Mulder coined this word in 1835 when he recognized that these nitrogen-containing organic compounds were essential for all life processes.

Proteins are copolymers made up of 20 different amino acids. Many lower animals can synthesize all of these amino acids, but as noted in Table 14.1, humans must obtain 8 of these from their diet. Fortunately, soybeans, which have served as a staple food for centuries, contain all the essential amino acids.

14.4 PROTEIN STRUCTURE

The structures of naturally occurring L-leucine and D-leucine are shown for comparative purposes:

$$\begin{array}{c} \mathrm{COOH} \\ | \\ \mathrm{H_2NCH} \\ | \\ \mathrm{HCH} \\ | \\ \mathrm{H_3CCCH_3} \\ | \\ \mathrm{H} \end{array} \qquad\qquad \begin{array}{c} \mathrm{COOH} \\ | \\ \mathrm{HCNH_2} \\ | \\ \mathrm{HCH} \\ | \\ \mathrm{H_3CCCH_3} \\ | \\ \mathrm{H} \end{array}$$

L-Leucine (Leu)
(naturally occuring)

D-Leucine (Leu)
(does not occur naturally)

The amino acids

$$\begin{array}{c} \mathrm{NH_2} \\ | \\ \mathrm{RCCOOH} \\ | \\ \mathrm{H} \end{array}$$

in proteins are arranged in characteristic head-to-tail order and joined through peptide linkages

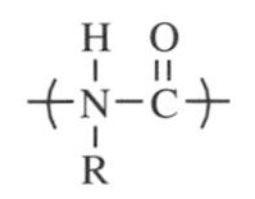

Peptide (or amide) linkage

A dipeptide (or dimer) is produced when two amino acids are joined together. The low-molecular-weight polymers containing relatively few amino acid residues or mers are called polypeptides. The high-molecular-weight polymers or macromolecules are also polypeptides but are usually called proteins.

It is of interest that the peptide linkages that connect the various amino acid units to form proteins are structurally similar to those connecting the synthetic nylons:

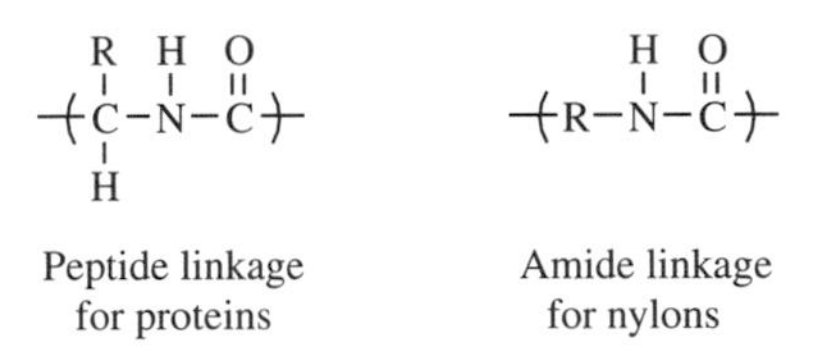

Peptide linkage for proteins

Amide linkage for nylons

We can represent the structures of proteins by using the symbols for the amino acids. The dipeptide made from the reaction between valine and glycine through reaction of the carboxylic acid function on the valine with the amine group on the glycine is represented as Val · Gly and given the name valylglycine. If the amine function of the valine were reacted with the acid function of glycine, then the representations are reversed, giving Gly · Val the name of glycylvaline. In each case the amino acid whose carboxyl group is involved in the formation of the amide linkage is placed first. This convention is biochemical shorthand for representing complex structures.

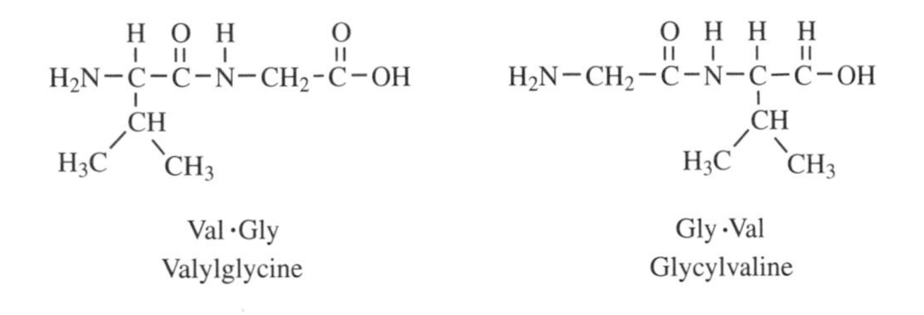

Val·Gly
Valylglycine

Gly·Val
Glycylvaline

When a protein is boiled in hydrochloric acid, hydrolysis of many of the amide repeating units occurs, and a mixture of dimers, monomers, trimers, and so on, is produced. Because of the extreme importance of proteins, techniques have been developed to allow the sequence of amino acids to be reconstructed from such data. Molecular weight measurements and electron and x-ray diffraction measurements provide information on the actual shape and size of the protein. This process is tedious, and hence it has been used for only a few of the more important and simpler proteins, but today with modern techniques many more structures are being determined.

Lysome vasopressin is composed of nine amino acid units. It is excreted by the pituitary gland, is employed clinically as a hypertensive agent, and was the first naturally occurring hormonal polypeptide synthesized in the laboratory. The Nobel prize was awarded in 1955 to V. du Vigneaud for this synthesis. The chemical and abbreviated structure for bovine vasopressin is

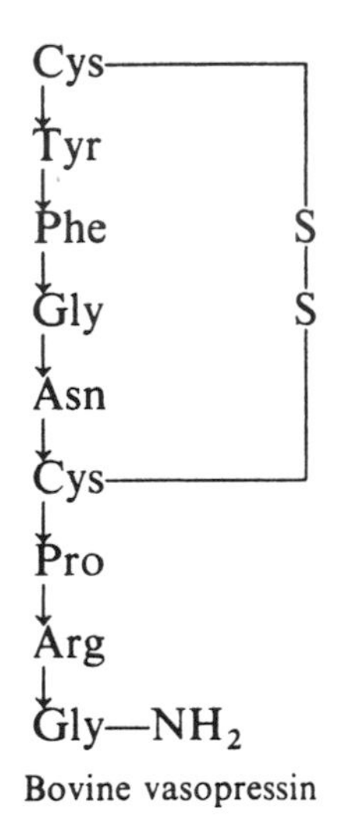

Bovine vasopressin

Proteins can be divided according to function, as noted in Table 14.2, or according to overall shape as fibrous or globular. Fibrous proteins can be likened to a strand of rope—they are largely linear and nonbranched. Globular proteins exhibit a great deal of twisting and turning, with the overall shape dependent on the specific sequence of amino acid units, and are often "held in shape" through cross-links that "lock in" a specific geometry.

Linus Pauling received the Nobel prize in 1954 for his work on protein structure. The sequence of amino acid units joined by peptide linkages in the polypeptide chain is called the primary structure. Because of essentially free rotation around the covalent bonds in this chain, the macromolecules may assume an infinite number of shapes or conformations. However, Pauling showed that certain conformations are preferred because of intramolecular and intermolecular hydrogen bonding forming the so-called secondary structure.

The shape, size, and specific sequence of amino acid units composing a protein determine the specific function of the protein. For example, fibrous proteins are typically found as connective tissue, including tendons, bones, and cartilage, whereas globular proteins such as hemoglobin and myoglobin are often involved with transport. Because the structure of proteins is so important, great effort has been made to describe the structure of specific proteins in an effort to understand particular structure–property relationships. Four classifications are employed to describe the structure of a protein.

We will now briefly look at these four levels of structure for proteins. Remember that the primary structure drives the secondary structure and the secondary

Table 14.2 Protein classification according to function

Function	Example	Description
A. Structural proteins	Collagen	Major component of connective tissue in animals, including bones, cartilage, and tendons.
	Keratins	Comprise most protective coverings of animals: hair, hoofs, claws, feathers, beaks, nails.
B. Regulator proteins		
1. Enzymes	Chymotrypsin	Involved in digestive process, cleaves polypeptides excreted by pancreas.
	Lysozyme	Involved in digestion, cleaves polysaccharide chains; found in many natural sources such as egg whites.
2. Hormones	Bradykinin	Regulates blood pressure; in blood plasma.
	Insulin	Required for normal glucose metabolism.
C. Transport proteins		
	Hemoglobin	Responsible for oxygen transport from lungs to the cells and for removal of waste carbon dioxide from cells; found in red blood cells.
	Myoglobin	Responsible for binding oxygen, which it obtains from hemoglobin, and storing it until needed; found in muscle tissue.

structure, in turn, drives the tertiary structure and the tertiary drives the quaternary structure.

Primary Structure. Primary structure describes the specific sequence of amino acid units composing the protein. Typically, only the primary bonding is considered when describing the primary structure of a protein. The diagram of bovine vasopressin represents one of the ways commonly employed to show primary structures.

As the number of amino acid units increases, interaction between the amino acid units within a single chain becomes possible. The major "driving force" fixing preferred geometrics is secondary bonding forces, which are primarily hydrogen bonds. Two major secondary structures are observed, that is, the helix (Figure 14.4) and the sheet (Figure 14.5). Within these two major categories there exist variations. For instance, the particular carbon atoms involved with bonding and the number of amino acids within a complete circle of the spiral lead to names such as alpha helix and beta helix. Within such a straightforward division of secondary structure, further variations can arise. Thus, wool consists of helical protein chains connected to give a "pleated" sheet.

Figure 14.4. Alpha-helix conformation of proteins.

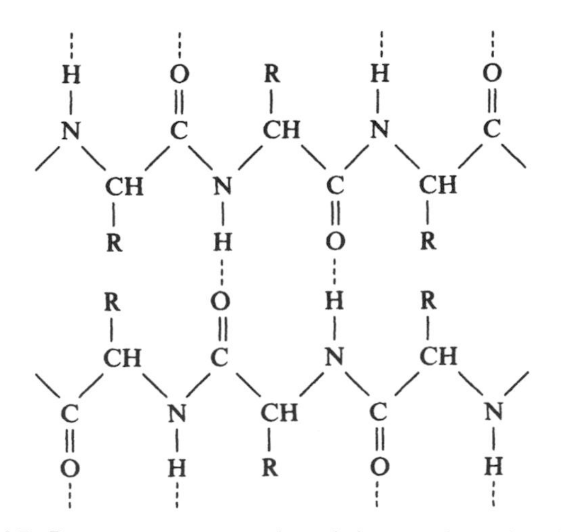

Figure 14.5. Beta arrangement or pleated sheet conformation of proteins.

In Nature, extended helical conformations appear to be utilized in two major ways: to provide linear systems for the storage, duplication, and transmission of information (DNA, RNA) and to provide inelastic fibers for the generation and transmission of forces (F-actin, myosin, and collagen). Examples of the various helical forms found in nature are single helix (messenger and ribosomal DNA), double helix (DNA), triple helix (collagen fibrils), and complex multiple helices (myosin). Generally, these single and double helices are readily soluble in dilute aqueous solution. Often solubility is only achieved after the inter- and intrahydrogen bonding is broken.

The structures of proteins generally fall into two groupings: fibers and globular. The structural proteins such as the keratines, collagen, and elastin are largely fiberous. A recurring theme with respect to conformation is that the preferential secondary structures of fiberous synthetic and natural polymers approximate that of a pleated sheet or skirt or helix. The pleated sheet structures in proteins are referred to as beta arrangements. In general, proteins with bulky groups take on a helical secondary structure while those with less bulky groups exist as beta sheets.

Keratines. As noted above, two basic "ordered" secondary structures predominate in synthetic and natural polymers. These are helices and the pleated sheet structures. These two structures are illustrated by the group of proteins called the keratines. It is important to remember that hydrogen bonding is critical in both structures. For helices the hydrogen bonding occurs within a single strand, whereas in the sheets the hydrogen bonding occurs between adjacent chains.

Hair and wool are composed of helical keratine called alpha-keratine. A single hair on our head is composed of many strands of keratine. Coiled, alpha-helix chains of alpha-keratine intertwine to form protofibrils that in turn are clustered with other protofibrils forming a microfibril. Hundreds of these microfibrils, in turn, are embedded in a protein matrix, giving a macrofibril that in turn combines, thereby producing a human hair.

While combing will align the various hairs in a desired shape, after a while, the hair will return to its "natural" shape through the action of the sulfur cross-links pulling the hair back to its original shape.

Stronger secondary bonding is involved in forming the helical structures of keratines, but the various bundles of alpha-keratine in our hair are connected by weak secondary interactions that allow the bundles to readily slide past one another. This sliding or slippage along with the "unscrewing" of the helices allows our hair to be flexible.

Some coloring agents and most permanent waving of our hair involves breakage of the sulfur cross-links and a reforming of the sulfur cross-links at new sites to "lock in" the desired hair shape.

Fingernails are also composed of alpha-keratin, but here the keratin has a greater amount of sulfur cross-links, thereby producing a more rigid material. In general, increased cross-linking leads to increased rigidity.

The other major structural feature is pleated sheets referred to as beta-keratine. The silk produced by insects and spiders is beta-keratine. This sheet structure is partially responsible for the "softness" felt when we touch silk. While silk is not easily elongated because the protein chains are almost fully extended, beta-keratin is flexible because of the low secondary bonding between sheets, allowing the sheets to flow past one another.

The beta-keratine structure is also found in the feathers and scales of birds and reptiles.

Wool, while naturally existing in the helical form, forms a pleated-skirt sheetlike structure when stretched.

Collagen. Collagen is the most abundant single protein in vertebrates, making up to one-third of the total protein mass. Collagen fibers form the matrix or cement material in our bones where mineral materials precipitate. Collagen fibers constitute a major part of our tendons and act as a major part of our skin. Hence, it is collagen that is largely responsible for holding us together.

The basic building block of collagen is a triple helix of three polypeptide chains called the tropocollagen unit. Each chain is about 1000 residues long. Collagen fibers are strong. In tendons, the collagen fibers have a strength similar to that of hard-drawn copper wire. Much of the toughness of collagen is the result of the cross-linking of the tropocollagen units to one another. The formation of cross-links continues throughout our life, resulting in our bones and tendons becoming less elastic and more brittle. Again, a little cross-linking is essential, but more cross-linking leads to increased fracture and brittleness.

Collagen is a major ingredient in some "gelation" materials like Jello™. Here, collagen forms a triple helix for some of its structure while other parts are more randomly flowing single collagen chain segments. The bundled triple helical structure acts as the rigid part of the polymer, while the less ordered amorphous chains act as a soft part of the chain.

Elastin. Collagen is found where strength is needed, but some tissues, such as arterial blood vessels and ligaments, need materials that are elastic. Elastin is the protein of choice for such applications. Elastin is rich in glycine, alanine, and valine and it is easily extended and flexible. Its conformation approaches that of a random coil so that secondary forces are relatively weak, allowing elastin to be readily extended as tension is applied. The structure also contains some lysine side chains that are involved in cross-linking. The cross-linking is accomplished when four lysine side chains are combined to form a desmosine cross-link. This cross-link prevents the elastin chains from being fully extended and causes the extended fiber to return to its original dimensions when tension is removed.

One of the areas of current research is the synthesis of polymers with desired properties based on natural analogues. Thus, elastin-like materials have been synthesized using glycine, alanine, and valine and some cross-linking. These materials approach elastin in its elasticity.

Tertiary Structure. Discussions related to the tertiary structure of a protein focus on the overall folding—that is, the turning of the protein chains. Although it is important to remember that such chains may exist in a helix or sheetlike secondary structure, the tertiary structure concerns only the overall, gross shape of the protein chain. Both secondary and primary forces in the form of cross-links are important in determining the tertiary structure.

Globular Proteins. As noted above, protein structures generally fall into two groupings, fiberous and globular. There is a wide variety of so-called globular proteins. Many of these have varieties of alpha and beta structures embedded within the overall globular structure. Beta sheets are often twisted or wrapped into a "barrel-like" structure. They contain portions that are beta sheet structures and portions that are in an alpha conformation. Furthermore, some portions of the globular protein may not be conveniently classified as either an alpha or beta structure.

Globular proteins take on this shape so as to offer a different "look" or polar nature to its outside than is present in its interior. Hydrophobic portions are generally found in the interior, while hydrophilic portions are found on the surface interacting with the hydrophilic water-intense external environment. (This theme is often found for synthetic polymers that contain polar and nonpolar portions. Thus, when polymers are formed or reformed in a regular water-filled atmosphere, many polymers will favor the presence of polar moieties on their surface.)

Globular proteins act in maintenance and regulatory roles—functions that often require mobility and thus some solubility. Included within the globular grouping are enzymes, most hormones, hemoglobin, and fibrinogen that is changed into an insoluble fibrous protein fibrin that causes blood clotting.

Denaturation is the irreversible precipitation of proteins caused by heating, such as the coagulation of egg white as an egg is cooked, or by addition of strong acids, bases, or other chemicals. This denaturation causes permanent changes in the overall structure of the protein and because of the ease with which proteins are denatured, it makes it difficult to study protein structure. Nucleic acids also undergo denaturation.

Quaternary Structure. Protein chains can group together to give a larger, more intricate structural arrangement described as a quaternary structure. Thus, hemoglobin is composed of four protein chains, and each protein chain is described in terms of a tertiary structure and the sum are described in terms of a quaternary structure. Just as there are primary, secondary, tertiary, and quaternary structures for biological giant molecules, there are analogous structures for the somewhat simpler, more regular synthetic polymers. Illustrations are given in Figure 14.6.

Protein chains can also interact with other protein chains. Here we will consider only one type of such interactions, those forming multiple helices. Nature employs the triple helix as a building block; it has greater strength than a simple helix composed of a single protein chain, possesses greater flexibility than simple sheet proteins, and results from the intertwining of three alpha helices. Disulfide (—S—S—) bonds act as cross-links and hydrogen bonding provides the cohesive forces holding the protein chains together. This triple helix is also called a protofibril.

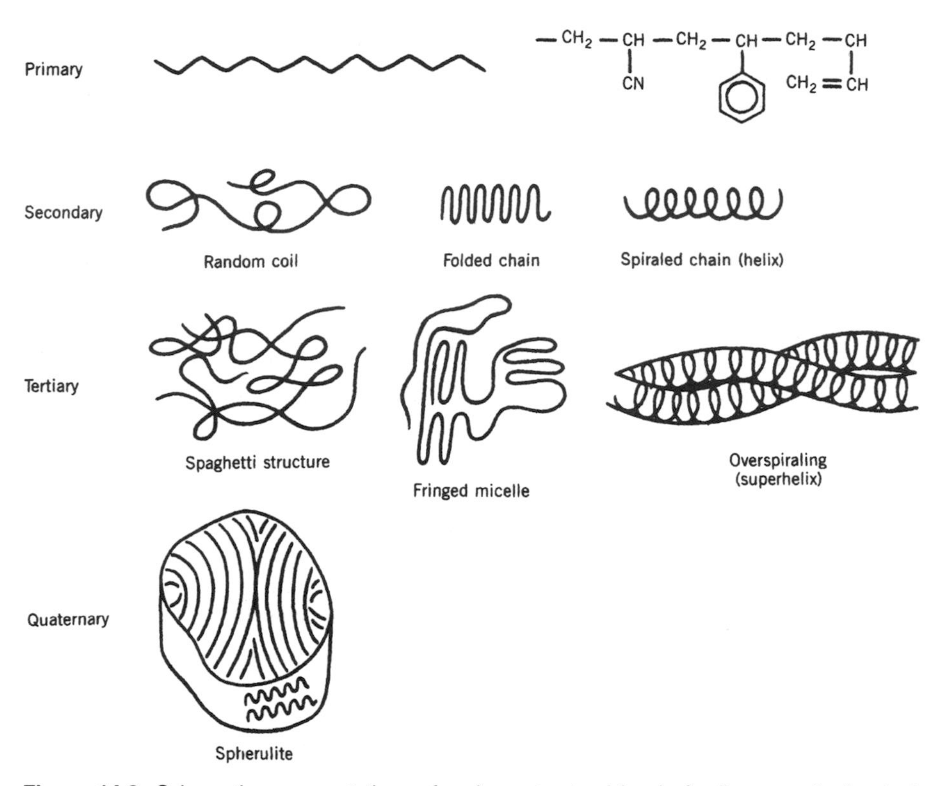

Figure 14.6. Schematic representations of various structural levels for linear synthetic giant molecules.

Human hair is an example of a macrofibril, with the various triple helices, microfibrils, and finally the macrofibril held together by hydrogen bonding and sulfide cross-links. Moths excrete an enzyme that breaks such disulfide cross-links in the deterioration of wool. These cross-links are also cleaved and reformed in the cold waving of hair.

The preferred structure of fibrous proteins, such as hair, wool, and silk, is a sheet in which the hydrogen atoms on the nitrogen atom of the amide group (CONH) in one chain are attracted to the oxygen atoms of the amide group on an adjacent chain. In contrast, the preferred structure of nonfibrous proteins is a right-handed alpha helix, which results from intramolecular hydrogen bonding—that is, within the same molecule. Polymers of both L and D amino acids form helices with 3.6 units per turn. However, the winding direction is right for the helices of the L polymer and left for the helices of the D polymer.

Fibrillar proteins, such as keratin of the hair and nails, collagen of connective tissue, and myosin of the muscle, are strong, water-insoluble polymers. In contrast, globular proteins, such as enzymes, hormones, hemoglobin, and albumin, are usually weaker and more water-soluble polymers.

14.5 ENZYMES

In addition to serving as food and structures for the animal kingdom, proteins, in the form of enzymes (from the Greek word *enzymos*, which means soup dough), perform specific functions by acting as catalysts in biological reactions. These specific proteins are responsible for reactions varying from eye movement, to the maintenance of body temperature, to the production of blood cells, to the digestion of food.

The "lock and key" concept proposed by Nobel laureate E. Fischer in the early 1900s is the most widely accepted theory for the specificity of enzymatic reactions. According to this theory, the chiral atoms in the amino acids provide a geometric pattern that permits specific reactions to occur while the reactant is locked in place. The products are released rapidly after the extremely fast reaction occurs, and the mechanism is then repeated numerous times. Nobel laureate J. Sumner isolated the crystalline enzyme urease in 1926. However, the leading organic chemists and biochemists of that era maintained that it was impossible to isolate crystalline proteins. It is now believed that specificity of an enzyme molecule is related to its interaction with three groups in the substrate. Many enzymes have approximately rounded overall shapes. Since the spherical shape requires less energy than rod- or coil-shaped polymers, it facilitates easy transport of enzymes. The spherical or globular shape is also a convenient way to exist in two environments. The external environment interfaces with a water and polar exterior, while the interior is less polar and more hydrophobic (not liking water).

14.6 WOOL

Wool from sheep was woven into fabrics by the ancient inhabitants of Egypt, Nineveh, Babylon, and Peru. The first factory in America using water power to weave wool was established at Hartford, Connecticut, in 1788. About 40 thousand tons of wool is used annually in the United States, and the worldwide consumption is 3 million tons annually.

The polymer chains in wool consist of parallel polypeptide alpha helices joined by disulfide (S—S) bonds. When wool is ironed with a wet cloth, immersed in an alkaline detergent, or subjected to tension in the direction of the helical axis, the hydrogen bonds parallel to the axes and the disulfide linkages are broken, and the structure can be elongated to nearly 100% of its possible length.

The opening and closing of these disulfide cross-linking groups allows the curling and uncurling of wool and of human hair. For humans, the wavesetting lotions sodium bisulfite ($NaHSO_3$) and ammonium thioglycolate ($HSCH_2COONH_4$) in hot- and cold-hair waving, respectively, open disulfide linkages and disrupt hydrogen bonds. The hair is curled and a neutralizer is added to reform the disulfide linkages, thus setting the hair in a curled or straight manner as desired.

14.7 SILK

Sericulture, that is, the culture of the silkworm (*Bombyx mori*), and the weaving of the silk filaments produced by the mulberry silkworm were of prime importance over 5000 years ago. In 2640 B.C., the Empress HSi-Ling-Shi developed the process of reeling by floating the cocoons on warm water. This process and the silkworm itself were monopolized by China until about A.D. 550 when two missionaries smuggled silkworm eggs and mulberry seeds from China to Constantinople (Istanbul). The crystalline silk fiber is three times as strong as wool.

Because of its high cost, only a small amount of silk (55 thousand tons) is woven worldwide annually. Both the discrete fibers of wool and the continuous filaments of silk are made up of macromolecules in which the repeating units consist of about 20 different amino acids.

The composition within a spider web is not all the same. We can look briefly at two of the general types of threads. One is known as the network or frame threads, also called the dragline fabric. It is generally stiff and strong. The second variety is the catching or capture threads that are made of viscid silk that is strong, stretchy, and covered with droplets of glue. The frame threads are about as stiff as nylon 6,6 thread and on a weight basis stronger than steel cable. Capture thread is not stiff but is more elastomeric-like and on a weight basis about one-third as strong as frame thread. While there are synthetic materials that can match the silks in both stiffness and strength, there are few that come near the silk threads in toughness and their ability to withstand a sudden impact without breaking. Kevlar, which is used in bullet-resistant clothing, has less energy-absorbing capacity in comparison to either frame or capture threads. In fact, when weight is dropped onto frame silk, it adsorbs up to 10 times more energy than Kevlar. On impact with frame thread, most of the kinetic energy dissipates as heat which, according to a hungry spider, is better than transforming it into elastic energy which might simply act to "bounce" the pray out of the web.

The frame threads are composed of two major components: Highly organized microcrystals compose about one-quarter of the mass, and the other three-quarters are composed of amorphous spaghetti-like tangles. The amorphous chains connect the stronger crystalline portions. The amorphous tangles are dry and glass-like, acting as a material below its T_g. The amorphous chains are largely oriented along the thread length as are the microcrystals giving the material good longitudinal strength. As the frame threads are stretched, the tangles straighten out, allowing it to stretch without breaking. Because of the extent of the tangling, there is a lessening in the tendency to form micro-ordered domains as the material is stretched, though that also occurs. Frame thread can be reversibly stretched to about 5%. Greater stretching causes permanent creep. Thread rupture does not occur until greater extension, such as 30%. By comparison, Kevlar fibers break when extended only 3%.

The capture threads are also composed of the same kinds of components, but here the microcrystals compose less than 5% of the thread with both the amorphous and microcrystalline portions arranged in a more random fashion within the thread.

A hydrated glue that coats the thread acts as a plasticizer imparting to the chains greater mobility and flexibility. It stretches several times its length when pulled and is able to withstand numerous shocks and pulls appropriate to contain the prey as it attempts to escape. Furthermore, most threads are spun as two lines so that the resulting thread has a kind of build in redundancy.

The spinning of each type of thread comes from a different emission site on the spider; also, the spider leaves little to waste, using unwanted and used web parts as another source of protein.

Cloning of certain spider genes have been included in goats to specify the production of proteins that call for the production of silk-like fibroin threads that allow the production and subsequent capture of spider-like threads as part of the goat's milk.

14.8 NUCLEIC ACIDS

Nucleic acids, which are found in the nucleus of all living cells, are responsible for the synthesis of specific proteins and are involved in the generic transmission of characteristics from parent to offspring. Nucleic acids were discovered in 1867 by J. Miesher, who isolated these materials from the remnants of pus from cells. Since nucleic acid was found in the nucleus of cells, he called this material "nuclein," but this name has been changed to nucleic acid. It should be noted that nucleic acids are also found in the cytoplasm as well as in the nucleus of cells.

P. Levene, who discovered D-deoxyribose in 1929, showed that D-riboses were present in pure nucleic acid. Avery, MacLeod, and McCarty showed that deoxyribonucleic acid (DNA) was the basic genetic component of chromosomes in 1944. Nobel laureate Alexander R. B. Todd synthesized adenosine diphosphate (ADP) and adenosine triphosphate (ATP) in 1947. These compounds are not only important components of nucleic acids but are also involved in biological energy transfer.

The terms DNA and RNA are derived from the specific sugar moiety present. Deoxyribose is the sugar present in deoxyribose nucleic acids (DNAs), whereas ribose sugar is present in ribose nucleic acids (RNAs). The difference between the two sugars is the absence of one of the hydroxyl (OH) groups in the deoxyribose.

The three components of a nucleotide are a purine or a pyrimidine base, a pentose, and phosphoric acid (Figure 14.7). As already noted, the difference between DNA and RNA is the pentose component and the particular bases that are present. Both DNA and RNA contain adenine, guanine, and cytosine, whereas RNA also contains uracil and DNA also contains thymine and 5-methylcytosine. A third difference is the tendency of RNA to be single-stranded and not to possess a regular helical structure, whereas DNA can be double-stranded and typically forms a regular helical structure.

The purine and pyrimidine bases, called adenine (A), guanine (G), cytosine (C), and thymine (T), are held together by hydrogen bonds in the parent cell. The pyrimidine molecules are smaller than the purine molecules, and one of each of these bases can fit between the strands in the DNA double helix, as shown in

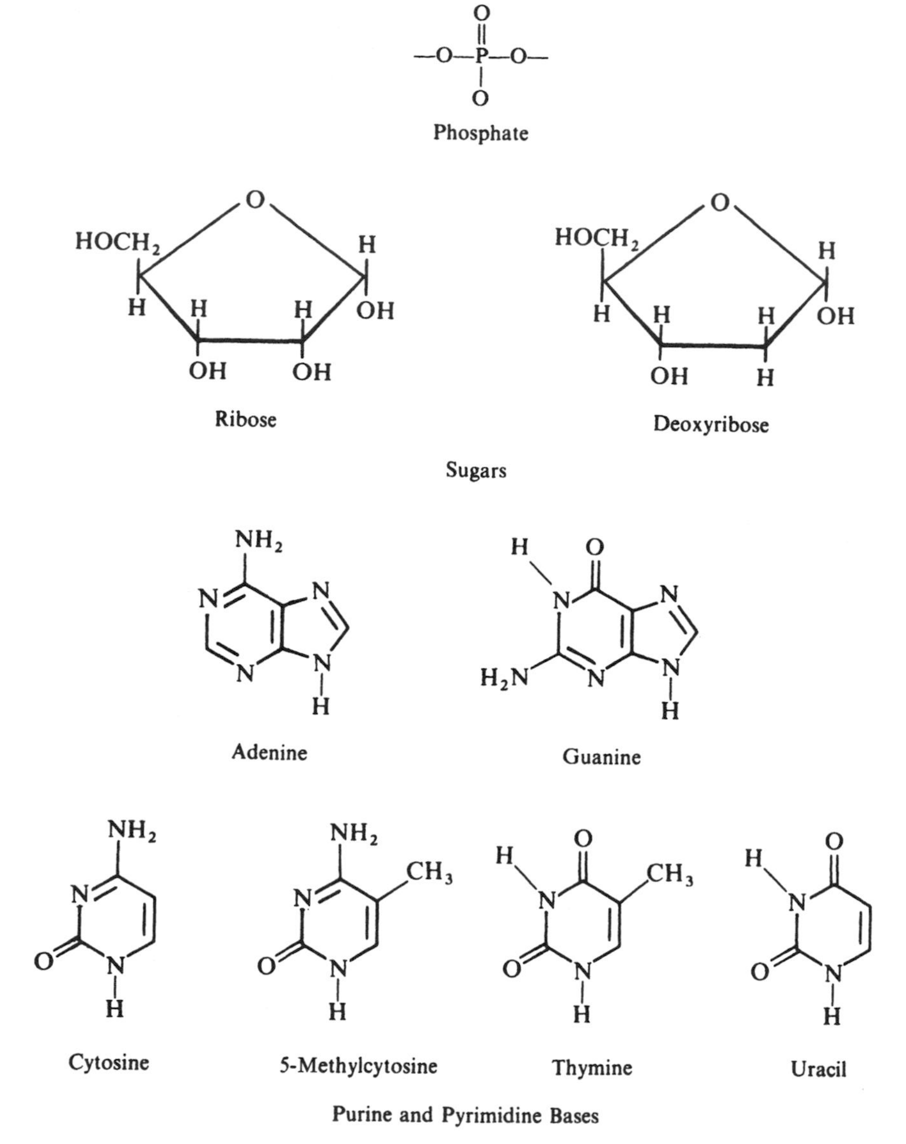

Figure 14.7. Chemical structures of components of nucleic acids.

Figures 14.8 and 14.9. This formation of base pairs may be remembered from the mnemonic expression Gee-CAT. The pentose molecules are joined together through the phosphate units, and the base portions are present as substituents on the pentose molecule. A portion of a DNA chain is shown in Figure 14.10.

Nobel laureates James Watson and Francis Crick correctly postulated the double-stranded helical structure in 1953; this was confirmed by x-ray diffraction

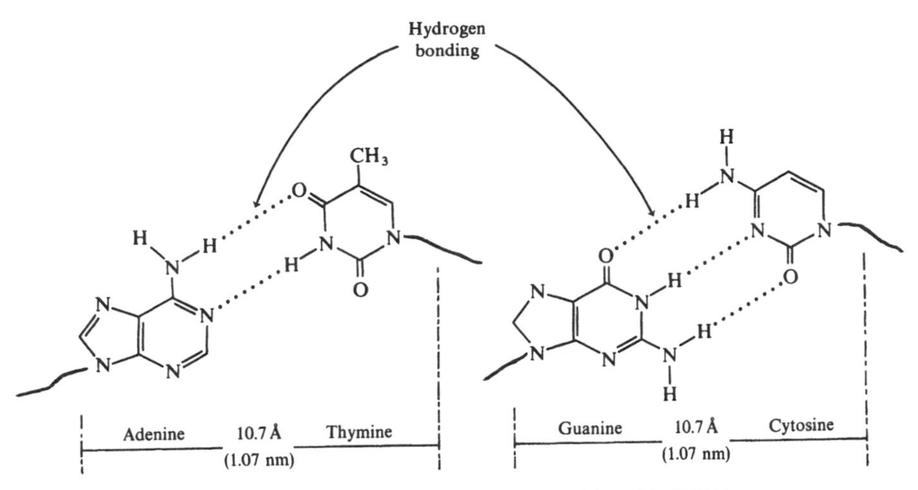

Figure 14.8. Allowable base pairs in nucleic acids (DNA).

in 1973. The complete biochemical synthesis of a biologically active DNA of a virus was accomplished by utilizing two enzymes discovered by Nobel prize winner Arthur Kornberg in 1967. The enormity of such a DNA molecule is easier to understand when one considers that a single DNA molecule laid lengthwise would be about a centimeter in length.

The human genome is composed of Nature's most complex, exacting, and important macromolecule. It is composed of nucleic acids that appear complex in comparison to simpler molecules such as methane and ethylene, but simple in comparison to their result on the human body. Each unit is essentially the same, containing a phosphate, a deoxyribose sugar (below),

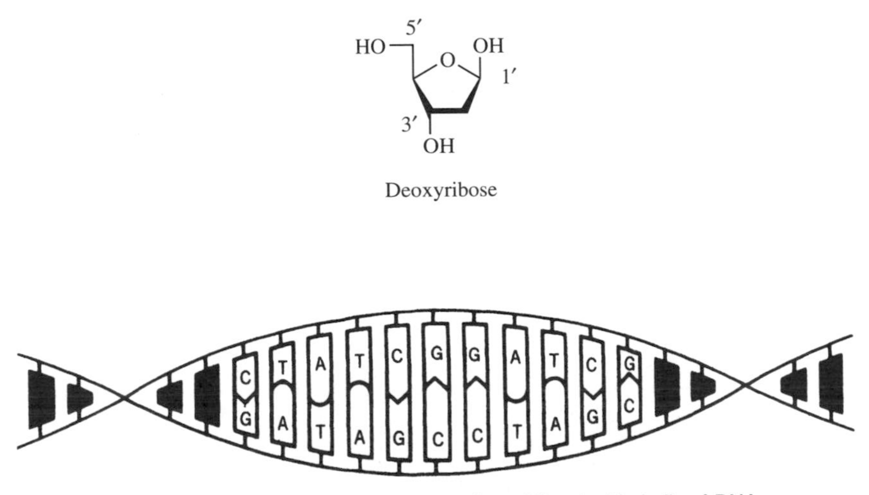

Figure 14.9. A schematic representation of the double helix of DNA.

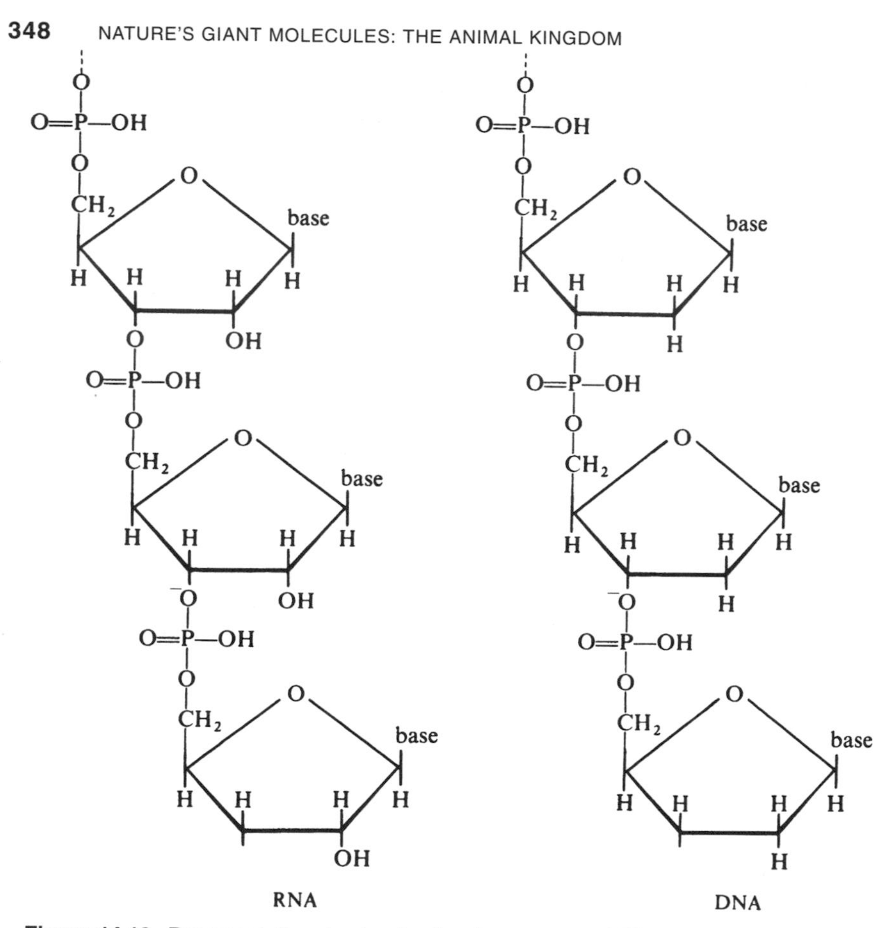

Figure 14.10. Representative structural units of a segment of RNA (*left*) and DNA (*right*).

and one of four bases (Figure 14.7), with each base typically represented by the capital of the first letter of their name, G, C, A, and T. In fact, the complexity is less than having four separate and independent bases because the bases come in matched sets—they are paired. The mimetic Gee CAT allows an easy way to remember this pairing. The base, sugar, and phosphate combine, forming nucleotides such as adenylic acid, and adenosine-3′-phosphate shown below and represented by the symbols A, dA, and dAMP.

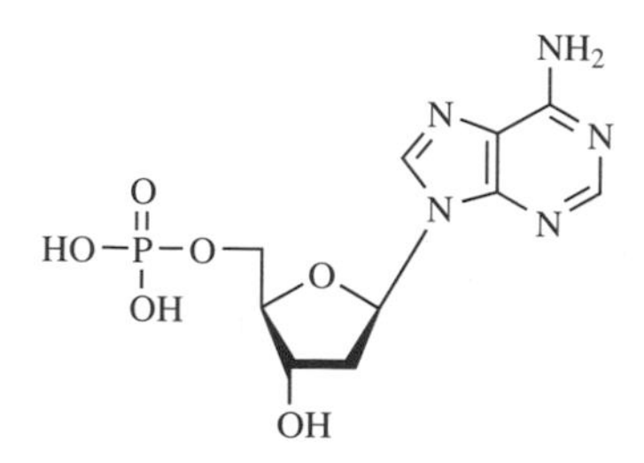

The backbone of nucleic acids is connected through the 3′ and 5′ sites on the sugar with the base attached at the 1′ site. Because the sugar molecule is not symmetrical, each unit can be connected differently but there is order (also called sense or directionality) in the sequence of this connection so that phosphodiester linkage between units is between the 3′ carbon of one unit and the 5′ carbon of the next unit. Thus nucleic acids consist of units connected so that the repeat unit is a 3′–5′ (by agreement we consider the start to occur at the 3′ and end at the 5′, though we could just as easily describe this repeat as being 5′–3′) linkage. Thus, the two ends are not identical: One of them contains an unreacted 3′ and the other an unreacted 5′ hydroxyl.

A shorthand is used to describe sequences. Following is a trimer containing in order the bases cytosine, adenine, and thymine:

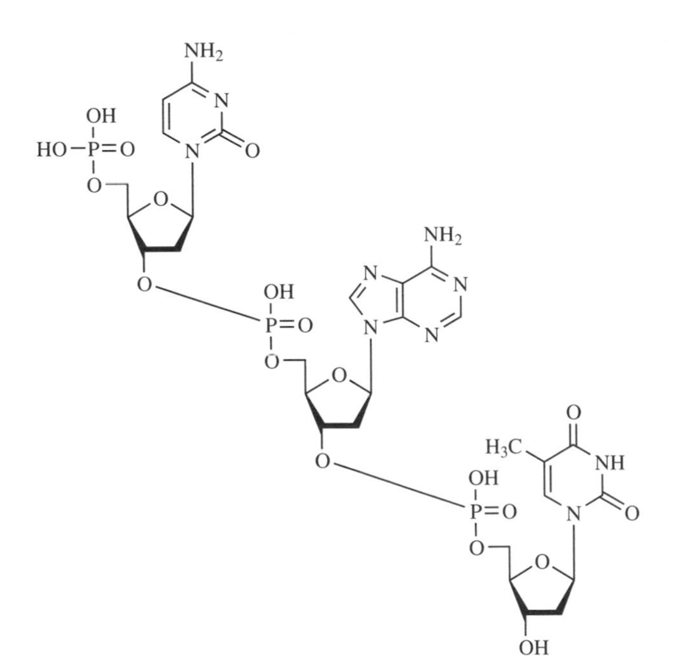

This sequence is described as p-5′-C-3′-p-5′-A-3′-p-5′-T-3′ or pCpApT or usually as simply CAT.

Secondary Structure. Watson and Crick correctly deduced that DNA consists of a double-stranded helix in which a pyrimidine base on one chain or stand was hydrogen-bonded to a purine base on the other chain.

The combination AT has two hydrogen bonds while the combination GC has three double bonds, contributing to making the GC a more compact structure as seen above (Figure 14.8). This results in a difference in the twisting resulting from the presence of the AT or GC units, and combinations of these units result in structures that are unique to the particular combination. It is this twisting, and the particular base sequence, that eventually results in the varying chemical and subsequently biological activities of various combinations.

In solution, DNA is a dynamic, flexible molecule. It undergoes elastic motions on a nanosecond time scale most closely related to changes in the rotational angles of the bonds within the DNA backbone. The net result of these bendings and twistings is that DNA assumes a compact shape. The overall structure of the DNA surface is not that of a reoccurring "barber pole"; but rather because of the particular base sequence composition, each sequence will have its own characteristic features of hills, valleys, bumps, and so on.

Supercoiling. Electron microscopy shows that individual DNA chains consist of two general structures: linear and circular. The chromosomal DNA in bacteria is a closed circle, a result of covalent joining of the two ends of the double helix, but the DNA within eukaryotic cells, like our cells, is believed to be linear.

The most important secondary structure is supercoiling. Supercoiling simply is the coiling of a coil or in this case a coiling of the already helical DNA. The typical relaxed DNA structure is the thermally stable form. Two divergent mechanisms are believed responsible for supercoiling. The first, and less prevalent, is illustrated by a telephone cord. The telephone cord is typically coiled and represents the "at rest" or "unstressed" coupled DNA. As I answer the telephone I have a tendency to twist it in one direction and after answering and hanging up the telephone for awhile it begins forming additional coils. Thus, additional coiling tends to result in supercoiling. The second, and more common, form involves the presence of less than normal coiling. Thus, underwinding occurs when there are fewer helical turns than would be expected. Purified DNA is rarely relaxed.

Compaction. Essentially all of human DNA is chromosomal, with a small fraction found within the cell's energy-producing "plant," the mitochondria. The contour length, the stretched-out helical length, of the human genome material in one cell is about 2 meters in comparison with about 1.7 meters for *E. coli*. An average human body has about 10^{14} cells, giving a total length that is equivalent in length to traveling to and from the earth and sun about 500 times or 1000 one-way trips.

Bacterial DNA appears as a loose, open arrangement of the closed-loop DNA that exhibits supercoiling (Figure 14.11, right). These supercoiled DNA molecules are generally circular, are usually right-handed in a supercoiled DNA, and tend to be extended and narrow rather than compacted, with multiple branches.

By comparison, our DNA is present in very compacted packages (Figure 14.11, left). One of the major compacting comes in the form of supercoiling. As noted before, our DNA is linear, but because of their large size they act as though they are looped, forming coils about specific proteins. Subjection of chromosomes to treatments that partially unfold them show a structure where the DNA is tightly wound about "beads of proteins," forming a necklace-like arrangement where the protein beads represent precious stones embedded within the necklace fabric. This combination forms the nucleosome, the fundamental unit of organization

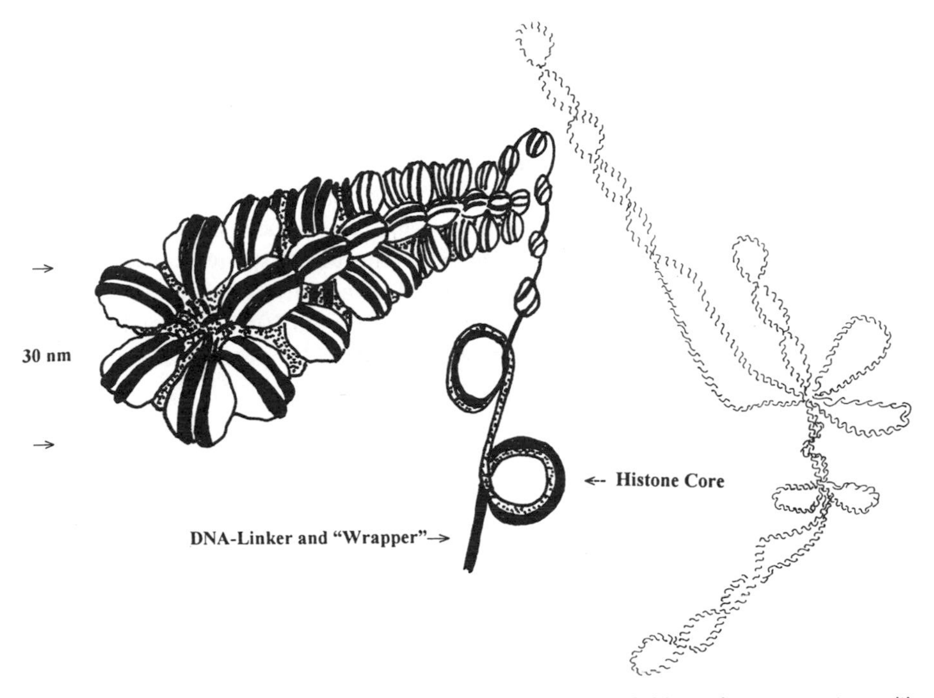

Figure 14.11. Typical bacterial chromosome showing the usual thin and open structure with supercoils, branch points, and intersections (*right*) and the compact DNA found in eukaryotic cells (*left*). Note the regularly spaced nucleosome necklace-like linker and "wrapper" combination folding into a 30-nm superthread.

upon which higher-order packing or folding occurs. Histone proteins are small proteins with molecular weights between 11,000 and 21,000.

Wrapping of DNA about a nucleosome core compacts the DNA length about sevenfold. The overall compacting though is about 10,000-fold. Additional compacting of about 100-fold is gained from formation of so-called 30-nm fibers or threads. These fibers contain one histone for each nucleosome core. The name "30-nm fibers" occurs because the overall shape is of a fiber with a 30-nm thickness (Figure 14.11, left). The additional modes of compaction are just beginning to be understood but may involve scaffold-assisting—that is, DNA-containing segments wrapped about or within protein-containing units.

The scaffold contains several proteins, especially histone in the core and topoisomerase II. Both appear important to the compaction of the chromosome. In fact, the relationship between topoisomerase II and chromosome folding is so vital that inhibitors of this enzyme can kill rapidly dividing cells, and several drugs used in the treatment of cancer are topoisomerase II inhibitors.

Replication. Replication occurs with a remarkably high degree of fidelity such that errors occurs only about once per 1000 to 10,000 replications, or an average single missed base for every 10^9 to 10^{10} bases added. This highly accurate reproduction occurs because of a number of reasons including probably some that are as yet unknown. As noted before, the GC group has three hydrogen bonds while the AT has two. Thus, A that wants to form two hydrogen bonds would likely not bind to G that has three hydrogen bonding sites. This is believed to give a precision of about 10^4—that is, only one error in 10,000. Some mistakes are identified and then corrected. This process is very precise. If a wrong base has been added, this enzyme prevents addition of the next nucleotide removing the mispaired nucleotide and then allowing the polymerization to continue. This activity is called proofreading, and it is believed to increase the accuracy another 10^2- to 10^3-fold. Combining the accuracy factors results in one net error for every 10^6 to 10^8 base pairs, still short of what is found. Thus, other factors are at work.

14.9 THE GENETIC CODE

Flow of Biological Information. Nucleic acids, proteins, some carbohydrates and hormones are informational molecules. They carry directions for the control of biological processes. With the exception of some hormones, these are macromolecules. In all these interactions, secondary forces such as hydrogen bonding and van der Waals forces, along with ionic bonds and hydrophobic/hydrophilic character, play critical roles. *Molecular recognition* is the term used to describe the ability of molecules to recognize and interact bond-specifically with other molecules. This molecular recognition is based on a combination of these interactions just cited and on structure.

In general, the flow of biological information can be mapped as follows:

$$\text{DNA} \rightarrow \text{RNA} \rightarrow \text{Proten} \rightarrow \text{Cell structure and function}$$

The total genetic information for each cell, called the *genome*, exists in the coded two-stranded DNA. This genetic information is expressed or processed through duplication of the DNA so it can be transferred during cell division to a daughter cell, or it can be transferred to manufactured RNA that in turn transfers the information to proteins that carry out the activities of the cell. *Transcription* is the term used to describe the transfer of information from the DNA to RNA. The process of "moving" information from the RNA to the protein is called *translation*. The ultimate purpose of DNA expression is protein synthesis.

Duplication of double-stranded DNA is self-directed. The DNA, along with accessory proteins, directs the *replication* or construction of two complementary strands forming a new, exact replicate of the original DNA template. As each base site on the DNA becomes available through the unraveling of the double-stranded helix, a new nucleotide is brought into the process held in place by hydrogen bonding and van der Waals forces so that the bases are complementary. It is then covalently bonded through the action of an enzyme called DNA polymerase. After duplication, each DNA contains one DNA strand from the original double-stranded helix and one newly formed DNA strand. This is called *semiconservative replication* and increases the chance that if an error occurs, the original base sequence will be retained.

How is DNA suitable as a carrier of genetic information? While we do not entirely understand, several features are present in DNA. First, because of the double-stranded nature and mode of replication, retention is enhanced. Second, DNA is particularly stable within both cellular and extracellular environments, including a good stability to hydrolysis within an aqueous environment. Plant and animal DNA have survived thousands of years. Using polymerase chain reactions, we can reconstruct DNA segments allowing comparisons to modern DNA.

The genome is quite large, on the order of a millimeter in length if unraveled, but within it exists coding regions called genes. Transcription is similar to DNA replication, except ribonucleotides are the building units instead of deoxyribonucleotides; the base thymine is replaced by uracil; the DNA:RNA duplex unravels, releasing the DNA to again form its double-stranded helix and the single-stranded RNA; and the enzyme linking the ribonucleotides together is called RNA polymerase.

Many viruses and retroviruses have genome that are single-stranded RNA instead of DNA. These include the AIDS virus and some retroviruses that cause cancer. Here, an enzyme called reverse transcriptase converts the RNA genome of the virus into the DNA of the host cell genome, thus infecting the host.

The transcription of the DNA involves three kinds of RNA: ribosomal, messenger, and transfer. The most abundant RNA is ribosomal RNA, rRNA. Most rRNA is large and is found in combination with proteins in the ribonucleoprotein complexes called ribosomes. Ribosomes are subcellular sites for protein synthesis.

DNA controls the synthesis of RNA, which in turn controls the synthesis of proteins. The general steps can be outlined as follows: First DNA synthesizes smaller RNA molecules called messenger RNA (m-RNA). The DNA determines the sequences of bases that will be presented in the messenger RNA through pairing with a partially untwisted portion of a DNA chain.

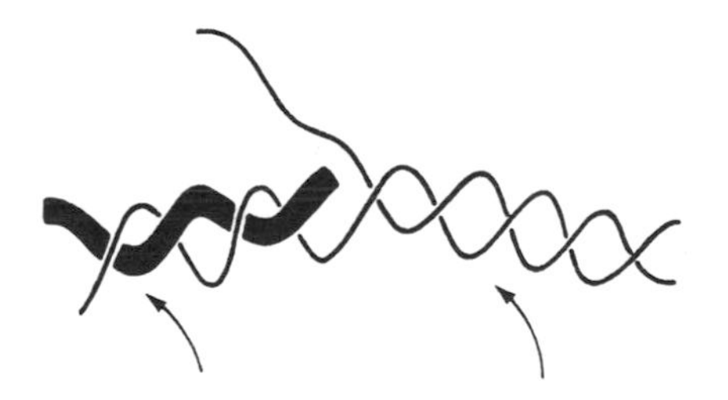

Another kind of RNA, called transfer RNA (t-RNA), serves as the "traffic officer" for amino acids; it selects only certain amino acids and transports these amino acids to the m-RNA template. The molecular weight of transfer RNA is much lower than that of m-RNA, and it is more soluble and more mobile in the cell fluids. There are at least 20 different forms of t-RNA (soluble RNA), each being specific for a given amino acid.

The anticodons, which consist of a specific sequence of bases, allow the t-RNA to bind with specific sites, called the codons of m-RNA. The order in which amino acids are brought by t-RNA to the m-RNA is determined by the sequence of codons. This sequence constitutes what is referred to as a genetic message. Individual units of that message (individual amino acids) are designated by triplets of nucleic acid units.

The m-RNA can be visualized as a template with indentions that have specific spatial (geometric, steric) and electronic characteristics.

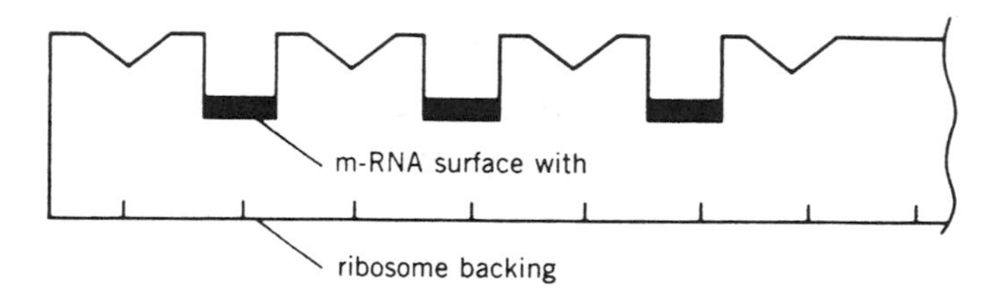

Briefly, instructions to replicate a DNA molecule can be visualized as a long sentence containing about 10 letters, one for each nucleic acid unit. There are only four letters in this alphabet—A, C, G, and T—but since the t-RNA operates on a 3-base code, 4^3 or 64 different sequences are possible. This number is far more than is needed to recall only 20 amino acids, and more than one sequence may be available to recall a specific amino acid. Sequences are also available for the messages of "start here" and "end here." After much experimentation, each of these 64 sequences has been identified with a specific amino acid or other message.

Each cell contains a chromosome, which in turn contains a DNA molecule. Reproduction occurs on signal when a new cell is needed as described in the foregoing.

The amount, presence, or absence of a particular protein is generally controlled by the DNA in the cell. Protein synthesis can be signaled external to the cell or within the cell. Growth factors and hormones form part of this secondary messenger service.

The translation and transcription of DNA information is polymer synthesis and behavior, and the particular governing factors and features that control these reactions are present in the synthesis and behavior of other macromolecules—synthetic and biological.

For the human genome there exists so-called coding or active regions called *exons* and noncoding regions called *introns*. The average size of an exon is about 120 to 150 nucleotide units long or coding for about 40 to 50 amino acids. Introns vary widely in size from about 50 to over 20,000 units. About 5% of the genome is used for coding. It was thought that the other 95% was silent or junk DNA. We are finding that the introns regions play essential roles. Interestingly, introns are absent in the most basic prokaryotes, only occasionally found in eukaryotes, but common in animals.

14.10 GENETIC ENGINEERING

A gene is a section of DNA in a chromosome. Genetic engineering consists mainly of uniting pieces of spliced DNA from different sources to produce a recombinant DNA. The potential applications of genetic engineering sometimes appear in fiction, where the possibility of cloning human beings has fascinated novelists, playwrights, and comic book writers. The term clone comes from the Greek word *klon*, meaning a cutting used to propagate a plant. Cell cloning is the production of identical cells from a single cell. Similarly, gene cloning is the production of identical genes from a single gene. During gene cloning, genes from different organisms are often joined together to form one artificial molecule known as a recombinant DNA.

The chemical reactions used in gene cloning are analogous to those encountered in elementary organic chemistry. The first of these, the "cutting" of a gene, is actually the hydrolysis of DNA. The second type of reaction involves joining DNA molecules together by a dehydration reaction. Most procedures require a vector (plasmids) and restriction enzymes. Vectors are the name given to the material that carries the recombinate DNA into a cell so that it will be accepted in the new environment and allowed to reproduce. Today the usual vectors for bacterial cells are plasmids, which are small, free-floating ringlets of DNA that are present in most cells and carry genetic information concerning resistance to antibiotics.

Restriction enzymes cut double-stranded DNA at predictable places, leaving the DNA with uneven ends containing a short segment of only one of the strands. These ends are called sticky ends since they can combine with other sticky ends through a process in which the bases are paired. These restriction enzymes cut DNA strands only at locations where there is a palindrome base structure. A palindrome is an arrangement of letters that reads the same way forward and backward, such as

DOG, OTTO, GOD

or, for the four bases (*G*ee-*CAT*),

GAATTAAG

would be a palindrome as would

AGATTAGA

Thus for a paired sequence

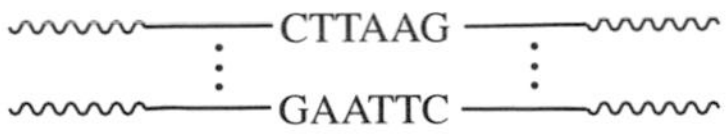

```
~~~~~———CTTAAG ———~~~~~
        :       :
~~~~~———GAATTC ———~~~~~
```

we see a palindrome in the pair where the top pair reads in reverse direction to the lower pair. A restriction enzyme may then split this pair as

```
~~~~~———C            TTAAG ———~~~~~
        :                  :
~~~~~———GAATT            C———~~~~~
```

producing two sticky ends—one containing the uncoupled TTAA portion and on the lower strand an AATT uncoupled portion.

A second DNA sample can now be "cut" by employing the same restriction enzyme to produce complementary sticky ends:

These DNA portions from the two different genes are mixed along with another enzyme (DNA lipase) that will chemically bind the complementary sticky ends:

CTTAAG ---- ---- CTTAAG
GAATTC ---- ---- GAATTC

The new recombinate DNA then enters a cell, carrying with it the new information.

New techniques and sequences are continually being developed and new applications are being studied. The goal of genetic engineering is to couple desired characteristics. For instance, if the synthesis of a specific protein is desired but its natural production occurs slowly and only as a trace material, a genetic engineer might couple a gene portion that dictates production of that specific protein with a gene portion that rapidly reproduces itself.

14.11 DNA PROFILING

DNA profiling is also called DNA fingerprinting and DNA typing. It is used in paternity identification, criminal cases, and so on. DNA is particularly robust and thousand year old DNA has been analyzed using DNA profiling techniques. Here we will concentration on its use in criminal cases where questions of law become intertwined with statistical arguments and chemical behavior.

The stability of the DNA is due to both internal and external hydrogen bonding as well as ionic and other bonding. First, the internal hydrogen bonding is between the complementary purine–pyrimidine base pairs. Second, the external hydrogen bonding occurs between the polar sites along exterior sugar and phosphate moieties and water molecules. Third, ionic bonding occurs between the negatively charged phosphate groups situated on the exterior surface of the DNA and cations such as Mg^{+2}. Fourth, the core consists of the base pairs, which, along with being hydrogen-bonded, stack together through hydrophobic interactions and van der Waals forces.

DNA profiling relies on the very small differences that exit between individual's (except for identical siblings) DNA. Samples can be obtained from saliva, skin, hair, blood, and semen. Two major types of DNA profiling are the polymerase chain reaction (PCR) approach and the restriction fragment length polymorphism (RFLP) approach. PCR utilizes a sort of molecular copying process. A specific DNA region is selected for study. PCR requires only a few nanograms of DNA as a sample.

RFLP focuses on sections of the DNA called restriction fragment length polymorphisms. The DNA is isolated from the cell nucleus and broken into pieces by restriction enzymes that cleave DNA at specific locations. Because we each have some differences in our DNA, some DNA fragments of different lengths result. The fragments are separated and transferred to a backing material where radioactive probes that bind to only certain sites are added. The radioactive probes allow identification using x-ray film that is sensitive to the decay of the radioactive probes, allowing the identification of the location of the binding sites on the particular DNA fragments. These locations are then compared to the test sample for a match/not match.

There are a number of different restriction enzymes that cut the DNA at different specific sites. Each sequence of using a different restriction enzyme and locating the binding sites increases the probability of differentiating between individuals. For instance, the use of one restrictive enzyme might give a 1-in-100 match; that is, it is 1% confident that another individual might have the same match. Another restriction enzyme might have a 1-in-50 match—that is, a 2% confidence rate. A third might have a 1-in-1000 match, or a 0.1% confidence. The use of all three is then about $0.01 \times 0.02 \times 0.001 = 0.0000002$ or 0.00002% or 1 part in 50,000,000 or 1 in 50 million.

The RFLP requires a sample about 100 times the size required for PCR but with repeated sequences using different restriction enzymes RFLP is more precise.

14.12 MELANINS

Light is continuous, ranging from wavelengths smaller than 10^{-14} m (gamma radiation) to those greater than 10^6 m. Radiation serves as the basis for the synthesis of many natural macromolecules via photosynthesis. Radiation is used commercially to increase the wood pulp yield through cross-linking and grafting of lignin and other wood components onto cellulosic chains. Radiation is also used in the synthesis and cross-linking of many synthetic polymers.

Radiation is also important in the synthesis and rearrangement of important "surface" macromolecules. Tanning of human skin involves the activation of the polypeptide hormone beta MSH that in turn eventually leads to the phenomena of tanning. Exposure to higher-energy light from about 297 to 315 nm results in both tanning and burning, whereas exposure to light within the 315- to 330-nm region results in mainly only tanning. UV radiation activates enzymes that modify the amino acid tyrosine in pigment-producing cells, the melanocytes. The concentration of tyrosine is relatively high in skin protein. These modified tyrosine molecules undergo condensation forming macromolecules known as melanins (Figure 14.12). Melanins have extended chain resonance where the pi electrons are associated with the growing melamine structure. As the melanine structure grows, it becomes more colored, giving various shades of brown color to our

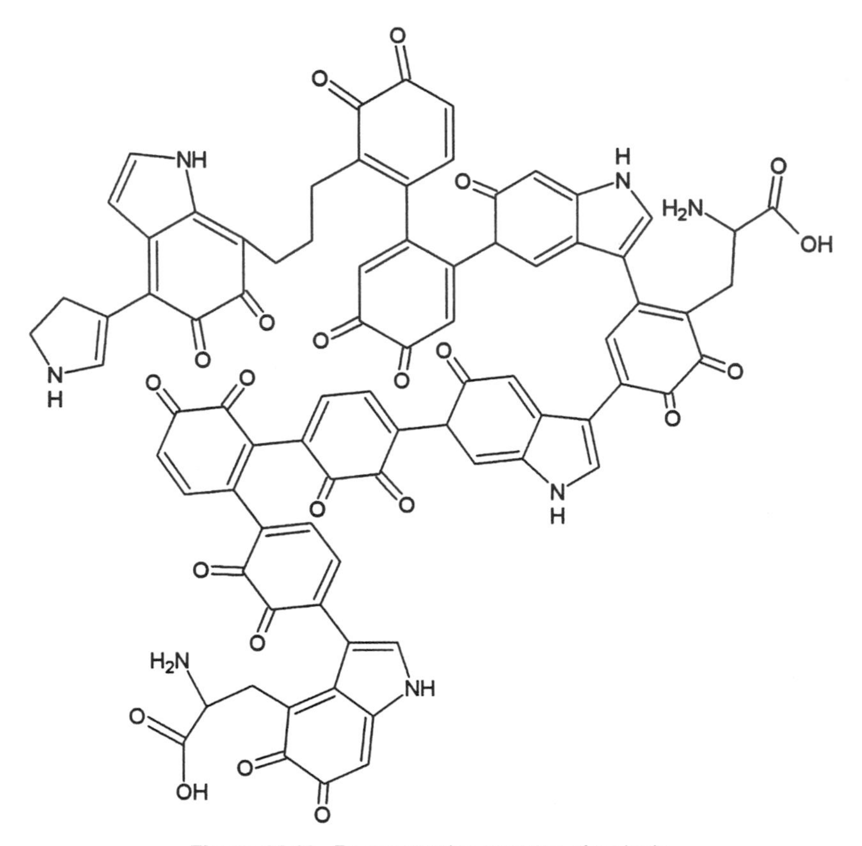

Figure 14.12. Representative structure of melanin.

skin. This brown coloration acts to help protect deeper skin elements from being damaged by the UV radiation. The absence of the enzyme tyrosinase that converts tyrosine to melanin can lead to albinism.

At least two colored melanins are formed: a series of black melanins and a series of so-called red melanins. Our skin pigmentation is determined by the relative amounts of these red and black melanins in our skin.

The concentration of melanin also contributes to the color of our hair (except for redheads where the iron-rich pigment trichosiderin dominates). The bleaching of hair, generally achieved through exposure to hydrogen peroxide, is a partial result of the oxidation of the melanin. A side reaction of bleaching is the formation of more sulfur cross-links leading to bleached hair being more brittle because of the increased cross-linking leading to a decrease in hair flexibility.

Melanin also provides a dark background in our eye's iris, is involved in animal color changes (such as the octopus and chameleon), is formed when fruit is bruised, and is partially responsible for the coloration of tea.

GLOSSARY

Adenosine phosphate: A purine base.

Amino group: $-NH_2$.

Anticodon: A specific sequence of purine and pyrimidine bases that permits t-RNA to bond with a specific set.

Bonds, secondary: Bonds based on van der Waals forces—that is, London dispersion bonds, dipole–dipole interactions, and hydrogen bonds.

Carboxyl group: $-COOH$.

Chiral atom: A carbon atom with four different groups.

Conformation: Different shapes of a molecule.

D-Ribose: A five-carbon carbohydrate (pentose).

Deoxynucleic acid: DNA.

Deoxyribose: A ribose in which a hydrogen atom replaces a hydroxyl group.

Desmosine: A skin protein.

Disulfide bond: S—S.

DNA: Deoxyribonucleic acid.

Elastin: A skin protein.

Enzyme: A biological catalyst.

Essential amino acid: Amino acids required for good health.

Fibrillar protein: Hairlike, water-insoluble protein structures resulting from intermolecular hydrogen bonding.

Genetic engineering: Any artificial process used to alter the genetic composition.

Globular protein: Water-soluble protein structures with intramolecular hydrogen bonds.

Gly: Abbreviation for glycine; other abbreviations for amino acids are shown in Table 14.1.

Glycine: The first member of the α-amino acid homologous series ($CH_2(NH_2)COOH$).

Hydrophilic: Water-loving.

Hydrophobic: Water-hating.

Insulin: A hormone that controls glycogenesis and glycogenolysis.

Intermolecular bonding: Secondary bonding between atoms in two different molecules.

Isoelectric point: The pH value at which the negative and the positive charges on an amino acid are equal. This point is characteristic for each amino acid.

Isomers: Molecules with identical structures, that is, the same formulas.

Kwashiorkor: Chronic disease caused by deficiency of essential amino acids.

L-Amino acid: Levo isomer that is the typical isomer for naturally occurring amino acids.

Lock and key concept: Theory of enzymatic activity that requires a specific fitting of a reactant in the enzyme structure that permits a specific reaction to occur.

Lysine: An essential amino acid (see Table 14.1).

Milk of magnesia: $Mg(OH)_2$.

Molarity: A measure of the number of moles present in a liter of solution. A 1 molar solution of sodium hydroxide (NaOH) contains 40 g of sodium hydroxide.

Mole: 6.023×10^{23} particles.

Molecular biology: Science applied at the molecular level in biological systems.

Mutation: A mistake in coding transfers.

Nucleotide: The repeating unit in nucleic acids.

Optical isomer: An isomer that rotates the plane of polarized light.

Pellagra: A chronic disease caused by a deficiency of lysine.

Peptide linkage:

```
  H O
  | ||
-(N-C)-
  |
  R
```

pH: An acidity scale in which 7.0 is the neutral point. pH values less than 7.0 are acidic, and those greater than 7.0 are alkaline.

Polypeptide: Name commonly used for low-molecular-weight amino acid polymers.

Polysome: A complex formed from RNA and ribosomes.

Primary structure: Structure resulting from primary bonding of atoms in proteins.

Prosthetic group: A nonprotein group, such as glucose, joined to a protein molecule.

Protein: A polymer made up of repeating units of α-amino acids.

Purine: A heterocyclic molecule with two fused rings.

Pyrimidine: A heterocyclic molecule consisting of one nitrogen-containing ring.

R: An alkyl group ($H(CH_2)_n$—).

Ribonucleic acid: RNA.

RNA: Ribonucleic acid.

m-RNA: Messenger RNA.

t-RNA: Transfer RNA.

Secondary structure: Conformations in proteins resulting from intermolecular and intramolecular attractions.

Sericulture: The culture of the silkworm and weaving of silk filaments.

Sodium hydroxide: NaOH.

Tertiary structure of proteins: Three-dimensional shape of protein molecules based on foldings of protein chains.

Tufting: Splitting of fibers.

Zwitterion: A polar molecule made up of an anion and a cation. In amino acids, the anion is the carboxyl ion ($-COO^-$) and the cation is the ammonium ion ($-NH_3^+$).

REVIEW QUESTIONS

1. What does DNA stand for?
2. What does RNA stand for?
3. How many different repeating units are present in proteins?
4. What does R stand for?
5. How does glycine differ from all other amino acids?
6. What does Gly stand for?
7. Are naturally occurring amino acids D or L?
8. What is the name given to a molecule like an amino acid that contains both an acidic and a base group?
9. What is the pH of distilled water?
10. Is the pH of an acid less than or greater than 7.0?
11. Define isoelectric point.
12. Give two major properties of an enzyme.
13. What is the repeating linkage in a protein?
14. What is the term used for hydrogen bonds between atoms in the same molecule?
15. Is the primary structure in proteins dependent on configurations or conformations?
16. Is the secondary structure in proteins dependent on configurations or conformations?
17. What type of hydrogen bonds are present in fibrillar proteins?
18. What type of hydrogen bonds are present in globular proteins?
19. Which is water repellent: hydrophilic or hydrophobic?
20. What is silk culture called?
21. What is the difference between D-ribose and D-deoxyribose?
22. What is the repeating unit in nucleic acid called?

23. The mnemonic expression Gee-CAT stands for what bases?

24. What is the general name for the bases in DNA?

BIBLIOGRAPHY

Atala, A., Mooney, D., Vacanti, J., and Langer, R. (1997). *Synthetic Biodegradable Polymer Scaffolds*, Springer, New York.

Carraher, C., and Gebelein, C. (1982). *Biological Activities of Polymers*, ACS, Washington, D.C.

Bloomfield, V., Crothers, D., and Tinoco, I. (2000). *Nucleic Acids: Structure, Properties and Functions*, University Science Books, Sausalito, CA.

Gebelein, C., and Carraher, C. (1985). *Biocative Polymeric Systems*, Plenum, New York.

Davies, K. (2001). *Cracking the Genome*, Free Press, New York.

Drauz, K, Waldmann, H., and Roberts, S. (2002). *Enzyme Catalysis in Organic Synthesis*, 2nd ed., Wiley, New York.

Dugas, H. (1995). *Bioorganic Chemistry*, Springer, New York.

Ferre, F. (1997). *Gene Quantification*, Springer, New York.

Harsanyi, G. (2000). *Sensors in Biomedical Applications*, Technomics, Lancaster, PA.

Kaplan, D. (1994). *Silk Polymers: Materials Science and Biotechnology*, ACS, Washington, D.C.

Kay, L. (2000). *Who Wrote the Book of Life*, Stanford University Press, Stanford, CA.

Keller, E. F. (2000). *Century of the Gene*, Harvard University Press, Cambridge, MA.

Moldave, K. (2000). *Progress in Nucleic Acid Research and Molecular Biology*, Academic Press, New York.

Nature (2001). Human Genome, Vol. 409, February 15 issue.

Ridley, M. (1999). *Genome*, HarperCollins, New York.

Rosenberg, I. (1996). *Protein Analysis and Purification*, Springer, New York.

Science (2001). Human Genome, Vol. 291, Number 5507, February issue.

ANSWERS TO REVIEW QUESTIONS

1. Deoxyribonucleic acid.

2. Ribonucleic acid.

3. About 20 different amino acid residues (repeating units).

4. In general, R can represent almost any moiety that is unspecified.

5. It does not contain a chiral carbon atom, therefore has no optical isomers.

6. Glycine.

7. L (levo).

8. Zwitterion.

9. 7.0.

10. Less than 7.

11. The pH at which the negative charge equals the positive charge in an amino acid. This is the neutral point.

12. Enzymes are catalysts and are specific.

13. The peptide linkage

$$-\!\!\left(\begin{matrix}\mathrm{H}\\ |\\ \mathrm{N}\end{matrix}-\mathrm{CO}\right)\!\!- \quad \text{or} \quad -\!\!\left(\begin{matrix}\mathrm{R}\\ |\\ \mathrm{C}\\ |\\ \mathrm{H}\end{matrix}-\begin{matrix}\mathrm{H}\\ |\\ \mathrm{N}\end{matrix}-\mathrm{CO}\right)\!\!-$$

14. Intramolecular hydrogen bonds.

15. Configurations—that is, primary bonds.

16. Conformations.

17. Mainly Intermolecular hydrogen bonds.

18. Mainly Intramolecular hydrogen bonds.

19. Hydrophobic.

20. Sericulture.

21. D-Ribose contains one more oxygen atom than does D-deoxyribose.

22. Nucleotide.

23. Guanine, cytosine, adenine, and thymine.

24. Purine and pyrimidine bases.

15

DERIVATIVES OF NATURAL POLYMERS

15.1 INTRODUCTION

In addition to their use as food, shelter, and clothing, polymeric carbohydrates, proteins, and nucleic acids are also important commercial materials.

Giant Molecules: Essential Materials for Everyday Living and Problem Solving, Second Edition,
by Charles E. Carraher, Jr.
ISBN 0-471-27399-6

15.2 DERIVATIVES OF CELLULOSE

The hydroxyl groups are not equivalent. The two hydroxyl (OH) groups on the ring are more reactive than the external hydroxyl. The two ring hydroxyl groups react first. If reaction continues, then the "off-ring" hydroxyl group is reacted.

In the cellulose regenerating process, sodium hydroxide is initially added such that approximately one hydrogen, believed to be predominately one of the two attached to the ring hydroxyl groups, is replaced by the sodium ion. This is followed by treatment with carbon disulfide, forming cellulose xanthate that is eventually re-changed back again, regenerated, to cellulose. This sequence is shown below.

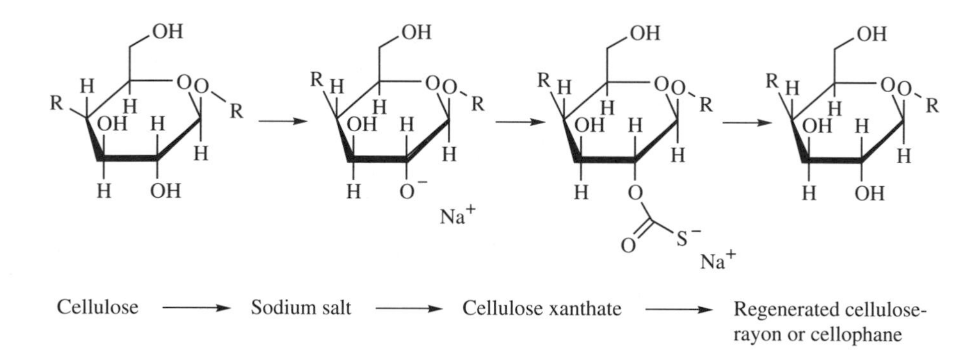

C. F. Schönbein accidentally produced explosive cellulose nitrate in 1846 when he used his wife's cotton apron to wipe up a spilled mixture of nitric acid (HNO_3) and sulfuric acid (H_2SO_4) and dried the wet apron by heating it on the stove. Gun cotton or cellulose trinitrate is produced when concentrated acids are used, but the more polar and less expensive cellulose dinitrate is obtained when dilute solutions of the acids are used. A segment of the polymer chain of cellulose dinitrate has the structure

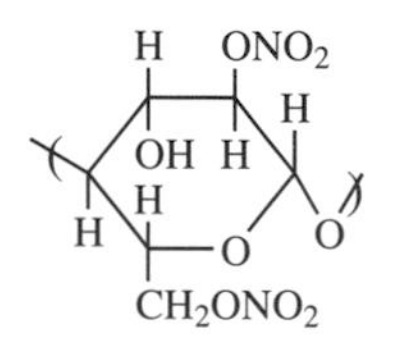

Cellulose dinitrate

For all of these polyfunctional materials, substitution occurs such that a repeating unit may contain, for example, one, two, or three substituents for each repeat unit. Thus cellulose dinitrate will contain, on the average, two nitrate substituents per unit, but some units will contain three, some two, some one, and some zero nitrate groups.

Nitrated cellulose, in a mixture of ethanol (C_2H_5OH) and ethyl ether ($(C_2H_5)_2O$), called collodion, is still used as a liquid court plaster (Nu/Skin).

It was modified by J. N. Hyatt to produce a plasticized cellulose dinitrate, which he called celluloid.

A plasticizer is a nonvolatile compound that enhances the segmental motion of the polymer chains. Hyatt, who was a printer and inventor of the roller bearing, used camphor to plasticize cellulose dinitrate. He added pigments to this plasticized polymer to produce billiard balls, and for his efforts he won a prize of $10,000. These balls had previously been carved out of the tusks of elephants. A competitive flexibilized product called Parkesine was produced by A. Parkes in England in 1861. In spite of these inventions, over 100,000 elephants were killed annually for their tusks in the nineteenth century.

Brush handles, baby rattles, shirt fronts, and collars were made from celluloid, and phonograph records were made by Thomas A. Edison from shellac in the nineteenth century. However, because of the combustibility of celluloid and the high cost of shellac, the use of these plastics was limited.

The fact that cellulose nitrate is highly flammable and explosive should not be surprising since most highly nitrated compounds, such as TNT and dynamite (glyceryl nitrate), are highly flammable and explosive.

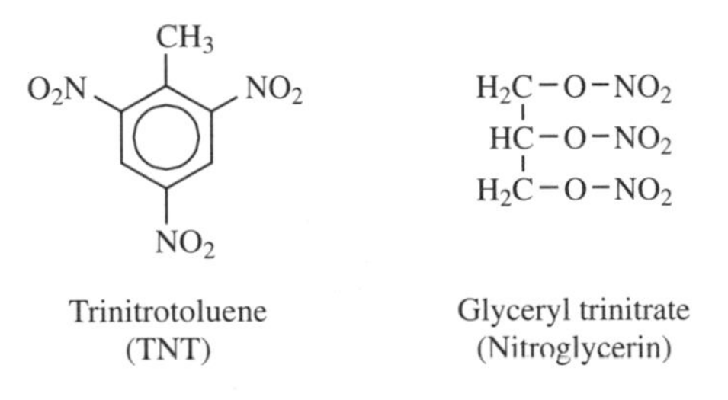

Trinitrotoluene (TNT) Glyceryl trinitrate (Nitroglycerin)

When all the hydroxyl groups in cellulose are nitrated, the material is known as nitrated cotton or gun cotton. When a mixture of gun cotton and lesser nitrated cellulose is mixed with ethanol (ethyl alcohol, CH_3CH_2OH) and diethyl ether ($CH_3CH_2OCH_2CH_3$), it produces a jellylike mass. This can be rolled into strips and dried to a consistency of dry gelatin. When ignited in small quantities in the open, this smokeless powder burns readily without exploding. Under the confinement of a rifle or gun cylinder, decomposition after ignition is rapid and much heat and gas are evolved. From the following balanced equation, it can be shown that from a 1-g (about the size of a small pea) charge of gun cotton, gas at about room pressure would occupy a volume of one quart. This expanding gas propels the shell from the gun. Since there is no solid residue left by the decomposition of gun cotton, the barrel chamber is left clean.

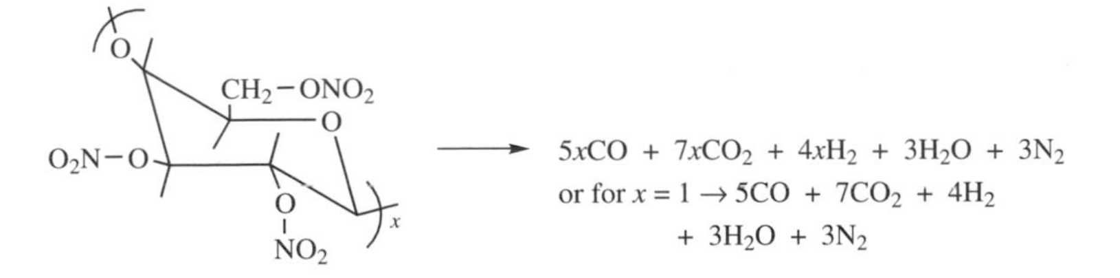

P. Schutzenberger reacted cellulose to produce cellulose triacetate in 1869, but the large-scale use of this cellulosic derivative was delayed because expensive solvents, such as chloroform, were required to dissolve this cellulose derivative. G. Miles solved this dilemma in 1902 by removing one of the acetyl groups from each repeating unit to produce cellulose diacetate in a process called partial saponification. The chemical equations for acetylation of cellulose and partial deacetylation by saponification are

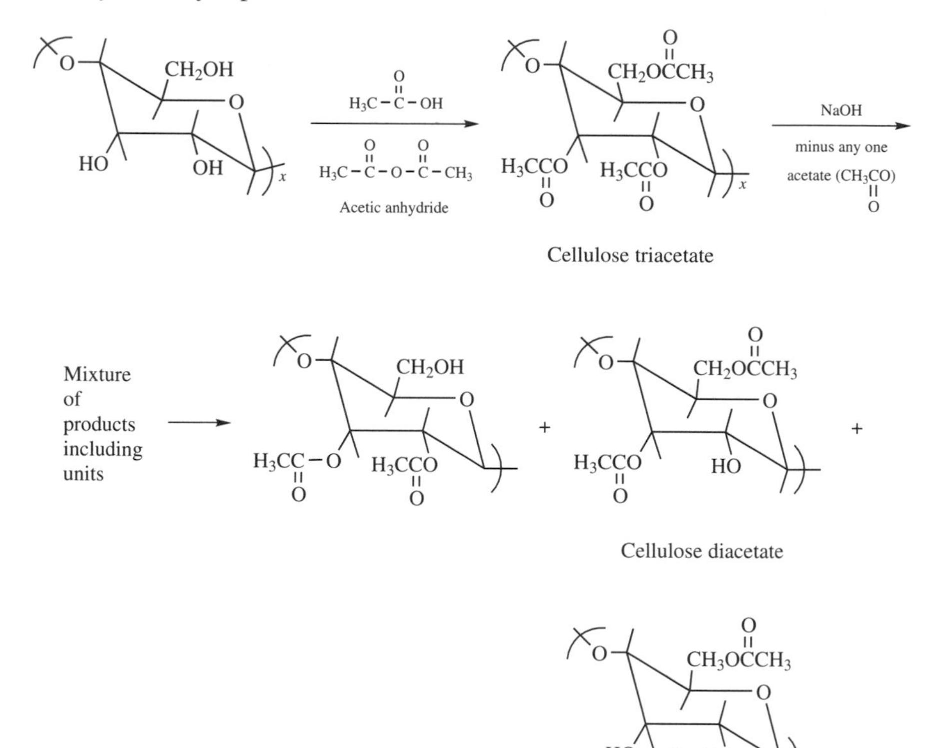

Henry and Camile Dreyfus developed the fiber-making process for cellulose acetate. This fiber, which is now called acetate rayon, has been produced in the United States by Celanese Corporation (now Hoechst-Celanese) since 1924. Cellulose triacetate is employed as the fabric for many tricot fabrics and sportswear. Today's triacetate is shrink- and wrinkle (permanent press modified)-resistant and easily washed.

J. Brandenberger made films of cellulose acetate in 1908 when he sprayed a solution of cellulose acetate on a tablecloth in an attempt to waterproof it. He found that he could peel the cellulose acetate from the cloth to produce a thin, transparent film. Because it was less flammable, cellulose acetate film was used as a replacement for celluloid film. Cellulose acetate is also used as a lacquer, called "dope," which was used to coat cloth for World War I airplanes. It is still used as a fingernail polish.

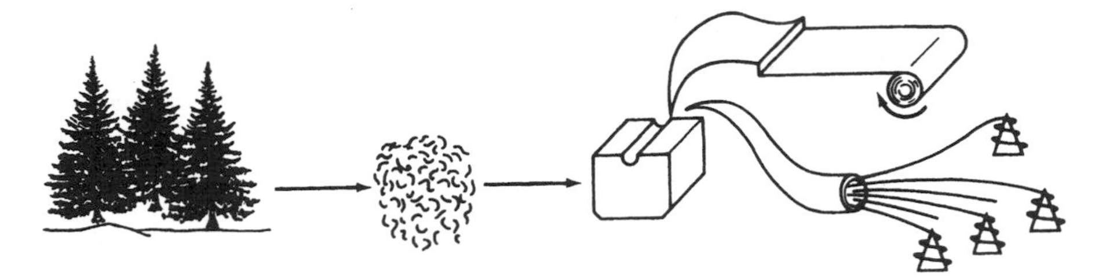

Figure 15.1. Cellophane and viscose rayon are produced from cotton or wood pulp that is matted, shredded, dissolved in basic solutions, and pumped through a slit (*top*) or spinneret (*bottom*) into an acid bath, where the cellophane or rayon precipitates to form sheets and threads.

Transparent films, such as cellophene, are produced by passing a solution of cellulose xanthane, called viscose, through a slit and precipitating this sheet in an acid solution. In this process, only an average of one hydroxyl group for each cellulose ring forms a xanthate as shown above. The annual production of cellophane film was about 500 million pounds in the 1960s but because of its high cost compared to other films, its annual production has decreased to below 150,000 pounds.

Although Robert Hooke suggested the possibility of making "artificial silk" in 1665, the first man-made fiber was not produced until two centuries later by Hilaire de Chardonnet. These regenerated cellulosic filaments were produced by passing collodion through small, uniformly sized holes, called spinnerets, evaporating the ethanol and ether solvent, and then removing the nitrate groups from the cellulose nitrate filaments (Figure 15.1).

Three principal methods have been employed in the synthesis of rayon. These are the viscose process, the cuprammonium process, and the acetate process. Viscose and cuprammonium rayons have similar chemical and physical properties. Both are dyed easily and lose their strength when wet (disruption of hydrogen bonds). Acetate rayon is easily softened when ironed and boiling water decreases its luster, but it is a finer texture and more readily dyed than viscose or cuprammonium rayons.

Reaction on each cellulose ring occurs differently since there are two "types" of hydroxyl groups (as noted before), the two ring hydroxyls and the methylene hydroxyl. In the typical formation of esters such as cellulose acetate, the ring hydroxyl groups are acetylated initially. Under the appropriate reaction conditions, reaction continues to almost completion with all three of the hydroxyl groups esterified (formation of an ester group). In triacetate products, only small amounts (on the order of 1%) of the hydroxyls remain unreacted.

$$\mathrm{R-\overset{\overset{\displaystyle O}{\|}}{C}-O-R'}$$

An ester

As in all large-scale industrial processes, the formation of the cellulose esters involves recovery of materials. Acetic anhydride is generally employed in the formation of cellulose esters. After reaction, acetic acid and solvent is recovered. The recovered acetic acid is employed in the production of additional acetic anhydride. The recovered solvent is also reintroduced after treatment.

Cellulose esters are used as plastics for the formation by extrusion of films and sheets and by injection molding of parts. They are thermoplastics and can be fabricated employing most of the usual techniques of (largely compression and injection) molding, extrusion, and casting. Cellulose esters plastics are noted for their toughness, smoothness, clarity, and surface gloss.

Acetate fiber is the generic name of a fiber that is partially acetylated cellulose. They are also known as cellulose acetate and triacetate fibers. They are nontoxic and generally nonallergic, so they are ideal from this aspect as clothing material.

While acetate and triacetate differ only moderately in the degree of acetylation, this small difference accounts for differences in the physical and chemical behavior for these two fiber materials. Triacetate fiber is hydrophobic, and application of heat can bring about a high degree of crystallinity that is employed to "lock-in" desired shapes (such as permanent press). Cellulose acetate fibers have a low degree of crystallinity and orientation even after heat treatment. Both readily develop static charge, and thus antistatic surfaces are typically added to clothing made from them.

For clothing applications there are a number of important performance properties that depend on the form of the textile. These properties include wrinkle resistance, drape, strength, and flexibility. These properties are determined using ASTM tests that often involve stress/strain behavior. Thus, the ability of a textile to resist deformation under an applied tensile stress is measured. As with any area of materials, specialty tests are developed to measure certain properties. Some of these are more standard tests like the aforementioned stress/strain test, while others are specific to the desired property measured for a specific application. For instance, resistance to slightly acidic and basic conditions is important for textiles that are to be laundered. Again, these are tested employing standard test procedures. In general, triacetate materials are more resistant than acetate textiles to basic conditions. Both are resistant to mild acid solutions but degrade when exposed to strong mineral acids. Furthermore, behavior to various dry cleaning agents is important. As the nature of dry cleaning agents change, additional testing and modification in the fabric treatments are undertaken to offer a textile that stands up well to the currently employed cleaning procedures. Again, both are stable to perchloroethylene dry cleaning solvents but can soften when exposed to trichloroethylene for extended treatment. Their stability to light is dependent upon the wavelength, humidity present, and so on. In general, acetates offer a comparable stability to light as that offered by cotton and rayon.

While cellulose acetates are the most important cellulose esters, they suffer by their relatively poor moisture sensitivity, limited compatibility with other synthetic resins, and a relatively high processing temperature.

A major effort is the free-radical grafting of various styrenic, vinylic, and acrylic monomers onto cellulose, starch, dextran, and chitosan. The grafting has been achieved using a wide variety of approaches including ionizing and ultraviolet/visible radiation, charge-transfer agents, and various redox systems. Much of this effort is aimed at modifying the native properties such as tensile-related (abrasion resistance and strength) and care-related (crease resistance and increased soil and stain release) properties, increased flame resistance, and modified water absorption. One area of emphasis has been the modification of cotton and starch in the production of superabsorbent material through grafting. These materials are competing with all synthetic cross-linked acrylate materials that are finding use in diapers, feminine hygiene products, wound dressings, and sanitary undergarments.

15.3 DERIVATIVES OF STARCH

Starch obtained from rice, wheat, potatoes, and corn is used worldwide as a source of food. Starch obtained from corn and potatoes is used industrially for adhesives and sizing for paper and textiles in the United States. Most starches contain about 25% amylose and 75% amylopectin. However, waxy corn is almost 100% amylopectin. Over 125 million tons of corn are used annually as a source of amylopectin in the United States.

The hydroxyl groups on each glucose residue in starch undergo the same reactions as the hydroxyl groups in cellulose. At high degrees of substitution (DS), starch becomes a thermoplastic. However, unplasticized starch triacetate tends to be brittle, and the higher esters are soft and weak.

A wide variety of graft copolymers of starch have been made, largely through addition of vinyl monomers such as acrylonitrile and methyl acrylate to free radical sites generated by irradiation or chemical treatment. The so-called Super Slurper, which absorbs hundreds of times its own weight of water, is a graft copolymer.

Soluble starch is produced by treating starch with 7.5% hydrochloric acid for 1 week at room temperature. The adhesiveness of starch is improved by oxidation with hydrogen peroxide (H_2O_2) or sodium hypochlorite (NaOCl) (bleach). Partially degraded starch (dextrin) is used as an adhesive on postage stamps. Starch nitrate is used as an explosive.

15.4 LEATHER

Leathermaking, which was the first man-made polymer reaction, involves the introduction of cross-links into the protein molecule. The tanners who produced this most historic of materials were some of our earliest craftsmen. That this ancient process was chemically sound has been demonstrated by the permanence of the tanned products. Sandals, garments, and water bags made over 2500 years ago can be found in museums today. The original vegetable tanning process has been displaced, to some extent, by the chrome tanning process, and leather itself has also been partially displaced by synthetic sheeting, but the tanning art is still practiced widely throughout the world.

15.5 REGENERATED PROTEINS

Proteins, such as those in animal blood, hides, and milk, are readily available and have been used as food and adhesives for many years. The proteinaceous gluten in wheat flour has been used as an adhesive in dough formation for centuries. The first animal glue factory was built in Holland in 1690, and a comparable plant was built in the United States in 1808.

Casein is obtained by acidification or by the addition of rennet to skimmed milk. E. Childs patented casein as a plastic in 1885, and W. Krische and A. Spitteler used this proteinaceous material to produce a commercial moldable plastic (artificial horn, Galalith) in 1897. Zein from maize and protein from soybeans and peanuts have been used to a limited extent for molded plastics, but these products are not competitive with general-purpose synthetic plastics.

In addition to food and adhesives, the widest use of proteins is for regenerated fibers. A. Ferretti produced casein fiber (Lanital) by forcing an alkaline solution of casein through small holes (spinnerets) into an acid bath. Other casein fibers are called Aralac and Merinova.

Regenerated protein fibers are also made from soybeans, peanuts (Ardil), corn (Zein), and chicken feathers. All of these regenerated protein fibers must be immersed in formalin to prevent bacterial attack. Unlike polyamide fibers, such as nylon, some of these bacteria-resistant polyamide fibers, given the generic name Azlon, are attacked by moths.

15.6 NATURAL RUBBER

In 1839, Charles Goodyear cross-linked natural rubber by heating it with a small amount of sulfur. Since the rubbery chains are held together by just a few sulfur cross-links, the "vulcanized" product is flexible and is said to have a low cross-linked density. His brother, Nelson Goodyear, produced hard rubber or ebonite by heating rubber with a large amount of sulfur to produce a plastic product with high cross-linked density.

The cross-linking of rubber by the formation of sulfur cross-links is a complex chemical reaction. However, Charles Goodyear carried out his first vulcanization reactions on his wife's kitchen stove, and many competitors followed his simple procedure and ignored his patents. He spent several hundreds of thousands of dollars contesting infringements on his patents, and because of his indebtedness, he spent considerable time in a debtor's prison. Notably, his attorney in "The Great India Rubber Patent Infringement Case" in 1852 was Daniel Webster. The court ruled in favor of Goodyear, but he died in 1860 before he could earn enough money from his royalties to pay his debts.

Charles Goodyear was the first to convert a linear elastomer to a cross-linked elastomer, and Nelson Goodyear was the first to convert an elastomer to a thermoset. The invention of the pneumatic tire by Dunlop in 1888 and the fountain pen by Waterman in 1884 were dependent on the inventions of Charles and Nelson Goodyear, respectively.

Chapter 10 contains additional information about rubber.

15.7 DERIVATIVES OF NATURAL RUBBER

Vulcanized rubber, of course, is a derivative, but this product is just called rubber by English-speaking consumers. Nevertheless, there are many other derivatives of natural rubber that were produced prior to the development of the synthetic polymer industry.

In 1927, Fisher isomerized rubber in the presence of acids, such as *p*-toluene sulfonyl chloride ($H_3CC_6H_4SO_2Cl$). These thermoplastics, or "thermoprenes," were used as adhesives for bonding rubber to metal. The most widely used product, called Pliolite, was produced by heating rubber in the presence of chlorostannic acid (H_2SnCl_6). This cyclized rubber continues to be used to a limited extent, but because of high costs it has been displaced by synthetic adhesives and coatings.

Since rubber is an unsaturated polymer, it adds chlorine readily. This chlorination reaction was known for many years but was not commercialized until 1925, when Peachey patented the chlorinated thermoplastic product under the trade name of Duroprene and Alloprene. Chlorinated rubber was produced in Germany under the trade name of Tornesit and later in the United States under the trade name of Parlon. Chlorinated rubber has been used as a protective coating for swimming pools.

Rubber hydrochloride (Pliofilm) is produced by the reaction of natural rubber with gaseous hydrogen chloride (HCl). Films of rubber hydrochloride, cast from solutions of this derivative in benzene, have been used as transparent packaging materials.

15.8 MODIFIED WOOL

Permanent-press wool is machine-washable, retains imparted creases or pleats, and requires minimum ironing after washing and drying. In the CSIRO solvent–resin–steam process, a reactive polyurethane is applied to the garment, which is pressed and hung in a steam oven. The polymer is cured on the garment in the required shape. The International Wool Secretariat's permanent-press process employs a thiol-terminated prepolymer.

Keratins have a high cystine content, approximately 500 μmol/g for wool. These cystine residues cross-link with adjacent protein chains via disulfide bonds, thus restricting their conformational motion. Treatment of wool with reducing agents converts these disulfide cystine groups to two cysteine residues:

$$\begin{array}{c} -NHCHCH_2-SS-CH_2CHCO- \\ \quad | \qquad\qquad\qquad | \\ -CO \qquad\qquad\qquad NH- \end{array} \longrightarrow \begin{array}{c} -NHCHCH_2SH \\ | \\ -CO \end{array} + \begin{array}{c} HSCH_2CHCO \\ | \\ NH- \end{array}$$

15.9 JAPANESE LACQUER

Japanese lacquer is a naturally occurring, phenolic coating material obtained from the lacquer tree *Rhus vernicifera*. Its use has been dated back 5000–6000 years to continental China. The sap consists of urushiol (65–70%), a polysaccharide plant gum (5–7%), a copper-bearing lactose ($<1\%$), and water (20–25%). On drying, Japanese lacquer forms a cellular structure about 0.1 μm in diameter, with gum as the cell wall.

15.10 NATURAL POLYMERS THROUGH BIOTECHNOLOGY

Poly(3-hydroxybutyrate) (PHB) is produced by bacteria. It occurs in soil bacteria, estuarine microflora, blue-green algae, and microbially treated sewage. A good laboratory source is *Alcaligenes eutrophus*. These microorganisms can be grown in large tanks on a variety of substrates, including natural sugars, ethanol, or gaseous mixtures of carbon dioxide and hydrogen. Large quantities of PHB accumulate as discrete intercellular granules, which may constitute up to 80% of the cell's dry weight.

15.11 OTHER PRODUCTS BASED ON NATURAL POLYMERS

Gutta-percha and balata, which are trans isomers of polyisoprene, have been used as cable coatings and molding resins. Since their chemical formulas are identical to the cis isomer of polyisoprene, they undergo all of the reactions listed for natural rubber. However, none of these derivatives has been produced commercially.

Shellac is the secretion of the lac insect (*Laccifer lacca*), which feeds on the sap of trees in southeastern Asia. Thomas Edison used this product for molding gramophone records. Moldings of mica-filled shellac have also been used for electrical insulators. Alcohol solutions of shellac continue to be used as coatings, but oleoresinous coatings are more widely used.

The art of decorative painting was developed in prehistoric times, and the first successful human nonvocal communication was probably through paintings on cave walls. Varnishes based on natural materials, such as beeswax, were used over 3000 years ago.

All oil paints consist of finely ground pigments, a solvent or thinner, and a resin-forming component called a binder. The solution of the binder in the solvent is called the vehicle. Some modern paints contain natural resins, such as copals, kauri, manila, and others. However, these natural resins as well as the oleoresinous paints are being displaced by coatings based on synthetic resins, which are generally superior and more economical than the natural products.

GLOSSARY

Acetate rayon: Cellulose acetate fibers.

Alum: Aluminum sulfate.

Ardel: Peanut protein fiber.

Azlon: Protein fibers.

Binder: Resin-forming component in paint.

Cachuchu: South American Indian name for rubber.

Casein: Milk protein.

Cellophane: Regenerated cellulose film.

Celluloid: Cellulose nitrate plasticized by camphor.

Cellulose: A polycarbohydrate with repeating units of D-glucose joined by beta acetal linkages.

Cellulose acetate: A product of the reaction of cellulose and acetic anhydride.

Cellulose nitrate: A product of the reaction of nitric acid and cellulose (erroneously called nitrocellulose).

Cellulose xanthate: The reaction product of sodium cellulose and carbon disulfide (Cell-CSS^-, Na^+).

Chlorinated rubber: The reaction product of rubber and chlorine.

Cinnabar: HgS.

Collodion: A solution of cellulose nitrate in an equimolar mixture of ethanol (C_2H_5OH) and ethyl ether ($(C_2H_5)_2O$).

Cyclized rubber: A derivative of rubber with a cyclic (ring) structure.

Dextrin: Partially degraded starch.

Dope: A solution of cellulose acetate in acetone.

Ebonite: Hard rubber.

Formalin: 37% aqueous solution of formaldehyde.

Galalith (milkstone): Molded casein articles.

Gluten: Wheat flour protein.

Hematite: Fe_2O_3.

Hydrogen peroxide: H_2O_2.

Lanital: Casein fiber.

Lapis lazuli: Egyptian blue.

Leather: Cross-linked protein produced by a tanning process.

Linear elastomer: An elastomer with a continuous polymer chain.

Malachite green: $CuCO_3 \cdot Cu(OH)_2$.

Methylol group: $HOCH_2-$.

Ocher: Hydrated iron oxide pigment.

Orpiment yellow: As_2S_3.

Plasticizer: A flexibilizing additive.

Pyrolusite: MnO_2.

Rayon: Regenerated cellulose fiber.

Rubber hydrochloride: The reaction product of rubber and gaseous hydrogen chloride.

Saponification: Hydrolysis of an ester to produce an alcohol and an acid.

Schweitzer's solution: $Cu(NH_3)_4(OH)_2$.

Shellac: A natural resin obtained from the excreta of coccid insects that feed on twigs of trees in southeastern Asia.

Sizing: A coating on paper or textiles.

Sodium hypochlorite: NaOCl.

Spinnerets: Small, uniform-sized holes used in fiber production.

Starch: A polysaccharide with repeating units of D-glucose joined by alpha acetal linkages.

Tanning: The cross-linking of animal skins by tannic acid.

Thermoprene: Cyclized rubber.

Thermoset polymer: Cross-linked polymer.

Titanium dioxide: TiO_2.

Vehicle: Solution of a binder in a solvent.

Viscose: A solution of cellulose xanthate.

Zein: Corn protein.

REVIEW QUESTIONS

1. Why is the term nitrocellulose incorrect?
2. What is the solvent present in collodion?
3. What is the effect of camphor on cellulose nitrate?
4. Why is cellulose acetate less flammable than cellulose nitrate?
5. What is the original meaning of saponification?
6. What is the principal difference between rayon and acetate rayon?
7. Would you expect a viscose rayon plant to have a bad odor? Why?
8. What is the difference between rayon and cellophane?

9. What is the principal difference between amylose starch and cellulose?

10. Which has the higher degree of polymerization (DP): starch or dextrin?

11. What is the oldest man-made cross-linked polymer?

12. What famous chemist gave the name of rubber to *Hevea braziliensis*?

13. Why is casein no longer used as a molded plastic?

14. Which has the higher cross-linked density: soft vulcanized rubber or ebonite?

15. Is thermoprene elastic?

16. Is chlorinated rubber elastic?

17. What is the principal use for rubber hydrochloride?

18. What name is given to the resinous component of a paint?

19. What is the most widely used pigment?

20. What pigments are red, white, and blue?

BIBLIOGRAPHY

Craver, C., and Carraher, C. (2000). *Advanced Polymer Science*, Elsevier, New York.

Freitag, R. (2001). *Synthetic Polymers for Biotechnology and Medicine*, Eurekah Company, Georgetown, Washington, D.C.

Gebelein, C., and Carraher, C. (1995). *Industrial Biotechnological Polymers*, Technomic, Lancaster, PA.

Gebelein, C., and Carraher, C. (1985). *Bioactive Polymeric Systems*, Plenum, New York.

McGrath, K., and Kaplan, D. (1997). *Protein-Based Materials*, Springer, New York.

Meister, J. (2000). *Polymer Modification*, Marcel Dekker, New York.

Okano, T. (1998). *Biorelated Polymers and Gels*, Academic Press, Orlando, FL.

Ottenbrite, R., and Kim, S. (2001). *Polymeric Drugs and Drug Delivery Systems*, Technomic, Lancaster, PA.

Shonaike, G., and Simon, G. (1999). *Polymer Blends and Alloys*, Marcel Dekker, New York.

ANSWERS TO REVIEW QUESTIONS

1. Nitrate (NO_3) groups and not nitro (NO_2) groups are present.

2. An equimolar mixture of ethanol and ethyl ether.

3. It serves as a flexibilizer or plasticizer for the stiff cellulose nitrate.

4. Acetate groups are less flammable (less explosive) than nitrate groups.

5. Soap making—that is, the alkaline hydrolysis of animal fat to produce glycerol and sodium salts of fatty acids (soap).
6. Rayon is a regenerated cellulose fiber; cellulose acetate is an acetyl derivative of cellulose.
7. Yes, carbon disulfide (CS_2) is odiferous.
8. The physical form—rayon is a fiber, cellophane is a film; both are regenerated cellulose.
9. The arrangement of the acetal linkages. Cellulose has beta acetal linkages and amylose starch has alpha acetal linkages between the D-glucose repeating units.
10. Starch; dextrin is degraded starch.
11. Leather.
12. Joseph Priestley.
13. Too expensive, subject to degradation, and properties are inferior to those of many other thermoplastics. It is used as an adhesive.
14. Ebonite—the maximum cross-linked density is approached when large amounts of sulfur are reacted with natural rubber.
15. No. It is a plastic used as a coating and as an adhesive.
16. No. It is a hard thermoplastic used as a coating.
17. As a packaging film (Pliofilm).
18. Binder (resin-forming component).
19. Titanium dioxide (white pigment).
20. Pb_3O_4, TiO_2, and lapis lazuli. (There are many others.)

16

INORGANIC POLYMERS

Giant Molecules: Essential Materials for Everyday Living and Problem Solving, Second Edition, by Charles E. Carraher, Jr.
ISBN 0-471-27399-6

16.1 INTRODUCTION

Just as polymers are abundant in the world of organic chemistry, they also abound in the world of inorganic chemistry. Inorganic polymers are the major components of soil, mountains, cement (concrete), and glass.

The first man-made, synthetic polymer was probably alkaline silicate glass, used in the Badarian period in Egypt (about 12,000 B.C.) as a glaze that was applied to steatite after it had been carved into various animal or ornamental shapes. Faience, a composite containing a powdered quartz or steatite core covered with a layer of opaque glass, was employed from about 9000 B.C. to make decorative objects. The earliest known piece of regular (modern type) glass has been dated at 3000 B.C. and is a lion's amulet found at Thebes and now housed in the British Museum. It is a blue opaque glass partially covered with a dark-green glass. Transparent glass appeared about 1500 B.C. Several fine pieces of glass jewelry were found in Tutankhamen's tomb (about 1300 B.C.), including two birds' heads of light-blue glass incorporated into the gold pectoral worn by the Pharaoh.

16.2 PORTLAND CEMENT

Portland cement is the least expensive and most widely used synthetic inorganic polymer. It is employed as the basic nonmetallic, nonwoody material in construction. Concrete highways and streets span our countryside, and concrete skyscrapers crowd the urban skyline. Less spectacular uses include sidewalks, fence posts, and parking bumpers.

The name portland is derived from the cement having the same color as the natural stone quarried on the Isle of Portland, south of Great Britain. The word cement comes from the Latin word *caementum*, which means pieces of rough, uncut stone. Concrete comes from the Latin word *concretus*, meaning to grow together. Common cement consists of anhydrous, crystalline calcium silicates, the major ones being tricalcium silicate (Ca_3SiO_5) and β-dicalcium silicate (Ca_2SiO_4); lime (CaO, 60%) and alumina (a complex aluminum-containing silicate, 5%) are also present.

When anhydrous (without water) cement mix is mixed with water, the silicates react by forming hydrates (compounds containing water and calcium hydroxide, $Ca(OH)_2$). Hardened portland cement contains about 70% cross-linked calcium silicate hydrate and 20% crystalline calcium hydroxide.

$$2Ca_3SiO_5 + 6H_2O \rightarrow Ca_3Si_2O_7 3H_2O + 3Ca(OH)_2$$

$$2Ca_2SiO_4 + 4H_2O \rightarrow Ca_3Si_2O_7 3H_2O + Ca(OH)_2$$

Calcium silicate trihydrate — Calcium hydroxide

The manufacture of portland concrete consists of three basic steps: crushing, burning, and finish grinding. Most cement plants are located near limestone

($CaCO_3$) quarries, since this is the major source of lime. Lime may also come from oyster shells, chalk, and a type of clay called marl. The silicates and alumina are derived from clay, silica sand, shale, and blast-furnace slag.

The powdery mixture is fed directly into rotary kilns for burning at 2700°F. These cement kilns are the largest pieces of moving machinery used in industry and can be 25 ft in diameter and 750 ft long. The heat changes the mixture into particles called clinkers, which are about the size of a marble. The clinkers are cooled and reground, with the final grinding producing portland cement that is finer than flour. The United States produces over 60 million tons of portland cement a year.

16.3 OTHER CEMENTS

There are a number of cements that are specially formulated for specific uses. Air-entrained concrete contains small air bubbles that were formed by addition of soaplike resinous materials to the cement or to the concrete when it is mixed.

Lightweight concrete may be made through use of lightweight fillers such as clays and pumice in place of sand and rocks or through the addition of chemical foaming agents, which produce air pockets as the concrete hardens.

Reinforced concrete is made by casting concrete about steel bars or rods. Most large cement structures such as bridges and skyscrapers employ reinforced concrete.

Prestressed concrete is typically made by casting concrete about steel cables stretched by jacks. After the concrete hardens, the tension is released, resulting in the entrapped cables compressing the concrete. Steel is stronger when tensed and concrete is stronger when compressed. Thus prestressed concrete takes advantage of both of these factors. Archways and bridge connections are often made from prestressed concrete.

There are non-portland cements as well. Calcium-aluminate cement has a much higher percentage of alumina than portland cement. Its active ingredients are lime and alumina. In the United States, it is manufactured under the trademark of Lumnite. Its major advantages are the rapidity of hardening and high strength within a day or two.

Magnesia cement is largely composed of magnesium oxide (MgO). In practice, the magnesium oxide is mixed with fillers and rocks and an aqueous solution of magnesium chloride. This cement sets up (hardens) within 2 to 8 h and is employed for flooring.

Gypsum, or hydrated calcium sulfate ($CaSO_4 \cdot 2H_2O$), serves as the base of a number of products, including plaster of Paris (also known as molding plaster, wall plaster, and finishing plaster), Keen's cement, Parisian cement, and Martin's cement.

16.4 SILICATES

Silicon is the most abundant metal-like element in the Earth's crust. It is seldom present in pure elemental form, but rather is present in a large number of polymers

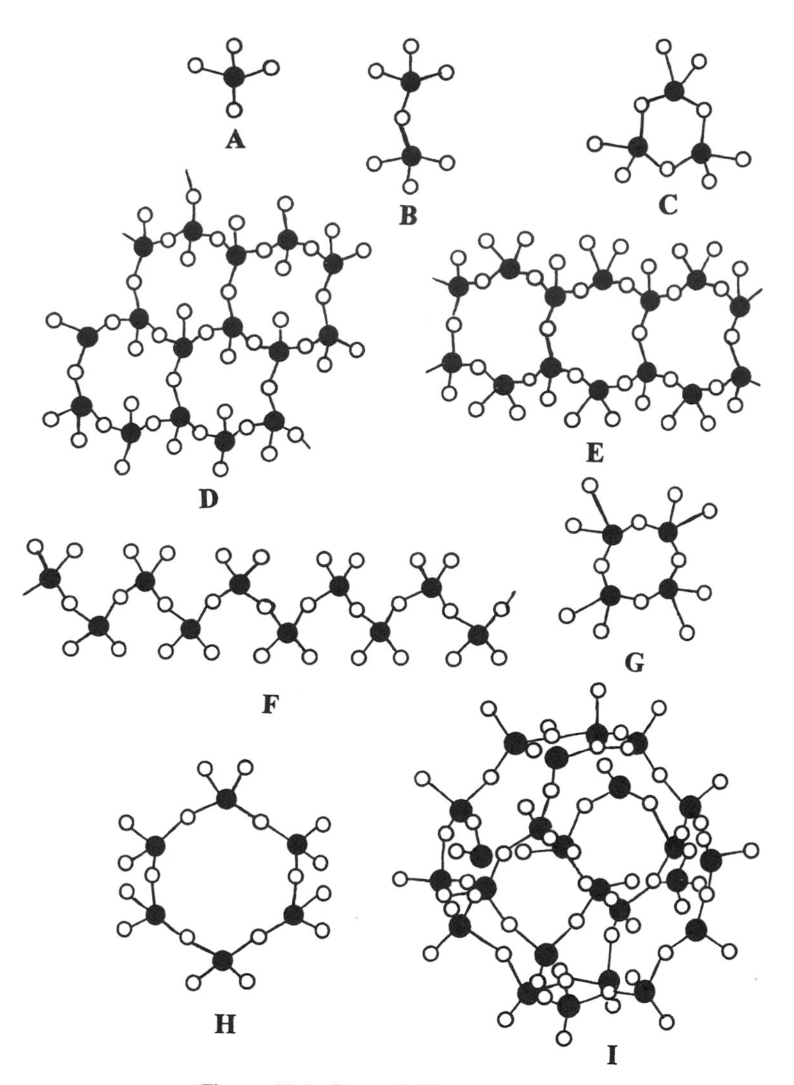

Figure 16.1. General silicate structures.

largely based on the polycondensation of the orthosilicate anion, SiO_4^{4-}. Central geometric forms are given in Figure 16.1, and examples are given in Table 16.1.

Each of these structures is based on a tetrahedral silicon atom attached to four oxygen atoms. The complexity and variety of natural occurring silicates is due to two major factors: first, the ability of the tetrahedral SiO_4^{4-} unit to be linked together often giving polymeric structures; second, the substitution of different metal atoms of the same approximate size as that of Si often occurs giving many different materials.

In the structures cited in Table 16.1, except for pure SiO_2, metal ions are required for overall electrical neutrality. These metal ions are positioned in

Table 16.1 Inorganic polymeric silicates as a function of common geometry

Basic Geometric Unit	Structural Geometry	General Silicate Formula[a]	Examples[a]
Tetrahedron	A	SiO_4^{4-}	Granite olivine–$(Mg,Fe)_2SiO_4$ Fosterite–Mg_2SiO_4, topaz
Double tetetrahedron	B	$Si_2O_7^{6-}$	Akermanite–$Ca_2MgSi_2O_7$
Triple ring	C	$Si_3O_9^{6-}$	Wollastonite
Tetra ring	D	$Si_4O_{12}^{8-}$	Neptunite
Six ring	E	$Si_6O_{18}^{12-}$	Beryl–$Al_2Be_3Si_6O_{18}$
Linear chain	F	$Si_4O_{12}^{8-}$	Augite, enstatite–$MgSiO_3$ Diopside–$CaMg(SiO_3)_2$ Chrysotile–$Mg_6Si_4O_{11}(OH)_6$
Double-stranded ladder	G	$Si_4O_{11}^{6-}$	Hornblende
Parquet (layered)	H	$Si_4O_{10}^{4-}$	Talc–$Mg_3Si_4O_{10}(OH)_2$ Mica–$KAl_3Si_3O_{10}(OH)_2$ Kaolinite–$Al_2Si_2O_5(OH)_4$ (Condensed) silicic acid–$H_2Si_2O_5$
Network	I	SiO_2	Quartz, feldspar (orthoclase)–$KAlSi_3O_8$

[a]The formulas given are for the most part simplified.

tetrahedral, octahedral, and so on, positions in the silicate-like lattice. Sometimes they replace the silicon atom. Kaolinite asbestos has aluminum substituted for silicon in the Gibbosite sheet.

Furthermore, sites for additional anions, such as the hydroxyl anion, are available. In ring, chain, and sheet structures the neighboring rings, chains, and sheets are often bonded together by metal ions held between the rings. In vermiculite asbestos the silicate sheets are held together by non-silicon cations.

For sheet-layered compounds, the forces holding together the various sheets can be substantially less than the forces within the individual sheets. Similar to graphite, such structures may be easily cleaved parallel to the sheets. Examples of such materials are mica, kaolinite, and talc.

Bonding occurs through a combination of ionic and covalent contributions just as are present in organic polymers except that the ionic character is a little higher.

Most silicate-like polymers can be divided into three major classes: the network structures based on a three-dimensional tetrahedral geometry (such as quartz), layered geometries with stronger bonding occurring within the "two-dimensional" layer (such as talc), and linear structures.

Network Structures. Quartz is an important network silicate (Section 16.6). A number of these tetrahedral silicate-like materials possess some AlO_4 tetrahedra substituted for the SiO_4 tetrahedra. Such structures offer a little larger "hole" in comparison to the entirely SiO_4 structures, thereby allowing alkali and alkaline-earth cations to be introduced. Feldspar (orthoclase) is such a mineral. The aluminosilicate networks are almost as hard as quartz. The feldspars are a similar network silicate to quart and widely distributed and compose almost two-thirds of all igneous rocks.

Some of the network structures exhibit a framework sufficiently "open" to permit ions to move in and out. The zeolite minerals used for softening water are of this type.

Zeolites include a whole group of aluminosilicates with an approximate formula of Si,AlO_4. Many of the so-called framework variety have an open hollow-like tubular structure. This tubular structure can be varied in size and used to filter out objects larger than the pore size of the hollow opening. They can also be used to "capture" atoms, holding them in a nanosized vice. Such captured atoms have been used as very specific catalysts.

Molecular sieves was the name first given to framework zeolites dehydrated by heating in vacuum to about 350°C because of their ability to capture and remove water and certain other species. Today, other materials, such as microporous silicas and aluminum phosphate, are three-dimensional structures that are also employed as molecular sieves.

Zeolites and related ordered clay-associated materials have been suggested to be involved in the initial primeval synthesis of basic elements of life.

Layer Structures. Layered structures typically conform to the approximate composition $Si_4O_{10}{}^{4-}$ or $Si_2O_5{}^{2-}$. For most of these, three of the oxygen atoms of each tetrahedron are shared by other tetrahedra, and the fourth oxygen is present on one side of the sheet.

In talc and kaolinite the layers are neutral. Thus, the layers slide over one another, easily imparting to these minerals a softness and ease in cleavage. In other minerals the layers are charged and held together by cations. In mica, the alumino silicate layers are negatively charged and cations, generally K^+, are present between the layers, giving the entire system of layers electronic neutrality. The ionic attractive forces between the layers result in mica being much harder than talc and kaolinite. Even so, these intersheet bonding forces are less than the "within-the-sheet" bonding forces, permitting relatively easy and clean cleavage of mica. Mica is used as an insulator for furnaces and electric equipment. Montmorillonite, another layered structure silicate-like material, is an important ingredient in soils and is employed industrially as a catalyst in the conversion of straight-chain hydrocarbons to more branched hydrocarbons.

Vermiculites are formed by the decomposition of mica. They contain layers of water and magnesium ions in place the potassium ions. When heated to 800–1100°C, vermiculite expands because of the conversion of the water to a gas. The expanded vermiculite has a low thermal conductivity and density and is used as a thermal and sound barrier and as an aggregate in lightweight concrete. It is also used as a moisture-retaining soil conditioner in planting.

A number of clays are layered silicate-like materials. Most clays contain finely divided quartzs, micas, and feldspars. Iron oxide-rich clays are employed to make pottery and terra cotta articles. Clays containing iron oxide and sand are used to make bricks and tiles. Clays rich in calcium and magnesium carbonate are known as marls and are used in the cement industry (Section 16.2).

Kaolinite is the main constituent in china clay used to make porcelain. The layers are largely held together by van der Waals' forces. Bentonite is used in cosmetics, as a filler for soaps, and as a plasticizer, and it is used in drilling-muds as a suspension stabilizer.

Asbestos also has a layered structure (Section 16.7).

Chain Structures. Both single- and double-stranded chains are found. The most important members of single chains are the pyroxenes and includes diopside. The most important double-chained minerals are the amphiboles. Some of these contain hydroxyl and fluoride ions, bonded directly to the metal cation and not to the silicon atom.

Jade, which has been valued in carving by eastern Asians for centuries, is generally one of two minerals: pyroxene or jadeite, $NaAl(SiO_3)_2$, and the amphibole nephrite. The nephrite has triple chains. Because the interchain bonding between the chains is weaker than the Si–O backbone bonding, these chain structures can generally be easily cleaved between the chains, allowing the jade to be carved.

Silicon dioxide plays a critical role in the electronics industry. The silicon used to produce silicon chips is derived from silicon dioxide. Semipure silicon dioxide (to about 99%) is prepared from thc rcaction of silicon dioxide with coke (a poor grade of graphite) using high temperature and an electronic arc.

$$SiO_2 + C \rightarrow Si + CO_2$$

Even so, this level of purity falls far short of the purity needed to produce the chips used in computers. The purity required is about 99.9999996, or a level of impurity of about one part in a billion. This is achieved through multistep processes.

16.5 SILICON DIOXIDE (AMORPHOUS)—GLASS

Silicon dioxide (SiO_2) is the repeating general structural formula for most rock and sand, as well as for the material we refer to as glass. The term glass can refer to many materials, but here we will use the ASTM definition: Glass is an inorganic product of fusion that has been cooled to a rigid condition without crystallization. We will consider silicate glasses and the common glasses used for electric light bulbs, window glass, drinking glasses, glass bottles, glass test tubes and beakers, and glass cooking ware.

Glass has many desirable properties. It ages (changes chemical composition and/or physical property) slowly, typically retaining its fine optical and hardness-related properties for centuries. Glass is referred to as being a supercooled liquid, or a very viscous liquid. Indeed it is a slow-moving liquid as confirmed by sensitive measurements carried out in many laboratories. This information is corroborated by the observation that the old stained-glass windows adorning European cathedrals are a little thicker at the bottom of each small, individual piece than at the top of the piece. For our purposes, though, we should treat glass as a brittle solid that shatters on sharp impact.

Glass is mainly silica sand and is made by heating silica sand and powdered additives together in a specified manner and proportion, much as one bakes a cake. This recipe describes the items to be included, amounts, mixing procedure (including sequence), oven temperature, and heating time. The amounts and nature of additives all affect the physical properties of the final glass.

Typically, cullet—recycled or waste glass—is added (from 5% to 40%) along with the principal raw materials. The mixture is thoroughly mixed and then added to a furnace, where the mixture is heated to near 2725°F to form a viscous, syrup-like liquid. The size and nature of the furnace corresponds to the intended uses. For small, individual items the mixture may be heated in small clay (refractory) pots.

Most glass is melted in large (continuous) tanks that can melt 400 to 600 tons a day for the production of glass products. The process is continuous, with raw materials fed into one end as molten glass is removed from the other. Once the process (called a campaign) begins, it is continued indefinitely, night and day, often for several years until the demand is met or the furnace breaks down.

A typical window glass will contain 95–99% silica sand, with the remainder being soda ash (Na_2CO_3), limestone ($CaCO_3$), feldspar, borax, or boric acid, along with the appropriate coloring, decolorizing, and oxidizing agents.

Processing of glass includes shaping and retreatments of glass. Since shaping may create undesirable sites of amorphous structure, most glass objects are again heated to *near* their melting point. This process is called annealing. Since many materials tend to form more ordered structures when heated and recooled slowly, the effect of annealing is to "heal" these sites of major dissymmetry.

Four main methods are employed for shaping glass: drawing, pressing, casting, and blowing. Most flat glass is shaped by drawing a sheet of molten (heated so it can be shaped but not so it freely flows) glass onto a surface of molten tin. Since the glass literally floats on the tin, it is called "float glass." Its temperature is carefully controlled. The glass from a "float bath" typically has both of its sides quite smooth with a brilliant finish that requires no polishing.

Glass tubing is made by drawing molten glass around a rotating cylinder of appropriate shape and size. Air can be blown through the cylinder or cone to make glass tubing like that used in laboratories. Fiberglass is made by drawing molten glass through tiny holes, with the drawing process helping to align the tetrahedral clusters.

Pressing is accomplished by simply dropping a portion of molten glass into a form and then applying pressure to ensure that the glass takes the form of the mold. Lenses, glass blocks, baking dishes, and ashtrays are examples of objects commonly made by pressing.

The casting process involves filling molds with molten glass, much the same way cement and plaster of Paris molded objects are produced. Art glass pieces are typical examples of articles produced by casting.

Glassblowing is one of the oldest arts known to human culture. For art or tailor-made objects, the working and blowing of the glass are done by a skilled worker who blows into a pipe intruded into molten glass. The glass must be maintained at a temperature that permits working but not free flow, and the blowing must be at a rate and force to give the desired result. Mass-produced items are made using machine blowers, which often blow the glass to fit a mold, much like the blow molding of plastics.

In the following we give brief summaries describing different kinds of glass. The type and properties of glass can be readily varied by changing the relative amounts and nature of ingredients. Soda-lime glass is the most common of all glasses, accounting for about 90% of the glass made. Window glass, glass for bottles and other containers, glass for light bulbs, and many art glass objects are all soda-lime glass. Soda-lime glass typically contains 72% silica, 15% soda (sodium oxide, Na_2O), 9% lime (calcium oxide, CaO), and 4% minor ingredients. Its relatively low softening temperature and thermal shock resistance limit its high-temperature applications.

Vycor, or 96% silicon, glass is made using silicon and boron oxide. Initially, the alkali–borosilicate mixture is melted and shaped using conventional procedures. The article is then heat-treated, resulting in the formation of two separate phases—one high in alkali and boron oxide and the other phase containing 96% silica and 3% boron oxide. The alkali–boron oxide phase is soluble in strong acids and is leached away by immersion in hot acid. The remaining silica–boron oxide phase is quite porous. This porous glass is again heated to about 2200°F, resulting in a 14% shrinkage as the remaining portion fills the porous voids. The best variety is "crystal" clear and is called fused quartz. The 96% silica glasses are more stable and exhibit higher melting points (2725°F) than soda-lime glass. Crucibles, ultraviolet filters, range-burner plates, induction-furnace linings, optically clear filters and cells, and super-heat-resistant laboratory ware are often 96% silicon glass.

Borosilicate glass contains about 80% silica, 13% boric oxide, 4% alkali, and 2% alumina. It is more resistant to heat shock than most glasses because of its unusually small coefficient of thermal expansion (typically between 2 and 5×10^{-6} cm/cm °C; the value for soda-lime glass is about $8\text{–}9 \times 10^{-6}$ cm/cm °C). Borosilicate glass is better known by trade names such as Kimax and Pyrex. Bakeware and glass pipelines are often made of borosilicate glass.

Lead glasses (also called heavy glass) are made by replacing some or all the calcium oxide by lead oxide (PbO). Very high amounts of lead oxide can be incorporated—up to 80%. Lead glasses are more expensive than soda-lime glasses, but they are easier to melt and work with. They are more easily cut and engraved and give a product with high sparkle and luster (due to higher refractive indexes). Fine art glass and tableware are often made of lead glass.

Glazes are thin, transparent coatings (colored or colorless) fused on ceramic materials. Vitreous enamels are thin, normally opaque or semi-opaque, colored coatings fused on metals, glasses, or ceramic materials. Both are special glasses but may contain little silica. They are typically low melting and often are not easily mixed in with more traditional glasses.

There are also special glasses for specific applications. Laminated automotive safety glass is a sandwich form made by combining alternate layers of poly(vinyl butyral) (containing about 30% plasticizer) and soda-lime glass. This sticky organic polymer layer acts both to absorb sudden shocks (like hitting another car) and to hold broken pieces of the glass together. Bulletproof or, more correctly stated, bullet-resistant glass is a thicker, multilayer form of safety glass.

Tempered safety glass is a single piece of specially heat-treated glass often used for industrial glass doors, laboratory glass, lenses, and side and rear automotive windows. Because of the tempering process, the material is much stronger than normal soda-lime glass. Optical fibers are glass fibers that are coated with a highly reflective polymer coating such that light entering one end of the fiber is transmitted through the fiber (even around curves and corners, as when inserted into a person's stomach) to emerge from the other end with little loss of light energy. Optical fibers can also be made to transmit sound and serve as the basis for the transmission of television and telephone signals over great distances.

16.6 SILICON DIOXIDE (CRYSTALLINE)—QUARTZ

Silicon crystallizes in mainly three forms—quartz, tridymite, and cristobalite. After the feldspars, quartz is the most abundant mineral in the Earth's crust, being a major component of igneous rocks and one of the commonest sedimentary materials in the form of sandstone and sand. Quartz can occur as large (several pounds) single crystals but is normally present as much smaller components of many of the common materials around us. The structure of quartz is a three-dimensional network of 6-membered Si–O rings (three SiO_4 tetraheda) connected such that every six rings enclose a 12-membered Si–O ring (six SiO_4 tetrahedra).

16.7 ASBESTOS

Asbestos has been known and used for over 2000 years. Egyptians used asbestos cloth to prepare bodies for burial. The Romans called it *amiantus* and used it as a cremation cloth and for lamp wicks. Marco Polo described its use in the preparation of fire-resistant textiles in the thirteenth century. Asbestos is not a single mineral but rather a grouping of materials that yield soft, threadlike fibers. These materials are examples of two-dimensional sheet polymers containing two-dimensional silicate ($Si_4O_{10}^{4-}$) anions bound on either one or both sides by a layer of aluminum hydroxide ($Al(OH)_3$, gibbsite) or magnesium hydroxide ($(Mg(OH))_2$, brucite). The aluminum and magnesium are present as positively charged ions, that is, cations. These cations can also have a varying number of water molecules (waters of hydration) associated with them. The spacing between silicate layers varies with the nature of the cation and the amount of its hydration.

These fibrous silicates are generally divided into the serpentine and amphibole groups of minerals. Chrysotile is the most abundant and widely used type of asbestos and is a member of the serpentine mineral group. It consists of alternate sheets of magnesia ($Mg(OH)_2$) and silica with an overall empirical formula of $Mg_3Si_2O_5(OH)_4$. The chrysotile fibers are coiled and exist as bundles of hollow tubes called fibrils. Chrysotile is mined in Canada, the United States, South Africa, Zimbabwe, and the U.S.S.R., and accounts for 90% of the world asbestos market.

Asbestosis is a disease that blocks the lungs with thick, fibrous tissue, causing shortness of breath and swollen fingers and toes. Bronchogenic cancer, or cancer of the bronchial tubes, is prevalent among asbestos workers who also smoke cigarettes. Asbestos also causes mesothelioma, a fatal cancer of the lining of the abdominal chest. These diseases may lie dormant for many years after exposure.

The exact causes of these diseases are unknown but appear to be characteristic of particles (whether asbestos or other particulants) about 5 to 20 μm in length (about 2×10^{-4} in.), corresponding to the approximate sizes of the mucous membrane openings in the lungs. Furthermore, the sharpness of asbestos fibers intensifies their toxicity since these fibers actually cut the lung walls; even though the walls heal, the deposited asbestos, if not flushed from the lungs, will again cut the lung walls when the individual coughs, thus causing more scar tissue to form. The cycle continues until the lungs are no longer able to function properly.

16.8 POLYMERIC CARBON—DIAMOND

Just as carbon serves as the basic building element for organic materials, it is also a building block in the world of inorganic materials. Elemental carbon exists in two distinct crystalline forms: graphite and diamond.

Although diamonds may not truly be "a girl's best friend," they are an important industrial mineral because they are the hardest naturally occurring substance. Hardness is a relative term. In the world of rocks and minerals, hardness is measured on the basis of a 10-point scale (Table 16.2) that is dependent on the ability of a material to scratch a mark on a member of this scale. Each mineral on the scale scratches the ones with lower values. The hardness is more accurately measured today using an instrument called a sclerometer, which records the force required to scratch the material with a diamond.

Diamonds are almost pure carbon that occurs in a tetrahedral structure in which each carbon atom is at the center of a tetrahedra composed of four other carbon

Table 16.2 Mohs' scale of hardness[a]

1. Talc	6. Feldspar
2. Gypsum	7. Quartz
3. Calcite	8. Topaz
4. Fluorite	9. Corundum
5. Apatite	10. Diamond

[a]Hardness of other materials: fingernail, 2; copper penny, 3; knife blade, 5.5; window glass, 5.5.

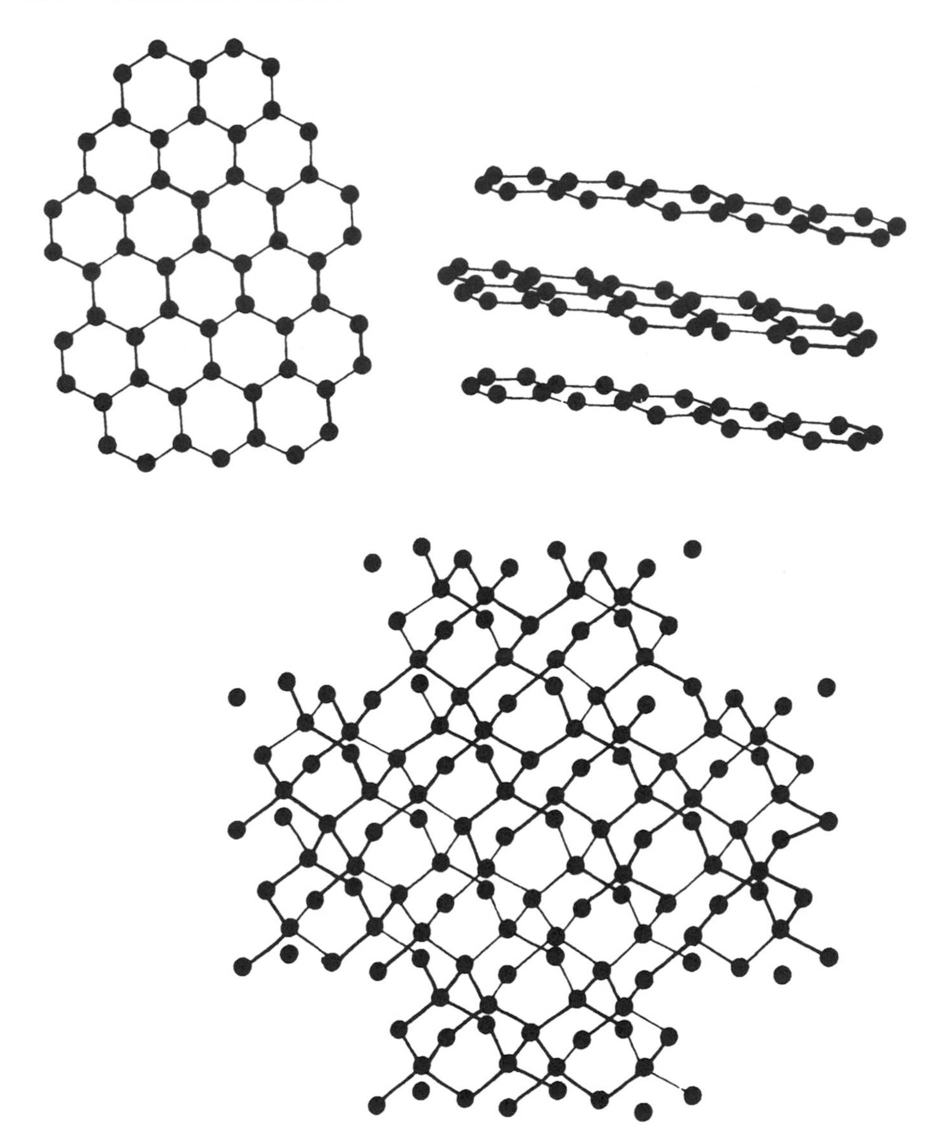

Figure 16.2. Structures of graphite (*top*) emphasizing the sheet nature (*left*) and emphasizing the layered nature (*right*), and diamond (*bottom*).

atoms (Figure 16.2). Natural diamonds were formed millions of years ago when concentrations of pure carbon were subjected by the earth's mantle to great pressure and heat. The majority of diamonds (nongems) are now man-made. The first synthetic diamonds were made by a team of scientists in 1955 at the General Electric Research Laboratory by compressing pure carbon under extreme pressure and heat. The majority of the synthetic diamonds are no larger than a grain of common sand. By 1970, G.E. was manufacturing diamonds of gem quality and size. These diamonds, available at a cost much higher than that of natural diamonds, are used for research. For instance, it was discovered that the addition of small amounts of boron to the diamonds causes the diamonds to become semiconductors. Today such doped diamonds are used to make transistors.

The major uses of diamonds are in industry as shaping agents to cut, grind, and bore (drill) holes in metals and ceramics. Most turntable cartrides employ a diamond needle to transmit differences in the record grooves into sound.

16.9 POLYMERIC CARBON—GRAPHITE

Although diamonds are the hardest naturally occurring material, the most common form of crystalline carbon is the softer graphite. Graphite occurs as sheets of hexagonally fused benzene rings (Figure 16.2) or "hexa-chicken wire." The bonds holding the fused hexagons together are traditional primary, covalent bonds. In contrast, the bonding between the sheets of fused hexagons consists of a weak overlapping of pi electrons and is considerably weaker than the bonding within the sheet. Thus, graphite exhibits many properties that are dependent on the angle with which they are measured. Graphite has some strength when measured along the sheet but very little strength if the layers are allowed to "slide over one another." Furthermore, the fused hexagons are situated such that the atoms in each layer lie opposite to the centers of the six-membered rings in the next layer. This arrangement further weakens the overlapping of pi electrons between layers such that the magnitude of layer-to-layer attraction is on the order of ordinary secondary forces. The "slipperiness" of the layers relative to one another accounts for graphite's ability to be a good lubricant.

As with diamonds and most other natural materials, graphite's discovery and initial use by humankind is lost in antiquity. Graphite was long confused with molybdenite (MoS_2), and at different times it was known as plumbago ("like lead"), crayon noir, silver lead, black lead, or carbo mineralis. In 1789 Werner first named it graphite, from the Greek *graphein*, meaning to write.

Although graphite has been extensively mined in China, Mexico, Austria, North and South Korea, Russian, and Madagascar, the majority of graphite used in the United States is manufactured from coke.

Graphite's properties led directly to its many uses in today's society. Because of its tendency to mark, hardened mixtures of clay and graphite are the "lead" in today's lead pencils. Graphite conducts electricity and is not easily burned; thus many industrial electrical contact points (electrodes) are made of graphite. Graphite is a good conductor of heat and is chemically quite inert even at high temperatures, and so many crucibles for melting metals are graphite-lined. Graphite has good stability to even strong acids, thus it is employed to coat acid tanks. It also is effective at slowing down neutrons, and thus composite bricks and rods (often called carbon rods) are employed in nuclear generators to regulate the progress of the nuclear reaction. Its "slipperiness" accounts for its use as a lubricant for clocks, door locks, and hand-held tools. Graphite is also the major starting material for the synthesis

Table 16.3 Comparison of properties of diamond and graphite

Property	Diamond	Graphite
Density (g/cc)	3.5	2.3
Electrical resistance	Increases with temperature	Decreases with temperature
Mohs' hardness	10	0.5–1.5
C–C bond length (Å)	1.54	1.42
Stability temperature (°F)	to ~4000	6300

of synthetic diamonds. Dry cells and some types of alkali storage batteries also employ graphite.

At ordinary pressures and temperatures, both graphite and diamonds are stable. At high temperatures (about 3250°F), a diamond is readily transformed to graphite. The reverse transformation of graphite to a diamond occurs only with application of great pressure and high temperatures. Thus, the naturally more stable form of crystalline carbon is not diamond but rather graphite. A comparison of some physical properties of diamond and graphite appear in Table 16.3.

Another somewhat distinct form of carbon is found in carbon black. Carbon black is normally impure and is often formed from incomplete burning of natural gas and other petroleum products. It contains chains of variable numbers of carbon atoms. Carbon black is used in tires, paints, drinking water purification, batteries, and inks.

16.10 POLYMERIC CARBON—NANOTUBES

Carbon nanotubes (CNTs) have probably been made in small amounts since the first fires reduced trees and organic material to ashes. It was not until recently, as part of the so-called nano revolution, that we first recognized the existence of these nanotubes. While there are many different nanotubes, we will focus on only carbon nanotubes.

Carbon nanotubes are carbon allotropes; that is, they are composed of the same carbon materials but in different structures, that have attracted much attention. Some have suggested that carbon nanotubes will be one of the most important twenty-first-century materials because of the exceptional properties and ready abundance of the feedstock, carbon.

CNTs are generally classified into two groups. Multiwalled carbon nanotubes (MWCNTs) are comprised of 2 to 30 or more concentric graphitic layers with diameters ranging from 10 to 50 nm with lengths that can exceed 10 μm. Single-walled carbon nanotubes (SWCNTs) have diameters ranging from 1.0 to 1.4 nm with lengths that can reach several micrometers.

An ideal CNT can be envisioned as a single sheet of fused hexagonal rings—that is, graphite—that has been rolled up, forming a seamless cylinder with each end "capped" with half of a fullerene molecule. Single-walled CNT can be thought

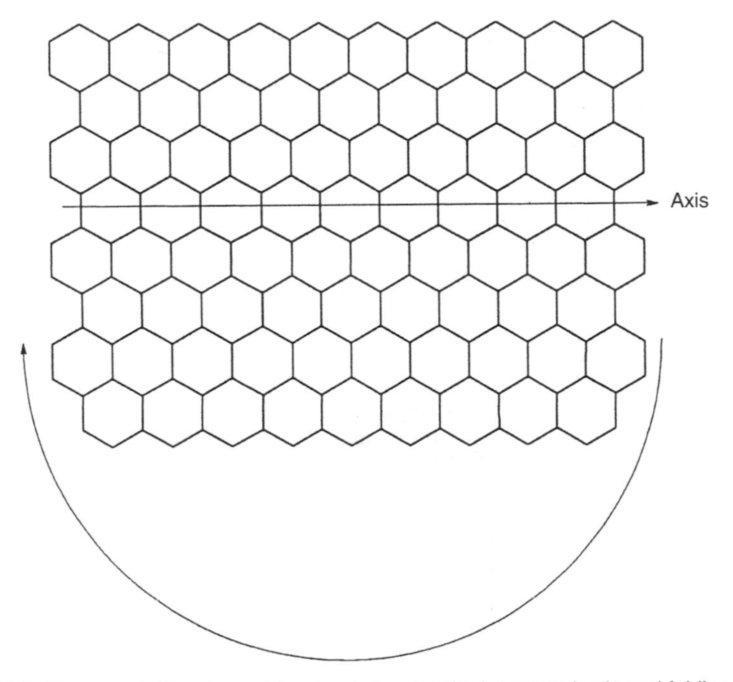

Figure 16.3. Representation of graphite sheet showing the hexagonal axis and folding to give a zigzag tube.

of as the fundamental cylindrical structure, with multiwalled CNTs simply being concentric tubes. They can also be conceived of as being the fundamental building block of ordered arrays of single-walled nanotubes called ropes.

Geometrically, there are three structures called armchair, zigzag, and helical. These three general structures can be conceptually described in terms of rolling a flat sheet (a.k.a. graphite sheet). Take a sheet of paper and draw a straight line in the middle from side to side. This will act as the axis line for a linear set of fused hexagonals. If the sheet is made into a cylinder so the line goes from one side to the other side of the open ends, it represents the armchair form. By comparison, formation of a cylinder so that the line runs through the middle of the cylinder gives a zigzag representation (Figure 16.3). Finally, formation of a cylinder so that the cylinder line is other than perpendicular or parallel gives a helical representation. Representations of these three major structural forms are given in Figure 16.4.

In real life, nothing is perfect, as is the case with CNTs. The defects are mainly inclusion of wrong-membered rings. Pentagonal defects—that is, the replacement of a hexagonal with a five-membered ring—results in a positive curvature causing the tube to curve inwards like a horseshoe. The closure of an open cyclindrical surface necessarily involves topological defects—often formation of pentagons. Heptagonal defects result in a negative curvature, with the lattice looking expanded around the defect.

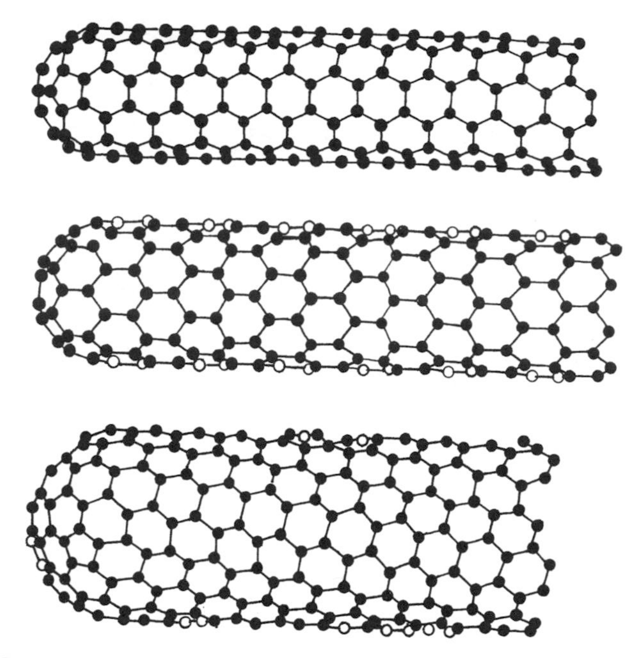

Figure 16.4. Representations of the three major structural forms of carbon nanotubes: Armchair (*top*), zigzag (*middle*), and helical (*bottom*).

One of the major reasons for the intense interest in CNT is their extreme and varied properties. Table 16.4 contains a comparison between single-walled CNTs and competitive materials/techniques.

Applications

Electrical. Nanotubes can be metallic or semiconducting, depending on their diameters and helical arrangement. Armchair tubes are metallic-like in that they are conductive. The other tubes (helical and zigzag) are typically semiconducting. Carbon nanotubes can in principle play the same role as silicon does in electronic circuits, but on a molecular scale where silicon and other standard semiconductors cease to work. Single carbon nanotube bundles have been used to construct elementary computing circuits.

Mechanical. Nanotubes have superior resilience and tensile strength. They can be bent and pressed over a large angle before they begin to ripple or buckle. Until the elastic limit is exceeded, the deformations are elastic with the deformation disappearing when the stress is removed. It is possible that buildings and bridges built from them may sway during an earthquake rather than fracturing and crumbling.

Field Emission. When stood on end and electrified, carbon nanotubes act as a lightning rod concentrating the electrical field at their tips. While a lightning rod

Table 16.4 Comparison of selected properties of SWCNTs with a typical competitive material

Property	Single-Walled Carbon Nanotubes	Comparison
Size	0.6–1.8 nm in diameter	Electron beam lithography can create lines 50 nm wide and a few nanometers thick
Density	1.33–1.40 g/cc	Aluminum has a density of 2.7 g/cc and titanium has a density of 4.5 g/cc
Tensile strength	~45 billion pascals	High-strength steel alloys break at about 2 billion pascals
Resilience	Can be bent at large angles and restraightened without damage	Metals and carbon fibers fracture at grain boundaries
Current-carrying capacity	Estimated at 1 billion A/cc	Copper wires burn out at about 1 million A/cc
Heat transmission	Predicted to be as high as 6000 W/m · K	Diamond transmits 3320 W/m · K
Temperature stability	Stable up to 2800°C in vacuum, 750°C in air	Metal wires in microchips melt at 600–1000°C

conducts an arc of electricity to a ground, a nanotube emits electrons from its tip at a rapid rate. Because the ends are so sharp, the nanotube emits electrons at lower voltages than do electrodes made from other materials, and their strength allows nanotubes to operate for longer periods without damage. Field emission is important in several industrial areas including lighting and displays.

Hydrogen and Ion Storage. While we can picture CNTs as being composed of hexagonal carbon atoms with lots of empty space between the carbons atoms, atoms "thrown" against them generally just bounce off. But in actuality, there is not empty space as shown in Figure 16.5. Even helium atoms do not readily penetrate the nanotubes. CNTs are then really membranes or fabrics that are one atom thick made of the strongest material that is also impenetrable. Thus, CNTs can be used for hydrogen storage in their hollow centers and release of the hydrogen can be controlled allowing the tubes to act as inexpensive and effective fuel cells.

Analytical Tools. Single-walled CNTs are being used as tips of scanning probe microscopes. Because of their strength, stability, and controllable and reproducible size, the tub probes allow better image fidelity and longer tip lifetimes. Nanotube-tipped atomic force microscopes can trace a strand of DNA and identify chemical markers that reveal DNA fine structure as well as detail the surface of most any other solid material.

Superconductors. Metallic CNTs are also high temperature superconductors.

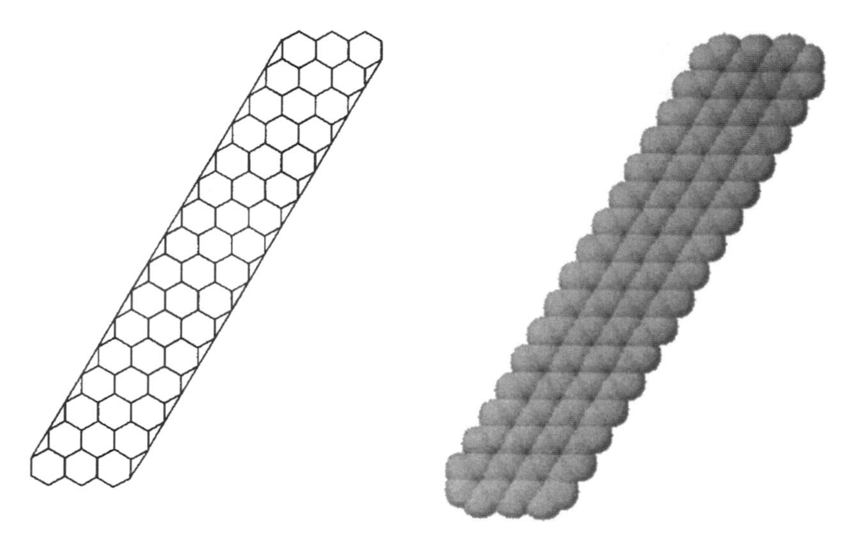

Figure 16.5. Sketal (*left*) and space-filling (*right*) models of a representative carbon-based nanotube.

16.11 CERAMICS

The term "ceramics" comes from the Greek word keramos which means "potter's clay" or "burnt stuff." While traditional ceramics were often based on natural clays, today's ceramics are largely synthetic materials. Depending on which ceramic and which definition is to be applied, ceramics have been described as inorganic ionic materials and as inorganic covalent (polymeric) materials. In truth, many ceramics contain both covalent and ionic bonds and thus can be considered as polymeric or not as polymeric materials. Many of the new ceramics such as the boron nitriles and the silicon carbides are polymeric without containing any ionic bonds.

Ceramics are typically brittle, strong, resistant to chemicals such as acids, bases, salts, and reducing agents, and high-melting. They are largely composed of carbon, oxygen, and nitrogen. They are largely made from silicates such as clay, feldspar, bauxite, and silica but now contain other materials such as borides, carbides, silicides, and nitrides.

They are generally made by two processes: sintering and fusing. In sintering, the starting material is reduced to a powder or granular form by a series of crushing, powdering, ball-milling, and so on. The ground pre-ceramic material is then sized, separated according to particle size, using different-sized screens.

Ceramic material is generally shaped by pressing it into a form or through extruding, molding, jiggering or slip-casting. Slip-casting uses a suspension of the pre-ceramic material in water. The mixture must be dilute enough to allow it to be poured. Deflocculants are often added to assist in maintaining the suspension. The "slip" is poured into a plaster of Paris mold that absorbs water, leaving the finished shape. The pre-ceramic material hardens next to the mold and surplus "slip" material poured off leaving a hollow item. At this point, the molded material is referred to as a "green body," which has little strength. Coffee pots and vases are formed using this technique.

In jiggering, machines press the pre-ceramic material into a rotating mold of desired shape. Dinnerware products are often made using jiggering.

Abrasives and insulators are formed from simply pressing the pre-ceramic material into a mold of desired shape. In extrusion, the pre-ceramic material is forced through an opening in a "shaping" tool. Bricks and drainpipes are formed using extrusion.

After the product has dried, it is heated or fired in a furnace or kiln. Modern ceramics generally require certain heating schedules that include the rate and duration of heating and under what conditions such as in the presence or absence of air. This is similar to procedures used to produce carbon fibers where the heating schedule is critical to the end products properties.

There are a number of other "non-oxygen" or non-oxide ceramic inorganic giant molecules including phosphonitric chlorides (PN backbone), boron nitriles (BN), aluminum nitriles (AIN), titanocarbosilanes (Si–Ti–C backbone), and silazanes (Si–C–N backbones).

Many ceramic products are coated with a glassy coating called a glaze. The glaze increases the resistance of the material to gas, and solvent permeability makes the surface smoother in art objects used for decoration.

16.12 HIGH-TEMPERATURE SUPERCONDUCTORS

As you have seen in this chapter, many inorganic compounds you find about you are giant molecules. In early 1986, George Bedorz and K. Alex Muller reported a startling discovery: A ceramic material containing lanthanum, barium, copper, and oxygen lost its resistance to electrical current at about −243°C. This was the first report of a so-called high-temperature superconductor. While this is very cold, this is much higher than temperatures needed to achieve superconductivity for other materials—thus the name high-temperature superconductor. What a superconductor does it to conduct electrical current without so-called line lose. As electricity is conducted to our homes, it loses some of its energy because of what is called electron friction; that is, when an electron travels through the wire, it creates friction, which creates heat and thereby results in loss in energy. Another thing that superconductors do is to act as though they suspend materials in mid-air. This is called the Meissner effect. It must be noted that below the superconducting temperature the material is superconducting, but above it the material is either nonconducting or only a little conducting.

The initial report was followed by a great flurry of activity, and in February 1987 the liquid nitrogen temperature barrier was broken with the report of a material that becomes superconducting at 180°C. Since liquid nitrogen is relatively safe, it does not explode but it does quickly freeze parts of the body that are exposed to it; also, it is and it was important to have a material that became superconductive when dipped in liquid nitrogen or exposed to an atmosphere made cold using liquid nitrogen. This new material had the formula $Y_1Ba_2Cu_3O_7$ and is referred to as the 123 superconductor material because of the subscripts 1, 2, 3.

The structure of the 123 compounds is related to an important class of minerals called perovskites. In general terms, the 123 materials contain polymeric layers of copper–oxygen atoms separated by ionic bonding by barium and yttrium atoms holding the copper–oxygen layers together. This theme of polymeric layers held together by ionic bonding to metal ions is found in may materials including many of the silicates.

High-temperature superconductors have been envisioned as bringing about "air travel" just above the ground, motors that go on forever, and so on; however, the actuality is something less, but nevertheless truly exciting with new applications in medicine, commerce, communication, electronics, and so on, occurring daily. Along with the nano revolution, the use of high-temperature superconductors only awaits your imagination.

16.13 VISCOELASTIC BEHAVIOR

James Wright, a researcher for General Electric, first discovered Silly Putty™ in 1943 during a search for synthetic rubber during World War II. Initially no practical use was found. By 1949 it found its way into a local toy store as a novelty item. Despite its good sales, the store dropped it after the first year. The next year Peter Hodson began packaging it in the now familiar plastic eggs, gave it the name "Silly Putty," and started the sales campaign. Today, it sells for about the same price that it did in 1950, at a rate of about 6 million eggs yearly equal to about 90 tons.

The first Silly Putty was made over 50 years ago from mixing together silicone oil with boric acid. That original formula has changed little, though colorants have been added, giving the material brighter colors and some the ability to "glow in the dark." It is a three-dimensional inorganic polymer. Silly Putty easily exhibits a wide variety of properties. Thus, when struck with a hammer it will shatter as a solid or when rapidly "snapping" it, it will act as a brittle material. The impact time is referred to here as an interaction time, and sharp blows such as hitting the Silly Puddy with a hammer will be called a short interaction time. Under a relatively long interaction time the molecular chains are able to yield and the material acts as a liquid. Under moderate interaction times there is segmental movement and the material acts as a rubber.

Giant molecules are called viscoelastic materials in that they behave under some conditions like liquids (the "visco" part of the name), under other conditions they behave as solids (the "elastic" part of the name), and finally under other conditions they behave as both liquids and solids. We can relate giant molecule behavior to temperature and interaction time and to chain movement.

The viscoelastic behavior can be divided into groupings with respect to molecular movement. At low temperatures and fast reaction times, chain segmental motion is restricted and changes involve mainly only bond bending and bond angle

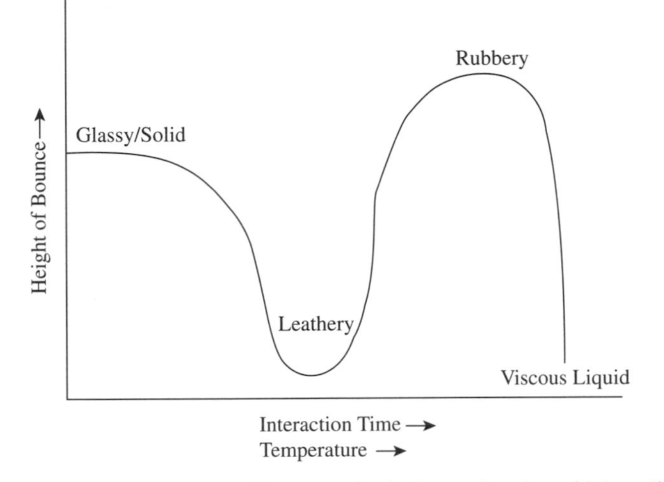

Figure 16.6. Idealized plot of height of bounce of a ball as a function of interaction time and temperature.

deformation. This region is called the viscous glass, Hookean (after Robert Hooke a seventh-century English physicist) glass, or simply glassy region where the material behaves as a glass like a glass window. For glass windows, flow occurs but it is very slow so that the segments of stained glass in the old European cathedrals are a little thicker at the bottom due to this slow flow.

As temperature increases and interaction time is a little shorter, side chains (e.g., for polyethylene it would be the short branches) and groupings on the main chain (e.g., for polypropylene it would be the methyl group) can move and some rotation of the main chain occurs. Here some dampening may occur. This region includes the leathery region of Figure 16.6.

Next, as temperature increases more and the interaction time decreases, there is enough energy (temperature) and/or time (interaction time) so that local segmental mobility occurs but not entire chain movement. This region is called the rubbery region. Finally, at high temperatures non-cross-linked materials melt, and for long interaction times (such as thousands of years for window glass) the material flows like a viscous liquid. Each of these regions is both temperature- and time-dependent, with this time dependency described in terms of the interaction time.

If we constructed a ball out of various materials and dropped it onto a solid floor, varying the temperature and interaction time and measuring the height of the bounce, we would get something like what appears in Figure 16.6. The particular real behavior will vary as to the material we made the ball out of, but this behavior is general for giant molecules. Silly Putty easily illustrates three of these behaviors. When struck rapidly, it shatters as a glassly solid; when dropped at a moderate rate it bounces well, acting as a rubber ball; and when allowed to lay in its container, it flows to occupy the container contour.

Other non-giant molecules also act in a similar manner. Thus, if we were to jump off a tall bridge into a river below, unless we were very lucky, it would be similar to hitting cement because the interaction time is too fast to allow the water molecules to give way and only the flexing of the H–O bond occurs. Water is acting like a solid here. Also, at low temperatures water is a solid. By comparison, if we were to jump into the neighborhood pool the interaction time is slow enough as to allow the individual water molecules to move past one another and the water acts as a liquid. At high temperatures, like room temperature, water is a liquid. Thus, water shows the same general behavior as giant molecules.

GLOSSARY

Alumina: Al_2O_3.

Anisotropic: Dependent on direction; directionally dependent.

Annealing: Subjecting materials to heat near their melting point.

Asbestos: Grouping of silica-intensive materials containing aluminum and magnesium that give soft, threadlike fibers.

Asbestosis: Disease that blocks lungs with thick, fibrous tissue.

Borosilicate glass: Relatively heat-shock-resistant glass with a small coefficient of thermal expansion (Kimax and Pyrex).

Calcium-aluminate cement: Cement with more alumina than portland cement.

Chrysotile: The most abundant and widely used type of asbestos.

Colored glass (stained glass): Glass containing coloring agents such as metal salts and oxides.

Concrete: Combination of cement, water, and filler material such as rocks and sand.

Diamond: Polymeric carbon in which the carbon atoms are at the centers of tetrahedra composed of four other carbon atoms; the hardest known natural material.

Feldspar: A derivative of silica in which one-half to one-quarter of the silicon atoms are replaced by aluminum atoms.

Fiberglass: Fibers of drawn glass.

Float glass: Glass made by cooling sheets of molten glass in a tank of molten tin; most common window glass is of this type.

Glass: An inorganic product of fusion that has been cooled to a rigid condition without crystallization; most glasses are based on amorphous SiO_2.

Glaze: Thin, transparent coatings fused on ceramic materials.

Graphite: Polymeric carbon consisting of sheets of hexagonally fused rings in which the sheets are held together by weak overlapping pi electron orbitals; anisotropic in behavior.

Gypsum: $CaSO_2 \cdot 2H_2O$; serves as the basis of plaster of Paris, Martin's cement, Keen's cement, and Parisian cement; shrinks very little on hardening; rapid drying.

Hole: Unoccupied site in a material.

Inorganic polymer: A polymer containing no organic portions.

Kaolinite: An important type of asbestos clay.

Lead glass (heavy glass): Glass in which some or all the calcium oxide is replaced by lead oxide.

Lime: $CaCO_3$; derived from oyster shells, chalk, and marl.

Magnesia cement: Cement composed mainly of magnesium oxide; rapid hardening.

Optical fiber: Glass fibers coated with highly reflective polymer coatings; allows light entering one end of the fiber to pass through to the other end with little loss of energy.

Piezoelectric material: Materials that develop net electronic charges when pressure is applied; sliced quartz is piezoelectric.

Portland cement: A major three-dimensional inorganic polymer construction material consisting of calcium silicates, lime, and alumina.

Precast concrete: Portland concrete cast and hardened prior to being taken to the site of use.

Prestressed concrete: Portland concrete cast around steel cables stretched by jacks.

Quartz: Crystalline forms of silicon dioxide; basic material of many sands, soils, and rocks.

Reinforced concrete: Portland concrete cast around steel rods or bars.

Safety glass: Laminated glass; sandwich form containing alternate layers of poly(vinyl butyral) and soda-lime glass.

Sandstone: Granular quartz.

Silicon glass: Glass made by fusing pure quartz crystals or glass sand; high melting.

Soda: Na_2O.

Soda ash: Na_2CO_3.

Soda-lime glass: Most common glass; based on silica, soda, and lime.

Tempered safety glass: A single piece of specially heat-treated glass.

Tempering: A process of rapidly cooling glass, resulting in an amorphous glass that is weaker but less brittle.

Vitreous enamel: Thin, normally somewhat opaque-colored inorganic coatings fused on metals.

Vycor: 96% silicon glass; made from silicon and boron oxide; best variety is called fused quartz.

REVIEW QUESTIONS

1. Compare window glass with organic thermoplastics.
2. Why is portland cement an attractive basic building material?
3. Name five important natural inorganic giant molecules.
4. Is window glass a thermoset polymer?
5. Why are specialty cements and concretes necessary?
6. What is meant by the comment that "glass is a supercooled liquid"?
7. Is quartz a thermoset polymer?
8. Why are specialty glasses important in today's society?
9. Where does sand come from?
10. Which is the most brittle: window glass, quartz, fibrous glass, asbestos fiber, polypropylene?
11. Compare the structures of graphite and diamond.

BIBLIOGRAPHY

Archer, R. (2001). *Inorganic and Organometallic Polymers*, Wiley, New York.

Brinker, C., and Scherer, D. (1990). *The Physics and Chemistry of Sol–Gel Processing*, Academic Press, Orlando, FL.

Brook, M. (1999). *Silicon in Organic, Organometallic, and Polymer Chemistry*, Wiley, New York.

Bruce, D. W., and O'Hare, D. (1997). *Inorganic Materials*, 2nd ed., Wiley, New York.

Bunsell, A., and Berger, M-H. (1999). *Fine Ceramic Fibers*, Marcel Dekker, New York.

Bye, G. (1999). *Portland Cement: Composition, Production and Properties*, Telford, London, UK.

Carraher, C., Sheats, J., and Pittman, C. (1978). *Organometallic Polymers*, Academic Press, NY.

Carraher, C., Sheats, J., and Pittman, C. (1981). *Metallo Organic Polymers*, Mer, Moscow, USSR.

Carraher, C., Sheats, J., Pittman, C. (1982). *Advances In Organometallic and Inorganic Polymer Science*, Marcel Dekker, New York.

Cunha, A., and Fakirov, S. (2000). *Structural Development During Processing*, Kluwer, New York.

Donnet, J. (1998). *Carbon Fibers*, Marcel Dekker, New York.

Jones, R., Andeo, W., and Chojnowski, J. (2000). *Silicon-Containing Polymers*, Kluwer, New York.

Mittal, K. L. (2000). *Silanes and Other Coupling Agents*, VSP, Leiden, Netherlands.

Mitura, S. (2000). *Nanomaterials*, Elsevier, New York.

Nanotechnology (2000). C & EN, Oct 16.

Ohama, T., Kawakami, M., and Fukuzawa, K. (1997): *Polymers in Concrete*, Routledge, New York.

Pittman, C., Carraher, C., Zeldin, M., Culbertson, B., and Sheats, J. (1996). *Metal-Containing Polymeric Materials*, Plenum, New York.

Ramachandran, V. S., and Beaudoin, J. J. (2001). *Handbook of Analytical Techniques in Concrete*, ChemTec, Toronto.

Schropp, R., and Zeman, M. (1998). *Amorphous and Microcrystalline Silicon Solar Cells*, Kluwer, Hingham, MA.

Schubert, U., and Husing, N. (2000). *Inorganic Materials*, Wiley, New York.

Sheats, J., Carraher, C., and Pittman, C. (1985). *Metal-Containing Polymeric Systems*, Plenum, New York.

Smith, R. (2001). Modern Drug Delivery, *Nanotechnology*, April, 33.

Tsuchida, E. (2000). *Macromolecular–Metal Complexes*, Wiley, New York.

Weller, M. (1995). *Inorganic Materials Cehmistry*, Oxford University Press, Cary, NC.

Wesche, R. (1999). *High-Temperature Superconductors: Materials, Properties and Applications*, Kluwer, Hingham, MA.

ANSWERS TO REVIEW QUESTIONS

1. Window glass is held together by directional covalent bonds; acts physically like many organic polymers that are above their glass transition temperature in being very viscous; acts like a solid on rapid impact but like a liquid on a much elongated time scale.
2. It is readily available on a large scale; inexpensive; relatively nontoxic; stands up well to most natural elements such as rain, cold, heat, and mild acids and bases; light and strong.
3. Quartz, asbestos, alumina, graphite, and diamond.
4. No.
5. To perform for a wide variety of conditions and applications.
6. It flows like a liquid, but the flow rate is very low.
7. Yes.
8. A wide variety of applications require materials that may possess glasslike properties; glasses have generally good resistance to natural elements, are easily shaped, polished, and cut, and many transmit light and can be colored.
9. From breakup of larger silicon dioxide-intense rocks.
10. Quartz.
11. Both are made from carbon, but graphite is a sheet and diamond is a three-dimensional tetrahedra.

17

SPECIALTY POLYMERS

Giant Molecules: Essential Materials for Everyday Living and Problem Solving, Second Edition,
by Charles E. Carraher, Jr.
ISBN 0-471-27399-6

17.1 WATER-SOLUBLE POLYMERS

Because of the nonpolar structure of some polymers, such as polystyrene and polyethylene, they repel water. Some polymers are polar, and a few are polar enough to be soluble in water. These polymers are said to be hydrophilic, that is, water-loving.

Most of us are familiar with water-soluble starch, which is used both as a food and as a textile and paper coating. However, few of us are aware of water-soluble derivatives of cellulose, such as hydroxyethylcellulose, methylcellulose, and carboxymethylcellulose, which are used to increase the viscosity (resistance to flow) of water and to provide water-soluble coatings. These and other water-soluble polymers, such as guar gum, have been used to enhance the recovery of oil from wells and thus increase oil production.

Aqueous solutions of other water-soluble polymers, such as polyacrylamide,

$$\left[CH_2-\underset{|}{\overset{COONH_2}{CH}} \right]_n$$

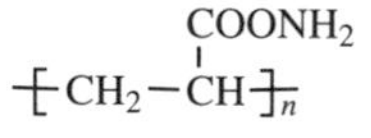

the sodium salt of polyacrylic acid,

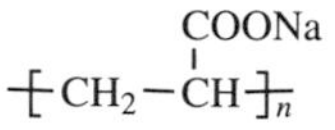

$$\left[CH_2-\overset{COONa}{\underset{}{CH}} \right]_n$$

and copolymers of vinyl acetate and maleic anhydride

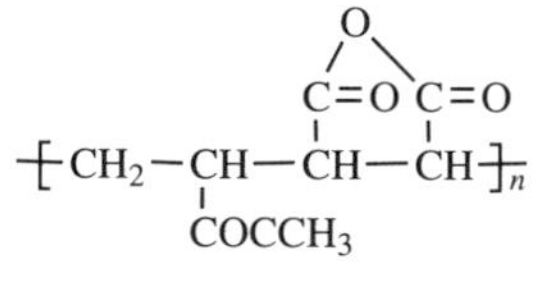

$$\left[CH_2-\underset{COCCH_3}{CH}-\overset{C=O}{CH}-\overset{C=O}{CH} \right]_n \quad \text{(the two C=O groups bridged by O)}$$

have also been used for enhanced oil recovery. Maleic anhydride copolymers must be hydrolyzed to produce a water-soluble salt.

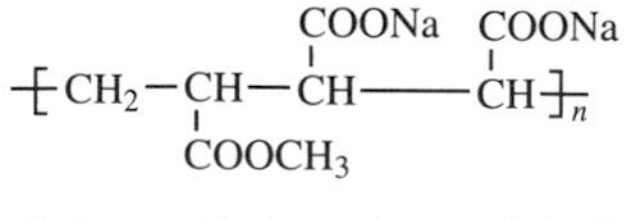

$$\left[CH_2-\underset{COOCH_3}{CH}-\overset{COONa}{CH}-\overset{COONa}{CH} \right]_n$$

Sodium salt of copolymer of vinyl acetate and maleic acid

Water-soluble polymers such as polyacrylamide are used as flocculants in papermaking. Salts of polyacrylic acid, polyacrylamide, polyvinyl alcohol, and polyethylene oxide are also used as flocculants to improve the efficiency of municipal waste treatment. Polyethylene oxide, $(CH_2CH_2-O)_n$, has been added to water to increase the flow when it is used to extinguish fires. Water-soluble polymers such as hydrolyzed maleic anhydride copolymers are also used to increase the utility of floor polishes.

17.2 OIL-SOLUBLE POLYMERS

Nonpolar polymers are used to assist the flow of crude oil in pipelines, and they are also used to control the viscosity of lubricating oils. The 10–40 oil that is used in automobiles contains a small amount of an oil-soluble polymer, such as polyisobutylene, polybutyl methacrylate, or polycyclohexylstyrene. Since these polymers are not particularly soluble in cold oil, their chains tend to form tight coils, which have little effect on the viscosity of the oil. However, these coils extend when the oil is heated, and these extended chains increase the viscosity of the oil so that it is more viscous at engine operating temperatures than at room temperature. This increased viscosity helps the oil reduce the friction between the moving parts in the engine.

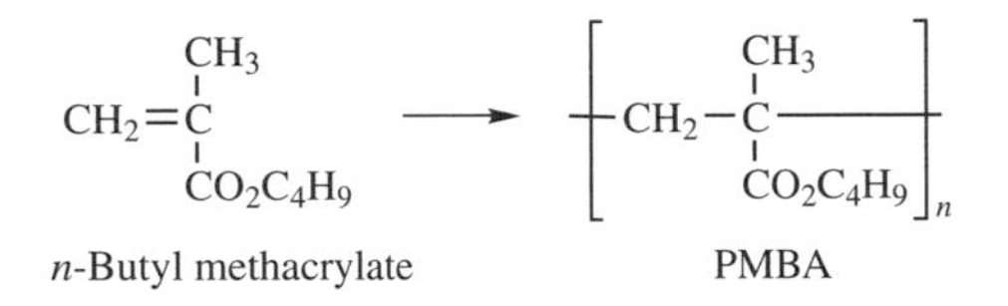

17.3 POLYMERIC FOAMS

The insulation properties of nonconductive polymers may be enhanced by the addition of hollow glass or polymer spheres to produce syntactic foams. Foams may also be produced by the addition of volatile liquids or gases to molten polymers. Thus, foamed polystyrene (PS, Styrofoam) is produced by the extrusion of a mixture of polystyrene and pentane.

The propellant pentane may also be incorporated into PS beads before molding. Coffee cups produced by molding these expandable beams have been widely used by fast-food establishments, but their use is being curtailed because of the alleged adverse environmental impact of these chemicals.

Most rigid foam is also made from polyurethane derived from methylenediphenyl isocyanate (MDI) and difunctional polyether polyols. The rigid foams are typically blown using fluorocarbons as the gas to produce a closed cell foam with outstanding insulation properties. Major applications of rigid foam include use in building and construction (60%), cryogenic (low-temperature) transport of materials, furniture, packaging, and molded structural parts such as solar panels.

Bedding and furniture (50%) account for the major end use of flexible polyurethane, PUR, foams. Most furniture and pillow filling (30%) is PUR foam, as is much of the automotive seating material. PUR foam is also employed as carpet underlay and in the packaging of breakable items.

17.4 POLYMER CONCRETE

The use of polymers for structural applications has usually been restricted because of the relatively high costs of these polymers. However, their lower specific gravity, good tensile strength, and the increasing cost of competitive products, such as hydraulic cement, have sparked considerable interest in polymer concrete.

The original polymer concrete was a filled resole phenolic resin that was converted to a hard polymer *in situ* by the presence of a compatible acid, such as *p*-toluene sulfonic acid,

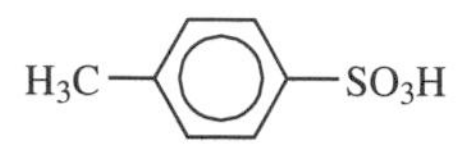

The phenolic polymer concrete has been displaced to some extent by a more alkaline-resistant cement based on furfuryl alcohol,

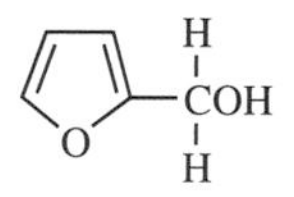

epoxy, and polyester resins.

The most widely used polymer cement is based on a filled polyester prepolymer and is called "cultured marble." It is used for casting bathroom sinks and tubs. Because of its high strength, the epoxy concrete is competitive with portland cement and is used for patching roads and surfacing bridges. Unlike patching with portland cement, where appreciable bonding does not occur between the patched portion and the newly deposited patch material, considerable bonding does occur between the polymer cement and the old section, thus making such patches more permanent.

17.5 XEROGRAPHY

Photoconductive materials form the basis for xerography. The two major photoconductive materials are selenium and polyvinylcarbazole, which are applied to a backing. The surface is made light-sensitive by electrostatic charging in darkness. The surface is then exposed to the desired image, which in turn is developed by application of a toner. The image is then transferred to a paper and set or fixed by heat, and the xerography paper emerges from inside the xerography machine. The drum is then automatically cleared and ready to accept another image. Improvements are continuing, with the giant molecule polyvinylcarbazole playing a major role.

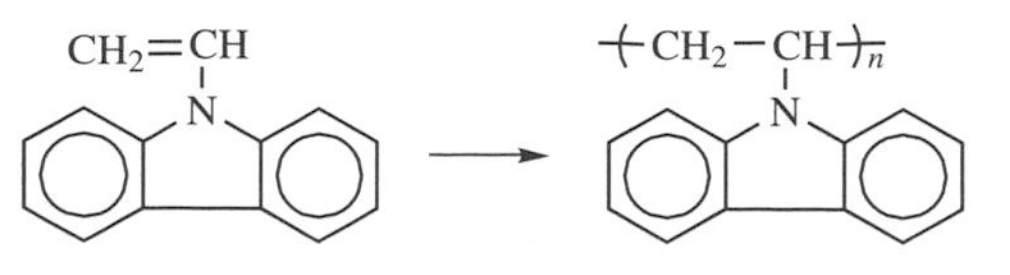

17.6 PIEZOELECTRIC MATERIALS

A slice of quartz develops a net positive charge on one side and a negative charge on the other side when pressure is applied. The same effect is found when pressure is applied by means of an alternating electric field. Such crystals are employed for quartz watches and clocks, for TVs, radios, hearing aids, ignition of flash bulbs, and in telephone receivers. Rochelle salt and tourmaline also exhibit piezoelectric properties.

Several organic polymers are also effective piezoelectric materials. The most widely used is polyvinylidene fluoride, which is employed in loudspeakers, fire and burglar alarm systems, microphones, and earphones. Nylon 11 is also piezoelectric and can be aligned when placed in a strong electrostatic field, giving films used in infrared-sensitive TV cameras, in underwater detection devices, and as part of electronic devices because nylon 11 films can be overlaid with printed circuits.

$$\text{-}[CH_2CF_2]_n\text{-}$$

Polyvinylidene fluoride

$$H_2N(CH_2)_{10}CO_2H \xrightarrow{\text{heat}} \left[NH(CH_2)_{10}\overset{O}{\overset{\|}{C}} \right]_n + H_2O$$

11-Aminoundecanoic acid   Nylon 11

17.7 CONDUCTIVE AND SEMICONDUCTIVE MATERIALS

In the past, most conductive materials have been composed of metals such as copper. Whereas most giant molecules have been employed because of their lack of electrical conductivity, today some are being considered for their conductive

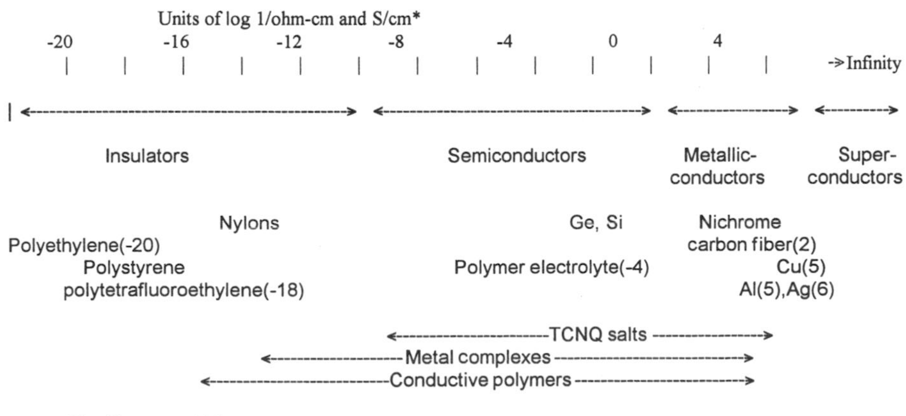

Figure 17.1. Electrical conductivity for various materials.

properties. Advantages of polymeric materials are the possibilities of flexible, finer, noncorrosive, inexpensive, and more easily processed materials. This search has increased to include, for some applications, nano-sized fibrils and tubes. The conductivity for general materials is noted in Figure 17.1.

A number of polymers have been found that, when partially oxidized or reduced (doped), form electrically conducting materials (Table 17.1). Almost all of these materials possess what is referred to as "whole-chain" resonance; that is, electrons can easily move from one end of the chain to the other end. Doping causes an electrical imbalance that allows electrons to flow when an electrical potential is applied. The "band gap" is the energy needed to promote an electron from the valence band to the empty energy or conductive band. Metals have zero band gaps, while insulators like polyethylene have large band gaps, meaning that a lot of energy is needed to promote an electron to an empty band.

One of the major polymers being investigated as an organic conductive material is polyacetylene. Polyacetylene, in order to conduct, must be doped with oxidizing agents such as the halides or arsenic pentafluoride or reducing agents such as metallic sodium. This chemical doping transforms the polyacetylene from an insulator or semiconductor to a conductor. Ordinary polyacetylene is composed of small fibers, fibrils, that are randomly oriented. Conductivity is decreased because of the nonconductive contacts between the various random fibrils. Two approaches are taken to align the polyacetylene fibrils. The first approach is to use a liquid crystal solvent for the acetylene polymerization and to form the polymer under external perturbation. The second approach is to mechanically stretch the polyacetylene material in order to align the fibrils.

Polyacetylene has good inert atmosphere thermal stability but easily oxides in the presence of air. The doped samples are even more sensitive to air. Polyacetylene films have a lustrous silvery appearance and have some flexibility.

Other conductive polymers (Table 17.1) include polyanilines, polypyrrole, polythiophene, poly(p-phenylene), and poly(sulfur nitride) (also known as polythiazyl) ($-(SN)-$).

Table 17.1 Structures of some conductive polymers

Name	Structure	σ (reciprocal ohm, cm)
Poly(p-phenylene)	$-[C_6H_4]_n-$	5000
Polythiazyl	$-[S=N]_n-$	3700
Polyacetylene	$-[CH=CH]_n-$	3000
Poly(phenylene sulfide) (PPS)	$-[C_6H_4-S]_n-$	1.0

While the amount of electricity that can be conducted by polymer films and "wires" is limited, on a weight basis the conductivity is comparable to that of copper. These polymeric conductors are lighter, some more flexible, and they can be "laid down" in "wires" that approach being one atom thick. They are being used as cathodes and solid electrolytes in batteries, and potential applications include use in fuel cells, "smart" windows, nonlinear optical materials, light-emitting diodes, conductive coatings, sensors, electronic displays, and electromagnetic shielding.

17.8 SILICON CHIPS

The silicon chip is essential for transmission, but the photoresist polymers supply the message. Thus, there are many sophisticated polymers that are coated on the chip and then selectively degraded to produce an effective chip.

Since poly(methyl methacrylate) (PMMA) is easy to thermally degrade, it was one of the first coatings used on the chip. However, PMMA has been displaced by other polymers designed specifically for this end use.

17.9 ION-EXCHANGE RESINS AND ANCHORED CATALYSTS

Just as the calcium ion forms an insoluble compound through reaction with the carbonate ion (namely, calcium carbonate), so also will it form a complex with a carboxyl group attached to a polymer. This concept forms the basis for many analysis, separation, and concentration techniques. Many of these are based on benzene and divinylbenzene, silicon dioxide, and dextran-based resins. These resins are almost always cross-linked and then functionalized; that is, functional groups are added onto the surface of the resin beads. Functional groups such as carboxylic acid and sulfonic acid attract and retain positively charged ions and are appropriately called cation-exchange resins. Functional groups such as amines attract anions and are called anion-exchange resins (Table 17.2).

Applications of these resins are extremely varied, including the purification of sugar, identification of drugs and biomacromolecules, concentration of uranium, and use as therapeutic agents for the control of bile acid and gastric acidity. In the latter use, a solid polyamide (Colestid) is diluted to be taken with orange juice to help in the body's removal of bile acids. Removal of bile acids causes the body to produce more bile acid from cholesterol, thus effectively reducing the cholesterol level. Recent catalysts and stereoregulating groups have been attached to the backbones of polymeric materials.

The Merrifield protein synthesis (using chloromethylated polystyrene) makes use of ion-exchange resins as do many of our industrial and home water purifiers. Water containing calcium, iron, or magnesium ions is called hard water. These ions

Table 17.2 Active functional groups on ion-exchange resins

Active Group	Structure
Cation-exchange resins	
Sulfonic acid	C_6H_4–SO_3H
Carboxylic acid	$-CH_2CH(COOH)CH_2-$
Anion-exchange resins	
Quaternary ammonium salt	C_6H_4–$[CH_2N(CH_3)_3]^+Cl^-$
Secondary amine	C_6H_4–CH_2NHR
Tertiary amine	C_6H_4–CH_2NR_2

are normally from natural sources; for example, calcium ions are generally derived from the passage of water over and through limestone ($CaCO_3$). Hardness in water is objectionable since the metal ions generally form insoluble salts when the water is cooled or when soap is added. The precipitate (formation of insoluble materials) may be deposited in the pipes and water heater, forming boiler scale. The ions also lower the efficiency of added soaps, and the precipitate forms "curds" that are often "captured" in the laundery. Furthermore, the precipitate forms bathtub rings, and it adheres to us as we are taking a shower or bath.

Most home water softeners are based on ion-exchange resins. The first ion-exchange materials used in softening water were naturally occurring polymeric aluminum silicates called zeolites ($NaAlSi_2O_6$), which exchanged their ions for calcium, iron, and magnesium ions. Synthetic zeolites were later developed. Today most ion-exchange materials are based on styrene and divinylbenzene (vinylstyrene) resins, which are then sulfonated. When the resin system is ready for use, sodium ions generated from rock salt (sodium chloride) are passed through the resin "bed," replacing the hydrogen ions (protons). Then water to be used for drinking, cooking, washing, and bathing is passed through the resin bed. The sulfonate functional groups have a greater affinity for calcium, iron, and magnesium ions; thus these displace the sodium ions, resulting in water that has few, if any, "hard ions," but with a few more sodium ions. Eventually, the sulfonate sites on the resin become filled and the resin bed must be recharged by adding large amounts of dissolved sodium ions from sodium chloride, which displace the more tightly bound, but overwhelmingly outnumbered, "hard ions"; after the system is flushed free of these "hard ions," the resin bed is again ready to deliver "soft water" for our use (Figure 17.2).

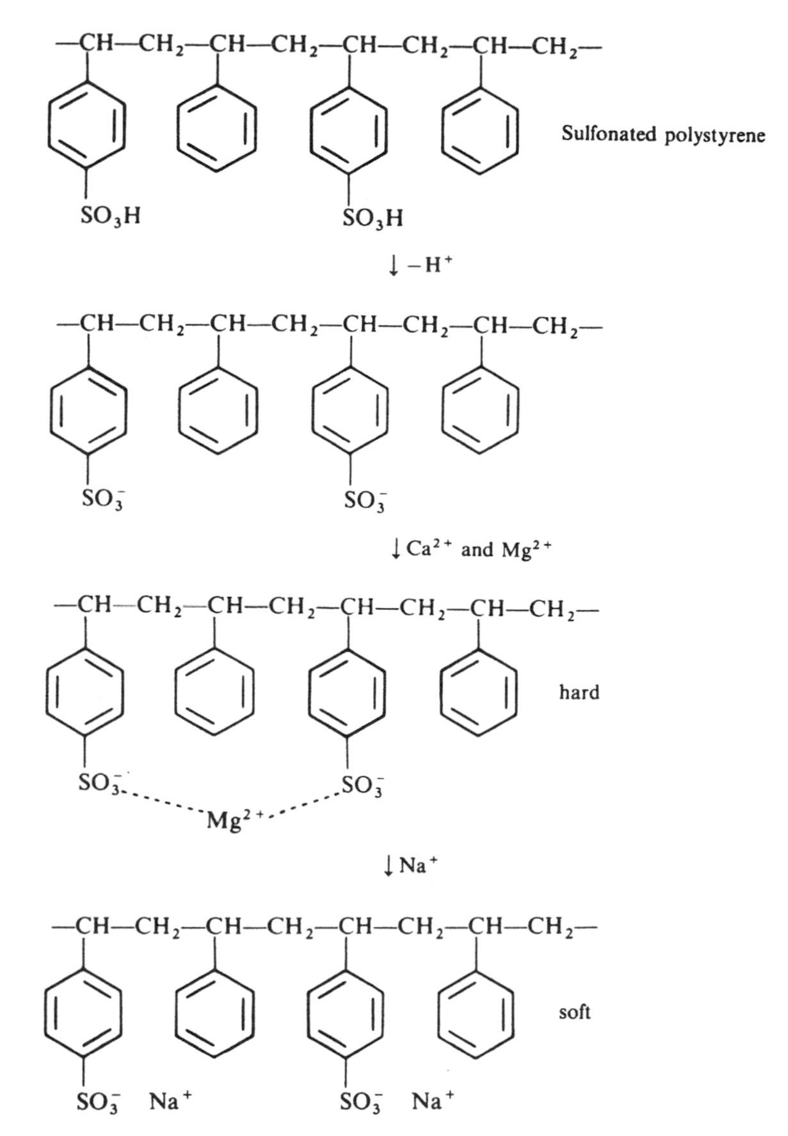

Figure 17.2. The ion-exchange cycle. *From top to bottom*: Uncharged resin, charged resin, resin with complexed "hard ions," and recharged resin. After initial use, the usual cycle will involve the two bottom steps.

17.10 PHOTOACTIVE MATERIALS

Photo-cross-linking and photopolymerization have been employed to produce permanent images. For DuPont's Dyoryl and Lydel systems, soluble linear giant molecules are made insoluble through cross-linking by application of ultraviolet radiation.

Three major approaches have been studied in developing photoactive materials for imaging. An Eastman–Kodak photo-cross-linkable system employs polyvinyl cinnamates. The photoactive group in this case is the cinnamate unit, which is attached to polyvinyl alcohol and is thus located as a pendant group. Another approach involves the photoactive group as part of the main chain. A third approach employs a difunctional reactant such as diazide, which, when activated, acts as a cross-linking agent. These photosensitive systems are called negative imaging systems, similar to the more traditional photographic systems. Since the exposed areas are cross-linked, these insoluble exposed areas are left after the unexposed areas are washed away, creating negative images. Positive images can be obtained through use of special solvent systems that render the cross-linked portions soluble through disruption of the cross-linking.

These systems are much superior to the classical silver halide systems in which the image sharpness was limited by the size of the silver halide grains, whereas the photocross-linking systems are limited only by the size of the individual giant molecules. Today, microcircuits and most of our newspapers are printed using photopolymers.

$$\underset{\beta\text{-Vinyloxyethyl cinnamate}}{CH_2{=}\underset{|}{C}H{-}OCH_2CH_2O\overset{O}{\overset{\|}{C}}CH{=}CH{-}C_6H_5} \xrightarrow[\text{toluene}]{BF_3\ \text{etherate}} \underset{\text{Poly }\beta\text{-Vinyloxyethyl cinnamate}}{{-}\!\!\left[CH_2\underset{|}{C}H\right]\!\!{-}_n \; OCH_2CH_2O\overset{O}{\overset{\|}{C}}CH{=}CH{-}C_6H_5}$$

17.11 CONTROLLED-RELEASE POLYMERS

Numerous examples of controlled-release systems are known. For many cases, the to-be-delivered group is incorporated within the polymer chain or as a side group. The controlled-release polymer is then allowed to work—in a book as a scratch and smell, in the body to treat drug addiction, on a flea collar to ward off fleas, in waterways to control undesirable plants, in chewing gum to prolong the flavor, on an adhesive bandage to extend the antiseptic period, in cattle to promote weight gain, or in a diabetic patient to maintain a balanced level of insulin.

17.12 DENDRITES

We are continuing to recognize that polymer shape is important in determining material property. Today, the polymer scientist can design giant molecules with various shapes such as stars, fans, and so on. Many of these can be included under the general grouping of dendrites. These molecules can act as "spacers," as "ball-bearings," and as building blocks for other structures. Usually, they are either wholly organic or they may contain metal atoms. They may or may not be copolymers, depending on the particular synthetic route employed in their synthesis.

Dendrites are highly branched, usually curved, structures. The name comes from the Greek word "dendron," meaning tree. Another term often associated with these structures is "dendrimers," describing the oligomeric nature of many dendrites. Because of the structure, dendrites can contain many terminal functional groups for each molecule that can be further reacted. Also, most dendrites contain "lots" of unoccupied space that can be used to "carry" drugs, fragrances, adhesives, diagnostic molecules, cosmetics, catalysts, herbicides, and other molecules.

The dendrite structure is determined largely by the "functionality" of the reactants. The dendrite pictured in Figure 17.3. can be considered as being derived from a tetra-functional monomer formed from the reaction of 1,4-diaminobutane and methyl acrylate. The resulting dendrimer has terminal acid groups that can be further reacted extending the dendrimer. In the case of Figure 17.3 the size of the dendrite is increased with each successive round of reaction between the amine and methyl acrylate. The number of "arms" doubles for each successive round. The resulting molecule is circular with some three-dimensional structure.

Numerous approaches have been taken in the synthesis of dendrites or dendrimers. These approaches can be divided into two groupings. In divergent dendrimer growth, growth occurs outward from an inner core molecule (Figure 17.3). In convergent dendrimer growth, various parts of the ultimate dendrimer are separately synthesized and then they are brought together to form the final dendrimer.

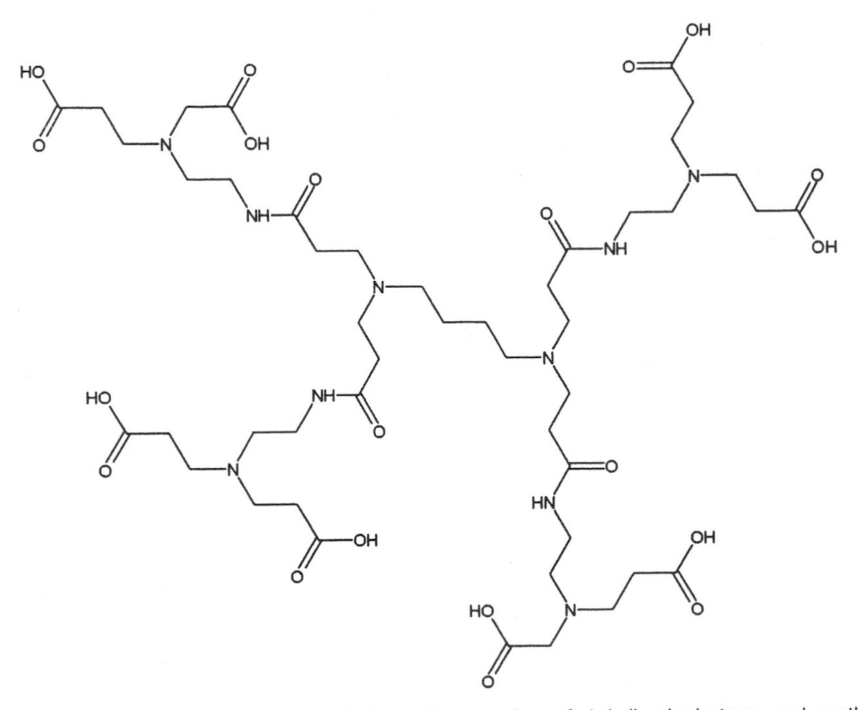

Figure 17.3. Initial dendrimer growth from the reaction of 1,4-diaminobutane and methyl acrylate.

The somewhat spherical shape of many dendimers gives them some different properties in comparison to more linear macromolecules. On a macroscopic level, dendimers act as ball-bearings and fans rather than strings. In solution, viscosity increases as molecular weight increases for linear polymers. With dendimers, viscosity also increases with molecular weight up to a point after which viscosity decreases as molecular weight continues to increase.

Dendimers are being used as host molecules, catalysts, and self-assembling nanostructures; as analogues of proteins, enzymes, and viruses; and in analytical applications.

17.13 IONOMERS

Ionomers are ion-containing copolymers typically containing over 90% (by number) ethylene units, with the remaining being ion-containing units such as acrylic acid. These "ionic" sites are connected through metal atoms. Ionomers are often referred to as processable thermosets. They are thermosets because of the cross-linking introduced through the interaction of the ionic sites with metal ions. They are processable or exhibit thermoplastic behavior because they can be reformed through application of heat and pressure.

As with all polymers, the ultimate properties are dependent upon the various processing and synthetic procedures that the material is exposed to. This is especially true for ionomers where the location, amount, nature, and distribution of the metal sites strongly determine the properties. Many of the industrial ionomers are made where a significant fraction of the ionomer is un-ionized and where the metal-containing reactants are simply added to the preionomer followed by heating and agitation of the mixture. These products often offer superior properties to ionomers produced from fully dissolved preionomers.

The metal–acid group bonding (salt formation) constitutes sites of cross-linking (Figure 17.4). It is believed that the "processability" is a result of the combination of the movement of the ethylene units and the metal atoms acting as "ball bearings." The "sliding" and "rolling" is believed to be a result of the metallic nature of of the acid–metal atom bonding. (Remember that most metallic salts are believed to have a high degree of ionic, nondirectional bonding as compared with typical organic bonds where there exists a high amount of covalent, directional bonding.) Recently, Carraher and co-workers have shown that the ethylene portions alone are sufficient to allow ionomers to be processed through application of heat and pressure.

Ionomers are generally tough and offer good stiffness and abrasion resistance. They offer good visual clarity, high melt viscosities, superior tensile properties, and oil resistance and are flame retarders. They are used in the automotive industry in the formation of exterior trim and bumper pads; in the sporting goods industry as bowling pin coatings and golf ball covers; and in the manufacture of roller skate wheels and ski boots. Surlyn (DuPont; poly(ethylene-co-methacrylic acid) is used in vacuum packaging for meats, in skin packaging for hardware and

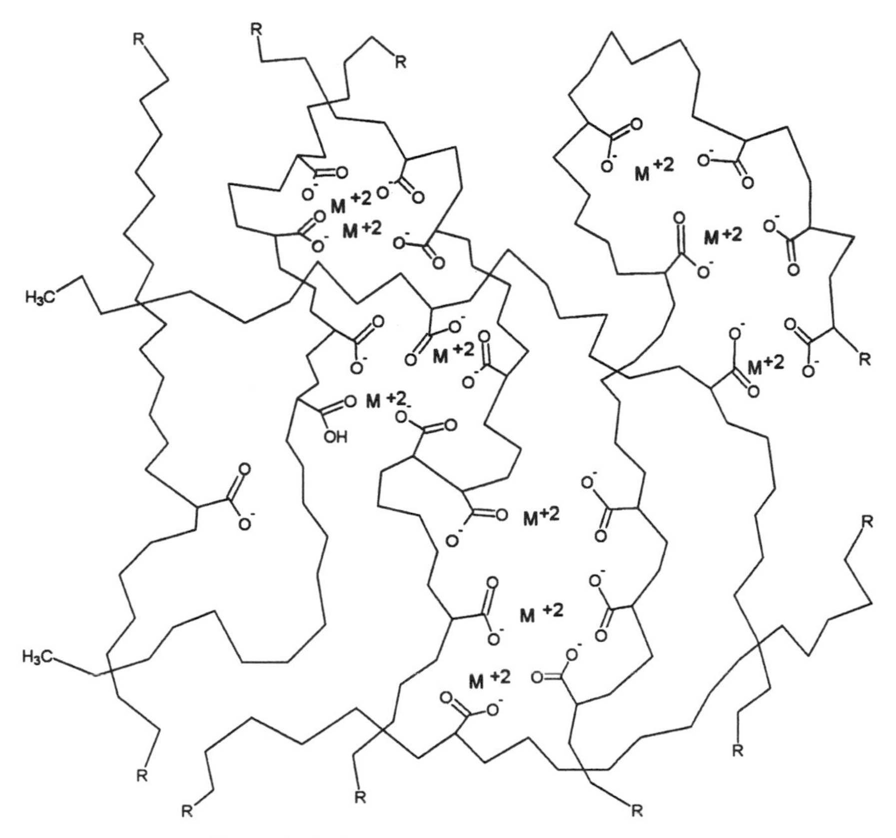

Figure 17.4. Representation of ionomer structure.

electronic items (such as seal layers and as foil coatings of multiwall bags), and in shoe soles.

Sulfonated ethylene–propylene–diene terpolymers (EPDM) are formulated to form a number of rubbery products including adhesives for footwear, garden hoses, and materials used in the formation of calendered sheets. Perfluorinated ionomers marketed as Nafion (DuPont) are used for membrane applications including chemical processing separations, spent acid regeneration, electrochemical fuel cells, ion-selective separations, electrodialysis, and use in the production of chlorine. It is also employed as a "solid"-state catalyst in chemical synthesis and processing. Ionomers are also used in blends with other polymers.

17.14 LIQUID CRYSTALS

We run across liquid crystals (LCs) each day. They are common in our computer monitors, digital clocks, high-end TVs, and so on. In 1888, Reintzer first reported liquid crystal behavior. In working with cholesteryl esters, he found that the ester formed opaque liquids which, on heating, turned clear. LCs are material that undergo physical reorganization whereby at least one of the rearranged structures involve molecular alignment along a preferred direction, causing the material to exhibit different physical properties dependent on the direction; that is, they show nonisotropic behavior or molecular asymmetry. While the first LCs were composed of small molecules, LCs are often giant molecules.

LCs are typically composed of materials that are rigid and rodlike. In general, because of the high order present in LCs, especially within their ordered state, they have low void densities and, as such, exhibit good stability to most chemicals including acids, bleaches, common liquids, and so on; low gas permeability; relatively high densities; strength and stiffness.

We can look at one illustration of the use of LC materials (Figure 17.5). A typical LC display (LCD) may contain thin layers of liquid crystal molecules sandwiched between glass sheets. The glass sheets have been rubbed in different directions and then layered with transparent electrode strips. The outside of each glass sheet is coated with a polarizer material oriented parallel to the rubbing direction. One of the sheets is further coated with a material to make it a reflecting mirror. The liquid crystalline molecules preferentially align along the direction that the two glass surfaces have been rubbed. Because the two glass surfaces are put at 90° to one another, the liquid crystal orientation changes as one goes from one glass surface to the other, creating a gradual twist of 90°.

Ordinary light consists of electromagnetic waves vibrating in various planes perpendicular to the direction of travel. As light hits the first polarizer material, only light that vibrates in a single plane is allowed to pass through. This plane-polarized light then passes through the layers of liquid crystals that effectively twist the plane of the light 90°, allowing the "twisted" light to pass through the second polarized surface, striking the mirrored surface, and "bouncing back," being seen as a white background.

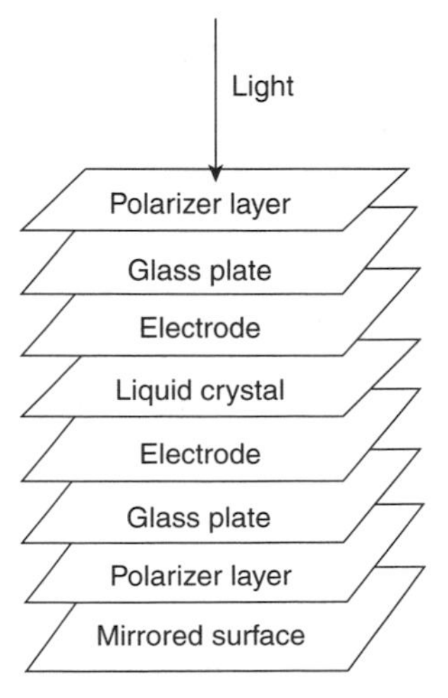

Figure 17.5. LCD illustration.

The LCD image is formed as a voltage is applied to an appropriate pattern of tiny electrodes that causes reorientation in the liquid crystals. Orientated liquid crystals no longer are at 90° to one another and thus are unable to transmit light through to the mirrored surface and thus appear as dark areas. This combination of dark and light surfaces then creates the LCD image.

Assemblies similar to the above have been employed to create images in clocks and watches and other similar LCD image-containing products.

17.15 RECYCLING CODES

Most of us have seen the triangle symbol with three arrows and realize that it has something to do with recycling. This symbol was originally developed by the Society of Plastics Industry (SPI) as a plastic identification code to assist in recycling these materials. While it was devised for use on plastic containers, today the "chasing-arrows" triangle sign is used as a universal symbol for recycling.

The SPI code uses numbers 1–7 and/or bold, capital letters as shown below to identify particular plastics.

Poly(ethylene terephthalate)—PET or PETE. PET is used to package soft drinks, some peanut butter jars, and other jars. About one-quarter of plastic bottles are PET. PET bottles are often clear, are tough, and hold carbon dioxide well.

HDPE

High-density polyethylene—HDPE. HDPE accounts for about 50% of the plastic bottle market. HDPE bottles are used to hold milk, juices, margarine, and some grocery snacks.

V

Poly(vinyl chloride)—PVC or V. PVC, or blends containing PVC, is used as PVC pipes for plumbing, food packaging film, and containers for window cleaners and solid detergents. It accounts for only about 5% of the container market.

LDPE

Low-density polyethylene—LDPE. LDPE is a major material for bread bags and inexpensive trash bags.

PP

Polypropylene—PP. PP is used for some screw-on caps, lids, yogurt tubs, margarine cups, straws, and syrup bottles, as well as grocery bags.

PS

Polystyrene—PS. PS is used for a variety of containers including those called "styrofoam" plates, cups, and dishes. Cups, yogurt containers, egg cartons, and meat trays are made of PS.

Other plastics. A wide variety of other plastics are used including blends, copolymers, alloys, and multilayered combinations.

17.16 SMART MATERIALS

Smart materials are materials that react to an externally applied force—electrical, stress/strain (including pressure), light, magnetic, chemical environment, and heat. A smart material is not smart simply because it responds to external stimuli, but it becomes smart when the interaction is used to achieve a defined engineering or scientific goal. For instance, polymers undergo a transitional phase change referred to as the glass transition temperature, T_g. A volume change accompanies the T_g. This volume change is used in multiple switching devices that detect and redirect electrical signals as a function of temperature.

Muscles contract and expand in response to electrical, thermal, and chemical stimuli. Other synthetic giant molecules also change shape on application of electrical current, temperature, and chemical environment. Piezoelectric materials such as poly(vinylene fluoride) emit an electric charge when pressure is applied. Windows made of giant molecules that change color and allow variable amounts of light to show through are part of the overall energy control system of industrial and residential buildings.

Smart giant molecules are being studied and used in many other applications including automotive, aerospace, communications, and biomedical.

GLOSSARY

Antioxidant: A stabilizer for polymers.

Benzophenone: Diphenyl ketone.

Cultured marble: A polymer cement usually based on a filled polyester.

Dopant: An additive that enhances the electrical conductivity of a polymer.

Guar gum: The endosperm of *Cyanopsis tetraganoloba*, cultivated in India. The water-soluble portion (85%) is called guaran.

Polyacetylene:

$$\left[\overset{\mathrm{H}}{\overset{|}{\mathrm{C}}}=\overset{\mathrm{H}}{\overset{|}{\mathrm{C}}}-\overset{\mathrm{H}}{\overset{|}{\mathrm{C}}}=\overset{\mathrm{H}}{\overset{|}{\mathrm{C}}} \right]_n$$

Polymer concrete: A composite produced by the *in situ* polymerization of a mixture of polymer and filler.

Poly(methyl isopropenyl ketone) (PMIK):

$$\left[\begin{array}{cccc} H & & H & H \\ | & & | & | \\ C - & C - & C - & C \\ | & \| & | & | \\ H & O & H & CH_3 \end{array} \right]_n$$

Resole phenolic resin: The reaction product of phenol and formaldehyde under alkaline conditions, usually used as a prepolymer.

Solubility parameter: A scale of solubility equal to the square root of the cohesive energy density, $(CED)^{1/2}$. The values range from about 4H (Hildebrands) to 22.5 H (for water).

Specialty polymer: A polymer that is used in applications other than molding or extrusion; or a polymer that is used as a fiber, coating, adhesive, or elastomer.

Structural foam: A molded cellular plastic with a solid surface.

Styrofoam: Trade name for cellular polystyrene.

Syntactic foam: A cellular product consisting of a polymer and a hollow glass bead filler.

REVIEW QUESTIONS

1. Why do some toys break when dropped on concrete surfaces?
2. Do the principles outlined in previous chapters apply to biopolymers as well as to elastomers, coatings, fibers, and plastics?
3. What is the requirement for a polymer to be water-soluble?
4. Will a hydrophilic polymer like polyvinyl alcohol be soluble in gasoline?
5. What happens if sodium chloride (rock salt) is not added to home ion-exchange water purifiers?
6. Is starch hydrophilic or hydrophobic?
7. Is polyethylene hydrophilic or hydrophobic?
8. Cellulose is not soluble in water but hydroxyethylcellulose is water-soluble. Why?
9. What is the difference in the shape of polyisobutylene in hot and cold lubricating oil?
10. Name one application in which biodegradable polymers are useful?
11. What is the advantage of a glyceryl nitrate controlled-release patch (nitro patch) over the glyceryl nitrate pill for control of angina pectoris?

12. Why are polymers essential for computers?

13. What is the advantage of a conductive polymer over a copper wire?

14. What is the principal use of flexible foams?

15. What is the principal use of rigid foams?

16. Which is more expensive: portland cement or polymer concrete?

17. What is the advantage of epoxy concrete over portland cement for patching cracks in highways?

18. Name some applications of synthetic polymers in the human body.

BIBLIOGRAPHY

Chandrasekhar, P. (1999). *Conducting Polymers: Fundamentals and Applications—A Practical Approach*, Kluwer, New York.

Chung, T. (2001). *Advances in Therotropic Liquid Crystal Polymers*, Technomic, Lancaster, PA.

Collings, P. J., and Hird, M. (1997). *Introduction to Liquid Crystals, Chemistry and Physics*, Taylor and Francis, London.

Frechet, J., and Tomalia, D. (2002). *Dendrimers and Other Dendritic Polymers*, Wiley, New York.

Gebelein, C., and Carraher, C. (1995). *Industrial Biotechnological Polymers*, Technomic, Lancaster, PA.

Kawazoe, X., Ohno, K., and Kondow, T. (2001). *Clusters and Nanomaterials*, Springer, New York.

Mishra, M., and Kobayashi, S. (1999). *Star and Hyperbranched Polymers*, Marcel Dekker, New York.

Newkome, G., Moorefield, C., and Vogtle, F. (2001). *Dendrimers and Dendrons*, Wiley, New York.

McCormick, C. (2000). *Stimuli-Responsive Water-Soluble Polymers*, ACS, Washington, D.C.

Rupprecht, L. (1999). *Conductive Polymers and Plastics*, ChemTec, Toronto.

Wallace, G., and Spinks, G. (1996). *Conductive Electroactive Polymers: Intelligent Materials Systems*, Technomic, Lancaster, PA.

ANSWERS TO REVIEW QUESTIONS

1. Toys molded from inexpensive brittle polymers will break but there are many tough plastics that can be used. Unfortunately, these tough plastics are more expensive.

2. Yes.

3. It must have polar pendant groups such as hydroxyl, amino, or carboxyl or have polar groupings in the polymer backbone.

4. No.
5. The resins will not be recharged and the resins will not be able to remove the "hardness" from the water.
6. Hydrophilic.
7. Hydrophobic.
8. The pendant group is a bulky group that reduces intermolecular hydrogen bonding.
9. Polyisobutylene is a tight coil in cold lubricating oil but is an extended chain in hot oil.
10. As a polyolefin agricultural mulch.
11. The glyceryl nitrate is released in small amounts from the patch over a long period of time.
12. The message is supplied by photoresist polymers, etc.
13. Lighter weight and usually more flexible.
14. Upholstery, mattresses, and filling.
15. Insulating and packaging.
16. Polymer concrete on a weight basis but not always on a performance basis.
17. The epoxy concrete bonds adhere to clean concrete; faster drying.
18. Lens implants, tooth filling, false teeth, wigs, body parts, such as ears and noses, and joint implants.

18

ADDITIVES AND STARTING MATERIALS

Giant Molecules: Essential Materials for Everyday Living and Problem Solving, Second Edition,
by Charles E. Carraher, Jr.
ISBN 0-471-27399-6

18.1 INTRODUCTION

Additives are added to modify properties, assist in processing, and introduce new properties to a material. They may be added as solids, liquids, or gases. Many additives for synthetic giant molecules have become a part of general formulations whose development is as much of an art as it is a science. For instance, a typical tire tread recipe has an accelerator activator, antioxidant, processing aid, retarder, vulcanizing agent, accelerator, antiozonate, softener, and finishing aid as additives. A general paint formulation may have China clay as an extender, titanium dioxide as the white pigment, and calcium carbonate as an extender, along with a fungicide, defoaming aid, coalescing liquid, and a surfactant-dispersing agent.

A listing of typical additives includes

Antiblocking agents	Antifoaming agents	Antifogging agents
Antimicrobial agents	Antioxidants	Antistatic agents
Blowing agents	Colorants	Coupling agents
Curing agents	Fillers	Flame retardants
Foaming agents	Impact modifiers	Low-profile agents
Lubricants	Mold-release agents	Odorants
Plasticizers	Preservatives	Reinforcement agents
Slip agents	Heat stabilizers	Radiation stabilizers
Viscosity modifiers		

Additives are an essential functional ingredient of giant molecules. Some natural polymers, such as wool, silk, or cotton fibers, and some natural coatings, such as shellac and gutta-percha, may be used without additives. However, plant leaves consist of cellulose and pigments, natural rubber contains stabilizers (antioxidants), and wood is a reinforced polymer consisting of a continuous phase (lignin) and a discontinuous phase (cellulose). Likewise, most synthetic plastics, elastomers, and coatings consist of polymers and functional additives.

The types and purposes of additives (the word "additives" is derived from "addition" and simply means material or materials added) are varied and the exact proportions and nature of the additives are as much an art as a science. It is important to remember that (a) the addition of additives often requires extra processing steps, thus increasing the cost of the item; (b) the additives themselves may vary in cost from clay fillers and sulfur, which cost pennies per pound, to bioactive additives to prevent rot and mildew, which may cost hundreds of dollars per pound; and (c) the "additive industry" is also a major contributor to the polymer industry and to our industrial complex.

18.2 FILLERS

According to the ASTM D-883, a filler is a relatively inert material added to a material to modify its strength, permanence, working properties, or other qualities or to lower costs.

Charles Goodyear patented the use of small quantities of carbon black as a pigment for natural rubber in the 1840s. However, the advantageous use of larger quantities (50%) of carbon black as a reinforcement for rubber tires was not recognized until 1920. Wood-flour-filled shellac (Florence Compound) was formulated by A. P. Critchlow in 1845. Wood flour is made up of fibrouslike wood particles obtained by the attrition grinding of wood.

Heming and Baekeland used asbestos as reinforcements for cold-molded bituminous composites and for phenolic resins, respectively, in the early 1900s. α-Cellulose was used as a filler in urea and melamine plastics in the early 1930s, and these formulations are still in use today. Over 2 million tons of fillers are used annually by the American plastics industry.

The most widely used inorganic filler is calcium carbonate, which is used at an annual rate of over 1 million tons. Asbestos continues to be used in moderate amounts (250,000 tons), but, because of its toxicity, it is being displaced by other fillers. Among the naturally occurring filler materials are cellulosics, such as wood flour, α-cellulose, shell flour, and starch, and proteinaceous fillers, such as soybean residues. Approximately 40,000 tons of cellulosic fillers are used annually by the American polymer industry.

Wood flour, which is produced by the attrition grinding of wood wastes, is used as a filler for phenolic resins, dark-colored urea resins, polyolefins, and PVC. Shell flour, which lacks the fibrous structure of wood flour, has been made by grinding walnut and peanut shells. It is used as a replacement for wood flour.

Cellulose, which is more fibrous than wood flour, is used as a filler for urea and melamine plastics. Melamine dishware is a laminated structure consisting of molded resin-impregnated paper. Starch and soybean derivatives are biodegradable, and the rate of disintegration of resin composites may be controlled by the amount of these fillers present.

Many incompatible polymers are added to increase the impact resistance of other polymers, such as polystyrene. Other comminuted resins, such as silicones or polyfluorocarbons, are added to increase the lubricity of some plastics. For example, a hot melt dispersion of polytetrafluoroethylene in polyphenylene sulfide is used as a coating for antistick cookware.

Carbon black, which was produced by the smoke impingement process by the Chinese over a thousand years ago, is now the most widely used filler for polymers. Much of the 1.5 million tons produced annually in the United States is used for the reinforcement of elastomers. The most widely used carbon black is furnace carbon black.

Carbon-filled polymers, especially those made from acetylene black, are fair conductors of heat and electricity. Polymers with fair conductivity have also been obtained by embedding carbon black in the surfaces of nylon or polyester filament reinforcements. The resistance of polyolefins to ultraviolet radiation is also improved by the incorporation of carbon black.

Although glass spheres are classified as nonreinforcing fillers, the addition of 40 g of these spheres to 60 g of nylon 6,6 increases the flexural modulus, compressive strength, and melt index. However, the tensile strength, impact strength, creep resistance, and elongation of these composites are less than those of the unfilled nylon 6,6.

Zinc oxide is used to a large extent as an active filler in rubber and as a weatherability improver in polyolefins and polyesters. Titanium dioxide is used as a white pigment and as a weatherability improver in many polymers.

The addition of finely divided calcined alumina, corundum, or silicon carbide produces abrasive composites. Alumina trihydrate (ATH) serves as a flame-retardant filler in plastics. Ground barytes ($BaSO_4$) yield x-ray-opaque plastics with controlled density. Zirconia, zirconium silicate, and iron oxide, which have specific gravities greater than 4.5, are also used to produce plastics with controlled densities.

Clay is used as a filler in making synthetic paper and rubber. Talc, a naturally occurring, fibrouslike, hydrated magnesium silicate, is used with polypropylene (Figure 18.1). Since talc-filled polypropylene is much more resistant to heat than PP, it is used in automotive accessories subject to high temperatures. Over 40 million tons of talc are used annually as a filler.

Silica, which has a specific gravity of 2.6, is used as naturally occurring and synthetic amorphous silica, as well as in the form of large crystalline particulates, such as sand and quartz. Diatomaceous earth, also called infusorial earth, fossil flour, or Fuller's earth, is a finely divided amorphous silica consisting of the skeletons of diatoms. Diatomaceous earth is used to prevent rolls of film from sticking to themselves (antiblocking) and to increase the compressive strength of polyurethane foams.

Pyrogenic or fumed silica is a finely divided filler obtained by heating silicon tetrachloride in an atmosphere of hydrogen and oxygen. This filler is used as a thixotrope to increase the viscosity of liquid resins. Finely divided silicas are

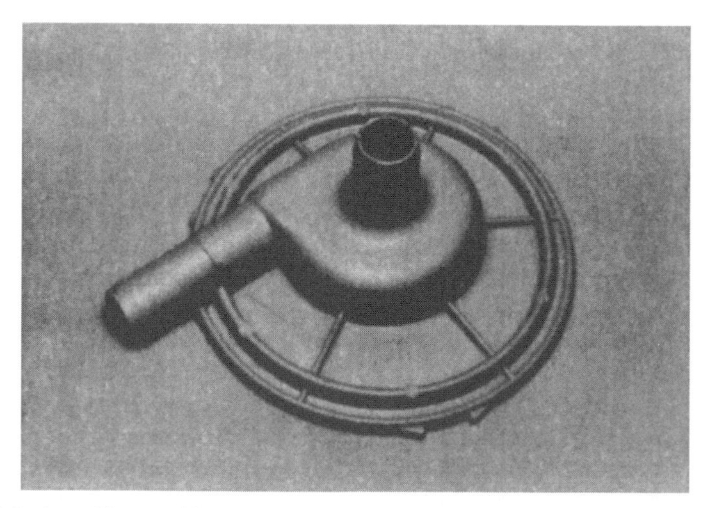

Figure 18.1. A washing machine pump housing made from 40% talc and 60% polypropylene, the latter containing 0.5% titanate.

Table 18.1 Fillers for polymers

I. ORGANIC
- A. Cellulosic products
  - 1. Wood
  - 2. Comminuted cellulose
  - 3. Fibers (cellulose, cotton, jute, rayon)
- B. Lignin-based
- C. Synthetic fibers
  - 1. Polyesters
  - 2. Nylons
  - 3. Polyacrylonitrile
- D. Carbon
  - 1. Carbon black
  - 2. Graphite whiskers and filaments
  - 3. Ground petroleum coke

II. INORGANIC
- A. Silicates
  - 1. Minerals (asbestos, mica, China clay (kaolinite), talc, wollastonite)
  - 2. Synthetics (calcium silicate, aluminum silicate)
- B. Silica-based
  - 1. Minerals (sand, quartz, diatomaceous earth, tripoli)
  - 2. Synthetics
- C. Metals
- D. Boron filaments
- E. Glass
  - 1. Solid and hollow glass spheres
  - 2. Milled fiber
  - 3. Flakes
  - 4. Fibrous glass (woven, roving, filament, yarn, mat, fabric)
- F. Metallic oxides
  - 1. Ground (zinc oxide, titania, magnesia, alumina)
  - 2. Whiskers (aluminum oxide, magnesium oxide, thorium oxide, zirconium oxide, beryllium oxide)
  - Polyfluorocarbons
- G. Calcium carbonate
  - 1. Limestone
  - 2. Chalk
  - 3. Precipitated calcium carbonate
- H. Other fillers
  - 1. Whiskers (nonoxides; aluminum nitride, boron carbide, silicon nitride, tungsten carbide, beryllium carbide)
  - 2. Barium sulfate
  - 3. Barium ferrite

also produced by the acidification of sodium silicate solutions and by the evaporation of alcoholic solutions of silicic acid.

Sharp silica sand is used as a filler in resinous cement mortars. Reactive silica ash, produced by burning rice hulls, and the lamellar filler novaculite, from the novaculite uplift in Arkansas, are also used as silica fillers in polymers.

Conductive composites are obtained when powdered metal fillers or metal-plated fillers are added to resins. These composites have been used to produce forming tools for the aircraft industry. Powdered lead-filled polyolefin composites have been used as shields for neutron and gamma radiation.

It is of interest to note that much of the theory on property enhancement by fillers is based on an equation developed by Einstein, which states that the viscosity (η) of a liquid is increased as the concentration (C) of spherical particles is increased in accordance with the equation

$$\eta = \eta_0(1 + K_E C)$$

The constant K_E is a universal constant that is equal to 2.5 for spherical particles—that is, those in which the ratio of length (l) to the diameter (d) is equal to 1. This constant increases as the aspect ratio (l/d) increases.

The stiffness (modulus) and other strength properties of composites (mixtures of polymers and fillers) are related to the viscosity (i.e., the resistance to flow). Many properties of the composites may be estimated from the rule of mixtures in which the volume percentage of the additive and that of the polymer are important factors in determining the properties of the composite. For example, the specific heat of a composite is equal to that of the polymer multiplied by its fractional volume plus the specific heat of the filler multiplied by its specific volume.

Fillers such as glass beads are spherical, those like wood flour and α-cellulose are fibrous, and those like mica are platelike. The properties of the composites are usually enhanced when the surface of the filler is treated with a coupling agent that makes the filler more compatible with the resin. See Table 18.1 for a listing of common fillers.

18.3 REINFORCEMENTS

By definition, reinforcements are fillers with aspect ratios (l/d) greater than 100. Fiberglass, which is the most widely used reinforcement, was produced commercially from molten glass by Slayter in 1938. In spite of the potential of this fiber as a reinforcement for unsaturated polyester resins, little enhancement in the properties of glass-reinforced plastics was noted until the glass surface was treated with coupling agents.

Alkylsilane ($R_2Si(OCH_3)_2$) coupling agents are widely used today for the surface treatment of fiberglass. Organotitanates and organozirconates have also been used successfully as coupling agents. The fiberglass may be used as chopped strands, continuous filaments, mats, or woven cloth.

Graphite is an excellent but expensive reinforcement for plastics. Aramid (aromatic polyamide), polyester, and boron filaments are also used as reinforcements.

Polyester resin-impregnated fibrous glass is used as a sheet molding compound and bulk molding compound. The former is used like a molding powder and the latter is hot-pressed in the shape of the desired object, such as one-half of a suitcase. Chopped fibrous glass roving may be impregnated with resin and sprayed, and glass mats may be impregnated with resin just prior to curing.

Table 18.2 Properties of reinforcements

Reinforcement	Specific Gravity	Tensile Strength (psi)[a]	Tensile Modulus (psi)	Specific Modulus (psi)[b]
E-glass	2.55	450,000	10,000,000	4,000,000
S-glass	2.48	650,000	12,500,000	5,000,000
Graphite—whiskers	1.74	400,000	40,000,000	23,000,000
Alumina	4.0	4,100,000	103,000,000	26,000,000
Silicon carbide	3.2	2,000,000	70,000,000	22,000,000
Silicon nitride	3.2	20,000,000	57,000,000	19,000,000
Potassium titanate	3.2	10,000,000	40,000,000	12,000,000

[a] 68,948 psi = kPa.
[b] Specific modulus = tensile modulus/specific gravity.

Strong composites are made from continuous filaments impregnated with resin before curing. These continuous filaments are wound around a mandrel in the filament winding process, gathered together, and forced through an orifice in the pultrusion molding process.

The first continuous filaments were rayon; these, as well as polyacrylonitrile fibers, have been pyrolyzed to produce graphite fiber. High-modulus reinforcing filaments have also been produced by the deposition of boron atoms from boron trichloride vapors on tungsten or graphite filaments.

Small single crystals, such as potassium titanate, are being used at an annual rate of over 10,000 tons for the reinforcement of nylon and other thermoplastics. These composites are replacing die-cast metals in many applications. Another microfiber, sodium hydroxycarbonate (called Dawsonite), also improves the physical properties and flame resistance of many polymers. Many other single crystals, called whiskers, such as alumina, chromia, and boron carbide, have been used for making high-performance composites.

Microfibers that have an aspect ratio of at least 60 to 1 also have good reinforcing properties in composites. The principal microfibers are processed mineral fiber (PMF, slag), Franklin fiber (gypsum, $CaSO_4$), Dawsonite, and Fybex (potassium titanate, K_2TiO_3). Over 1 million tons of reinforcements are used annually by the American plastics industry. The properties of reinforcing fibers are shown in Table 18.2.

18.4 COUPLING AGENTS

Although natural rubber bonds well to carbon black, in 1956, H. M. Leeper discovered that the adhesion could be enhanced by the addition of small amounts of Elastopar (*N*-4-dinitroso-*N*-methylaniline).

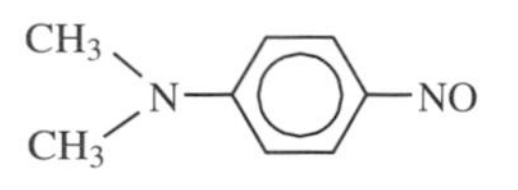

In addition to improving the properties of carbon black-filled butyl rubber, this additive also dramatically reduces the time of milling of the filler and butyl rubber.

There are many different proprietary alkylsilanes, organozirconates, and organotitanates that have been developed for use with specific composites. These coupling agents have two different functional groups, one that is attracted to the resin and the other that is attracted to the surface of the filler. For example, dialkyldimethoxysilanes are hydrolyzed to produce dialkyldihydroxysilanes *in situ*. As shown in the following equation, the hydroxyl groups bond with the filler surface and the alkyl groups are attracted to the resin.

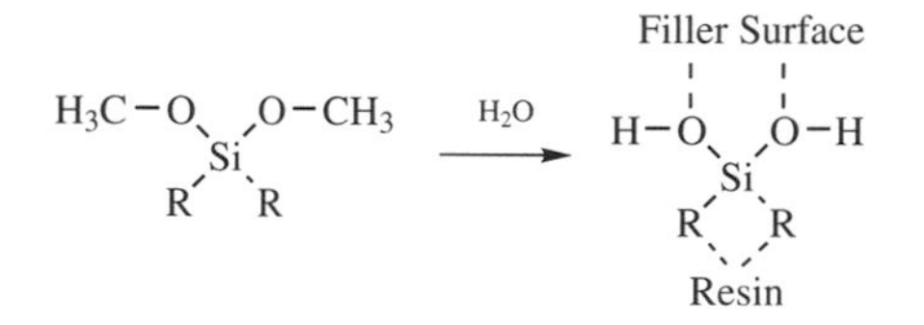

18.5 ANTIOXIDANTS

Polypropylene cannot be used out-of-doors unless a small amount of stabilizer (antioxidant) is present. It is known that the tertiary hydrogen atoms on every other carbon atom in the repeating units may be readily cleaved to form free radicals (R·):

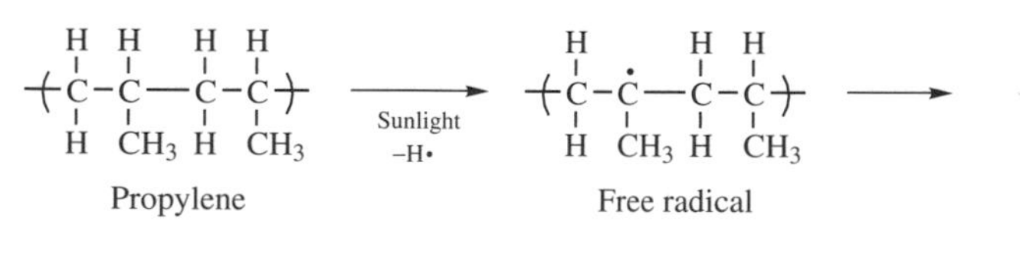

$$\text{—(CH=C(H)CH}_3\text{)} + \cdot\text{CH(H)—CH(CH}_3\text{)—}$$

Two chains

The deterioration resulting from the formation of these free radicals is lessened when antioxidants, such as hindered phenols, alkyl phosphites ($(ArO)_3P{=}O$), thioesters

$$(\text{RC}(=\text{S})-\text{SR})$$

or hindered amines, are present. The hindered phenols have relatively large alkyl groups adjacent to the hydroxyl group in phenols.

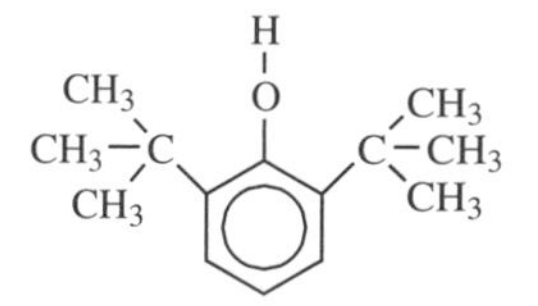

2,6-Di-*tert*-butylphenol

Aromatic amines, such as phenyl β-naphthylamine (Ar_2NH), are used as antioxidants in the rubber industry, but some tests have shown that these antioxidants cause cancer in rats. In contrast, some hindered phenols extend the life of rats.

18.6 HEAT STABILIZERS

As mentioned previously, polyvinyl chloride (PVC) is difficult to process since it decomposes at temperatures below the processing temperature. However, stabilizers such as barium, cadmium, calcium, and zinc salts of moderately high molecular weight carboxylic acids, organotin, and organoantimony compounds are effective stabilizers. Presumably, these heavy metal cations react with hydrogen chloride to produce insoluble salts. The general reaction may be represented by

$$-\!\!\left[CH_2\!-\!CHCl \right]\!\!- \xrightarrow{\Delta} -\!\!\left[CH\!=\!CH \right]\!\!- + HCl$$

$$2HCl + M(OOCR)_2 \longrightarrow 2HOOCR + MCl_2$$

18.7 ULTRAVIOLET STABILIZERS

Sunlight, which has an energy corresponding to 72–100 kcal, may break bonds having similar energy. This degradation of polymers may be minimized when compounds that absorb this high-UV energy are present. Since the reaction is equivalent to that which occurs in the sunburning of skin, the additives present in sunburn lotions, such as phenyl salicylate, are effective UV stabilizers for polymers. Phenyl salicylate, which has been used as a medicinal for years, rearranges to 2,2′-hydroxybenzophenone in the presence of ultraviolet light.

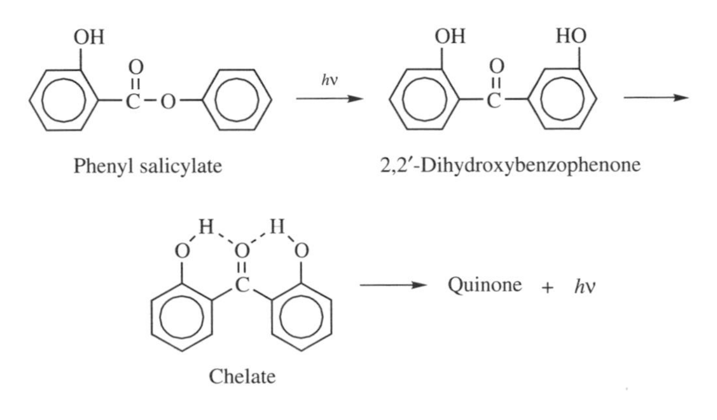

Stabilizers such as 2,2′-hydroxylbenzophenone produce cyclic compounds (chelates) that absorb the UV energy and release it at a lower, less destructive energy level. A chelate is a five- or six-membered ring that may be formed by intramolecular attraction of a hydrogen atom to an oxygen atom in the same compound. Thus, when the hydroxyl group is in the proper (2) position, it will form a chelate with the carbonyl oxygen. Hindered amine light stabilizers (HALS), which, in the presence of sunlight, produce nitroxyl radicals (=NO•), are excellent UV stabilizers.

18.8 FLAME RETARDANTS

General-purpose plastics, such as polyethylene and polystyrene, are readily combustible. However, when polymers are used for furniture or in construction, it is essential that they not be combustible under conditions that may exist in a burning building. Polymers such as polytetrafluoroethylene (PTFE) are usually considered to be flame-resistant. However, they will burn in the high-oxygen/low-nitrogen atmosphere present in aerospace capsules.

Other halogen (Cl, Br) and phosphorus-containing polymers are also flame-resistant in air. However, it is usually essential that flame retardants be added to most flame-resistant polymers. Alumina hydrate (ATH, $Al_2O_3 \cdot 3H_2O$) is a colorless filler that is readily used as a flame retardant. This filler releases water, which helps to quench the flames at the burning temperature.

Most flame retardants consist of mixtures of aliphatic chlorides (RCl) and antimony oxide (Sb_2O_3). The antimony chloride ($SbCl_3$) produced at the temperature of the burning plastic is the flame retardant.

Char, formed in some combustion processes, also shields the reactants from oxygen and retards the outward diffusion of volatile combustible products. Aromatic polymers tend to char, and some phosphorus and boron compounds tend to catalyze char formation.

Synergistic flame retardants, such as a mixture of antimony trioxide and an organic bromo compound, are much more effective than single flame retardants. Thus, whereas a polyester containing 11.5% tetrabromophthalic anhydride burned without charring at high temperatures, charring but no burning was noted when 5% antimony oxide was added.

Since combustion is subject to many variables, tests for flame retardancy may not predict flame resistance under unusual conditions. Thus, a disclaimer stating that flame-retardant tests do not predict performance in an actual fire must accompany all flame-retardant polymers. Flame retardants, like many other organic compounds, may be toxic or they may produce toxic gases when burned. Hence, extreme care must be exercised when using fabrics or other polymers treated with flame retardants.

18.9 PLASTICIZERS

Water has been called the plasticizer of life, and it is extensively used as a plasticizer in nature. It permits "Nature to be flexible"; that is, it allows leaves and trees to bend in the wind, and enables the nucleic acids and proteins to operate. Fats and some proteins also act as plasticizers.

W. Semon lowered the processing temperature of PVC by the addition of tricresyl phosphate as a plasticizer in the early 1930s. This somewhat toxic plasticizer has been replaced by esters of phthalic acid, such as diethylhexylphthalate (DOP, DEHP). It is believed that plasticizers weaken the intermolecular attractions between molecules in PVC and allow the semicrystalline polymer to flow at lower-than-normal temperatures. It should be noted that tests with massive doses of DEHP have shown it to be toxic to laboratory animals.

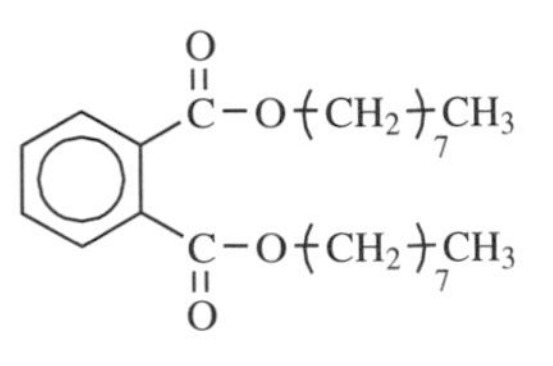

Dioctyl phthalate (DOP)

Synthetic plasticizers are fairly large molecules. Natural plasticizers, such as some proteins, are also large but molecular water is small—yet it probably acts as a larger unit through its hydrogen bonding. Most plasticizers work on the basis of solubilizing polymer units to permit segmental movement but not wholesale chain movement.

The annual worldwide production of plasticizers is 3.2 million tons, and the U.S. production is in excess of 1 million tons. In fact, plasticizers are major components of a number of polymer-containing products. For instance, automobile safety glass is composed mainly of poly(vinyl butyral) and about 30% plasticizer.

The development of plasticizers has been plagued with toxicity problems. For example, the use of highly toxic polychlorinated biphenyls (PCBs) has been discontinued. Phthalic acid esters, such as DOP, may be extracted by blood stored in plasticized PVC blood bags and tubing. These aromatic esters are also extracted from PVC tubing and are distilled from PVC upholstery in closed automobiles in hot weather. These problems have been solved by using oligomeric polyesters instead of DOP as nonmigrating plasticizers.

Many copolymers, such as poly(vinyl chloride-*co*-vinyl acetate), are internally plasticized because of the flexibilization brought about by the change in structure of the polymer chain. In contrast, DOP and others are said to be external plasticizers. The presence of bulky groups on the polymer chain increases segmental motion. Thus, the flexibility increases as the size of the pendant group increases. However, linear bulky groups with more than 10 carbon atoms will reduce flexibility because of side chain crystallization when the groups are regularly spaced.

Plasticizer containment still remains a major problem, particularly for periods of extended use. For instance, most plastic floor tiles become brittle with age, mainly due to the leaching out of plasticizer. This may be overcome through many routes, including treatment of polymer product surfaces to reduce porosity and use of branched polymers that can act as plasticizers to themselves. These highly branched polymers are slow to leach because of physical entanglements within the total polymer matrix.

18.10 IMPACT MODIFIERS

Polystyrene (PS) is brittle and not suitable for applications that require good resistance to impact. This objection has been overcome, to some extent, by blending PS with SBR to produce high-impact polystyrene (HIPS). HIPS was patented by Seymour in the early 1940s. The impact resistance of PS and other brittle polymers may be overcome by the addition of ABS or polyalkyl vinyl ethers. Of course, specific copolymers, such as those of methyl methacrylate and ethyl acrylate, have better resistance to impact than poly(methyl methacrylate) (PMMA).

18.11 COLORANTS

Since cosmetic effects are important to the consumer, colorants are added to plastics, fibers, and coatings to meet aesthetic requirements. Some polymeric objects, such as rubber tires, are black because of the presence of high proportions of carbon black filler. Many other products, including some paints, are white because of the presence of titanium dioxide, the most widely used inorganic pigment. Over 50,000 tons of colorants are used annually by the American polymer industry.

Pigments are classified as organic or inorganic. The former are brighter, less dense, and smaller in size than the more widely used, more opaque inorganic colorants. Iron oxides or ochers, available as yellow, red, black, brown, and tan, are the most common pigments.

Carbon black is the most widely used organic pigment, but phthalocyanine blues and greens are available in many different shades and are also common. Other organic pigments are the azo dyestuffs, such as the pyrazolone reds, diarylide yellows, dianisidine orange, and tolyl orange; quinacridone dyestuffs, such as quinacridone violet, magenta, and red; the red perylenes; acid and basic dyes, such as rhodamine red and victoria blue; anthraquinones, such as flavanthrone yellow; dioxazines, such as carbazole violet; and isoindolines, available in the yellow and red range.

18.12 CATALYSTS AND CURING AGENTS

By definition, a catalyst is an additive that accelerates the velocity of reaction but remains unchanged; thus most so-called catalysts used in polymer reactions are actually initiators and not catalysts. One of the most important of this type of additive is 2-mercaptobenzothiazole (Captax), which accelerates the rate of cross-linking of rubber with sulfur. The first rubber accelerator was developed by G. Oenslager in 1906.

Hexamethylenetetramine, which is produced by the condensation of ammonia and formaldehyde, is used as a source of formaldehyde in the curing of phenolic (novolak) resins. Large quantities of peroxides, such as benzoyl peroxide (BPO), and azo compounds, such as azobisisobutyronitrile (AIBN), are used as initiators for the polymerization of vinyl monomers.

Cyclic anhydrides, such as phthalic anhydride, are used for the elevated-temperature curing of epoxy resins. These prepolymers are cured at ordinary temperatures in the presence of secondary amines. Tertiary amines and organic tin compounds are used as curing agents for polyurethanes.

18.13 FOAMING AGENTS

Nitrogen gas, produced by the decomposition of azo compounds such as azodicarbonamide (ABFA), is used to form cellular plastics (foam). In addition to these chemical blowing agents (CBA), physical blowing agents (PBA) such as nitrogen, pentane, and volatile fluorocarbons have been used for foam formation.

18.14 BIOCIDES

Biocides are added to certain polymers such as paints to retard degradation by microorganisms. For example, wood is readily attacked under moist, humid conditions by bacteria and fungi, which cause mildew and rot. Antimicrobials such as tin, arsenic, antimony, and copper-containing organometallic salts and oxides are employed in the polymer industry to prevent or discourage the attack of bacteria or fungi on both the natural polymer and, in the case of coatings, the coated material. It must be remembered that most synthetic polymers are not attacked by microorganisms.

Tin-containing compounds have been widely employed as additives in coatings and sealants to inhibit mildew- and rot-causing microorganisms. The monomeric tin compounds have recently been outlawed for use in marine paints since such compounds dissolve in the water and kill marine plants and fish. Organotin-containing polymers have not been banned, are only slightly soluble, and are thus more environmentally acceptable.

18.15 LUBRICANTS AND PROCESSING AIDS

Lubricants, such as calcium or lead stearates, paraffin wax, or fatty acid esters, serve as lubricants and processing aids for plastics. Polybutylene and polystyrene promote flow of plastics in extrusion and injection molding; and thixotropes, such as colloidal silica, reduce the flow of prepolymers.

18.16 ANTISTATS

Since most polymers are nonconductors of electricity, they tend to store electrostatic charges and attract dust. These electrostatic charges can be reduced by the addition of organic compounds such as hydroxylated amines, which attract water, which in turn dissipate the charge. Since electromagnetic interference (EMI) is a problem with business machines with plastic housings, metallic flakes such as aluminum are added to solve this electrostatic problem.

It is important to note that most of these additives are used in small amounts (1%). However, their presence is essential for the attainment of optimum properties in plastics.

18.17 STARTING MATERIALS

In 1925, Phillips Petroleum Company was only one of dozens of small oil companies in Oklahoma. The only distinction was the large amount of natural gasoline (or naphtha), the lightest liquid fraction Table 18.3, found in its crude oil. As was customary, Phillips, and most of the other oil companies of the time, employed a distillation process to isolate the butane and propane. Even so, they were sued for the use of this distillation process, probably because they were small, had no real research capacity of their own, and had no real legal defense team. Frank Phillips elected to fight, supposedly including an argument that the ancient Egyptians had used a similar process to create an alcoholic equivalent to an Egyptian alcoholic drink. Phillips won the suit but became convinced that if the company was to remain successful, it would need to have a research effort. In 1935, during the early dust bowl years, they established the oil industry's first research team in Bartlesville, OK. George Oberfell, hired by Phillips to fight the lawsuit, planned the initial research efforts that involved three main initiatives: first, to develop technology to use light hydrocarbons in new ways as motor fuels; second, to develop markets for butane and propane; and finally, to find new uses for the light hydrocarbons outside the fuel market. All three objectives were achieved.

Table 18.3 Typical hydrocarbon fractions obtained from distillation of petroleum materials

Boiling Range, °C	Average Number of Carbon Atoms in Chain	Name	Typical Uses
<30	1–4	Gas	Heating, cooking (propane, butane)
30–180	5–10	Gasoline (Naphtha)	Automobile fuel
180–230	11,12	Kerosene	Heating, jet fuel
230–300	13–17	Light gas oil	Diesel fuel, heating
300–400	18–25	Heavy gas oil	Heating

Frederick Frey and Walter Shultze were instrumental early researchers. Frey was among the first to dehydrogenate paraffins catalytically to olefins and then dehydrogenate the olefins to diolefins that serve as feedstocks to the production of many of today's polymers. In competition with Bakelite, he discovered the preparation of polysulfone polymers made from the reaction of sulfur dioxide and olefins, creating a hard Bakelite-like material. Frey and Schultz also developed a process that allowed the production of 1,3-butadiene from butane, which allowed the synthesis of synthetic rubber.

Probably Frey's most important invention involved the use of hydrogen fluoride to convert light olefins produced as byproducts of a catalytic cracker into high-octane motor and aviation fuels. This process is still widely used. It came at a critical time for America's World War II efforts, allowing fuel production for the Allied forces. This fuel allowed aircraft faster liftoffs, more power, and higher efficiency.

The interest in polymers involves many aspects, with two of these aspects being the source of (a) inexpensive starting materials and (b) the vast amount of readily available starting materials. Starting materials are often referred to as feedstocks. Most of the starting materials, monomers, employed in the synthesis of synthetic polymers like polystyrene, polyethylene, and nylons are derived indirectly from fossil fuels. The synthesis of these feedstocks from fossil fuels is a lesson in inventiveness. The application of the saying that "necessity is the mother of invention" has truly been followed by the polymer industry leading to sequences of chemical reactions where little is wasted and byproducts from one reaction are employed in another reaction. These sequences were developed over many years and involved the discovering of the precise conditions of pressure, temperature, catalysts, and

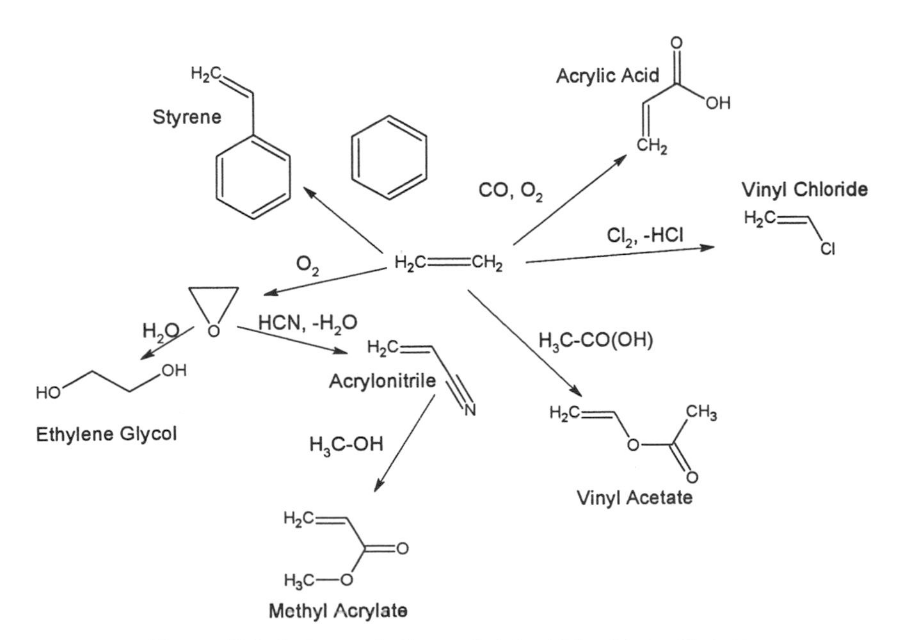

Figure 18.2. Polymer starting materials obtained from ethylene.

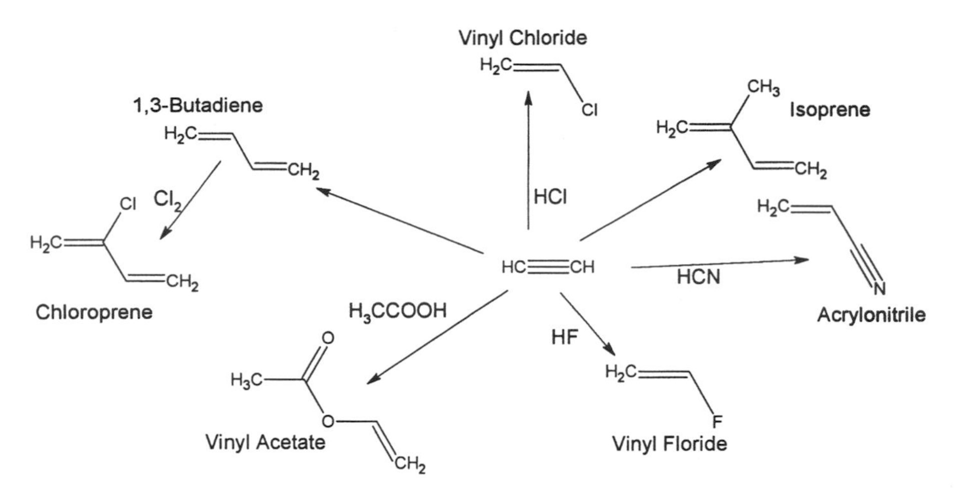

Figure 18.3. Polymer starting materials obtained from acetylene.

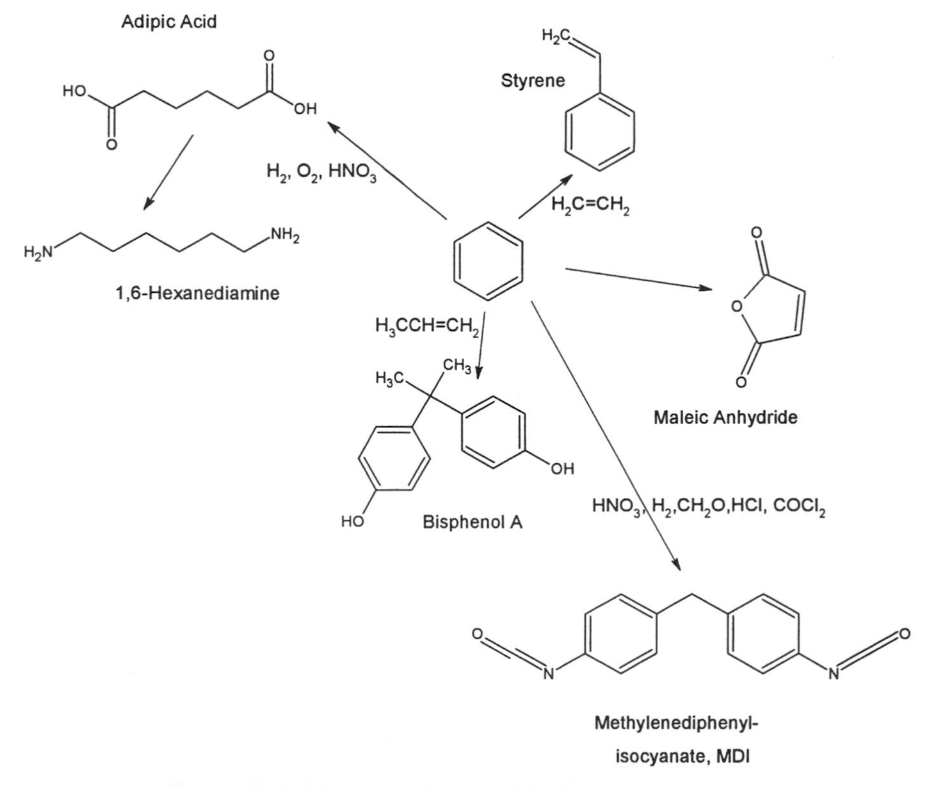

Figure 18.4. Polymer starting materials derived from benzene.

so on, that must be present as we go from starting materials, feedstocks, to monomers, to polymers.

The term fossil fuels refers to materials formed from the decomposition of once-living matter. Because these once-living materials contained proteins that contain sulfur, nitrogen, and metals such as iron and cobalt, these chemicals are present in the fossil fuels and must be removed.

The major one-carbon feedstock is methane. One major two-carbon feedstock is ethylene. From Figure 18.2 you can see that a number of the monomers are directly synthesized from ethylene. Again, while the "react" arrow goes directly from ethylene to the product, as noted above, it often took years to develop an economical procedure to obtain the product in essentially 100% yield. A number of polymers are derived from starting materials supplied by ethylene including polyacrylic acid, polyvinyl chloride, polystyrene, polyvinyl acetate, polyesters (in particular PET), and polymethyl acrylate.

Another major two-carbon feedstock is acetylene. As shown in Figure 18.3, a number of different monomers can be synthesized from acetylene. Many of these are the same as those from ethylene and illustrate some of the diversity industry uses to synthesize particular monomers. Acetylene is generally obtained from coal through reacting calcium carbide obtained from the coal with water.

Monomeric materials obtained from benzene are given in Figure 18.4. These giant molecule starting materials form the basis of many of the polyurethanes (MDI), nylons, polystyrene, polyesters (in particular PET), and so on.

Monomer synthesis continues to be improved with an emphasis on economics, saving energy, and safety. While the polymer industry is large, it reflects well less than 5% of the fossil fuel use.

GLOSSARY

α-Cellulose: High-molecular-weight cellulose, insoluble in 17.5% aqueous sodium hydroxide.

Accelerator: An additive that accelerates the cross-linking (vulcanization) of rubber by sulfur.

Additive: A substance added to a polymer.

AIBN: Azobisisobutyronitrile.

Alkylsilane: A coupling agent for polymer and filler.

Alumina hydrate (ATH): $Al_2O_3 \cdot 3H_2O$, a flame retardant.

Antioxidant: An additive that deters polymer degradation in air.

Antistat: An additive that aids in dissipating electrostatic charges in polymers.

Aramid: Aromatic polyamide.

Aspect ratio: *l/d* of a fiber.

Biocide: An additive that deters attack on polymers by bacteria, fungi, and so on.

BPO: Benzoyl peroxide.

Calcium carbonate: $CaCO_3$.

Carbon black: Finely divided carbon produced by the incomplete combustion of natural gas or liquid hydrocarbons.

Catalyst: A substance that hastens the attainment of equilibrium. The term is often misused and applied to initiators.

CBA: Chemical blowing agent.

Chelate: A cyclic compound formed by intramolecular attractions, such as hydrogen bonding between atoms.

Colorant: Pigment or dye.

Continuous phase: The polymeric phase in a composite.

Coupling agent: A compound that contains resin-attracting and fiber-attracting functional groups, such as organosilane.

DEHP: Diethylhexylphthalate plasticizer.

Discontinuous phase: The noncompatible additive phase in a composite.

E-glass: Electrical-grade fiberglass.

Einstein, Albert: Nobel laureate who developed equations for the property enhancement by fillers.

Einstein equation: $\eta = \eta_0(1 + K_E C)$, where η = viscosity of composite, η_0 = viscosity of polymer, $K_E = 2.5$ for spherical particles, and C = concentration of filler.

Elastopar: *N*-4-Dinitroso-*N*-methylaniline, a coupling agent.

Electromagnetic interference (EMI): Interference resulting from storage of electrostatic charges by nonconductive polymers.

Filler: Originally considered a cost-reducing additive but has been found to be a functional additive that enhances the physical properties of polymers in many instances.

Flame retardant: An additive that retards combustion of polymers.

Franklin fiber: Calcium sulfate microfiber.

Fybex: Potassium titanate microfiber.

Graphite fiber: Carbon fiber obtained by the pyrolysis of polyacrylonitrile fibers or pitch.

HALS: Hindered amine light stabilizer.

Heat stabilizer: An additive that deters degradation of polymers, such as PVC, at elevated temperatures.

Impact modifier: An additive that reduces the brittleness of polymers.

Microfiber: A fiber with an aspect ratio (*l/d*) of at least 60 to 1.

Modulus: Stiffness.

Novolak: A resin obtained by the condensation of phenol with an inadequate amount of formaldehyde under acidic conditions.

Organotitanates: Coupling agents for polymers and fillers.

PBA: Physical blowing agent.

Plasticizer: A flexibilizing additive for polymers.

PMF: Slag microfiber.

Reinforcing fiber: One in which $l/d > 100$.

Rule of mixtures: Each component contributes in accordance to its concentration.

S-glass: A high-tensile-strength fibrous glass.

Tertiary amine: Trisubstituted amine, R_3N.

Ultraviolet light stabilizer: An additive that deters degradation of polymers in sunlight.

Viscosity: Resistance to flow.

Whiskers: Finely divided single crystals.

Wood flour: Fibrouslike wood particles produced by the attrition grinding of debarked wood.

REVIEW QUESTIONS

1. Name a commercial polymer that does not contain an additive.
2. Which is the continuous phase in a polymeric composite?
3. What function does carbon black have in addition to its reinforcing property?
4. What is the difference between wood flour and sawdust?
5. How is α-cellulose separated from lower-molecular-weight cellulose?
6. What is the aspect ratio of a sphere?
7. Name an antioxidant.
8. Name an ultraviolet light stabilizer.
9. Why are flame retardants important?
10. How does ATH function as a flame retardant?
11. Name a widely used rubber accelerator.
12. Why should an antistat be added to the polymer in plastic bottles?

BIBLIOGRAPHY

Connell, N., and Baker, E. (1999). *Surfaces of Nanoparticles and Porous Materials*, Marcel Dekker, New York.

Craver, C., and Carraher, C. (2000). *Applied Polymer Science*, Elsevier, New York.

Datta, S., and Lohse, D. (1996). *Polymeric Compatibilizers*, Hanser-Gardner, Cincinnati.

Lutz, J., and Grossman, R. (2000). *Polymer Modifiers and Additives*, Marcel Dekker, New York.

Wypych, G. (2000). *Handbook of Fillers*, ChemTech, Toronto.

Zweifel, H. (2001). *Plastics Additives Handbook*, Hanser-Gardner, Cincinnati.

ANSWERS TO REVIEW QUESTIONS

1. If you had trouble, it's understandable since almost all commercial polymers have additives. Nondyed textile yarn is one.

2. The resinous phase.

3. It is a black pigment.

4. Wood fiber is more fibrous than sawdust.

5. Lower-molecular-weight cellulose is soluble in 17.5% aqueous sodium hydroxide.

6. 1.

7. 2,6-Di-*tert*-butylphenol.

8. Phenyl salicylate (salol).

9. Because polymers used in buildings, automobiles, ships, and airplanes should not ignite in the presence of flames.

10. ATH releases water when heated.

11. Captax, 2-mercaptobenzothiazole.

12. To prevent dust from being attracted to the bottle during storage.

19

THE FUTURE OF GIANT MOLECULES

19.1 THE AGE OF GIANT MOLECULES

We are living in the age of giant molecules which began when the first living organism appeared on Earth. Giant molecules are the key to life and living, to thinking, to breathing, to being. They serve as our covering (skin and hair) and as our inner works (muscles, tendons), and they are involved in the creation of our energy (food, photosynthesis). Without giant molecules there is no life as we know it. Giant molecules are also the fabric of today's society as essential materials in housing, communication, local motion, medicine, and so on.

The polymer industry is an employer of well over a million workers in the United States, with sales in the realm of $200 billion. The annual production of synthetic polymers has grown from less than 100 million pounds in the 1940s to about 60 billion pounds in 1988 and to about 110 billion pounds in 2000. The growth of general-purpose polymers will continue to outpace that of the gross national product, but the most important growth will occur in speciality materials

Giant Molecules: Essential Materials for Everyday Living and Problem Solving, Second Edition,
by Charles E. Carraher, Jr.
ISBN 0-471-27399-6

in such diverse areas as communications, electronics, aerospace, biomedical, automotive, and building. They will be essential in the containment of fuel and energy costs. It will become more important to recycle these important resources.

Because of its strength, flexibility, ease in molding, low cost, and resistance to chemical degradation including rust formation, lightweight traditional polymers will continue to be exploited in the automotive industry for both interior and exterior applications. Almost all construction employs giant molecules in various forms as foundation, exterior and interior walls and siding, adhesive and sealant materials, coatings/paints, windows, doors, and so on. In fact, giant molecules account for well over 90% of the building materials and cost in home construction. The amount of giant molecules used in the construction of aircraft continues to increase again because of the combination of ease in molding, strength, flexibility, light weight, cost, and resistance of wear and corrosion.

Because of their low density and relatively high strength per weight ratio, polymers replace materials many times their weight when substituted for metals, They are also less energy-intensive, meaning that it requires less energy to produce products derived from them. This conservation of energy pervades all phases of polymer production, fabrication, and recycling when compared to metals.

Monsanto built a plastics "House of the Future" in Disneyland in the 1950s. This house withstood the wear and tear of millions of visitors until it was dismantled in the late 1970s. The success of this house did not catalyze the construction of comparable house, but it did act to catalyze the use of giant molecules in recreational vehicles, mobile homes, and permanent residential structures. General Electric built a "Living Environment" demonstration house in 1988, setting the stage for an increase in the use of giant molecules in construction. The basement of this house consists of precast reinforced concrete covered on both sides by plastic panels. The aboveground tongue-and-groove assembled walls consist of reinforced plastic with molded spaces for ducts and conduits. Thus, the builder of the future may assemble tongue-and-groove panels without the use of hammer, nails, and saws.

Similar simplification and decreases in size are found in circuit boards, where all of the components are included in a small area. These units are based on high-performance giant molecules. Today's electronics industry is dependent on giant molecules, and the future electronics industry will be even more dependent.

Since fuel accounts for a significant portion of aircraft operating costs, and this portion continues to increase, giant molecules are an essential component of aircraft. All of the commercial aircraft, both small-private and large, are increasing the use of giant molecules in their construction. This dependence on giant molecules is even greater for aerospace missions, satellites, and outerspace platforms. In fact, the majority by weight of outerspace platforms will be giant molecules.

In his "Of English Verse," written in the seventeenth century, Edmund Waller wrote: "Poets that lasting marble seek must come in Latin or in Greek." Although science came in German in the early years of the twentieth century, it now comes predominantly in English. More and more, an educated person in a developing country must be fluent in English. In fact, science has its own language, and that language has words such as atoms, molecules, electrons, bonds, and so on. That is the language you have been exposed to in this text.

The educated American of the seventeenth century was fluent in Latin and Greek, but this counterpart in the twenty-first century must be literate in at least some of the material about them. Since much of this material is composed of giant molecules, an understanding of giant molecules is an important step in achieving this goal.

If we are to maintain our position as world leaders and have an appropriately knowledgeable general population, we must increase scientific illiteracy. The present book will help.

19.2 RECYCLING GIANT MOLECULES

Many giant molecules are naturally recycled. This includes both plant (Chapter 13) and animal materials (Chapter 14). Others are more difficult to recycle, such as thermosets (Chapter 8) and cross-linked elastomeric materials (Chapter 10). The recycling of paper, glass, and many thermoplastic materials is well underway, but still large amounts are simply cast into the local dump, burnt as fuel, or collected with no plan to utilize these valuable materials. It is a price we pay when we live in a somewhat "throwaway" society. Difficulty increases with respect to recycling as the complexity of giant molecule-base products increases. Thus, even a simple jogging shoe may have about 10 different materials in it, and each different brand of shoe can have a different combination of speciality material present. Even so, techniques to recover synthetic giant molecules continue to improve.

Recycling is still in its infancy. Experiments continue in the use of synthetic biodegradable versus environmentally stable materials. Experiments also continue with respect to the recyclability of synthetic polymers. For thermoplastic materials, recyclability is more straightforward since these materials can be melted and reshaped. One major problem involves the presence of additives and their affect on the behavior of the pre-recycled material (for recyclable processing) and upon materials made from the recycled materials. As the use of additives for behavior modification increases, this problem will increase, but solutions will continue to be developed.

For thermoset materials, the range of options with respect to recyclability is more limited. Some of these recycled thermoset materials are employed in such low-technology applications as extenders and fillers in asphalt. Other avenues of recyclability involve chopping the thermoset materials into small particles and then adding binders to "hold" the "recycled" material together.

For both thermoplastic and thermoset materials, another area of work involves the degradation of the polymers into smaller molecules that can again be reformed to produce needed materials.

An additional problem involves the rapid identification and division of these materials into compatible groups for recycling. The use of "Recycling Codes" is

Table 19.1 Weight of selected recycled plastic materials

	Weight of Plastic Material Recycled (millions of pounds)			
Plastic	1988	1993	%-Recycled (1993)	1998[a]
Poly(ethylene terephthalate)	146	505	35	900
HDPE	76	475	4	960
LDPE	15	120	1	230
Polypropylene	115	220	2	340
Polystyrene	3	60	1	100
Poly(vinyl chloride)	3	30	0.3	60

[a] 1998 figures projected.

Source: Freedonia Group.

only one step in the process toward more efficient and effective recycling of these valuable materials.

In 1993, only about 2% of the synthetic plastic materials were recycled. By 2002 this increased to only about 4%. Even so, efforts continue with respect to the recycling of synthetic polymers. Table 19.1 contains current and projected recycling values for some of the more common plastics. Solutions will continue to emerge as consumers and producers become more responsible.

19.3 EMERGING AREAS

At least three major themes are helping drive synthetic organic polymer synthesis and use of polymers today. These involve

- Synthesis and assembling on an individual scale (nano level)
- Synthesis in confined spaces (selected inorganic zeolites and biological syntheses; stereoregular and multifunctional materials)
- Single and multiple-site catalysis (both selected biological and synthetic polymer synthesis)

Superimposed on this is the applications aspects including the human genome/biomedical, electronic/communications, and so on.

Biological polymers represent successful strategies that are being studied by scientists as avenues to different and better synthetic polymers and polymer structure control. Sample "design rules" and approaches that are emerging include the following:

- Identification of repeat sequences that give materials with particular biological and physical properties
- Identification of repeat sequences that key certain structural changes

- Formation of a broad range of materials with a wide variety of general/specific properties and function (such as proteins/enzymes) through a controlled sequence assembly from a fixed number of feedstock molecules (proteins—about 20 different amino acids; five bases for nucleic acids and two sugar units)
- Integrated, *in situ* (in cells) polymer production with precise nanoscale control
- Repetitive use of proven strategies with seemingly minor structural differences but resulting in quite divergent results (protein for skin, hair, and muscle)
- Control of polymerizing conditions that allow ready production far from equilibrium

We are beginning to understand better how we can utilize the secondary structure of polymers as tools of synthesis. One area where this is being applied is in an area known as "folded oligomers." Here, the secondary structure of the oligomer can be controlled through its primary structure and use of solvents. Once the preferred structure is achieved, the oligomers are incorporated into larger chains, eventually forming structures with several precise structures "embedded" within them. The secondary structure of these larger polymers can also be influenced by the nature of the solvent, allowing further structural variety. Addition of other species such as metal ions can assist in locking in certain structures, and they can also act to drive further structural modification.

It is important to remember that while polymer scientists are making great progress in designing giant molecules with specific properties through manipulation of polymer structure, we have not approached Nature in this, particularly with respect to tertiary and quaternary structures. Mimicking natural structures has become an important area of polymer research.

Nanomaterials are and will play a role in many areas. They will play a crucial role as we move to computer chips that operate essentially on a molecular level. This, in turn, will enable increased computer capacity and decreased computer time per operation. In turn, this will allow for faster and more complete structure–property relationships to be established that will allow for better specific design of materials for manufacture, medicine, communications, transport—"an upward spiral." Nanotubes and nanopolymers are being (considered and actually) used as electric switches; as specific site catalysts; for specific removal of toxins on a molecular level; for storage of energy; for identification and separation of biologically important materials on an individual chain level; as interfaces for energy and information transfer including use as biosensors; for better material property and structure control; as superstrength material (on a weight basis); as high-efficiency fuel cells; and so on. Study of the use of DNA and DNA-like circuitry will increase.

19.4 NEW PRODUCTS

A number of small-scale polymeric materials will continue to enter the marketplace on a regular basis. These include biomaterials and electronics materials where the cost per pound is high and the poundage is low, generally well less than a hundred tons a year. These are materials that fulfill specific needs.

The number of new larger-scale giant molecules that enter the marketplace will be small. It has been estimated that it takes about $1 billion to introduce and establish a new material. It is a daunting task with no guarantee of success. In the past, new giant molecules could be introduced that offered improvements in a number of areas and thus would attract a market share in a number of application areas. Today, there are already a wide range of materials for most application areas that compete for that particular market share, so it is difficult for any material to significantly break in to any market area. A new material needs a "flagship" property that a particular market needs.

DuPont and Shell have developed a new polyester, poly(trimethylene terephthalate) (PTT), that is structurally similar to poly(ethylene terephthalate) (PET), except that 1,3-propanediol (PDO) is used in place of ethylene glycol. The extra carbon in Sorona allows the fiber to be more easily colored, giving a material that is softer to the touch with greater stretch for textile use. Furthermore, it offers good wear and stain resistance for carpet use. The ready availability of the monomer PDO is a major consideration with efforts underway to create PDO from the fermentation of sugar through the use of biocatalysts for this conversion. Sorona and Lycra blends have already been successfully marketed. Sorona is also targeted for use as a resin and film.

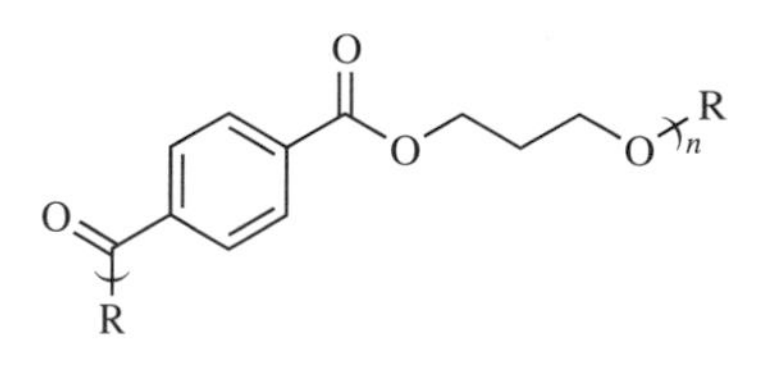

Poly(trimethylene terephthalate), PTT

In 1990, DMS introduced nylon 4,6 called Stanyl, based on the reaction between adipic acid and 1,4-diaminobutane. Stanyl can withstand temperature to 595°F, allowing it to create a niche between conventional nylons and high-performance materials. It was not able to break in to the film market and has only now begun to be accepted for tire cord applications. About 22 million pounds of Stanyl was produced in 2001.

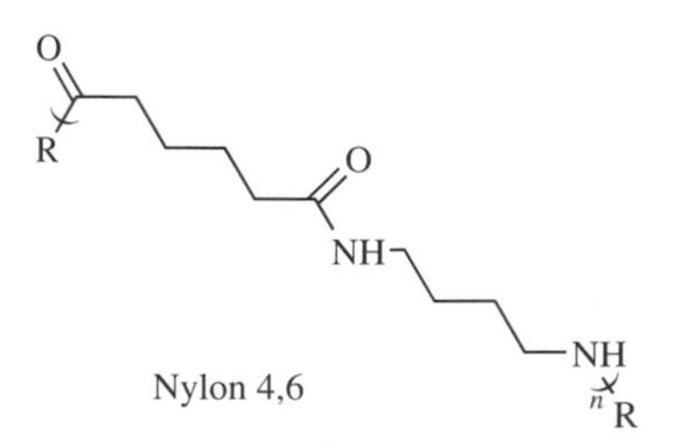

Nylon 4,6

In 1997, Dow introduced syndiotactic polystyrene under the trade name Questra. The technology for the production of Questra is based on relatively new technology and science involving soluble stereoregulating catalysts that produce polystyrene that has a fixed and repeating geometry as each new styrene monomer unit is added to the growing polystyrene chain. Targeted areas include medical, automotive, and electronic applications.

Several other produces have been developed based on the relatively new soluble stereoregulating catalysis systems. Index, an ethylene-styrene interpolymer, was introduced in 1998 and is intended to compete with block copolymers such as styrene–butadiene–styrene, flexible PVC, polyurethanes, and polyolefins. It is being used as a modifier for polystyrene and polyethylene. Dow is also developing soundproofing and packaging foam applications for Index. Hoechst Celanese (now Ticona) developed Topas, a cyclo-olefin copolymer, in the 1980s and in 2000 began commercial production of it. Topas has high moisture-barrier properties and is being considered for use in blister packaging for pharmaceuticals. It is also being used in resealable packages where it provides stiffness to the sealing strip. It is also being use in toner resin applications and is being blended with linear low-density and low-density polyethylene providing stiffness and to improve sealing properties.

A number of new materials are looking toward being involved in the upcoming move toward blue-light CDs. For any of these to become important materials in this area, they will need to improve on the present polycarbonate-based materials.

GE introduced in 2000 a new polyester carbonate based on resorcinol arylates it called W-4 and is now marketed as Sollx. Sollx does not need to be painted, it offers good weather, chip, scratch, and chemical resistance, and it is being used as the fenders for the new Segway Human Transporter. It is also aimed at automotive applications, including use as body panels. Sollx is coextruded into two layers, one clear and one colored, to simulate automotive paint. It is then thermoformed and molded into the finished product.

Several new ventures are based on using natural, renewable materials as the starting materials instead of petrochemicals. These products are known as "green" products since they are made from renewable resources and they can be composted. Along with the production of 1,3-propanediol by Shell and DuPont to produce nylon 4,6, Cargill Dow is making PLA, beginning with corn-derived dextrose. The polylactide, PLA, is made from corn-derived dextrose that is fermented, producing lactic acid. The lactic acid is converted into lactide, a ring compound, that is polymerized through ring opening.

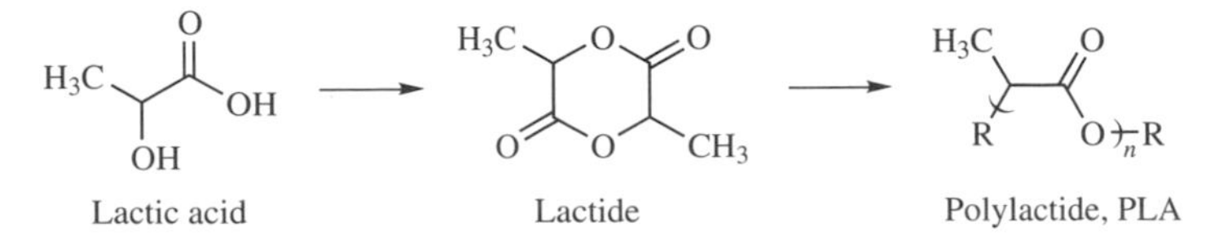

PLA looks and processes like polystyrene. It has the stiffness and tensile strength of PET, offers good odor barrier, and resists fats, oils, and greases. PLA is being considered for use in fibers and in packaging. As a film, PLA has good deadfold properties—that is, the ability to be folded and to stay folded. It is being used as a fiber for apparel and carpeting applications. It is being sold as a bridge between synthetic and natural fibers in that it processes like synthetic fibers but has the touch, comfort, and moisture-management of natural fibers.

BIBLIOGRAPHY

Allcock, H. R., and Lampe, F. W. (1990). *Contemporary Polymer Chemistry*, 2nd ed., Wiley, New York.

Brewer, D. (1987). *50 Years of Progress in Plastics*, HBJ Plastics Publishers, Denver.

Callister, W. (2000). *Materials Science and Engineering*, 5th ed., Wiley, New York.

Campbell, I. (2000). *Introduction to Synthetic Polymers*, Oxford University Press, New York.

Carraher, C. (2003). *Polymer Chemistry*, 6th ed., Marcel Dekker, New York.

Carraher, C., Swift, G., and Bowman, C. (1997). *Polymer Modification*, Plenum, New York.

Craver, C., and Carraher, C. (2000). *Applied Polymer Science*, Elsevier, New York.

Ehrenstein, G. (2001). *Polymeric Materials*, Hanser-Gardner, Cincinnati.

Elias, H. G. (1997). *An Introduction to Polymers*, Wiley, New York.

Elliott, E. (1986). *Polymers and People*, Center for History of Chemistry, Philadelphia.

Fried, J. R. (2002). *Polymer Science and Technology*, 2nd ed., Prentice-Hall, Upper Saddle River, NJ.

Gebelein, C., and Carraher, C. (1985). *Bioactive Polymeric Systems*, Plenum, New York.

Grosberg, A., and Khokhlov, A. R. (1997). *Giant Molecules*, Academic Press, Orlando, FL.

Hummel, R. E. (1998). *Understanding Materials Science: History, Properties, Applications*, Springer-Verlag, New York.

Mark, H. (1966). *Giant Molecules*, Time-Life Books, New York.

Nicholson, J. W. (1997). *The Chemistry of Polymers*, Royal Society of Chemistry, London.

Ravve, A. (2000). *Principles of Polymer Chemistry*, Kluwer, New York.

Rodriguez, F. (1996). *Principles of Polymer Systems*, 4th ed., Taylor and Francis, Philadelphia.

Salamone, J. C. (1998). *Concise Polymeric Materials Encyclopedia*, CRC Press, Boca Raton, FL.

Sandler, S., Karo, W., Bonesteel, J., and Pearce, E. M. (1998). *Polymer Synthesis and Characterization*, Academic Press, Orlando, FL.

Schwartz, M. (1996). *Emerging Engineering Materials*, Technomic, Lancaster, PA.

Seymour, R. (1982). *History of Polymer Science and Technology*, Marcel Dekker, New York.

Seymour, R., and Mark, H. (1988). *Applications of Polymers*, Plenum, New York.

Seymour, R., and Carraher, C. (1997). *Introduccion a la Quimica de los Polymeros*, Editorial Reverte, S. A., Barcelona, Spain.

Sorenson, W., Sweeny, F., and Campbell, T. (2001). *Preparative Methods in Polymer Chemistry*, Wiley, New York.

Sperling, L. (2001). *Introduction to Physical Polymer Science*, 2nd ed., Wiley, New York.

Tanaka, T. (1999). *Experimental Methods in Polymer Science*, Academic Press, New York.

Thrower, P. (1996). *Materials in Today's World*, 2nd ed., McGraw-Hill, New York.

Tonelli, A. (2001). *Polymers Inside Out*, Wiley, New York.

Walton, D. (2001). *Polymers*, Oxford University Press, New York.

APPENDIX 1

STUDYING GIANT MOLECULES

Studying about giant molecules is similar to studying any science. Following are some ideas that may assist you as you study.

Much of science is abstract. While much of the study of giant molecules is abstract, it is easier to conceptualize (i.e., make mind pictures) of what a polymer is and how it should behave than many areas of science. For linear giant molecules, think of a string or rope. Long ropes get entangled with themselves and other ropes. In the same way, giant molecules entangle with themselves and with chains of other giant molecules that are brought into contact with them. Thus, create mental pictures of the giant molecules as you study them.

Giant molecules are real and all about us. We can look at giant molecules on a micro or atomic level or on a macroscopic level. The PET bottles we have may be composed of long chains of poly(ethylene terephthate) (PET) chains. The aramid tire cord is composed of aromatic polyamide chains. Our hair is made up of complex bundles of fiberous proteins, again polyamides. The giant molecules you study are related to the real world in which we live. We experience these "giant molecules" at the macroscopic level everyday of our lives, and this macroscopic behavior is a direct consequence of the atomic-level structure and behavior. Make pictures in your mind that allow you to relate the atomic and macroscopic worlds.

At the introductory level we often examine only the primary factors that may cause particular giant molecule behavior. Other factors may become important under particular conditions. The giant molecules you study at times examines only the primary factors that impact polymer behavior and structure. Even so, these primary factors form the basis for both complex and simple structure–property behavior.

Giant Molecules: Essential Materials for Everyday Living and Problem Solving, Second Edition,
by Charles E. Carraher, Jr.
ISBN 0-471-27399-6

The structure–property relationships you will be studying are based on well-known basic chemistry and physical relationships. Such relationships build upon one another, and as such you need to study in an ongoing manner. Understand as you go along. Read the material *before* you go to class.

This course is an introductory-level course. Each chapter or topic emphasizes knowledge about one or more area. The science and excitement of giant molecules has its own language. It is a language that requires you to understand and memorize certain key concepts. Our memory can be short term or long term. Short-term memory may be considered as that used by an actor or actress for a TV drama. It really does not need to be totally understood, nor retained after the final "take." Long-term memory is required in studying about giant molecules since it will be used repeatedly and is used to understand other concepts (that is, it is built upon).

In memorizing, learn how you do this best—time of day, setting, and so on. Use as many senses as necessary—be active—read your assignment, write out what is needed to be known, say it, listen to yourself say it. Also, look for patterns, create mnemonic devices, avoid cramming too much into too small a time, practice associations in all directions, and test yourself. Memorization is hard work.

While knowledge involves recalling memorized material, to really "know" something involves more than simple recall—it involves comprehension, application, evaluation, and integration of the knowledge. Comprehension is the interpretation of this knowledge-making predictions, applying it to different situations. Analysis involves evaluation of the information, and comparing it with other information and synthesis has to do with integration of the information with other information.

In studying about giant molecules, please consider doing the following:

- Skim the text *before* the lecture.
- Attend the lecture and take notes.
- Organize your notes and relate information.
- Read and study the assigned material.
- Study your notes and the assigned material.
- Review and self-test.

Learning takes time and effort. Study daily skimming the text and other study material, think about it, visualize key points and concepts, write down important material, make outlines, take notes, study sample problems, and so on. All of these help, but some may help you more than others, so focus on these modes of learning—but not at the exclusion of the other aspects.

In preparing for an exam, consider the following:

- Accomplish the above. *Do not* wait until the day before the exam to begin studying; create good study habits.
- Study wisely: Study how *you* study best—time of day, surroundings, and so on.

- Take care of yourself; get plenty of sleep the night before the exam.
- Attend to last-minute details: Is your calculator working, is it the right kind, do I have the needed pencils, review the material once again, and so on.
- Know what kind of test it will be, if possible.
- Get copies of old exams if possible; talk to others that might have already had the course.

During the test

- Stay cool, *do not panic.*
- Read the directions; try to understand what is being asked for.
- In an essay or similar exam, work for partial credit; plan your answers.
- In a multiple choice or T/F exam, eliminate obviously wrong choices.
- Look over the entire exam; work questions that you are sure of; then go to less sure questions; check answers if time permits.

The study of giant molecules contains several types of content:

- *Facts.* The term *polymer* means "many" (poly) "units" (mers).
- *Concepts.* Linear polymers are long molecules like a string.
- *Rules.* Solutions containing polymer chains are more viscous (i.e., slower flowing) than solutions that do not contain polymers.
- *Problems.* What is the approximate molecular weight of a single polyethylene chain that has 1000 ethylene units in it?

These varied types of content are often integrated within any topic, but in this introduction to giant molecules, the emphasis is often on concepts but all the aspects are important.

APPENDIX 2

ELECTRONIC WEB SITES

The amount of giant-molecule-related information that appears on the World Wide Web, WWW, is rapidly increasing. In general, giant-molecule-related material can be obtained from "surfing" the "POLY.COM" and looking for specific topics. Sites are being renamed, dropped, created, and so on, in an ongoing fashion, so this may have happened to some of the following sites. The following site was developed with the cooperation of PolyEd for polymer education use. It is probably the most used giant-molecule-associated site aimed at the general public.

www.psrc.usm.edu/macrog/index.html

Following are other sites that may be of use. The author welcomes additions. Please send to carraher@fau.edu the site and short description.

PolyEd and IPEC

www.polyed.org General site for PolyEd, the joint polymer education for the ACS Divisions of Polymer Chemistry and Polymeric Materials: Science and Engineering.

www.ipeconline.org General site for IPEC, a joint society group that focuses on K-12 science education utilizing polymers as the connective material.

www.polymerambassadors.org General site maintained by the Polymer Ambassadors, which focuses on materials suitable for K-12 teachers.

Giant Molecules: Essential Materials for Everyday Living and Problem Solving, Second Edition,
by Charles E. Carraher, Jr.
ISBN 0-471-27399-6

For Biomacromolecules

http://biotech.icmb.utexas.edu Contains a glossary of over 6700 terms related to biochemistry and biotechnilogy.

http://www/gdb.org/Dan/DOE/prim1.html Contains a primer on molecular genetics.

http://web.indstate.edu/them/mwking/biomols.html Chemistry of nucleic acids.

http://moby.ucdavis.edu/HRM/Biochemistry/molecules.htm Molecular models of nucleic acids.

http://moby.ucdavis.edu/HRM/Biochemistry/molecules.htm Models of secondary and tertiary protein structures.

http://expasy.hcuge.ch/pub/Graphics/IMAGES/GIF Images of many proteins.

http://www.rcsb.org/pdb Protein data bank.

http://www.ncbi.nlmnih.gov/Entrez Protein data bank.

http://www.ilstu.edu/depts/chemistry/che242/struct.html Structures of many saccharides.

The unraveling of much of the human genome is one of the most important advances made since our dawning. An online tour of the human genome is found at a number of web sites allowing access to some of this valuable information. Some of these sites are

http://genome.ucsc.edu

www.nature.com/genomics/human/papers

www.sciencemag.org/contents/vol291/issue5507

www.ncbi.nlm.nih.gov/sitemap/index.html#humangenome

www.nhgri.nih.gov/data/

www.celera.com

www.ensembl.org/genome/central/

http://genome.ucsc.edu

http://genome.cse.ucsc.edu

http://genome.wustl.edu/gsc/human/mapping

The association of a particular disease with a particular gene or group of genes is rapidly increasing. A spot check of www.ncbi.nlm.nih.gov/omim, the online version of Mendelian Inheritance in Man, OMIN, gives an ongoing updated progress report of this activity. Currently, about 1500 disease-mutations have been entered.

Nano Materials

www.nano.gov NSF site for the National Nanotechnology Initiative.

www.nano.org.uk/index.html Institute of Nanotechnology in Stirling, Scotland.

http://cnst.rice.edu center for Nanoscale Science and Technology at Rice University.

http://ipt.arc.nasa.gov:80/index.html NASA Integrated Product Team on Devices and Nanotechnology.

Coatings

http://cage.rti.org/altern.htm Coating Alternative Guide, CAGE, information about different coatings.

http://www.execpc.com/~rustoleu/coatings.htm Corrosion and protective coatings and paint resources.

http://www.coatingsstech.org/ Federation of Societies for Coatings Technology, FSCT, home site.

http://www.iscc.org/ Inter-Society Color Council-organization that promotes application of coatings and color.

http://www.nace.org/ National Association of Corrosion Engineers, NACE, home web site.

http://www.paint.org/ National Paint and Coatings Association, NPCA, home web site.

http://www.paintcoatings,net/pcnmain.htm Paint coatings network giving directories, tradeshows, product news, and related articles.

http://www.pra.org.uk/index.htm Paint Research Association.

http://www.jvhltd.com/paintwebs/default1.html Paint web sites for suppliers.

http://protectivecoatings.com Protective coatings worldwide information resource site.

Others

Most products have individual sites that may contain useful general and specific product information.

www.pct.edu/prep Plastics Resources for Educators Plastics-Emphasizes polymer processing.

http://russo.chem.1su.edu/howto/HowTo.html Experimentation with polymer solutions.

http://www.msi.com Computer modeling of polymers-synthetic and natural -go to "WebLab Viewer Lite."

www.teachingplastics.org General interest articles about specific polymer applications.

http://www.asm-intl.org/index.htm ASM-Materials engineering society.

http://www.astm.org/ ASTM home site for technical standards.

http://www.plasticstrends.net Nice assembly of recent, general articles focusing on plastics and applications.

INDEX

Giant Molecules: Essential Materials for Everyday Living and Problem Solving, Second Edition,
by Charles E. Carraher, Jr.
ISBN 0-471-27399-6